Functional Nanostructured Membranes

Functional Nanostructured Membranes

edited by

Enrico Drioli
Lidietta Giorno
Annarosa Gugliuzza

PAN STANFORD PUBLISHING

Published by

Pan Stanford Publishing Pte. Ltd.
Penthouse Level, Suntec Tower 3
8 Temasek Boulevard
Singapore 038988

Email: editorial@panstanford.com
Web: www.panstanford.com

British Library Cataloguing-in-Publication Data
A catalogue record for this book is available from the British Library.

Functional Nanostructured Membranes

ISBN 978-981-4774-79-6 (Hardcover)
ISBN 978-1-351-13511-5 (eBook)

Contents

Preface

There is a worldwide demand for the development of new nanostructured materials with associated functions and high-performing properties, urging one to provide new advanced solutions for environment, water, energy, food, safety, and health needs. Membrane science is a part of this context because membranes can be envisaged as unique arrays where different and complementary elements and functions can be accommodated in a defined space according to chemical, physical, and biological rules. Desired functions and performance can be achieved and amplified through manipulation of materials providing new solutions and benefits and fitting with eco-sustainable economical development and societal growth.

This book contains insightful contributions from scientists with a long experience in membrane science and technologies and have been working at ITM-CNR in the field of nanostructured materials membranes, including selected materials and practices for manufacturing highly defined functional interfaces with the ability to transport mass, charge, and energy. It describes appropriate techniques for monitoring molecular-scale events that regulate the assembly of organic and inorganic materials on different length scales needed to realize the developed applications. It covers different fields, including concepts related to nanostructures, breakthrough membrane fabrication, and modeling as well as their connected use in environment, energy, water, textiles, biomedical, and bioengineering areas.

Chapter 1 is a general introduction to the concepts of nanostructured materials and membranes, with a special focus on challenges, opportunities, and future tasks deriving from a planned use of newly developed membrane processes. Chapter 2 describes theoretical and modeling approaches to investigate selective transport through nanostructured matrices. Chapter 3 is an overview of the featured techniques that are used for the fabrication of porous ordered membranes with a very high level of organization and structure at various length scales. Chapter 4 introduces the reader to the fabrication of ion-exchange membranes and microenvironment-related effects on the final permselective properties. Chapter 5 focuses on the fabrication and application of nanostructured zeolite membranes. Chapter 6 examines the potential of developing a new classes of 2D materials for the fabrication of ultrathin membranes that can be used in ion-filtering, gas separation, and energy production and storage. Chapter 7 describes the high performance of the new nanostructured membranes in gas separation, informing the reader about the shift in the state of the art by ladder and semi-ladder polymers. In context to the catalytic nanoreactors, Chapter 8 offers an outlook on the potential of inorganic nanostructured membranes in advanced catalytic processes. Chapter 9 focuses on the manipulation of surface properties to impart self-cleaning behavior to patterned membranes. Chapter 10 refers to the formulation of new coatings to improve development processes of water purification membrane. Chapter 11 opens new horizons on nanostructured membranes with smart functions for many-sided textile applications. Chapter 12 is a critical discernment on the key role played by nanostructured porous membranes in membrane distillation and membrane crystallization processes. Chapter 13 is a slice of a broader argumentation describing the potential attractive role of nanostructured membranes in the field of biosensors while Chapter 14 describes preparation, characterization, and conceivable applications of nano-engineered membranes in biotechnologies. The build-up of nanocapsules, nanomatrices, and nanoemulsions is the subject of Chapter 15 dedicated to controlled release, while Chapter 16 embraces the field of the engineering of organs and tissues wherein nanostructured membranes currently find a large application.

The intent of this book is to catalyze the process of integration between complementary disciplines in order to share knowledge and expertise on nanomaterials and nanotechnologies, which can draw researchers, investigators, graduated students, and end-users in the world of the future science and technology.

Enrico Drioli
Lidietta Giorno
Annarosa Gugliuzza
Summer 2018

Chapter 1

An Introduction to Nanostructured Membranes for Advanced Applications in Strategic Fields

Enrico Drioli, Lidietta Giorno, and Annarosa Gugliuzza
Institute on Membrane Technology of the National Research Council of Italy (ITM-CNR), Via P. Bucci, Cubo 17C, 87036 Rende (CS), Italy
a.gugliuzza@itm.cnr.it

This chapter is a brief introduction to concepts of nanostructured materials, including membranes. The role of bio-inspired nanostructures in nanoscience and nanotechnology has been discussed with particular emphasis on membrane nanotechnologies and related societal and economic impacts. Challenges, opportunities, and future tasks have been envisioned.

1.1 Nanosized Materials and Nanostructured Membrane Concepts

According to ISO Nanotechnologies Vocabulary (Nanotechnologies Vocabulary, 2010), the expression "nanotechnology" deals with

Functional Nanostructured Membranes
Edited by Enrico Drioli, Lidietta Giorno, and Annarosa Gugliuzza

ISBN 978-981-4774-79-6 (Hardcover), 978-1-351-13511-5 (eBook)
www.panstanford.com

"Structure-dependent properties and phenomena as distinct from those associated with individual atoms or molecules or with bulk materials." This means that the manipulation of atomic and molecular interactions leads to a modification of macroscopic properties causing a radical renovation of mechanical, electrical, optical and transport, and/or other material features. Macroscopic adjustments can be ascribed to small size effects, surface interface effects, quantum size effects, quantum tunneling effects, and so on (Cao, 2007; Denton et al., 1971).

In this context, it is useful to categorize some nano-based concepts according to ISO/TS 80004-1 rules (Nanotechnologies Vocabulary, 2010): (a) Nanosized materials are intended with at least one external dimension on nanoscale (1–100 nm) or with an internal structure or surface structure on nanoscale, while (b) nanostructures are designed for an assembly of inter-linked parts, in which one or more of those parts can be on nanoscale length. A system such as a nanostructured membrane can exhibit both the internal or surface nanostructure, because nano-scaled features can be into the bulk or throughout the surface (Fig. 1.1).

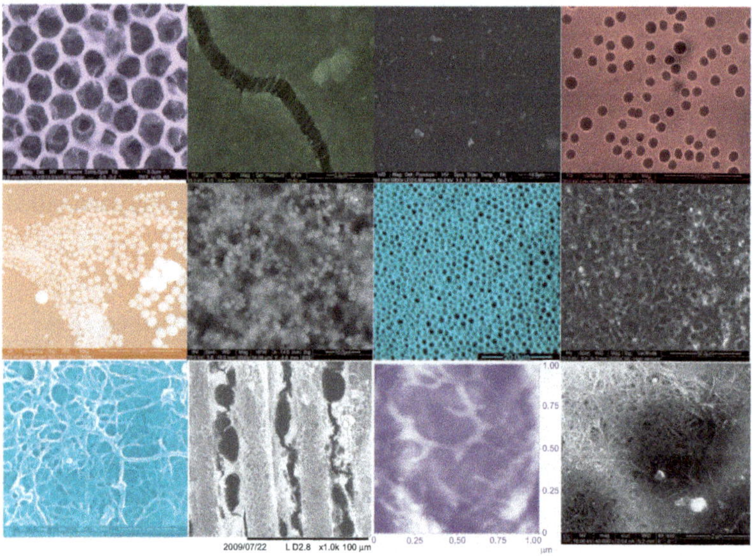

Figure 1.1 Examples of tailor-made nanostructured membranes.

Pores with size from 1 to 100 nm, through which particles and compounds can be separated on the basis of their own size, are typical structural elements that allow membranes to be classified as nanostructures. However, the inclusion of nanofillers in polymer matrices also makes the membranes categorized as nanostructures. The creation of hierarchically textured membrane surfaces is another requisite for falling into the class of nanostructures. Block copolymers together with nanotubes, catalytic nanoparticles, zeolites, or bio-hybrid compounds can be embedded in the polymer matrix or deposited onto the related surface in order to improve the productivity–efficiency trade-off, reducing capital costs and energy consumption (Fig. 1.1).

The substantial allocation of nano-objects and complementary components in confined spaces yields new tunable functions and advanced performances that depend on intrinsic relationships between local structure, transport and surface properties (Perrotta et al., 2017; Scott et al., 2011). On the other hand, the origin of transport phenomena controlling the membrane selectivity and permeability is based on the phenomena active at nanolevels.

Uniformity is, however, necessary to increase the structural sophistication of nanostructured membranes, since a large number of species need to arrange in complex organizations that depend on the previous level of assembly. This imposes organizational rules and precise control on different length scale. The chemical composition and conformation of involved species can play a crucial role in hierarchical organizations under particular boundary environmental conditions (O'Brien et al., 2016).

In this context, nature is regarded as an infinite source of nanostructures and organizational regulation, which inspire the work of many scholars and scientists dedicated to the exploration of ordering of natural building blocks and the understanding of mechanisms controlling complex spontaneous phenomena (Fig. 1.2).

Natural nanomaterials and nanostructures are everywhere (Barth, 2005); plants and animals exhibit singular structures that confer unique properties on them. Soil generates a very large amount of nanomaterials, and the ocean is their natural reservoir. It is surprising how self-cleaning, self-powered transport, energy storage, motion under a geomagnetic field observed in natural systems can be ascribed to intrinsic relationships between textures and functions precisely rearranged by nano-orders (Fig. 1.2).

Figure 1.2 Examples of natural nanostructures.

Frequently, engineered materials and structures are discovered to be identical to some natural systems. As in nature, the same basic mechanisms operating at nanolevels can characterize processes at any large-scale level. Indeed, the recent availability of advanced instrumentation and metrology equipment has allowed many researchers to discover and tune the structure and properties of many natural materials, propelling their reproduction in the laboratory (Kumar, 2009; Gugliuzza et al., 2016). The major concern is, however, the production on scale of some types of nanostructures, including nano-patterned membranes (Han et al., 2017). Despite that bio-inspired nanostructured membranes are still far from commercial reality, other categories of nano-hybrid composite membranes seem to be simpler to implement commercially.

Another concern is the potential risk to human health and environment (Oberdorster et al., 2005). In nano-porous membranes, the pores are considered integral constituent parts of the whole framework. This means that nano-porous membranes do not have associated intrinsic hazards such as toxicity or exposure like it occurs for traditional nanomaterials. Otherwise, membranes containing nanofillers inside the bulk or deposited on the surface could have risks comparable to those associated to nanomaterials (Read et al., 2013).

1.2 Viewpoint on Membrane Processes

Membrane operations are regarded as ideal for industrial production because they work in mild conditions of pressure and temperature, require low energy input, and are clean processes. Simple in concept and operation, membrane processes are modular and easy to scale-up, energy-saving, and with a great potential for rational utilization of raw materials and recovery and reuse of byproducts.

Membrane technologies respond efficiently to the requirements of so-called "process intensification," because they permit drastic improvements in manufacturing and processing, substantially decreasing the equipment–size/production–capacity ratio, energy consumption, and/or waste production and resulting in cheaper, sustainable technical solutions.

Today, membrane technology has well-established uses in numerous industrial processes, including water desalination, wastewater treatments, agro-food, gas separation, energy power and storage, artificial organs, and chemical and petrochemical industry. Membrane operations are already dominant technologies in molecular separations, but they are becoming of interest also as membrane reactors and membrane contactors.

The significant positive outcomes of various membrane systems are, however, still far from fully exploiting the potentialities of this technology. There are still difficulties related to the pre-treatment of streams, membrane lifetime, fouling, aging, and sealing as well as slowing down the growth of large-scale industrial use.

The design of new nanostructured membranes with improved properties and the identification of efficient transport-properties relationships are key challenges for further development of this technology. In the last years, the literature has received an increasing number of contributions on nanostructured materials, membranes, and nanotechnologies (Fig. 1.3).

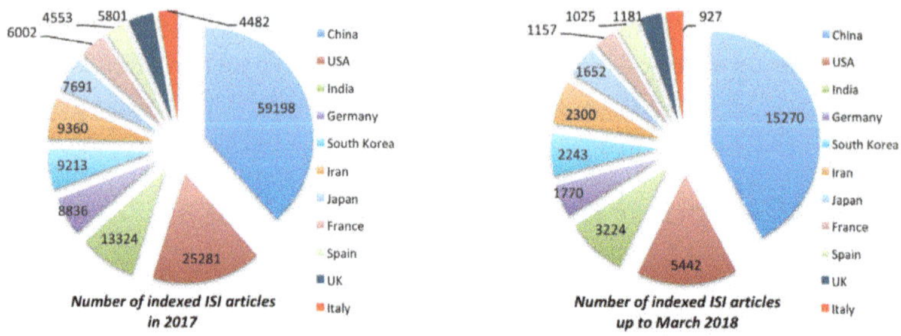

Figure 1.3 Number of nanotechnology articles indexed in ISI (Web of Science) year 2017 and up to March 2018. Data source: http://statnano.com/report/s29.

However, further intensive research activities both at academic and industrial levels are still necessary. Significant innovation toward large-scale diffusion of membrane technologies is expected in various sectors, and the role of membrane materials and engineering is crucial in this respect. New solutions and benefits may be achieved through the fabrication of new functional membrane structures and the integration of various membrane operations in the same industrial cycle, with important profits in terms of product quality, plant compactness, environmental impact, and energetic aspects. Existing non-membrane-based equilibrium-driven separation technologies (e.g., distillation, extraction, absorption, adsorption, ion exchange, stripping), which represent the core of the traditional desalination, chemical, and petrochemical industry, have significant limitations such as high energy consumption, inherent operational difficulties, lack of flexibility and modularity, slower rates, need for hazardous chemicals, high capital costs, need for large equipment volume and footprint. These shortcomings are aggravated by new separation demands (for example, environmental pollution control laws). New membrane-based separation concepts and technologies (e.g., membrane distillation, membrane crystallization, membrane condenser, membrane dryer, membrane emulsifier, etc.) do not suffer from many such deficiencies and are poised to invade increasingly the domain of traditional separation technologies. It is, then, trustworthy to assert that new widespread perspectives of nanostructured membranes associated with integrated membrane solutions for sustainable industrial growth are possible.

It is relevant to observe how the membranes differ widely in their structure, in their function, and the way they are operated. Furthermore, membrane properties can be tailored and adjusted on different length scale to specific separation tasks, while the membrane processes are often technically simple and are equally well suited for large-scale continuous operations as for batch-wise treatment of very small quantities.

At the heart of every membrane processes, there is an interface, which is clearly materialized by a nanostructured/functionalized thin barrier that controls the exchange between two phases, not only by external forces and under the effect of fluid properties but also through the intrinsic characteristics of the membrane material itself. A membrane may be biological or synthetic, solid or liquid, homogeneous or heterogeneous, isotropic or anisotropic in its structure. A membrane can be a fraction of a micrometer or several millimeters thick. Its electrical resistance can vary from millions of ohm to a fraction of an ohm.

Another characteristic property of a membrane is its permselectivity, which is determined by differences in the transport rates of various components in the membrane matrix. The permeability of a membrane is a measure of the rate at which a given component is transferred through the membrane under specific conditions of concentration, temperature, pressure, and/or electric field.

The structure of the membrane, the size of the permeating component, the chemical nature and electrical charge of the membrane material and permeating components together with driving force, i.e., concentration, pressure, or electrical potential gradient across the membrane, control the transport rate of a component through a membrane. The transport of certain components through a membrane may be facilitated by certain chemical compounds coupled to the transport of other components, or activated by a chemical reaction that takes place in the membrane. These phenomena are referred to as facilitated, coupled, or active transport. Another important characteristic is the driving force acting on the permeating components. Some driving forces such as concentration, pressure, or temperature gradients act equally on all components, in contrast to an electrical potential driving force, which is only effective with charged components. The use of different membrane structures and driving forces has resulted in a number of rather different membrane processes such as reverse osmosis, micro-, ultra-, and nanofiltration, dialysis, electrodialysis, Donnan dialysis, pervaporation, gas separation, membrane reactors, membrane contactors, membrane distillation, membrane emulsification, membrane crystallization, membrane condenser, membrane dryer, etc.

The design of new nanostructured functional membranes is expected to shift the productivity–efficiency trade-off for each single-membrane operations.

1.3 Perspectives of Nanostructured Membranes in Life Sciences

Nanostructured membranes with tuned surface properties and pore size are commonly used in biological systems. Interactions between biomolecules, cells, biocomponents and membrane materials, structures, and morphologies account for adhesion, repulsion, retention, and separation by means of charged-selective barrier and/or size- and shape-selective barrier.

Nanostructure and hierarchical organization play a key role in the structure–function relationship of system biomembranes. The phospholipid bilayer and the components present in it, including protein membrane channels, receptors, sensorial systems, are built according to an accurate multiscale hierarchical assembly that results in very highly defined and ordered supramolecular structure and morphology yielding extremely precise ion and molecular recognition as well as transport (Militano et al., 2016). Furthermore, individual cells bearing their own biomembrane functionalities are assembled in multilayer nanocomposite structures functioning as selective barrier, assuring high selectivity and permeability, as well as bearing components to assist functions such as repairing, healing, and cleaning.

An example of nanostructured hierarchical thin film is represented by the basal membrane of the glomerular capsule, the biofiltration unit of the kidney (Fig. 1.4). This basal membrane is a highly performing asymmetric nanostructured biomembrane formed of three major layers: (1) a layer of endothelial cells forming fenestrae of about 70 nm with negatively charged endothelial glycocalyx; (2) an intermediate layer, made of three negatively charged heparan sulfate layers that prevent passage of negative charged proteins, such as albumin; and (3) a layer of epithelial podocytes forming slits of about 20 nm. The first two layers form a charged-selective barrier, while the third one forms a size-selective barrier.

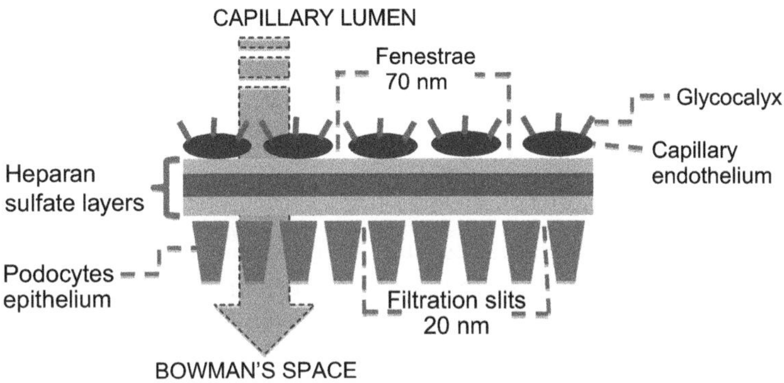

Figure 1.4 Example of nanostructured hierarchical thin film: basal membrane of the glomerular capsule as the biofiltration unit of the kidney.

Similar to nanostructures found in biosystem membranes, artificial membranes can be designed to compartmentalize, regulate transport between compartments, perform selective separation, adhesion, repulsion. Even though the performance of artificial membranes does not match the performance of biological membranes, they exhibit much higher mechanical properties and shelf-life, which make them suitable for large production and separation.

Numerous studies clarified that structures properly organized on a nanoscale improve the macro-scale properties and functions of membranes in contact with biocomponents. Therefore, much interest is oriented toward the development of methods to manipulate and tune nanostructured membranes as well as to visualize them.

Some of the future perspectives of nanostructured membrane in life sciences are outlined here.

To improve performance of artificial membranes and fulfil their great potential in areas such as biotechnology, bioengineering, biochemistry, and pharmacology, one of the major challenge is selectivity, and the current trend is to develop membranes with nano-precision pore size, which able to distinguish between macromolecules with similar size (molecular weight) and shape (dynamic diameter). Permeability is also needed to be kept in a range suitable for real use, but selectivity seems the actual bottleneck. Materials and nanostructures promoting antifouling properties and/or better fouling control are also heavily researched.

Nanostructured membranes able to host enzymes immobilized as catalytic monolayers and to promote transport of substrate and product so that bioconversion may only occur toward right are needed to break through the field of biocatalytic membrane reactors.

Oriented immobilization of antibodies to develop affinity membranes for separation, recognition, and biosensing is gaining renewed attention.

Nanocapsule membranes able to release payloads in places where they are needed, thus reducing the amount of chemicals necessary while making the treatment more efficient, are the forefront of controlled drug delivery.

Nano-patterned biocompatible, bioactive, and biodegradable membranes able to promote cell adhesion, guide cell growth, and promote transport of nutrients to and catabolites from the cell environment are used for tissue engineering. Autologous constructs for personalized medicine are in the research forefront.

The design of membranes aiming at reducing the size of membrane modules for dialysis is among the requests of this industrial sector.

References

Barth, J. V., Costantini, G., and Kern, K. (2005). Engineering atomic and molecular nanostructures at surfaces. *Nature*, **437**, p. 671.

Cao, M. S. (ed.) (2007). *Introduction to Nano-Materials*, 1st ed., Harbin Institute of Technology Press, Harbin.

Denton, R., Muhlschlegel, B., and Scalapino, D. J. (1971). Electronic heat capacity and susceptibility of small metal particles. *Phys. Rev. Lett.*, **26**, p. 707.

Gugliuzza, A., Perrotta, M. L., and Drioli, E. (2006). Controlled bulk properties of composite polymeric solutions for extensive structural order of honeycomb polysulfone membranes. *Membranes*, **6**(2), p. 27.

Han, Y. L., Giorno, L., and Gugliuzza, A. (2017). Photoactive gel for assisted cleaning during olive mill wastewater membrane microfiltration. *Membranes*, **8**, p. 66.

Militano, F., Poerio, T., Mazzei, R., Piacentini, E., Gugliuzza, A., and Giorno, L. (2016). Influence of protein bulk properties on membrane surface coverage during immobilization. *Colloids Surf. B: Biointerfaces*, **143**, p. 309.

Oberdorster, G., Oberdorster, E., and Oberdorster, J. (2005). Nanotoxicology: An emerging discipline evolving from studies of ultrafine particles. *Environ. Health Perspect.*, **113**, p.823e39.

O'Brien, M. N., Jones, M. R., and Mirkin, C. A. (2016). The nature and implications of uniformity in the hierarchical organization of nanomaterials. *Proc. Natl. Acad. Sci. USA*, **113**, p.11717.

Perrotta, M. L., Saielli, G., Casella, G., Macedonio, F., Giorno, L., Drioli, E., and Gugliuzza, A. (2017). An ultrathin suspended hydrophobic porous membrane for high-efficiency water desalination, *Appl. Mater. Today*, **9**, p. 1.

Scott, R. L., Datta, S., Guia, M., Cokera, E. L., Huggins, F. E., Daunertc, S., Bachasd, L., and Bhattacharyya, D. (2011). Reactive nanostructured membranes for water purification. *PNAS*, **108**, p.8577.

Kumar, C. S. S. R. (2009). *Nanomaterials for the Life Sciences. Nanostructured Oxides*, vol. 2. Weinheim, WILEY-VCH Verlag GmbH & Co. KGaA.

Read, S. A. K., Jimenez, A. S., Ross, B. L., Aitken, R. J., and von Tongeren, M. (2013). Nanotechnology and exposure scenarios. In: Vogel, U., Savolainen, K., Wu, Q., van Tongeren, M., Brouwer, D., and Berges, M. (Eds.), *Handbook of Nanosafety Measurement, Exposure and Toxicology*, Elsevier.

Nanotechnologies Vocabulary. (2010). 1st Edition, Core terms, International Organisation for Standardisation (ISO), ISO/TS 80004-1, Geneva, Switzerland.

Chapter 2

Nanostructured Membranes for Selective Separations: Modeling and Simulations

Elena Tocci, Carmen Rizzuto, Johannes Carolus Jansen, and Enrico Drioli
Institute on Membrane Technology of the National Research Council of Italy (ITM-CNR), Via P. Bucci, Cubo 17C, 87036 Rende (CS), Italy
e.tocci@itm.cnr.it

This chapter reports on the simulation techniques applied on the family of polymers of intrinsic microporosity (PIMs). The theoretical methods used to characterize morphology and gas transport in PIMs membranes will be described as well as the factors that control thermodynamic and kinetic properties in membranes.

2.1 Introduction

Modeling of structural and transport properties of polymers used in membrane fabrication has become one of the pillars of research in the field, especially in applications that use dense

Functional Nanostructured Membranes
Edited by Enrico Drioli, Lidietta Giorno, and Annarosa Gugliuzza

ISBN 978-981-4774-79-6 (Hardcover), 978-1-351-13511-5 (eBook)
www.panstanford.com

membranes. Many different polymers have been studied, of which PIMs will receive most attention in the present chapter. During the last decades, computational chemistry had a favorable impact in almost all branches of materials research, ranging from phase determination to structural characterization and property prediction (Baschnagel et al., 2000; Gubbins et al., 2011; Karger, Ruthven, and Theodorou, 2012; Maginn and Elliott, 2010; Sautet and Van Santen, 2009; Tafipolsky, Amirjalayer, and Schmid, 2010), as it allows dealing with different types of polymers as well as, for example, with thermal conductivity of composites (Ma et al., 2013), advanced batteries (Albertus et al., 2011; Ferrese et al., 2012), etc. Complementing experiment and theory, computer simulation gains greater importance in materials research. Simulations help to understand observed phenomena and even predict properties and scenarios in complex systems. Especially challenging is the quest for new materials and phenomena for which an experimental exploration without the knowledge from simulations would be prohibitive. Fundamental research in materials and structure design now considers numerical simulation, as a complement to theory and experimentation. The integration of physical testing, advanced computations, and system simulation would dramatically reduce the design and development time and costs. New materials are often developed not on rational considerations but by trial and error, due to the challenging time and length scales involved in the modeling of transport phenomena in polymeric membranes. However, the rapid progress in computational methodology and the development of new simulation tools have been gradually improving the understanding of different facets of gas transport in polymeric membranes for their effective use in materials design (ESF Programme n.d.; Sautet and Van Santen, 2009; Yampolskii, Pinnau, and Freeman, 2006).

The development of the nanostructured polymer-based membranes, formed by membranes with nanoscale pores, for membrane applications such as gas separation, is of paramount interest, providing a more advantageous combination of permeability and selectivity, i.e., the upper bound performance (Freeman, 1999; Robeson, 2008), than those currently used commercially. One direction of these researches has been to develop high free volume polymers or microporous organic polymers, which have exceptionally high gas permeabilities and very high glass transitions.

Their applications range in the area of gas storage (Tozawa et al., 2009; Yaghi et al., 2003), gas separation (Eddaoudi et al., 2002; Park et al., 2007), and low-dielectric applications (Long and Swager, 2003). Specific tailoring of the molecular structure can be regarded as a viable approach in obtaining an improvement on membrane permselectivity because of (i) the loose inter-segmental packing with a simultaneous inhibition of the intra-segmental (backbone) mobility, and (ii) a weakening of interchain interactions (reduction in charge transfer complexes) (Xiao et al., 2009). Rigidity in polymer chains can separate gas molecules of slightly different sizes (Guiver and Lee, 2013). Moreover, a depth analysis by Koros (Koros and Zhang, 2017) indicates the Achilles heel that reduces or eliminates the entropic factor advantage in the small-scale motion along the backbone at critical constrictions. For this reason, an accurate nanoscale control, in the range of sub-2 nm, is required for efficient and rapid molecular separation (Wang et al., 2017).

Recently developed highly permeable glassy polymers are basically constructed by rigid main chains such as ladder-like and contorted macromolecular structures or heteroaromatic structures and also contain flexible or rotatable domains in the main chains or side chains. In this context, PIMs have shown great promise for the future of gas separation technologies, as several PIMs are performing above the Robeson upper bound for certain gas pairs such as CO_2/light gas mixtures (Du et al., 2011). PIMs are a class of glassy polymers that create free volume as a result of the inefficient packing of adjacent polymer chains, yielding a significant amount of porosity below 2 nm (i.e., microporosity). This is the first reported study on the simulation techniques applied on the PIM family. After a brief discussion of some computational approaches used to cover different aspects of polymer membrane simulations, this chapter will describe theoretical methods used to characterize morphology and gas transport in PIM membranes, as well as the factors that control thermodynamic and kinetic properties in membranes.

2.2 Polymers of Intrinsic Microporosity

Since the first report by McKeown and Budd in 2004 (Budd et al., 2004), after previous works on network polymers (McKeown et al., 2002; McKeown and Budd, 2002), PIMs were reported to have intrinsic cavities in the polymer matrix by constructing ladder-like polymers with torsional domains. The intrinsic microporosity was defined as a "continuous network of interconnected intermolecular microcavities" less than 2 nm in diameter, which was formed as a consequence of the shape and rigidity of the polymer structure (Ilinitch et al., 1999; McKeown and Budd, 2011). PIM-1 (Fig. 2.1), the original material, was formed by a step polymerization based on a double-aromatic nucleophilic substitution mechanism from a spiro-containing tetrahydroxy monomer (5,5′,6,6′-tetrahydroxy-3,3,3′,3′-tetramethyl-1,1′-spirobisindane) and a tetrafluoro monomer (tetrafluoroterephthalonitrile), to form the dibenzodioxin linkage. In the case of PIM-1, a high molar mass polymer was readily obtained at temperatures in the range 60–80°C, but the molar mass was reduced at higher temperature (Budd, McKeown, and Fritsch, 2006).

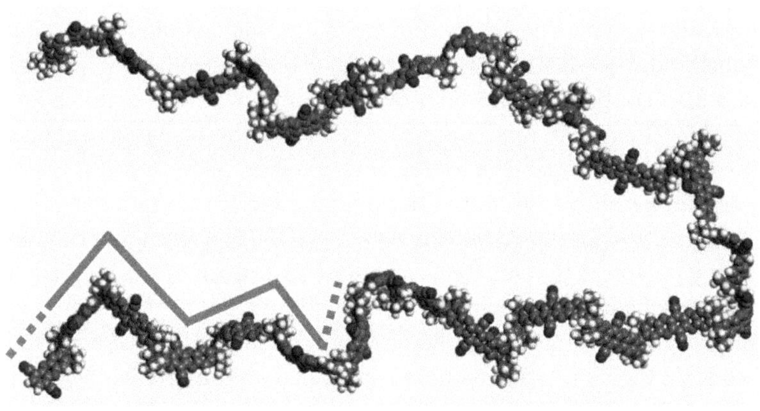

Scheme 2.1 Chemical structure of PIM-1 and polymer chain showing the rigid contorted structure.

PIMs have a structural combination of rigid ladder-like backbones with sites of contortion, which prevent intra- and/or interchain packing and create large free volume elements. Due to their fused ring structures, PIMs do not possess rotational freedom along the polymer backbone, which ensures that the macromolecular components cannot rearrange their conformation, so that their highly contorted shape is fixed during synthesis, providing some mechanical strength through entanglement (McKeown, 2012). Contortions in monomers are generally presented by spirocenters (i.e., a tetrahedral carbon shared by two rings) or other rigid non-planar structural units (McKeown and Budd, 2011). The microporous structures are not as highly dependent on process and thermal treatment history as for general glassy materials. PIMs are considered amorphous materials, since there is no evidence of crystallinity or a glass transition below the decomposition temperature. An investigation addressing the molecular mobility of PIM-1 was done by Böhning in 2016 (Konnertz et al., 2016) by using broadband dielectric spectroscopy. In PIM-1, a molecular relaxation process with Arrhenius behavior and an unusually high activation energy, denoted by β*, were observed together with significant conductivity in the glassy state. As expected, no α-relaxation (related to glass transition) was found. Both the β*-relaxation and the conductivity were explained with the formation of local intermolecular agglomerates due to the interaction of π-electrons in aromatic moieties of the polymer backbone (π–π stacking).

Figure 2.1 Picture of a PIM-EA(Me)-TB membrane. Courtesy: Mariolino Carta, University of Edinburgh, UK.

PIMs behave like molecular sieves and possess large internal surface areas—typically 300–3000 m^2/g, as measured by techniques based on the analysis of gas adsorption isotherms, such as the well-established Brunauer–Emmet–Teller (BET) model (Sing et al., 2013).

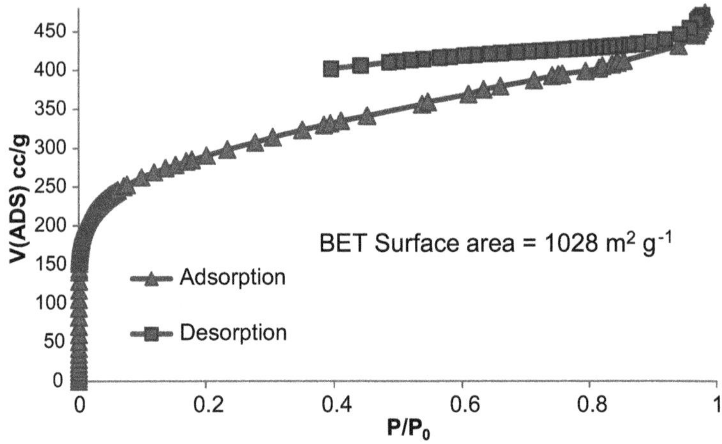

Figure 2.2 BET isotherm for ethanoanthracene TB-PIM treated with MeOH. Courtesy: Mariolino Carta, University of Edinburgh.

PIMs are unlike other polymers, which tend to enhance the attractive interactions between the component macromolecules, reducing the void space.

The high free volume in PIMs compared with other polymers increases gas solubility and diffusivity, resulting in improved permeability. The gas separation behavior of PIMs (Budd et al., 2008; Budd et al., 2005) for important gas pairs such as O_2/N_2 and CO_2/CH_4 was shown to surpass the upper bounds of performance as given in 1991 by Robeson (Robeson, 1991) and helped to define Robeson's revised 2008 upper bounds (Robeson, 2008). Subsequently, it was found that the permeability of PIM-1 was enhanced further just above the 2008 upper bounds for the CO_2/N_2 and CO_2/CH_4 gas pairs. Solvent cast films of PIM-1 were treated by methanol, which helps to flush out residual solvent and allows relaxation of the polymer chains (Budd et al., 2008). The upper bound performance of PIMs is attributed to high gas solubility, allowing increased permeability without sacrificing selectivity (Budd, McKeown, and Fritsch, 2005). A critical analysis of Robeson performed in 2016 compared the diffusion selectivity, solubility selectivity, diffusivity, and solubility coefficients of PIMs (and thermally rearranged polymers) (Robeson et al., 2016). The basic conclusion was that the upper bound performance for specific PIMs is due to a combination of higher diffusion selectivity, high diffusion coefficients, and high solubility coefficients, relative to other glassy polymers. The high solubility coefficients are a consequence of high free volume, combined with high glass transition temperatures, leading to significant dual-mode sorption, as noted previously (Robeson et al., 2014). Elevated solubility shifts the data to higher permeability values relative to other polymers at a given selectivity on the upper bound plots.

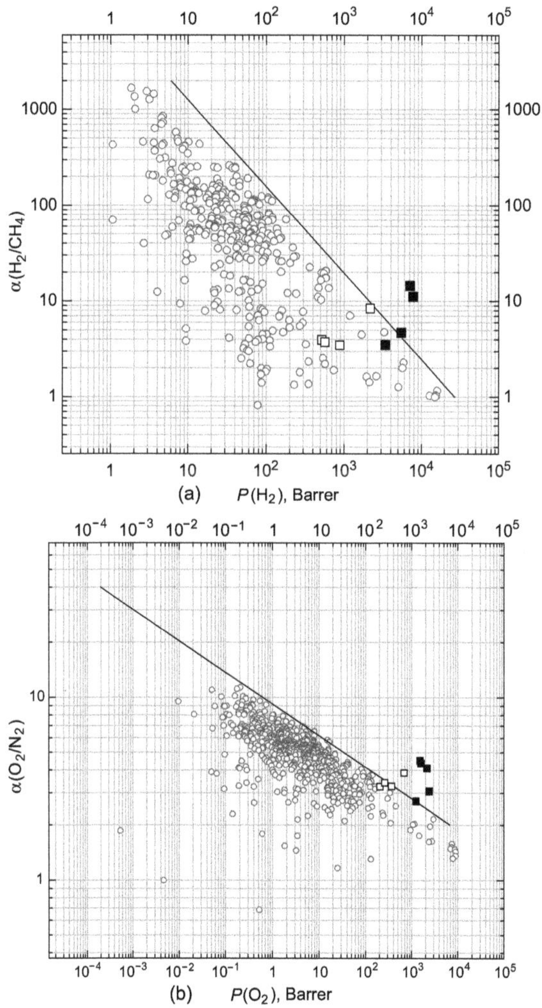

Figure 2.3 Robeson's diagrams: (a) H_2/CH_4 gas pair; (b) O_2/N_2 gas pair (Tocci et al., 2014). The line represents the upper bound of 2008 (Robeson, 2008). Reprinted with permission from Tocci et al. (2014). Copyright 2014 American Chemical Society.

There is a distinct tendency of this class of polymers to physical ageing, which is the major drawback with regard to practical membrane applications. PIM-based membranes exhibited accelerated "ageing" and thus leading to the loss of porosity, posing significant decrease in membrane performance over time. As shown in Fig. 2.4, the permeability of methanol-treated PIM-derivative membranes (PIM-EA-TB) decreased after the time lag analysis (Tocci et al., 2014). Nevertheless, PIMs contain interconnected regions of micropores with high gas permeability but with a level of heterogeneity that compromises their molecular selectivity. The PIM membrane displays extremely high gas permeability but relatively low gas selectivity.

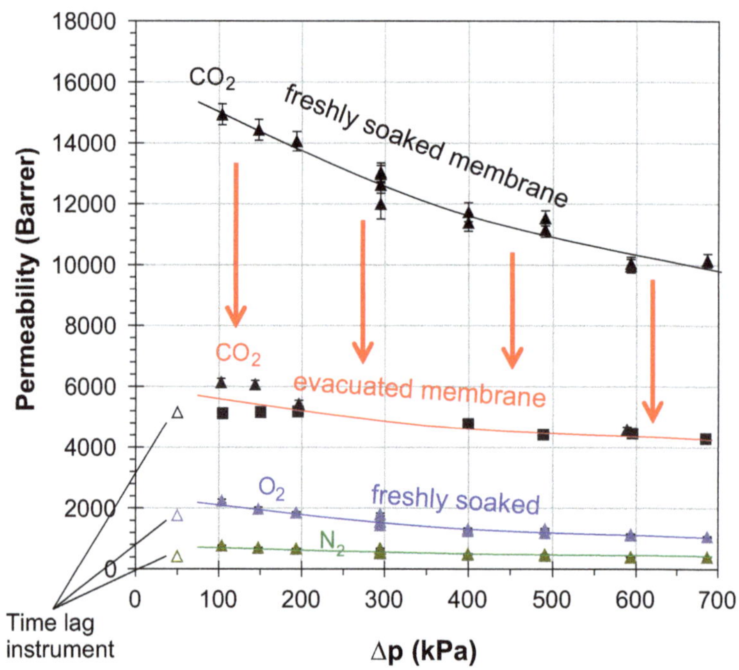

Figure 2.4 Permeability of pure CO_2 (black), O_2 (blue), and N_2 (green) gases measured on a MeOH-treated PIM-EA-TB membrane as a function of the feed pressure in the cross-flow cell. The red curve indicates the decrease in CO_2 permeability of the membrane after the analysis in the time lag instrument. Reprinted with permission from Tocci et al. (2014). Copyright 2014 American Chemical Society.

Several studies were pursued to modify the structure of the polymeric membranes by post-synthesis reactions centered on the nitrile group. These comprise hydrolysis groups to provide carboxylic acids (Du et al., 2009), reaction with P_2S_5 to provide thioamide (Mason et al., 2011), reaction of hydroxylamine to give amidoximes (Patel and Yavuz, 2012), and the reaction of sodium nitride to provide tetrazole functionality (Du et al., 2011). Also a chiral (+)-PIM material was prepared with an easy fabrication, showing both high and enantioselective permeability for a range of racemic compounds (Weng et al., 2015). New modifications are described in the following paragraphs.

2.2.1 PIM-PI, Ethanoanthracene (EA), and Tröger's Base–Based PIMs

In addition to the dibenzodioxin reaction, the more classical polymerization reaction of imide formation has been used to form PIMs termed PIM-PIs. Although the imide link is not composed of fused ring units, if suitable diamine aromatic monomers that contain methyl groups adjacent to the amine group are used, rotation about the C–N single bond is sufficiently restricted to prohibit conformation rearrangement and to generate significant intrinsic microporosity (McKeown, 2012).

These high free volume polyimides such as polyimides with intrinsic microporosity (Budd et al., 2004; Ghanem et al., 2008b, 2009, 2014; Rogan et al., 2014; Swaidan et al., 2014) were reported, some of which demonstrated high separation performance for several gas pairs exceeding Robeson's upper bounds.

The permselectivity can be tuned by incorporating molecular units of different lengths between the spirocenters, such as thianthrene (Du et al., 2010), 9,10-dimethyl-9,10-dihydro-9,10-ethanoanthracene (Emmler et al., 2010), ethanoanthracene (Ghanem et al., 2008a; Carta et al., 2014a; Carta et al., 2013) and with pyrazine (Budd et al., 2006).

PIMs with more rigid structure based on Tröger's base (TB) unit and polyimides containing TB (Zhuang et al., 2014; Wang, Wang, and Jin, 2014) (Fig. 2.5) have recently been reported and show improved permeabilities. Also the introduction of spirobifluorene (Fig. 2.6) as the main structural unit (PIM-SBF) results in a greater rigidity in the PIM structure and pushes the gas permeability above the 2008 Robeson upper bound for several gas pairs (Bezzu et al., 2012).

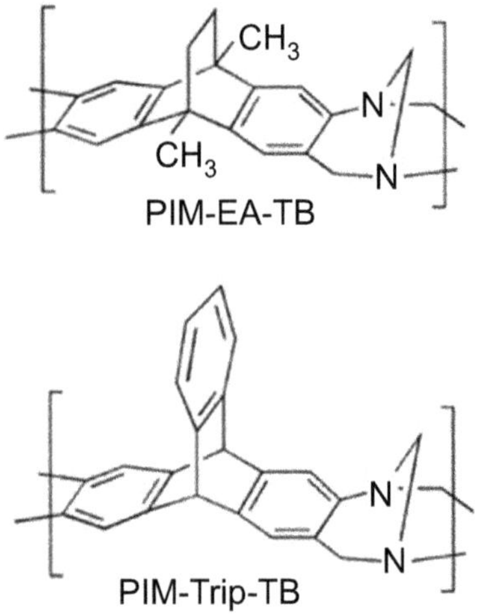

Figure 2.5 Ethanoanthracene (EA) and Tröger's base (TB) based PIMs (Carta et al., 2014).

Moreover, triptycene (Trip)-based PIMs prepared via TB formation show further enhancement of gas permeability data for O_2/N_2 and performance near the 2008 upper bounds for the CO_2/CH_4 and CO_2/N_2 gas pairs (Carta et al., 2014b).

Pendant substituents on the PIM backbone may also increase chain rigidity and act as interchain filling material, which effectively tunes the cavity shape and size, such as carboxylic acid groups (Du et al., 2009), sulfone-based groups (Du et al., 2009), and trifluoromethyl groups (Du et al., 2008).

2.2.2 Crosslinked PIMs

A further approach to improving PIM membrane performance is post-modification (Zhao et al., 2016) with crosslinks formed through multivalent metal ions.

Different approaches were utilized, including both carboxylated PIM-1 membranes by hydrolysis (Du et al., 2009) and PIM-1 membranes with CO_2-philic pendant tetrazole groups (Du et al., 2011) done by Guiver's research group.

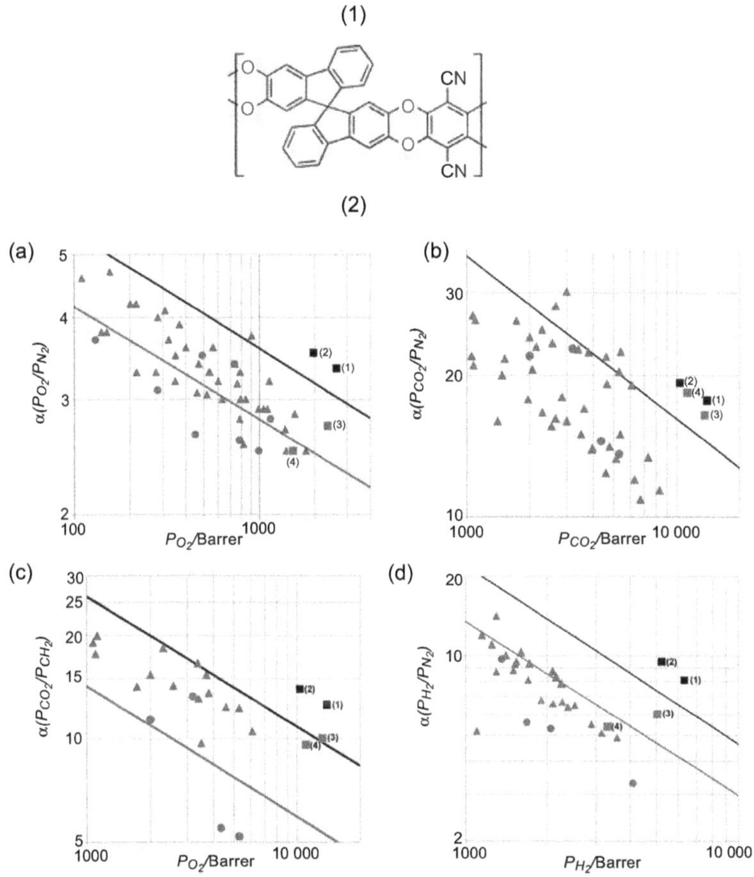

Figure 2.6 (1) Spirobifluorene unit; (2) Robeson plots for (a) O_2/N_2; (b) CO_2/N_2; (c) CO_2/CH_4; and (d) H_2/N_2 gas pairs showing the data for methanol-treated PIM-SBF with data points ▪(1) for a 180 μm film and ▪(2) for a 81 μm film. The black and gray lines represent the 1991 (Robeson, 1991) and 2008 (Robeson, 2008) upper bounds, respectively. Reprinted with permission from Bezzu et al. (2012). Copyright © 2012 WILEY-VCH Verlag GmbH & Co. KGaA, Weinheim.

Pure-gas permeation test results of UV-crosslinked PIM-1 membranes (Lesch, Estates, and Shumate, 2010) demonstrate that they exhibited CO_2/CH_4 performance well above Robeson's polymer upper bound trade-off curve for CO_2/CH_4 separation. Moreover, the amidoxime-PIM-1 membranes were synthesized by Yavuz's research group (Patel and Yavuz, 2012) and the high-temperature thermally self-crosslinked PIM-1 membranes were prepared by Chung's research group (Li et al., 2012a).

Song et al. (2014) prepared post-modified PIM-1 membranes with controllable thermal oxidative crosslinking by heat treatment in the presence of trace amounts of oxygen, producing covalently crosslinked networks that were thermally and chemically stable, mechanically flexible, and had remarkable selectivity at permeability of three orders of magnitude higher than commercial polymeric membranes.

Budd's research group prepared thioamide-modified PIM-1 membranes (Mason et al., 2011), amine-PIM-1 membranes (Mason et al., 2014), and the di/mono-ethanolamine-modified PIM-1 membranes (Satilmis, Alnajrani, and Budd, 2015). A strategy to increase CO_2 solubility in comparison to other gases is the incorporation of *N*-containing organic heterocyclic molecules such as pyridine, imidazole, and tetrazole into polymers. The interactions between CO_2 and such heterocyclic molecules (Vogiatzis et al., 2009) revealed that Lewis acid–Lewis base interactions are most important, as is hydrogen bonding with the negatively charged oxygen atoms of CO_2.

2.2.3 Mixed Matrix Membranes

Introducing a secondary filler phase to form mixed matrix membranes is one of the main approaches for improving PIM membrane performance and to aim at reducing physical ageing.

Their enhanced stability combined with perm-selective performances on or above Robeson's upper bound for several important gas pairs was improved by blending with inorganic materials such as silica nanoparticles (Ahn et al., 2010), zeolite imidazolate frameworks (ZIFs) (Bushell et al., 2013a), carbon nanotubes (CNTs) (Khan et al., 2013), few-layer graphene (Althumayri et al., 2016), porous organic cages (Bushell et al., 2013b), covalent–organic framework (Schiff base network) (Wu et al., 2017), β-cyclodextrins (Liu, Xiao, and Chung Tai-Shung, 2017), and porous aromatic framework (Lau et al., 2016).

In addition to the above materials, Cooper's group employed a cheap, acid-tolerant, nano-particulate hypercrosslinked polymer "sponge" as filler to prepare PIM membranes (Mitra et al., 2016).

As an alternative approach, Budd and coworkers recently prepared three-layer composite systems with a sandwich of PIM-1 between two MOF layers (Fuoco et al., 2017).

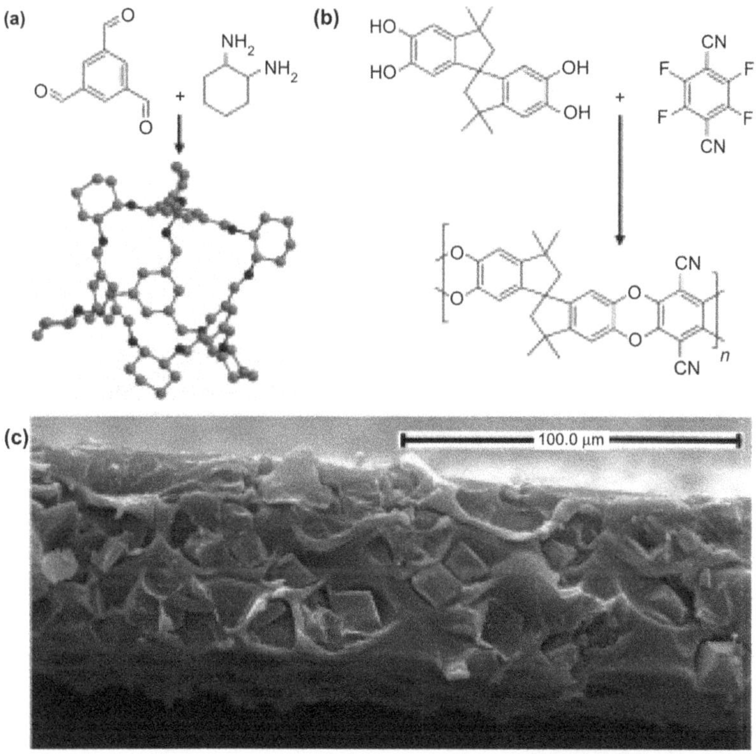

Figure 2.7 (a) Porous imine cage CC3 synthesized from 1,3,5-triformylbenzene and (R,R)-1,2-diaminocyclohexane by a condensation reaction. (b) PIM-1 is synthesized from 5,5′,6,6′-tetrahydroxy-3,3,3′,3′-tetramethyl-1,1′-spirobisindane and 1,4-dicyanotetrafluorobenzene by step polymerization involving a double-aromatic nucleophilic substitution. (c) SEM image of a cross section of a PIM-1/CC3 composite membrane (weight ratio 10:2). Reprinted from Bushell et al. (2013b). Copyright 2013 WILEY-VCH Verlag GmbH & Co. KGaA, Weinheim.

2.3 Simulation Methods

Molecular modeling can be used to describe a diversity of chemical systems with a wide range of complexity. A hierarchy of methods that refers to phenomena at different length and timescales (Theodorou, 2005; Gooneie, Schuschnigg, and Holzer, 2017) is necessary to address particular molecular systems and properties of interest. In this perspective, multiscale modeling methodologies, based on the information transfer between different scales, starting from the molecular level and ending up at the industrial scale, have been recently reviewed in the literature (Charpentier, 2016). They indicate a clear trend toward coupling of the design of chemical engineering equipment/units with nanoscale modeling.

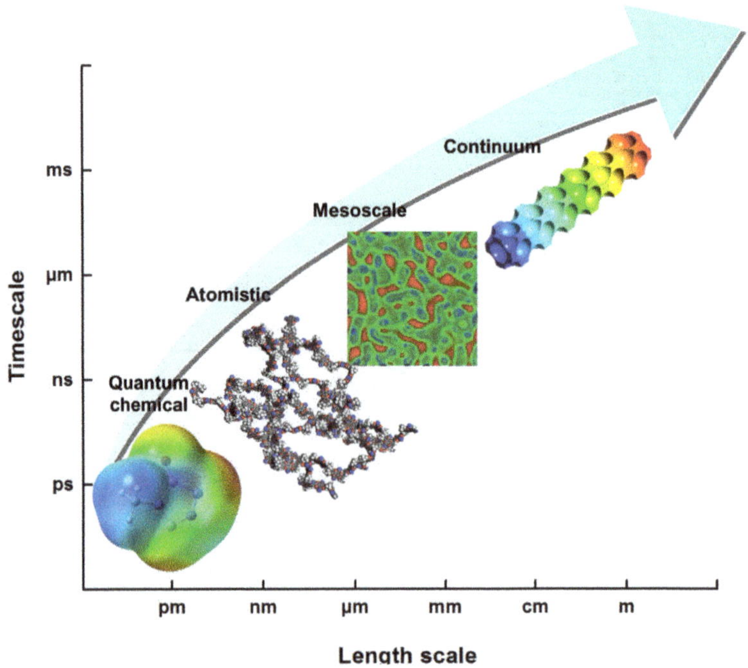

Figure 2.8 Time- and length scales relevant for materials and process development are sketched. Courtesy, for continuum model: Alessio Caravella, University of Calabria.

2.3.1 Ab initio

Ab initio methods (Friesner, 2005) refer to a collection of approximations to solve the Schrödinger equation, which is a many-body problem in its exact form, whose computational complexity grows exponentially with the number of electrons, and hence, a brute force solution is intractable (Dirac, 1929). The Hartree–Fock (HF) theory, a mean-field approach, produces reasonable results for many properties but is incapable of providing a robust description of reactive chemical events in which electron correlation has a major role. It neglects the correlated movement of the electrons—each electron experiences only the averaged potential of all other electrons (Head-Gordon, 1996). This leads to systematic errors that are not acceptable for calculations with chemical accuracy. Thus, a key problem has been the development of treatments of electron correlation that exhibit a tractable scaling in computational effort with the size of the system.

Although HF energies are not accurate enough by themselves, almost all higher correlated methods are based on the HF reference wave function via an inclusion of further (excited-state) determinants into the ground-state wave function. Density functional theory (DFT) methods (Kohn, 1999) are based on Hohenberg and Kohn's (Rajagopal and Callaway, 1973) theorem, which states that the exact ground-state energy of any molecular system can be computed from the one-particle electron density. With the knowledge of this functional, it would be possible to compute the exact ground-state energy. The functionals used within the local density approximation or the generalized gradient approximation (GGA) are guesses or approximations of this functional. First-principles DFT methods offer a high-level quantum mechanical description of the electronic structure at a cost that is affordable for the many thousands of structures that must be considered in the course of a reliable search.

2.3.2 Atomistic

The traditional simulation methods for many-body systems can be divided into two classes of stochastic and deterministic simulations, which are largely covered by the Monte Carlo (MC) method and the molecular dynamics (MD) method, respectively (Frenkel and Smit, 2001). Properties described at this scale range from thermodynamic properties to transport properties and phase equilibria. Gubbins and Moore (Law, 2010), Maginn and Elliott (Maginn and Elliott, 2010), and Theodorou (Theodorou, 2010) have recently published excellent reviews on MD and MC methods in the context of membrane and engineering applications.

2.3.2.1 Monte Carlo simulations

The MC technique is a stochastic simulation method that probes the configuration space by trial moves of particles (Theodorou, 2010). Within the Metropolis algorithm (Hastings, 1970), the energy change from step n to $n+1$ is used as a trigger to accept or reject the new configuration. Paths toward lower energy are always accepted; those to higher energy are accepted with a probability governed by Boltzmann statistics. In that way, properties of the system can be calculated by averaging over all MC moves (where one move means that every degree of freedom is probed once on average). Since its development, the MC technique has been valuable as a tool for testing statistical-mechanics-based theories for the properties of matter. The invention of new ensembles has greatly enhanced the possibility to understand and predict macroscopic thermodynamics based on molecular-level information. For the construction of polymeric membrane models, filling a basic cubic volume element under periodic boundary conditions, a rotational isomeric state (RIS) MC technique incorporating long-range interactions can be used.

The simulation of sorption properties of gas molecules in the amorphous cells of a glassy polymer can also be estimated using Grand Canonical Monte Carlo (GCMC) calculations. This needs as input the structural model of the amorphous cell and the force field describing sorbate–sorbent and sorbate–sorbate interactions. For the prediction of gas sorption in the generated amorphous cells, the interaction potential is the most important ingredient. A simplified interaction potential including only a dispersive–repulsive short-range potential, represented by a Lennard–Jones 6–12 potential, combined with electrostatic interactions between partial charges on the adsorbent and guest atoms is used.

2.3.2.2 Molecular dynamics

Molecular dynamics is an atomistic simulation method for studying a wide class of materials, such as polymers, metals, ceramics, and biomolecules, under ambient as well as extreme conditions. MD methods are governed by the system's Hamiltonian and Hamilton's equations of motion:

$$\dot{p}_i = -\frac{\partial H}{\partial q_i}, \; \dot{q}_i = \frac{\partial H}{\partial p_i} \tag{2.1}$$

where $H = H(q,p,t)$ is the Hamiltonian, which corresponds to the total energy of the system, indicated as a sum of kinetic (K) and potential (U) terms $H = K + U$; p_i are the positions and q_i are the momenta. Particles in MD move naturally under their own intermolecular forces and follow Newton's second law. The Hamilton equations of motion are integrated to move particles to new positions and to assign new velocities at these new positions. MD allows one to predict the time evolution of a system of interacting particles (e.g., atoms, molecules, etc.) and estimate the relevant physical properties. It generates such information as atomic positions, velocities, and forces from which the macroscopic properties (e.g., pressure, energy, heat capacities) can be derived by means of statistical mechanics. MD simulation usually consists of three constituents: (i) a set of initial conditions (e.g., initial positions and velocities of all particles in the system); (ii) interaction potentials to represent the forces among all the particles; (iii) evolution of the system in time by solving a set of classical equations of motion for all particles in the system. During the simulation, the configuration space as well as the phase space is explored, allowing to extract information on the dynamics of the system. The simulation of the gas diffusion in a polymeric membrane requires a force field representing the interactions between all the atoms of the system, composed of the polymer amorphous cell and penetrant molecules. The force field has to be validated against experimental results and theoretical constraints.

2.3.3 Physical Structure and Properties

Wide-angle X-ray diffraction (WAXD) is commonly used to determine the crystallinity and intermolecular space of a membrane material, via measurement of the d-spacing by Bragg's law (Bragg, 1912)

$$d = \frac{\lambda}{2\sin\theta} \tag{2.2}$$

where λ is the wavelength of the X-ray source (usually 1.789 Å corresponding to the Co K_α transition or 1.54 Å of the Cu K_α) and θ is the scattering angle.

The **fractional free volume** (FFV) of the membrane can be estimated using the equations:

$$\text{FFV} = \frac{V - V_0}{V} \tag{2.3}$$

where V is the cell volume and V_0 is the volume occupied by the polymer chains, calculated as:

$$V_0 = 1.3V_w \tag{2.4}$$

where V_w is the van der Waals volume. V_vdW is calculated from simulations without using the Bondi group's contribution method (Bondi, 1916–1968).

The FFV and free volume distribution (FVD) can be calculated using the Hofmann–Heuchel software (Hofmann et al., 2002). The free volume is determined by first superimposing a fine grid over the cubic packing. Then a spherical test particle, representing the penetrating molecule, is inserted at every point of the grid to determine if an overlap occurs with any atom of the polymer, represented also by a corresponding hard sphere. The result is a classification of grid points as "occupied" or "free." Then the connectivity of the "free" grid points is considered, and the connected "free" grid points are collected into groups, which represent individual holes. This can be done in two ways. In the first approach (named $V_{connect}$), the topological criterion is that every point of a group has at least one next neighbor, which is also a member of this group. This approach identifies holes, which may be of complex shape and of large size. In the second approach, for every grid point, the shortest distance to a polymer atom is used to group points, and among these distances, local maxima are defined by calculating the related gradient. Then each grid point of the free volume regions is assigned to its nearest local maximum. This approach is referred to as R_{max}.

Another method used to characterize free volume size distribution analyzes cross-sectional images of the membrane model in the *x*, *y*, and *z* directions using the Matlab program. In each direction, nine positions of the cubic membrane model to cleave are selected, and then nine cross-sectional surfaces are formed. These surfaces have a thickness of 1.5 Å. The cross-sectional images are collected and then digitalized into an image of 256 pixels × 256 pixels. These cross-sectional images then allow determination of the areas occupied by the free volume elements and by the polymers. The areas of free volume elements are calculated and then transformed into an equivalent diameter (D_{eq}). Gathering of all values of D_{eq} and their corresponding quantity yields the free volume size distribution.

For the **free accessible volume** (FAV) analysis, the accessible volume is probed using a hard spherical particle with a radius, in which the factor of 1.3 is not adopted. The van der Waals volume and the accessible volume of the membrane model can be obtained from the Visualizer Module in the Materials Studio software [BIOVIA (ex Material Studio 7.0) package. Classical simulation theory section. Visualizer User Guide, San Diego, 2013].

2.3.4 Thermodynamics and Transport in Polymeric Membranes

Sorption is composed of adsorption and absorption, with the former being related to the interaction of gas, liquid, or dissolved solids with the surface of the material, whereas the latter being related to the interaction of the species with its internal structure (i.e., related to the material volume or mass). A distinction between the two cases can be made by considering a dense material, i.e., no bulk phase can be recognized inside the structure, or a material with a certain degree of microporosity.

Solubility (S) is a direct measure of the efficiency of sorption, which is usually considered an equilibrium process, even though it is actually a dynamic process. The solubility, S_i, of a gas species i is defined as (Matteucci et al., 2006; Wang, Cao, and Chung, 2002):

$$S_i = \frac{C_i}{p_i} \tag{2.5}$$

where C_i [mol/kg] is the loading of the i-th species and p_i [Pa] is its partial pressure in the bulk phase, considering the whole process at equilibrium. The solubility can be directly calculated from sorption isotherms. In pure-gas conditions and at fixed temperature, the solubility is only a function of the partial pressure, whereas in mixtures, it is a function of the content of all species, as all of them generally affect the sorption of the single species. Experimentally, the sorption can be easily determined by volumetric or gravimetric methods.

The solubility coefficient can be calculated via simulations in the canonical ensemble in which the chemical potential is calculated using the Widom particle insertion method (Widom, 1963). The interaction energy of a gas particle inserted within the accessible free volume of the polymer matrix is calculated and the excess thermodynamic potential μ_{excess} can be estimated from (Eq. 2.6)

$$\mu_{\text{excess}} = RT \ln\langle \exp(-E_{\text{int}}/kT)\rangle \tag{2.6}$$

The solubility S is then obtained from the equation:

$$S = \exp(-\mu_{\text{excess}}/RT) \tag{2.7}$$

In highly rigid polymers such as PIMs, complications may exist if the timescale of the sorption analysis is shorter than that of the relaxations in the polymer. In this case, slow relaxation phenomena may lead to time-dependent transport parameters and dilation of the sample. In computer simulations, Henry's constant is usually calculated via MC statistical mechanics methods. A method used to perform such calculations requires the evaluation of the simulation-cell loading at several fixed pressures (Grand Canonical Ensemble).

2.3.5 Transition-State Theory

The prediction of diffusivity in polymer glasses and low-temperature rubbery polymers via direct MD simulations would require extremely long simulation times because the penetrant diffusion becomes too slow to be predictable by MD. On the other hand, the transition-state theory (TST) is a well-established methodology for the calculation of the kinetics of infrequent events in numerous physical systems. According to the TST method, the gas transport mechanism through a dense polymer system is described as a series of activated jumps. For each transition, a "reaction trajectory" leading from a local energy minimum to another through a saddle point in the configuration space is tracked, and the transition rate constant is evaluated.

The general approach followed to obtain the diffusivity is based on atomistic TST-based determination of rate constants for individual jumps, executed by the penetrant in the polymer matrix. Subsequent use of these rate constants within a kinetic MC simulation to track displacement at long times is a good example of hierarchical modeling. Three different approaches have been applied for describing the coupling between the jumping of a penetrant and the motions of nearby polymer chains. The original Gusev–Suter TST method (Gusev et al., 1993) uses a frozen polymer. All polymer chains are considered fixed in place, and TST-based rate constants are calculated from the energy barriers found for a penetrant to pass from one local potential energy minimum to another. This method is the most straightforward; however, in polymers it yields rate constants that lead to much lower diffusion coefficients (by factors of 10^3–10^6) than experimental values, because neglecting chain fluctuation contributions is physically unrealistic in a polymeric material. In an improved method, Gusev and Suter adapted the frozen polymer method, taking into account the thermal vibrations of the polymer matrix (Gusev and Suter, 1993). They assumed that the polymer atoms in a sorption site execute uncorrelated harmonic vibrations around their equilibrium positions to accommodate the guest molecules. The magnitude of the fluctuations is controlled by a parameter similar to the Debye–Waller factor in X-ray scattering. In the most recent method (Greenfield and Theodorou, 1998, 2001), referred to as the explicit polymer method, the dimensionality of the jump path was increased to include explicitly both penetrant displacements and polymer chain motions. Each resulting rate constant then captures the details of the particular chain motions that accompany the jump, rather than mean-field average motions, but at the expense of a much higher computational cost. This method has been applied to a few penetrants in polypropylene (Greenfield and Theodorou, 1998; Rallabandi and Ford, 2000), with reasonable agreement compared to experiment and calculation.

The TST approach permits the calculation of the solubility through the relation:

$$S = \frac{1}{kTV}\int_V \rho(r)dV \tag{2.8}$$

2.3.6 Diffusivity

The gas diffusivity can be estimated either from MD simulations or from TST theory by using the Einstein formulation:

$$D = \frac{1}{6N_a}\lim_{t\to\infty}\frac{d}{dt}\sum_{i=1}^{N_a}\left\langle [r_i(t)-r_i(0)]^2\right\rangle \tag{2.9}$$

where N is the total number of molecules and $r_i(t)$ is the unfolded position of the gas molecule i at time t. In binary mixtures, D is calculated by means of the Maxwell–Stefan expression:

$$D^{MS} = \frac{1}{6N\chi_1\chi_2}\left(\frac{M_2}{M_1}\chi_1 + \chi_2\right)^2 \times \lim_{t\to\infty}\frac{d}{dt}\left\langle\left[\sum_{i=1}^{N_2} r_i(0) - \sum_{i=1}^{N_2} r_i(t) - \right]^2\right\rangle \quad (2.10)$$

Here N is the total number of molecules, χ_i's are the mole fractions, M_i is the molecular mass of the i-th component, and $r_i(t)$ is the unfolded position of the molecule i at time t.

The diffusivity of small gas molecules in glassy polymeric membranes depends on their concentration and reaches a constant value at relatively high concentrations. In fact, since glassy polymers are not in a thermodynamic equilibrium state, their final "metastable" chain configuration depends on the processing history of the membrane. This makes the modeling of glassy polymeric membranes even more difficult due to the lack of experimental structural data for validating the computational approaches.

The predicted self-diffusivity depends principally on the quality of the force fields used to model the interactions between not only the penetrant and polymer matrix, but also intramolecular interactions between polymer chains. The role of chain relaxation and matrix fluctuations in explaining the diffusion mechanism of small gas penetrants as N_2 in rubber polymeric membranes was clearly demonstrated through MD calculations in which the polymer matrix was kept fixed (van der Vegt, 2000; Tocci et al., 2008).

2.4 Modeling of Polymers of Intrinsic Microporosity

Starting from the prototypical PIM-1 (Heuchel et al., 2008; Larsen et al., 2011a), many other PIMs were simulated, and with different modified structures (Larsen et al., 2011a; Hart et al., 2013; Larsen, Hart, and Colina, 2014; McDermott et al., 2014) and functional groups such as cyano, trifluoromethyl-phenylsulfone, and carboxyl (Fang, Zhang, and Jiang, 2011), PIM-7 (Fang, Zhang, and Jiang, 2010), and sulfur-containing PIMs (Hart et al., 2013a), with various bridged bicyclic units (Zhang et al., 2016), backbone lengths (Hart and

Colina, 2014). Also different PIM-PI polymers have been simulated (Chang et al., 2013; Tocci et al., 2014; Zhao et al., 2016; Zhou et al., 2014). Most studies are related to the morphological properties, with the analysis of the dihedral angle profile, FFV, FAV, cavity size distribution (CSD), and surface area and pore size distribution (PSD). These properties were studied in parallel to thermodynamics and transport properties, such as adsorption of different gases, solubility coefficient, and diffusivity of various gases. They are discussed in more detail in the following paragraphs.

2.4.1 Physical Structure and Properties

2.4.1.1 Torsion angle analysis

PIM-1 consists of flat ladder parts connected by "joints," i.e., the spiro atoms that represent also the "sites of contortion." For this reason, the dihedral angle analysis emphasizes the impact of the shape (i.e., kink angle), the rigidity (i.e., shape persistence), and three-dimensional character of the contortion site on the macromolecular rigidity, i.e., the packing efficiency of the polymer chains. Heuchel et al. (the first paper simulating PIMs in 2008) (Heuchel et al., 2008) analyzed in PIM-1 the rigidity of the spirocenter, calculating the dihedral angles in the bulk state (i.e., amorphous polymeric boxes). A broad bimodal distribution of dihedral angles with maxima at about −75° and −35° was found. On the other hand, the distribution of the ether bond indicated a certain oscillation of the ladder planes out of planarity. This suggests partial flexibility of the ladder part, which was already observed as "bent" segments. Freeman's theoretical analysis of the position of the upper bounds suggested further enhancement in gas selectivities by designing polymers with great shape persistence (Freeman, 1999; Guiver and Lee, 2013) and stimulated the creativity of different groups that started to synthesize PIM-derivatives with such features. In 2013, Carta et al. synthesized two highly rigid PIMs, with bridged bicyclic amine 2,8-dimethyl-6H,12H-5,11-methanodibenzo [b,f][1,5]diazocine, called TB units: a spirobisindane-based TB polymer (PIM-SBI-TB) and an ethanoanthracene-based TB polymer (PIM-EA-TB) (Carta et al., 2013). They modelled the dihedral angles of the bridged bicyclic ring systems and demonstrated an increase in the torsion energy compared with the typical components of PIMs known so far (Fig. 2.9).

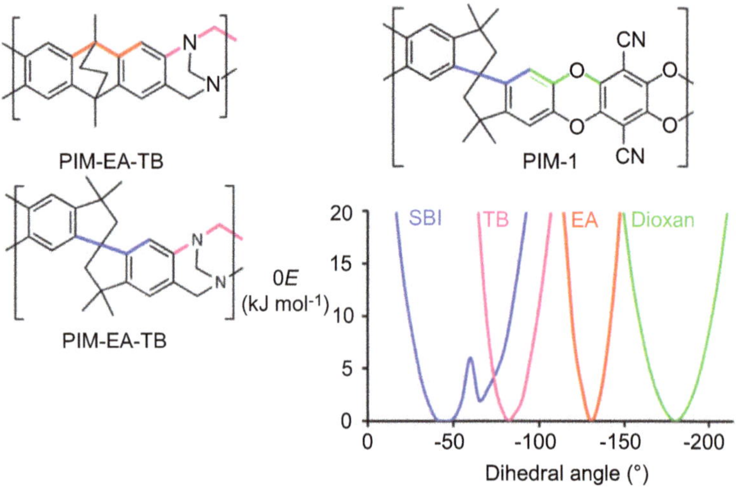

Figure 2.9 A plot showing the increase in energy associated with the deviation in the marked dihedral angle within the bridged bicyclic units of EA (red) and TB (purple) as compared with typical components of PIMs, such as the spirocenter of SBI (blue) and the dioxan linking unit (green). The narrower energy wells of the bridged bicyclic units TB and EA demonstrate their greater rigidity. Reprinted from Carta et al. (2013). Copyright 2013, American Association for the Advancement of Science.

Tocci et al. analyzed PIM-EA-TB (Tocci et al., 2014) while Zhou et al. modelled both PIM-EA-TB and PIM-SBI-TB (Zhou et al., 2014). In their analysis of the distribution of torsional angles, the bridged bicyclic EA was more rigid than the spirocenter SBI. Thus, PIM-EA-TB chains result stiffer than those of PIM-SBI-TB, explaining the experimentally observed high size selectivity of the polymer (Carta et al., 2013). The torsional angles of the PIM-PI-1 and PIM-PI-8 were studied by Chang et al. The spirocenters in the PI-derivative membranes stiffened the polymer chains and, as a result, they lowered the packing efficiency (Chang et al., 2013). Sulfur PIM (Hart et al., 2013a) exhibited a very different torsional distribution in comparison to PIM-1, owing to the electronic differences between sulfur atoms and the original oxygen, and to the Coulombic interactions of the strong S=O dipoles, maximizing distances between negative charges. The dihedral angle profiles of Trip-based-PIM, i.e., PIM-Trip-TB and KAUST-PI-1 were compared with those of PIM-PI-1 by Chen et al. (2016). The authors applied the ab initio calculation to obtain the charges of the simulated monomers. The torsion angle analysis demonstrated the high rigidity of the bridged Trip units in PIM-Trip-TB and KAUST-PI-1 membranes. Expanding the range of PIMs and PIM-PI, two solution-processable polyimides bearing 9,10-bridgehead-substituted triptycene were synthetized and characterized (Swaidan et al., 2014). The contortion center geometry was studied by computational methods, and the results were compared with the torsional angles of other bridged bicyclic and spiro-based PIMs. The results indicated that the development of microporosity depends largely on the ability of a contortion center to resist molecular deformation and to maintain the intramolecular rigidity that disrupts chain packing (Maier, 2013). Zhang et al. used an ab initio methodology for their studies on the rigidity of modified PIMs, synthesized by locking the spiro-carbon pivot point (Zhang et al., 2016). The spirobisindane structure obtained was an 8-membered dihydrooxocine ring (SBI). A full quantum mechanical approach, using DFT, was performed to quantify the rigidity of PIM-C1 and to compare it with other PIMs, using B3LYP functional with the 3-21G^{+} basis set. The equilibrium dihedral angle of the spirobisindane structure was found to be shifted to the left after locking (from $-41°$ for unlocked SBI to $-31°$ for locked SBI), so a reduced dihedral angle was found, meaning an increased rigidity. The curvature of the dihedral potential energy surface was calculated to quantify the difference in the structural rigidities of different moieties, fitting a harmonic model to the surface. The spring constant was taken as a rigidity parameter. Locking the SBI by adding an 8-membered ring substantially increased the rigidity parameter value of the spiro-carbon by 230% relative to the unlocked version. The authors found that the rigidity of the locked spiro-carbon was close to that of TB and ethano dihydroanthracene (EA) structures (21 kcal/mol rad^2 for EA and 24 kcal/mol rad^2 for TB).

2.4.1.2 WAXD and packing ability of PIMs

The first broad X-ray scattering data collected for PIM-1 were compared with WAXS patterns with structure factors from MD simulation by McDermott's work (McDermott et al., 2011).

Comparison of the X-ray patterns of the methanol-treated films with simulations indicated that the nearest-neighbor segments in nonporous amorphous polymers give rise to one or two amorphous halos against a relatively flat WAXS background. Instead, the inter-segmental distances in PIMs contributed to a scattering shoulder. The broad peaks observed, superimposed on the shoulder, represent characteristic distances between segments or sites of contortion on different chains. In summary, PIM-1 has at least three characteristic peaks, namely (A) the spirocenter–spirocenter and backbone–backbone correlations on the same chain (~12 Å), (B) the distance between loosely packed polymer chains (~7 Å), and (C) the distance between closely packed polymer chains (~4 Å).

Simulated scattering patterns for PIM-1c and PIM-1n showed similar characteristic peaks as PIM-1 (Larsen et al., 2011a). Another method to provide insight into the packing ability of PIMs was obtained considering the spirocenter–spirocenter radial distribution functions (RDFs) of non-swollen PIM-1, PIM-1c, and PIM-SBF-Me (Hart et al., 2013). The authors found that the initial peak was connected to the size of the spirocenter, as in previous papers (Larsen et al., 2011a; Larsen et al., 2014). The spirocenter of PIM-SBF-Me, with the largest group, creates a bulky functionality surrounding the spirocenter and forces closely associated chains further apart, while still allowing inter-spirocenter pockets of free volume to be present. In addition, scattering curves were calculated at different swelling percentages of PIM-1. The information gained indicated that during the swelling process, polymers closely packed together were relatively unaffected due to the favorable van der Waals interactions of the backbones, and chains containing void space were spread apart as the free volume (swelling) increases (Hart et al., 2013). The spirocenter–spirocenter RDFs of PIM-1f, a model structure of a hyperfluorinated polymer, supported the hypothesis that interchain spacing was influenced by spirocenter and pendant group functionality and size. This polymer has a larger free volume due to spirocenter functional groups than PIM-1, PIM-1c, and PIM-1n (Larsen et al., 2014). From wide-angle X-ray diffractions, the estimated *d*-spacing distances estimated via Bragg's law in PIMs with the functional groups cyano, trifluoromethyl-phenylsulfone, and carboxyl (Fang et al., 2011) agree well with available experimental results. However, with bulky trifluoromethyl-phenylsulfone, particularly the latter, the chain-to-chain distance showed a rather small increment. The plausible reason being that phenylsulfone acted as a filler, occupying the large voids. The chain-to-chain distance in carboxyl-PIM was the shortest among all membranes, indicating the formation of hydrogen bonds among carboxyl groups in the polymer chains. The carboxyl groups can act as a donor as well as an acceptor to form hydrogen bonds. Consequently, the carboxyl-PIM membrane was densely packed and possessed a higher density and a larger percentage of small voids. The WAXD analysis on the two PI-derived membrane models, PIM-PI-1 and PIM-PI-8, which contained spirocenters and several softer segments than PIM-1, indicated that PIM-PI-1 and PIM-PI-8 had larger *d*-spacing values due to the presence of the contortion sites (Chang et al., 2013).

The simulated WAXD patterns of PIM-EA-TB and PIM-SBI-TB indicated that the two membranes were essentially amorphous (Zhou et al., 2014). The small-angle region shows three typical peaks, representing the distance between neighboring stiff segments or micropores. The distance between the neighboring spirocenter SBIs was shorter than that of the neighboring bridged bicyclic EAs, and thus the rigidity (stiffness) of PIM-SBI-TB was less than that of PIM-EA-TB. The peak representing the chain-to-chain correlation between polymer backbones reveals a looser structure and a larger FFV in PIM-EA-TB than in PIM-SBI-TB. The TB-based PIM copolymer membranes (Wang et al., 2014) were also tested by WAXD measurements. As compared to PIM-1, the peaks around 11.3 Å for TB-based PIM copolymers move to lower angles and larger *d*-spacings. This result implies that the introduction of TB units interrupts the arrangement of polymer chains to a certain extent. It is mostly attributed to the more rigid TB structure instead of the more flexible spiro-unit in PIM-1, leading to a more outstretched and less coiled polymer chain.

2.4.1.3 Cavity size distribution

The geometric calculations of the simulated porosity provide density, pore volumes, and surface area. The CSD analysis was performed widely in PIM-1 and PIM-derivatives because the voids (microporosity) and their distributions govern sorption and diffusion. Heuchel et al. developed a method to analyze the void distribution in PIM-1 and found the size distribution function in the range of 5–15 Å with a peak at 3 Å, using the R_{max} FVD (Heuchel et al., 2008). Similarly, Fang et al. studied PIM-1 and PIM-7, revealing that both membranes exhibit a broad void size distribution with a diameter in the range of 0–9 Å and a peak around 2.2 Å (Fang et al., 2010). Instead, Colina's group used a new protocol to prepare realistic models of the amorphous structure of PIMs (Abbott, Hart, and Colina, 2013). They used the FFV results to derive the PSD by differentiation with respect to the pore width (Gelb and Gubbins, 1999)

$$\mathrm{PSD} = -\frac{dV_{\mathrm{pore}}(r)}{dr} \tag{2.11}$$

where r is the probe radius. The simulated PSDs were compared with the experimental PSDs derived from the Horvath–Kawazoe (HK) method (Budd et al., 2005) and Positron Annihilation Lifetime Spectroscopy (PALS) (Staiger et al., 2008). For PIM-1, the authors found a PSD with a peak at 4.3 Å, consistent with the experimental results (Larsen et al., 2011b). In another paper, they compared the PSD data of PIM-1 with that of two different PIM-1-like polymers, PIM-1c and PIM-1n. The similar structures of PIM-1, PIM-1c, and PIM-1n produced similar PSDs with peaks at 3.2, 2.5, and 3.2 Å, respectively. The force field chosen to simulate PIM-1 is essential to capture the pore topology (Hart, Abbott, and Colina, 2013).

The introduction of cyano, trifluoromethyl-phenylsulfone, and carboxyl groups (Fang et al., 2011) reduced the percentage of large voids in comparison with the precursor PIM-1, especially with the bulky trifluoromethyl-phenylsulfone because they occupied partially the large voids. When the effect of residual solvent in PIM-1 was studied (Zhang, Fang, and Jiang, 2011), the large-size voids in PIM-1/solvent (CH_3OH, $CHCl_3$, and H_2O) decreased, following the order $CH_3OH > CHCl_3 > H_2O$ in accordance with the measurements by PALS (Budd et al., 2008). Nitrogen adsorption isotherms at 77 K (for five PIMs: PIM-1, PIM-1c, PIM-SBF, PIM-SBF-Me, and sPIM-1)(Hart et al., 2013) indicated that the PSD shifted to larger pores due to swelling. The size of the voids increased with the simulated degree of swelling, while the closely packed chains remained associated. A newly proposed hypothetical polymer (Larsen et al., 2014), PIM-1f, i.e., a hyperfluorinated PIM, with a large volume spirocenter functional group exhibited a relatively low PSD, although it displayed a high surface area. This highlighted the fact that although a higher porosity is necessary for increased adsorption capacity, a higher surface area alone is not a sufficient criterion for the evaluation of adsorbent performance. With the introduction of the bulky trimethylsilyl group in four different PIMs, each of which altering spirocenter size or backbone functionality or sulfur moiety or backbone length, the PSD correspondingly increased (Larsen et al., 2014). The CSDs of PIM-PI-1 and PIM-PI-8 (Chang et al., 2013) were comparable in size to those of PIM-1. However, the TB-based PIM copolymers synthetized by Wang et al. (2014) evidenced a pore width of the TBPIMs smaller than PIM-1. The free volume element size distribution of PIM-EA-TB (Tocci et al., 2014) was quantified by the R_{max} and $V_{connect}$ approach, setting the probe radius to the size of an o-positronium particle (radius = 1.1 Å). The $V_{connect}$ distribution indicated peaks for equivalent spheres with apparent "radii" in a range up to about 15 Å, while the maximum of the R_{max} distribution is at about 2.6 Å, which agreed well with the PALS results (de Miranda et al., 2007) with a superposition of the small radius at 2.52 Å with the average R_{max} peak (Fig. 2.10).

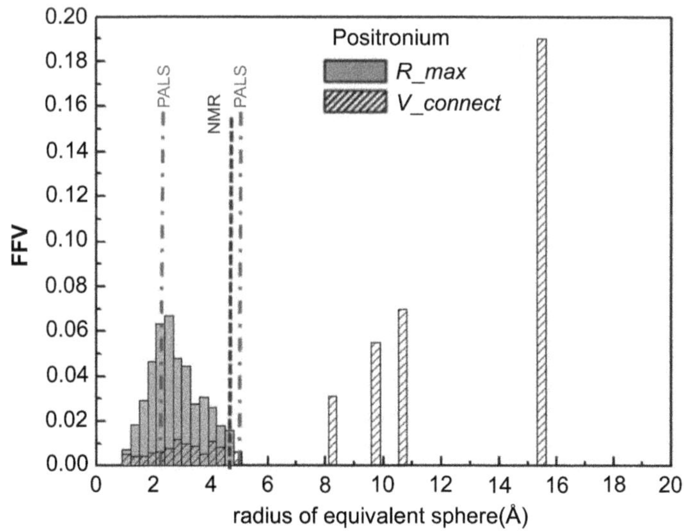

Figure 2.10 Averaged size distribution functions using positronium (r = 1.1 Å) as probe with the Hofmann–Heuchel method (Hofmann et al., 2002). Reprinted with permission from Tocci et al. (2014). Copyright 2014 American Chemical Society.

The structural variations in Trip-based PIM-PIs KAUST-PI-1 and KAUST-PI-1, prepared with different alkyl bridgehead substituents, were helpful to elucidate the effect of 9,10-bridgehead-substitution on the microstructures on the PSD of PIMs (Swaidan et al., 2014). The PSDs showed a bimodal distribution in microporosity, with a significant increase in the fraction of smaller, molecularly dimensioned pores, concomitant with a slight increase in the fraction of larger pores. Hypothesizing that the larger pores are interconnected through the smaller pores, this auspicious introduction of finer ultramicroporosity (<7 Å) along with some larger pores induced by the short and bulky branched substituent should enhance diffusion-based gas transport. Chen et al. (2016) used image analysis algorithm to analyze the cavity size of PIM-Trip-TB and KAUST-PI-1 compared to PTMSP and PIM-PI-1 membranes. They found that the Trip units enhanced the large PSD. The cavity size intensity was found to increase in the order of PIM-PI-1 < KAUST-PI-1 < PIM-Trip-TB < PTMSP at the large cavity size region.

2.4.1.4 Surface area

Molecular simulations provide a direct geometric measurement of the surface area. The solvent-accessible surface area was indicated to be most appropriate for comparison to BET surface areas (Düren et al., 2007). Experimental BET surface areas for PIM-1 have been reported from 750 to 860 m^2/g (Ghanem 2008a; Budd et al., 2004). The simulated solvent accessible of PIM-1 was of 448.99 ± 85.18 m^2/g, using a nitrogen size probe of 3.681 Å and is similar to the GCMC nitrogen adsorption based calculation of 435 m^2/g from Heuchel (Heuchel et al., 2008). Surface areas of PIM-1c and PIM-1n were of 319.28 ± 116.63 and 479.73 ± 98.47 m^2/g, respectively. The trend in surface areas for PIM-1 and other four PIMs (PIM-1c, PIM-SBF, PIM-SBF-Me, and sPIM-1) indicated that sPIM-1 < PIM-1c $\lesssim$ PIM-SBF $\lesssim$ PIM-1 < PIM-SBF-Me (Hart et al., 2013). The trends in the simulated data directly correlate with experimental BET surface areas: PIM-1c < PIM-SBF < PIM-1. The hypothetical PIM-SBF-Me showed the highest BET surface area. The geometric accessible surface areas for PIM-1, PIM-1c, PIM-1n, and PIM-1f (Larsen et al., 2014) indicated that PIM-1f and PIM-1n showed the highest (pore volumes and) surface areas, although the two differed widely in density, i.e., for PIM-1n, d = 0.959 ± 0.021 g/cm^3 and for PIM-1f, d = 1.177 ± 0.034

g/cm^3. Similar to the experimental trend, the simulated BET surface area of PIM-Trip-TB and KAUST-PI-1 compared to PTMSP and PIM-PI-1 membranes (Chen et al., 2016) increased following the order: PIM-PI-1 < KAUST-PI-1 < PIM-Trip-TB < PTMSP. In conclusion, the simulated surface areas give qualitatively very similar trends as the experimental values, and since surface area is an important index for permeability, this analysis is a useful tool to anticipate the membrane performance in terms of permeability.

2.4.1.5 Fractional free volume and fractional accessible volume

The FFV and FAV analyses show the membrane morphology and connective pathways, respectively. Heuchel et al. used the $V_{connect}$ and R_{max} methods to determine FV elements in the prototypal PIM-1 (Heuchel et al., 2008). A probe of the dimension of a positronium particle (r = 1.1 A) revealed, for the first time, a feature not found in high performance polyimides (Hofmann et al., 2002), namely peaks for equivalent spheres with apparent radii of about 16 Å. This was a clear indication of the presence of high interconnected free volume regions in PIM-1, and a confirmation of their intrinsic microporosity. The peaks indicated a kind of continuous hole phase, additional to standard free volume elements organized in isolated holes, with radii up to 6 Å. The same analysis performed with a probe radius similar to nitrogen (r = 1.82 Å) and oxygen (r = 1.73 Å) gave a smaller distribution of free volume elements with less interconnectivity in comparison to the smaller positronium probe), and relatively similar for both gases.

In another study, with a probe size equal to zero, the maximum FFVs in PIM-1, trifluoromethyl-phenylsulfone-PIM (TFMPS-PIM), and carboxyl-PIM (CX-PIM) (Fang et al., 2011) were about 45.2%, 42.1%, and 38.7%, respectively. Using Eq. (2.3), the FFVs in PIM-1, TFMPS-PIM, and CX-PIM were estimated to be 28.8%, 24.7%, and 20.3%, respectively. These values agreed quantitatively well with the FFVs calculated by Du et al. using the group contribution method (Du et al., 2008). The FFV decreased following PIM-1 > TFMPS-PIM > CX-PIM, which was consistent with void morphologies and void size distributions as discussed in previous paragraphs. In particular, the hydrogen bonding observed among carboxyl groups contributed to the lowest FFV in CX-PIM. The FFVs in PIM-1 and PIM-7, as calculated by Fang et al., showed unlikely high values of 47.7% and 46.6%, respectively (Fang et al., 2010). The authors related the FFVs to that of common polyimide membranes whose FFV is about 30–38% (Heuchel, Hofmann, and Pullumbi, 2004). The FFVs of PIM-1 and PIM-7 are higher than that of polyimide but lower than that in metal–organic frameworks (MOFs) and covalent–organic frameworks (COFs) (Babarao and Jiang, 2008; Babarao et al., 2008). The comparison between PIM-1 and PIM-7 indicated that PIM-1 had a larger FFV and a slightly higher percentage of voids with a diameter greater than 6.5 Å. This was attributed to the cyano groups in PIM-1, which lead to a steric hindrance and inefficient packing of the PIM-1 membrane. The FFVs of PIM-1 after contact with solvents ($CHCl_3$, CH_3OH, and H_2O) (Zhang et al., 2011) decreased from 47.7 to 44–45%. The comparison with dry PIM-1 in the absence of solvent [76] indicated that the density increased from 1.03 to 1.07–1.09 g/cm^3 and the FFV diminished. This is because the volume of the membrane remained almost constant in the absence/presence of solvent molecules and solvent molecules tended to occupy the voids in the membrane; consequently, the density increased slightly.

The mobility of PIM-1 chains was found to be reduced but was then facilitated by solvent due to the cooperative interactions between polymer and solvent. The FFVs and large-size voids in PIM-1/solvent membranes are observed to decrease in the order CH_3OH > $CHCl_3$ > H_2O, consistent with PALS measurements. Simulated ultraviolet (UV)-rearranged PIM-1 showed considerable reduction in FFV. The reason was due to the fact that polymer chains of PIM-1 underwent 1,2-migration reaction and transformed to close-to-planar like rearranged structure after UV radiation (Li et al., 2012b). The FFVs of PIM-PI-1 and PIM-PI-8 membranes, analyzed by Chang et al., were found to be comparable with that of the PIM-1 membrane (Chang et al., 2013). The FFV of the PIM-PI-8 membrane was greater than that of the PIM-PI-1 membrane because of the bulkier 3,3-dimethly-naphthidine moiety in PIM-PI-8, which reduces its packing efficiency. In order to compare the effective free volume for various species, the FAV was analyzed using hard spherical probes with different radii. The FAV values decreased as the probe radius increased. Interestingly, the FAV values approached zero when the probe radius was larger than approximately 2 Å in the case of PIM-1, PIM-PI-1, and PIM-PI-8. This phenomenon is in contrast to the experimental membrane cavity characterization (Budd et al., 2005; Budd et al., 2008; Ghanem et al., 2008b; Hofmann et al., 2002; Wang et al., 2005). The PIM-1, PIM-PI-1, and PIM-PI-8 membranes were found to have cavity widths larger than approximately 6 Å, and the PTMSP membrane had a cavity width larger than 7 Å. The authors explained this mismatch of the simulated FAV and the experimental cavity width by the nature of the simulation method: The FAV was analyzed from the static structure (at different simulation timeframes) and for this reason, the connectivity of the cavities would have been underestimated. Hölck et al. studied the sorption and dilation characteristics of three polymers—polysulfone (PSU), polyimide 6FDA-TrMPD, and PIM-1—upon contact with CO_2 and CH_4 (Hölck et al., 2013). Free volume analyses were performed by test particle insertion, and this procedure indicated that the free volume of all polymers increased during swelling. In particular, for PIM-1 the rise was in the ratio of 1.55. Moreover, calibration of the test particle radius to a standard gave good qualitative agreement to an estimation of the free volume by the Bondi's method. A visual, qualitative representation of the FVD was given for PIM-PI-EA (Rogan et al., 2014). Figure 2.11 shows the free volume in a PIM-PI-EA model, packed and equilibrated at the measured film density of 1.1 g/cm^3. The small o-positronium probe could access a highly interconnected void structure. The H_2 molecule could access less free volume, but more than the larger N_2, and for H_2, the free volume was also more interconnected than for the N_2 probe.

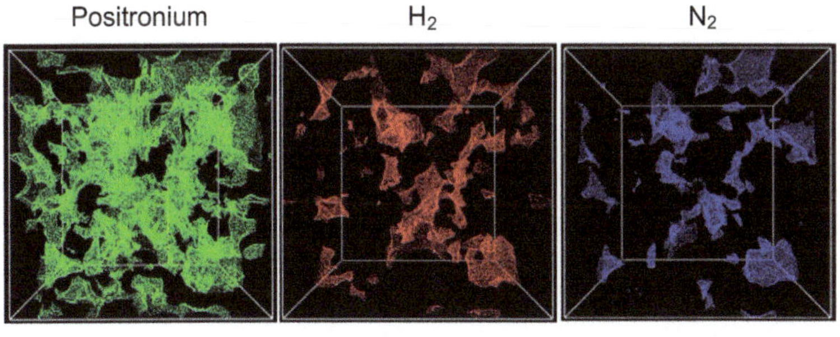

Figure 2.11 Representation of the free volume distribution in a 3D box model of PIM-PI-EA, packed down to a density of 1.1 g/cm^3. The blue, red, and green shaded areas highlight the accessible free volume when probing with o-positronium, hydrogen, and nitrogen probes, respectively. Reprinted from Rogan et al. (2014) under Creative Commons Attribution Licence, Copyright 2014, The Royal Society of Chemistry.

The FFV of PIM-EA-TB (Tocci et al., 2014), simulated using the conventional Bondi's method, was found to be 0.276, which is just slightly larger than the value of 0.24–0.26, estimated by Guiver et al. for PIM-1. The interconnectivity of the free volume nanostructure and the available FFV was found to be greater for oxygen than for nitrogen, explaining the relatively high permselectivity, $\alpha(O_2/N_2)$, and diffusion selectivity, $\alpha_D(O_2/N_2)$, for this gas pair. The FAVs in PIM-EA-TB and PIM-SBI-TB were probed, using a probe of carefully selected and representative radius (Zhou et al., 2014). PIM-EA-TB possesses a larger FAV than PIM-SBI-TB. With increasing radius, the FAVs decreased and approached zero when the radius was larger than 2.5 Å. At a radius of 1.0 Å, the FAVs were approximately 50.0% and 48.3% in PIM-EA-TB and PIM-SBI-TB. The bridged Trip units in the PIM- Trip-TB and KAUST-PI-1 (Chen et al., 2016) membranes lead to higher FFV than that of the PIM-PI-1 membrane, demonstrating the ability of the bridged Trip units to disrupt the polymer chain packing. The PIM-Trip-TB membrane exhibited a slightly higher FFV because TB units in the PIM-Trip-TB membrane are bulkier than the TMPD (PI) units in the KAUST-PI-1 membrane. The FAV of the KAUST-PI-1 membrane is twofold higher than that of the PIM-PI-1 membrane, because the substitution of spirocenter units with triptycene unit leads to higher internal free volume between three benzene arms.

2.4.2 Thermodynamics and Transport Properties

2.4.2.1 Adsorption of different gases and solubility coefficient

Heuchel calculated solubility coefficients for eight gases (He, H_2, Ar, O_2, N_2, CH_4, Xe, and CO_2) by the Gusev–Suter TST method (Heuchel et al., 2008). The simulated data agreed well with the experimental results for small penetrants, except for CO_2 and methane [14]. Solubilities obtained from GCMC sorption isotherms at 0.1 bar gave loading values that corresponded well with the experimental solubility coefficients for CO_2 and CH_4 (Hölck, 2008). Adsorption isotherms of N_2 were simulated at 77 K, under the assumption that the polymer matrix was fixed, using PCFF force field (Sun and Rigby, 1997). The solubility coefficients of H_2, O_2, CO_2, and CH_4 in PIM-1 and PIM-7 at 300 K were studied by Fang (Fang et al., 2010). PIM-1 and PIM-7 have the largest affinity for CO_2 and the least for H_2. The solubility coefficients of the four gases increased in the following order: H_2, O_2, CH_4, and CO_2. The predicted solubility coefficients matched better with the experiment than those evaluated by the coarse-grained TST (Heuchel et al., 2008). The solubility of PIM-1 was slightly higher than that of PIM-7 owing to the presence of cyano groups that enhance the affinity for gas molecules. RDFs between CO_2 and atoms of the cyano groups (carbon and nitrogen atoms) and the oxygen atoms in the backbone showed that CO_2 and the atoms of the cyano groups had the strongest interactions, confirming that the cyano groups are the favorable sites in PIM-1 for sorption. Combining molecular simulations and quantum chemical calculations, Fang et al. examined the permeation and separation of CO_2/N_2 in three PIM membranes: PIM-1, trifluoromethyl-phenylsulfone-PIM (TFMPS-PIM), and PIM modified by carboxyl groups (CX-PIM) (Fang et al., 2011). The strategy used was that to quantify how functional groups interact with the gas species. Ab initio calculations revealed that the interaction energies between the functional groups and CO_2 decreased in the order carboxyl > phenylsulfone > cyano > trifluoromethyl; consistently, the simulated solubility coefficient of CO_2 was the largest in CX-PIM and larger than in PIM-1 and TFMPS-PIM. The authors concluded that strong CO_2-philic groups should be chosen to functionalize PIM membranes for high-efficiency CO_2/N_2 separation. This is consistent with the largest binding energy estimated from ab initio calculations due to the formation of a hydrogen bond between CO_2 and carboxyl. However, experimental results show that too strong interaction might inhibit the permeation (Mason et al., 2014), and thus a compromise in polymer/penetrant affinity must be found. Larsen performed GCMC simulations of methane adsorption in PIM-1 at 20°C, using a united-atom model with Lennard–Jones parameters obtained from the TraPPE potential and partial charges calculated from ab initio methods (Larsen et al., 2011b). Their results qualitatively reproduced the shape of the experimental isotherm.

Using an all atom model, Larsen (Larsen et al., 2011a) calculated the adsorption isotherms of methane in PIM-1 and in two different PIM-1-like polymers, PIM-1c and PIM-1n. The adsorption isotherm for PIM-1, recalculated with corrected torsion parameters, qualitatively and quantitatively reproduced the corresponding experimental isotherm. Moreover, the only simulations using non-equilibrium molecular dynamics (NEMD) were reported for pure helium and carbon dioxide on a slice of PIM-1 polymer (Frentrup et al., 2015). The theoretical results were also in good agreement with experimental data. The adsorption isotherms and isosteric heats of adsorption of CO_2 and CH_4 were evaluated for four sulfur-containing PIMs (Hart, et al., 2013a). It was found for single-gas isotherms that CO_2 adsorption (calculated at 20°C) and isosteric heat were greatly increased in sulfonyl-functionalized PIMs. In addition, the 1:1 mixed-gas adsorption selectivity values of sulfonyl PIMs were shown to be enhanced by the highly polar functionality. In particular, while the sulfur PIMs exhibited little to no improvement compared to PIM-1, the subsequent oxidation to sulfonyl PIMs had a two- to three-fold CO_2/CH_4 selectivity increase due to the strong interaction of the polar groups preferentially adsorbing CO_2, without affecting the CH_4 sorption capacity. The adsorption capacities of the swollen and non-swollen samples of PIM-1, PIM-1c, and PIM-SBF were calculated using GCMC simulations at 77 K (Hart et al., 2013). This investigation evidenced that the amount of swelling observed during low-temperature nitrogen adsorption was strongly correlated with the interaction strength of the polymer matrix. The experimental isotherm was matched to the loading capacities of the artificially swollen samples, allowing for the comparison of the changes in isotherm capacity with the degree of swelling.

By comparing the total estimated swelling with the non-bonded interaction energy of the framework, PIMs that exhibit a higher energy of association were less susceptible to sorption-induced plasticization under these conditions. Therefore, to reduce the effect of plasticization of nitrogen at 77 K, linear polymer frameworks with strongly associating groups (e.g., hydrogen bond donors/acceptors) should be strategically incorporated in a way that does not reduce the amount of free volume, possibly on the spirocenter group. Larsen (Larsen et al., 2014) analyzed the adsorptive properties of the hyperfluorinated functionalized PIMs, PIM-1c, PIM-1n, and PIM-1f, and PIM-1 for CH_4, CO_2, N_2, and O_2. The solubility and solubility selectivity were obtained at 1 bar. The highest solubility was found for PIM-1c, having a narrower PSD, which allows for adsorbed molecules to interact with more pore walls simultaneously. Local dipoles from the carbonyl functionality on the spirocenter had the highest solubility selectivity for CO_2/CH_4 and CO_2/N_2, which is likely a result of the strong electrostatic interactions between the quadrupole of CO_2 and the C=O functionality, facilitating a higher capacity relative to the non-polar CH_4 and weakly quadrupolar N_2. Colina et al. studied 14 hypothetical PIMs of four different series, each of which systematically altered one characteristic of the PIM structure: spirocenter size, backbone functionality, sulfur moiety, and length of the backbone unit (Hart and Colina, 2014). They simulated adsorption properties, permeability, and permselectivity of several industrially relevant gases, CO_2, CH_4, and N_2, and compared their results with experimental data. The screening of the molecular models indicated that the structure and functionality of the polymer had a large effect on the resulting porosity, including (1) a polar sulfonyl functionality effectively increases gas solubility and permselectivity; (2) extending the length of the conjugated aromatic backbone units improves gas diffusivity; and (3) increasing the size of the spirocenter unit results in gas separation performance beyond the Robeson upper bound for both CO_2/CH_4 and CO_2/N_2. It was also shown that PIMs display a higher permeability coefficient with respect to FFV than traditional glassy polymers as a result of the consistently higher gas solubility. Thus, a modified empirical permeability model was proposed and validated with experimental data of functionalized PIMs.

Hölck et al. studied the swelling of PIM-1/gas systems via modeling and experimental sorption and dilation analysis for CO_2 and CH_4 in PIM-1, at 35°C (Hölck et al., 2013). The method to simulate sorption isotherms used experimental input of two reference states (unswollen and swollen). Models for the latter were constructed in agreement with experimental data (pressure, temperature, gas concentration, and volume dilation). Concentration–pressure isotherms of each polymer/gas system, obtained with GCMC simulations for both reference states (depleted of gas molecules), were in good agreement with the experimental data in the respective pressure range. Due to the simulation technique used, the isotherms are merely based on hole-filling in a static host matrix and do not represent the sorption behavior over a broader range of gas pressures, which might involve significant structural rearrangements as well as swelling and relaxational phenomena. An interesting observation was performed by Zhao in their study on adsorption behavior of PIM-1 compared to silicalite-1 (Zhao et al., 2012). The sorption isotherms of CH_4 in PIM-1 and silicalite-1 at 77 K and 298 K, respectively, were simulated using COMPASS (Sun, 1998) and PCFF (Sun and Rigby, 1997) force fields. PCFF was apparently more reliable for hydrogen measurements. PIM-1 showed favorable adsorption properties for CH_4 and H_2 at low temperature. At 77 K, the amount of CH_4 adsorbed onto PIM-1 was 60–68% higher than that onto conventional silicalite-1 zeolites. However, this high adsorption performance could not be maintained at room temperature. Adsorption of CH_4 onto PIM-1 was not as stable as that onto silicalite-1. This indicated that PIM-1 was characterized by a large specific surface area that could provide more adsorption sites than silicalite-1, although the integration stability of adsorbate molecules was poor. With increasing temperature, adsorption sites on PIM-1 increasingly disappeared, unlike those on the completely rigid silicalite-1. This difference was ascribed to the ordered microporous structure of silicalite-1 and the amorphous pores of PIM-1. For hydrogen, the difference was even greater and the stability was low. The solubility coefficients of H_2 in PIM-1/solvent ($CHCl_3$, CH_3OH, and H_2O) membranes (Zhang et al., 2011) indicated a good agreement between the predicted and experimental data. The solubility decreased, following PIM-1/CH_3OH > PIM-1/$CHCl_3$ > PIM-1/H_2O, accordingly with the decreasing order of FFV in the three membranes, thus implying that the solubility is proportional to the free volume.

The solubility coefficients were smaller compared with those in a dry PIM-1 membrane in the absence of solvent. This is because solvent molecules bind/block sorption sites and decrease free volume, thus leading to a reduction in both sorption and diffusion. The gas sorption behavior of CO_2, CH_4, O_2, and N_2 in the two PI-derived membrane models, PIM-PI-1 and PIM-PI-8, studied by Chang et al. was shown to be dominated not only by the high membrane free volume, but it was also controlled by the nature of the gas species (Chang et al., 2013). The gas transport for He, H_2, O_2, N_2, CO_2, and CH_4 in PIM-EA-TB (Tocci et al., 2014) was analyzed considering the TST and GCMC approaches, to determine the solubility coefficients. Both in as-cast samples and in the methanol-treated samples, a higher gas uptake was found in PIM-EA-TB than in PIM-1. The increase in gas sorption in methanol-treated PIM-EA-TB is most significant for CO_2 and the bulkier CH_4 molecules. This suggests that the methanol treatment opened up a fraction of the free volume, which would otherwise be inaccessible to these gases. This is in agreement with the observations discussed above, where the presence of solvent in PIM-1 reduces the solubility (Zhang et al., 2011). The high infinite dilution solubility for CO_2 confirms the much stronger interaction of CO_2 with the PIM-EA-TB polymer matrix as compared to N_2. The solubilities in PIM-EA-TB are rather similar and only slightly higher than those in PIM-1, but there is a significant discrepancy between the sorption data and the solubilities determined indirectly from the gas permeation measurements, owing to anomalous or pressure-dependent transport of gases. With respect to PIM-1, there is a modest increase in solubility in PIM-EA-TB after methanol treatment. The comparison of permeation properties of PIM-EA-TB and PIM-SBI-TB, studied by Zhou et al. (2014), indicated that PIM-EA-TB possesses larger solubility coefficients for five gases (CO_2, CH_4, O_2, N_2, and H_2) examined. Except for H_2, the simulated solubilities agreed well with experimental data. The solubility coefficients were governed by the affinity between the gas and the membrane, decreasing in the order $CO_2 > CH_4 > O_2 > N_2 > H_2$. Furthermore, it is found that the separation for CO_2/N_2 and CO_2/CH_4 is primarily driven by sorption.

The simulations of PIM-1 loaded with graphene (Gonciaruk et al., 2015) were performed to understand how graphene could affect polymer packing and how it is reflected in structural properties and adsorption of carbon dioxide compared to pristine PIM-1 adsorbent. The results indicated similar adsorption isotherms of pristine PIM-1 and the PIM-1/graphene composite suggesting that there is no significant effect on affinity toward CO_2 caused by the presence of graphene. This suggests that graphene can increase PIM-1 membrane permeability without affecting its selectivity. The sorption in PIM-PI-1, PIM-Trip-TB, KAUST-PI-1, and PTMSP decreased following the order: $CO_2 > CH_4 > O_2 > N_2$ (Chen et al., 2016). The gas sorption of the four membranes was found to be affected not only by the surface area and free volume but also by the affinity between a gas and a membrane. In conclusion, the bridged triptycene units in the PIM-Trip-TB and KAUST-PI-1 membranes proved to enhance the gas permeabilities and permselectivities of dense membranes for gas separation. The PIM-Trip-TB and KAUST-PI-1 membranes are composed of rigid bridged Trip units and are potential candidate membranes in gas separation processes.

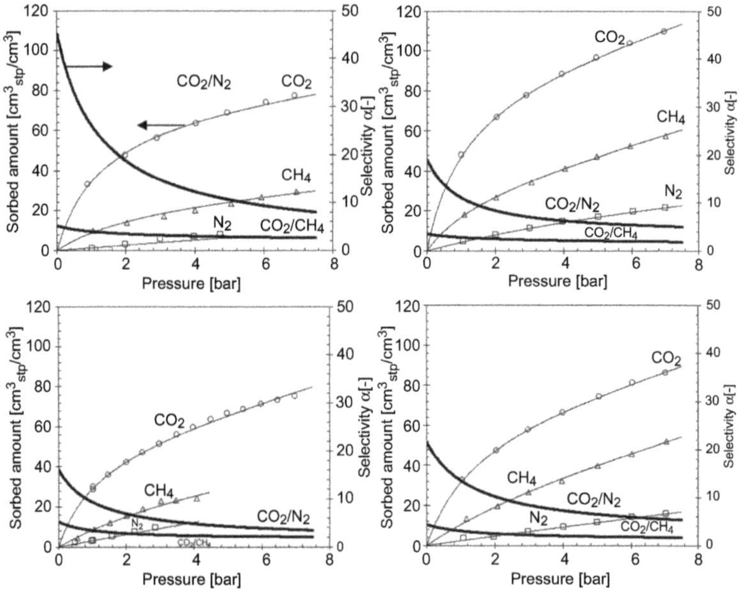

Figure 2.12 Gravimetric sorption isotherms at 25°C for CO_2 (○), CH_4 (Δ), and N_2 (□) in as-cast PIM-EA-TB (top left), MeOH-treated PIM-EA-TB (top right) and as-cast PIM-1 (bottom left) and MeOH-treated PIM-1 (bottom right) membranes. Reprinted with permission from Tocci et al. (2014). Copyright 2014 American Chemical Society.

2.4.2.2 Modeling of diffusion coefficients

Modeling of diffusion coefficients first requires the construction of reliable molecular models of the polymers and then analysis of the gas mobility. The comparison between theoretical and experimental diffusion coefficients calculated with the Gusev–Suter TST method by Heuchel agreed within a factor of three to the experimental values. With increasing size of penetrant, D decreases in PIM-1 over five orders of magnitude (Heuchel et al., 2008). Fang et al. estimated the diffusion coefficients of H_2, O_2, CO_2, and CH_4 in PIM-1 and PIM-7 at 300 K using the Einstein relationship (Fang, Zhang, and Jiang, 2010). The experimental diffusion coefficients in both membranes increased in the same order as the simulations: $CH_4 < CO_2 < O_2 < H_2$. The diffusion coefficients in PIM-1 were found to be slightly higher than in PIM-7 owing to the cyano groups in PIM-1, which increased the *d*-spacing and FFV of PIM-1. The simulated diffusion coefficients overestimated the experimental results approximately five-fold, except for CH_4, since the model membranes had 3–4% lower densities compared to the real samples. The overall permeability selectivities of three gas pairs CO_2/H_2, CO_2/O_2, and CO_2/CH_4 are dominated by the solubility selectivities. The diffusion coefficients of CO_2 and N_2 were studied by Fang et al. in PIM-1 and modified PIMs and decreased in the order PIM-1 > trifluoromethyl-phenylsulfone-PIM > PIM-modified by carboxyl groups, following the reducing order of FFV (Fang et al., 2011). The diffusion selectivity remained nearly constant; consequently, the permselectivity followed the same hierarchy as the solubility selectivity. The diffusion coefficients in the PIM-1/solvent membranes decreased in the hierarchy of PIM-1/CH_3OH > PIM-1/$CHCl_3$ > PIM-1/H_2O (Zhang et al., 2011). Also in this case, the trend was consistent with the decreasing order of FFV.

The presence of solvent molecules reduced free volume and diffusion coefficient. Particularly, H_2O clusters occupied the large-size voids in PIM-1/H_2O, causing the largest reduction in free volume and diffusion. As for solubility coefficients, diffusivity is lower compared with that in a dry PIM-1 membrane in the absence of solvent. This appeared because solvent molecules bind/block sorption sites and decrease free volume amount, thus leading to a reduction in both sorption and diffusion. The two PI-derived membrane models, PIM-PI-1 and PIM-PI-8, studied by Chang et al. (Chang et al., 2013) indicated great gas diffusivity due to the increase in the number of effective and continuous pathways for gas transport from one site to another, owing to their large free volume. The trend of gas mobility in the membrane followed the order $O_2 > N_2 > CO_2 > CH_4$, which means that the trend was primarily influenced by the size of the gases. Among the different membranes, gas diffusion dominated gas transport because the larger free volume effectively increased gas permeability; among the various gas species in each membrane, an increased gas solubility obviously facilitated gas permeation even when the species had a larger molecular volume.

The diffusion coefficients of He, H_2, O_2, N_2, CO_2, and CH_4 in PIM-EA-TB (Tocci et al., 2014) were simulated by using the TST method, and the mean square displacement of each gas molecule was determined through MD simulation runs. The comparison between theoretical and experimental diffusivity indicated a similar trend. The *D* values were rather similar to those of PIM-1, and methanol treatment of PIM-EA-TB resulted in a strong, 2–3-fold increase in diffusivity, also analogous to earlier observations in PIM-1 (Budd et al., 2008). Diffusivity of small gases was higher in PIM-EA-TB than in PIM-1. The difference became smaller or was reversed with increasing gas molecular dimensions, indicating the stronger size-sieving behavior of PIM-EA-TB, which possessed also higher diffusion coefficients than PIM-SBI-TB for the five gases (CO_2, CH_4, O_2, N_2, and H_2) examined (Zhou et al., 2014). The diffusion coefficients largely followed the kinetic diameter of the gases, but in the opposite direction, with *D* increasing in the order $CH_4 < CO_2 < N_2 < O_2 < H_2$. The performance of CO_2 was due to its strong interaction with the membrane. The information obtained from NEMD simulations for pure helium and carbon dioxide on a slice of PIM-1 polymer (Frentrup et al., 2015) was in good agreement with experimental results and shed light on the diffusion mechanism. In this case, a deviation from a straightforward gate-hopping mechanism was observed. The diffusion of carbon dioxide is slowed down by frequent and complex interactions inside the polymer. Both molecules within the polymer matrix spend a considerable amount of time in "random walks" throughout the extent of the available free volume, i.e., the mechanism deviates from a simple "pore hopping" trajectory expected for a dense polymer as a result of exhibiting highly interconnected porosity. The simulated diffusivities in the four membranes, PIM-PI-1, PIM-Trip-TB, KAUST-PI-1, and PTMSP (Chen et al., 2016), increased in the following order: $CH_4 < CO_2 < N_2 < O_2 < H_2$ similar to the experimental result (Ghanem et al., 2008b; Ghanem et al., 2009; Ghanem et al., 2014; Carta et al., 2014b; Merkel et al., 2000; Swaidan et al., 2015). Diffusivities in the PIM-Trip-TB and KAUST-PI-1 membranes were larger than those in the PIM-PI-1 membrane because the structures of the PIM-Trip-TB and KAUST-PI-1 membranes are looser.

2.5 Conclusion and Perspectives

Polymer rigidity is one of the key parameters to improve membrane gas transport performances, which stems from an intrinsic microporosity with high free volume elements. Future developments of microporous membranes for gas separation will require enhanced sorption of target gas pairs while maintaining the extraordinary gas permeability due to their high free volume elements. Tuning the cavity sizes is quite essential for target gas pairs in order to increase membrane performance. The trend for diffusivity was consistent with the decreasing order of FFV. Solvent molecules bind/block sorption sites and decrease free volume amount, thus leading to a reduction in both sorption and diffusion. The theoretical data were generally in good agreement with experimental data, as can be seen in Table 2.1.

The gas sorption is determined not only by the surface area and free volume of a membrane, but also by the affinity between a gas and a membrane. In particular, as in the case of PIM-1f, a higher surface area than PIM-1 did not translate to a sufficiently greater PSD and the newly proposed PIM was the lowest performing adsorbent for all gases reported (Larsen et al., 2014). This highlights the fact that although a higher degree of porosity is necessary for increased adsorption capacity, a higher surface area alone is not a sufficient criterion for the evaluation of adsorbent performance. Moreover, in the strategy of inserting/modifying the chemical structure of the archetypal PIM-1, Larsen et al. (2014) concluded that (1) chain spacing was influenced by spirocenter functionalization; (2) the effect of any functional modification on pore filling should be considered; (3) in order to maximize interactions with polar gases, the electronic structure should be designed to create local dipoles and quadrupoles; and (4) future PIMs should consider using lightweight constituent elements in monomeric design to achieve both gravimetric and volumetric storage targets.

To improve the gas separation performance of porous polymers, it was found (Hart and Colina, 2014) that (a) for CO_2/CH_4 gas separations, PIM-like porous polymers should be designed with dipolar functionalities and bulky spirocenter groups to increase both CO_2 enthalpy of adsorption and free volume, by which solubility and permeability are increased; (b) for the CO_2/N_2 gas pair, the gas

Table 2.1 Simulated diffusion coefficients and solubility coefficients in PIM membranes

	Gas	D_{exp} 10^{-8} cm^2/s	D_{sim} 10^{-8} cm^2/s	S_{exp} cm^3_{STP}/cm^3 bar	S_{sim} cm^3_{STP}/cm^3 bar	Reference
CX-PIM	CO_2		9.8 ± 2.1		78.1 ± 8.0	Fang, Zhang, and Jiang, 2011
	N_2		19.35 ± 10.66		6.47 ± 0.72	
PIM-1	CH_4	6.8 ± 0.5*	7.8 ± 3.8	13.7 ± 1.4*	30 ± 2	Heuchel et al., 2008
	CO_2	26 ± 2*	0.22 ± 0.10	69 ± 7*	345 ± 35	
	N_2	22 ± 2*	16 ± 6	3.2 ± 0.3*	6.2 ± 0.4	
	O_2	81 ± 6*	52 ± 24	3.5 ± 0.4*	7.7 ± 0.4	
	H_2	1700 ± 200*	3500 ± 230	0.58 ± 0.06*	0.42 ± 0.01	
	He	2700 ± 300*	6700 ± 2000	0.18 ± 0.02*	0.18 ± 0.01	
	Ar	40 ± 3*	67 ± 28	3.8 ± 0.4*	6.5 ± 0.3	
	Xe	0.46 ± 0.03*	0.49 ± 0.22	91 ± 9*	285 ± 35	
	CH_4	6.8*	112 ± 27	13.7*	14.2 ± 3.1	Fang, Zhang, and Jiang, 2010
	CO_2	26*	151 ± 47	66.9*	50.7 ± 9	
	O_2	81*	346 ± 100	2.35*	4.01 ± 0.26	
	H_2	1700*	6630 ± 215	0.58*	0.46 ± 0.03	
	CO_2	26* 120**	102 ± 27	66.0* 22.2**	47.1 ± 7.6	Fang et al., 2011
	N_2	22* 120**	223 ± 62	3.2* 1.54**	6.74 ±0.53	

	Gas	D_{exp} 10^{-8} cm^2/s	D_{sim} 10^{-8} cm^2/s	S_{exp} cm^3_{STP}/cm^3 bar	S_{sim} cm^3_{STP}/cm^3 bar	Reference
	CH_4			163***	6.39 ± 0.6	Chang et al., 2013
	CO_2			700***	31.9 ± 2.03	
	N_2			37***	2.41 ± 0.22	
	O_2			39***	4.13 ± 0.3	
	CO_2		190		4.53	Wang et al., 2014
	N_2		129		0.455	
	CH_4			11.77		Larsen, Hart, and Colina, 2014
	CO_2			50.51		
	N_2			3.67		
	O_2			4.66		
	CH_4		17		12.7	Hart and Colina, 2014
	CO_2		53		53.76	
	N_2		34		4.21	
PIM-1c	CH_4			16.93		Larsen et al., 2014
	CO_2			79.18		
	N_2			5.42		
	O_2			6.60		

(*Continued*)

Table 2.1 (*Continued*)

	Gas	D_{exp} 10^{-8} cm²/s	D_{sim} 10^{-8} cm²/s	S_{exp} cm^3_{STP}/cm^3 bar	S_{sim} cm^3_{STP}/cm^3 bar	Reference
PIM-1cf	CH_4		27		11.6	Hart and Colina, 2014
	CO_2		78		42.2	
	N_2		28		3.76	
PIM-1ch	CH_4		17		11.2	Hart and Colina, 2014
	CO_2		54		38.9	
	N_2		35		3.7	
PIM-1f	CH_4			6.70		Larsen et al., 2014
	CO_2			22.05		
	N_2			2.15		
	O_2			2.61		
PIM-1h	CH_4		54		15.0	Hart and Colina, 2014
	CO_2		137		65.8	
	N_2		95		4.8	
PIM-1n	CH_4			14.62		Larsen et al., 2014
	CO_2			52.38		
	N_2			4.10		
	O_2			5.16		

	Gas	D_{exp} 10^{-8} cm^2/s	D_{sim} 10^{-8} cm^2/s	S_{exp} cm^3_{STP}/cm^3 bar	S_{sim} cm^3_{STP}/cm^3 bar	Reference
PIM-1n2	CH_4		14		12.4	Hart and Colina, 2014
	CO_2		44		50.7	
	N_2		28		4.1	
PIM-1r7	CH_4		108		14.1	Hart and Colina, 2014
	CO_2		225		58.8	
	N_2		168		4.7	
PIM-1r9	CH_4		98		14.7	Hart and Colina, 2014
	CO_2		207		57.4	
	N_2		154		4.8	
PIM-1tb	CH_4		284		10.7	Hart and Colina, 2014
	CO_2		502		48.8	
	N_2		399		3.8	
PIM-1tms	CH_4		1099		12.6	Hart and Colina, 2014
	CO_2		1515		54.9	
	N_2		1318		4.1	
PIM-7	CH_4	5.1*	60 ± 6	9.1	1.16 ± 0.23	Fang et al., 2010
	CO_2	21*	130 ± 13	39.5	3.80 ± 0.37	
	O_2	62*	346 ± 100	2.35	0.30 ± 0.02	
	H_2	1100*	6860 ± 689	0.61	0.033 ± 0.001	

(Continued)

Table 2.1 (*Continued*)

	Gas	D_{exp} 10^{-8} cm²/s	D_{sim} 10^{-8} cm²/s	S_{exp} cm^3_{STP}/cm³ bar	S_{sim} cm^3_{STP}/cm³ bar	Reference
PIM-EA-TB	CH_4	12.9[a]	28 ± 5	15.12[a]	32.25 ± 5.25	Tocci et al., 2014
		22[b]	6 ± 3	5.17[b]	6.15 ± 0.68	
		9.94[c]		9.88[c]		
		36.0[d]		14.8[d]		
		31.9[e]		35.14[e]		
	CO_2	38.6[a]	1.4 ± 0.4	45.07[a]	283 ± 1.5	
		40.3[b]	7 ± 3	26.03[b]	24.0 ± 1.50	
		22.8[c]		33.68[c]		
		87.0[d]		57.0[d]		
		87.3[e]		83.95[e]		
	N_2	36.2[a]	79 ± 3	3.68[a]	6.60 ± 2.25	
		35.7[b]	42 ±6	1.61[b]	4.20 ± 0.53	
		25.3[c]		1.88[c]		
		99.5[d]		4.7[d]		
		89.3[e]		7.21[e]		
	O_2	125.5[a]	227 ± 5	4.11[a]	8.25 ± 3.0	
		77[b]	150 ± 50	2.52[b]	2.48 ± 0.23	
		68.2[c]		2.25[c]		
		318[d]		6.0[d]		
		310[e]		7.26[e]		

	Gas	D_{exp} 10^{-8} cm²/s	D_{sim} 10^{-8} cm²/s	S_{exp} cm^3_{STP} /cm³ bar	S_{sim} cm^3_{STP} /cm³ bar	Reference
	H_2	> 2986 [a]	11300 ± 50	< 0.55[a]	8.25 ± 3.0	
		349[b]	9790 ± 70	1.22[b]	24.48 ± 0.23	
		> 1200[c]		< 0.3[c]		
		> 7000[d]		< 0.8[d]		
		> 5000[e]		<1.1[e]		
	He	> 5028 [a]		< 0.12[a]		
		409[b]	19700 ± 30	0.41[b]		
		> 2000[c]		< 0.07[c]		
		> 10000[d]		< 0.2[d]		
		> 6000[e]		< 0.3[e]		
	CH_4	36	60	14.8	16.42	Zhou et al., 2014
		12		35.5		
		14.9		25.2		
	CO_2	87	152	57.0	61.64	
		41		92		
		60.7		58.5		
	N_2	99.5	154	4.7	4.84	
		40.5		7.0		
		49.1		5.7		
	O_2	318	485	6.0	6.72	
		177		6.9		
		210		6.0		

(Continued)

Table 2.1 (*Continued*)

	Gas	D_{exp} 10^{-8} cm²/s	D_{sim} 10^{-8} cm²/s	S_{exp} cm^3_{STP} /cm³ bar	S_{sim} cm^3_{STP} /cm³ bar	Reference
	H_2	>7000	69967	<0.8	0.64	
		>5000		<1.1		
		>5000		<1.1		
PIM-PI-1	CH_4			110	7.44 ± 1.12§	Chang et al., 2013
	CO_2			620	26.0 ± 2.8§	
	N_2			24	2.6 ± 3.0 §	
	O_2			28	3.9 ± 0.4§	
	CH_4	7§	52	11.0§	9.3	Chen et al., 2016
	CO_2	17§	127	62.0§	29.1	
	N_2	20§	123	2.4§	2.7	
	O_2	56§	389	2.8§	3.4	
	H_2	1200§	34600	0.4§	0.15	
PIM-PI-8	CH_4			180§	8.04 ± 0.6	Chang et al., 2013
	CO_2			810§	26.5 ± 1.42	
	N_2			39§	3.01± 0.15	
	O_2			41§	4.13 ± 0.22	
PIM-PI-EA	CH_4	24		14.2		Rogan et al., 2014
		8.4		13.5		

	Gas	D_{exp} 10^{-8} cm^2/s	D_{sim} 10^{-8} cm^2/s	S_{exp} cm^3_{STP}/cm^3 bar	S_{sim} cm^3_{STP}/cm^3 bar	Reference
	CO_2	95		57.8		
		48		50.1		
	N_2	84		3.29		
		32		3.3		
	O_2	270		3.83		
		144		3.4		
	H_2	≥ 3360		≤ 0.94		
		3581		0.60		
	He	≥ 5070		≤ 0.23		
		5740		0.15		
PIM-SBI-TB	CH_4	31.5	30	10.7	11.84	Zhou et al., 2014
		19.1		10.5		
	CO_2	74	65	29.4	31.18	
		66		30.7		
	N_2	75.2	73	2.3	3.75	
		70.1		2.3		
	O_2	201	211	2.7	4.82	
		187		2.6		
	H_2	3500	31.333	0.47	0.51	
		>3000		<0.5		

(Continued)

Table 2.1 (*Continued*)

	Gas	D_{exp} 10^{-8} cm^2/s	D_{sim} 10^{-8} cm^2/s	S_{exp} cm$^3_{STP}$ /cm^3 bar	S_{sim} cm$^3_{STP}$ /cm^3 bar	Reference
PIM-Trip-TB	CH_4	49$^@$	111	18.5$^@$	20.1	Chen et al., 2016
	CO_2	111$^@$	218	87.4$^@$	57.7	
	N_2	135$^@$	254	4.7$^@$	6.3	
	O_2	462$^@$	777	5.9$^@$	8.5	
	H_2	>7800$^@$	41700		0.3	
soPIM-0	CH_4		1		14.1	Hart and Colina, 2014
	CO_2		7		69.3	
	N_2		4		5.03	
soPIM-1	CH_4		3		15.6	Hart and Colina, 2014
	CO_2		14		74.6	
	N_2		8		5.7	
sPIM-0	CH_4		1		10.7	Hart and Colina, 2014
	CO_2		3		37.1	
	N_2		2		3.5	
sPIM-1	CH_4		1		13.8	Hart and Colina, 2014
	CO_2		6		52.0	
	N_2		3		4.4	

	Gas	D_{exp} 10^{-8} cm^2/s	D_{sim} 10^{-8} cm^2/s	S_{exp} cm$^3_{STP}$ /cm^3 bar	S_{sim} cm$^3_{STP}$ /cm^3 bar	Reference
TBPIM25	CO_2		120		27.8	Wang et al., 2014
	N_2		115		17.1	
TBPIM33	CO_2		90		36.3	Wang et al., 2014
	N_2		110		1.63	
TFMPS-PIM	CO_2		49.8 ± 17.8		48.4 ± 3.8	Fang et al., 2011
	N_2		107.9 ± 5.7		5.58 ± 0.31	
CX-PIM	CO_2		9.8 ± 2.1		78.1 ± 8.0	Fang et al., 2011
	N_2		19.35 ± 10.66		6.47 ± 0.72	
KAUST-PI-1	CH_4	—	52	—	13.3	Chen et al., 2016
	CO_2	—	127	—	42.1	
	N_2	53§	123	3.2§	3.9	
	O_2	226§	389	3.6§	5.0	
	H_2	—	34600		0.15	
PIM-0	CH_4		5		9.7	Hart and Colina, 2014
	CO_2		22		33.3	
	N_2		13		3.3	

Experimental data: *(Budd et al., 2005); **(Staiger et al., 2008); ***(Heuchel et al., 2008); §(Ghanem et al., 2008); @(Carta et al., 2014). In (Tocci et al., 2014): **a** (membrane as cast, 70 kDa), **b** (membrane as cast, 150 kDa), **c** (membrane as cast, 190 kDa); **d** (membrane MeOH-treated, 70 kDa), **e** (membrane MeOH-treated, 190 kDa).

separation performance was most effectively improved by increasing the FFV within the PIM-like matrix, resulting in increased CO_2 permeability, without sacrificing the high polymer–gas solubility coefficient. These principles are exemplified by the PIM-1 with bulky trimethylsilyl groups hypothetical structure, which increased the free volume within the framework and performed well above the Robeson upper bound for both gas pairs as a result of the bulky trimethylsilane spirocenter functional groups. Most importantly, the hypothetical PIM materials proposed here did not exhibit traditional "Robeson behavior" due to the high gas solubility and consistent solubility selectivity values even at large amounts of free volume. As such, significant effort should be placed toward increasing the free volume within amorphous polymer frameworks as a means to move beyond the Robeson upper bound. These design principles are expected to aid membrane scientists toward creating improved polymeric materials tailored for specific gas separations.

Appendix

Abbreviations

BET	Brunauer–Emmet–Teller (model)
CNTs	carbon nanotubes
COF	covalent organic frameworks
COMPASS	condensed-phase optimized molecular potentials for atomistic simulation studies
CSD	cavity size distribution
CX-PIM	carboxyl-PIM
Deq	equivalent diameters
DFT	density functional theory
FAV	free accessible volume
FFV	fractional free volume
FVD	free volume distribution
FVEs	free volume elements
GCE	grand canonical ensemble
GCMC	grand canonical Monte Carlo (method)
GGA	generalized gradient approximation
H	Hamiltonian
HF	Hartee–Fock (theory)

HK	Horvath–Kawazoe (method)
K	kinetic energy
KAUST-PI-1	triptycene-based-PIM
MC	Monte Carlo (method)
MD	molecular dynamics (method)
MMM	mixed matrix membranes
MOF	metal-organic frameworks
N	total number of molecules
NEMD	non-equilibrium molecular dynamics
PALS	positron annihilation lifetime spectroscopy
p_i	position of the of the *i*-molecule
PIM	polymers of intrinsic microporosity
PIM-1c	carbonyl-functionalized PIM
PIM-1f	hyperfluorinated PIM
PIM-1n	hyperfluorinated PIM
PIM-7	hyperfluorinated PIM
PIM-EA-TB	ethanoanthracene-based TB polymer
PIM-PI	polyimides-based PIM polymer
PIM-PI-1	PIM-polyimide (PI)-derived membranes
PIM-PI-8	PIM-polyimide (PI)-derived membranes
PIM-SBF	spirobifluorene-PIM
PIM-SBF-Me	methyl-functionalized spirobifluorene-based PIM
PIM-SBI-TB	spirobisindane-based TB polymer
PIM-Trip-TB	triptycene-based-PIM
PSU	polysulfone polymers
q	scattering vector
qi	momenta of the of the *i*-molecule
RDF	radial distribution function
RIS	rotational isomeric state
$r_i(t)$	unfolded position of the *i*-molecule at time *t*
sPIM-1	sulfur-based PIM
TB	Tröger's base
TFMPS-PIM	trifluoromethyl-phenylsulfone-PIM
TST	transition-state theory
U	potential energy
UP	upper bound
V	cell volume
VW	van der Waals volume
WAXD	wide-angle X-ray diffraction

ZIFs	zeolite imidazolate frameworks

Notation

C_i	loading of the *i*-th species (mol kg^{-1})
D	diffusion coefficient (cm^2 s^{-1})
M_i	molecular mass (g mol^{-1})
p_i	partial pressure (mol kg^{-1})
PSD	pore size distribution (cm^3 g^{-1} nm^{-1})
SA	surface area (m^2 g^{-1})
S_i	solubility (mol kg^{-1} Pa^{-1})

Symbols

α	permeselectivity
α_D	diffusion selectivity
β^*	activation energy
θ	scattering angle
λ	wavelength
μ_{excess}	thermodynamic potential
x_i	mole fraction

References

Abbott, L. J., Hart, K. E., and Colina, C. M. 2013. Polymatic: A generalized simulated polymerization algorithm for amorphous polymers. *Theoretical Chemistry Accounts*, **132**(3):1–19.

Ahn, J., Chung, W.-J., Pinnau, I., Song, J. et al. 2010. Gas transport behavior of mixed-matrix membranes composed of silica nanoparticles in a polymer of intrinsic microporosity (PIM-1). *Journal of Membrane Science*, **346**(2):280–287.

Albertus, P., Girishkumar, G., McCloskey, B., Sánchez-Carrera, R. S. et al. 2011. Identifying capacity limitations in the Li/oxygen battery using experiments and modeling. *Journal of The Electrochemical Society*, **158**(3):A343–351.

Althumayri, K., Harrison, W. J., Shin, Y., Gardiner, J. M. et al. 2016. The influence of few-layer graphene on the gas permeability of the high-free-volume polymer PIM-1. *Philosophical Transactions of the Royal Society A: Mathematical, Physical and Engineering Sciences* **374**(2060):20150031. Retrieved (http://rsta.royalsocietypublishing.org/content/374/2060/20150031.

abstract%5Cnhttp://www.scopus.com/inward/record.url?eid=2-s2.0-84956684284&partnerID=40&md5=294e9f7d33a096cbb546ac837f0d7a02%5Cnhttp://rsta.royalsocietypublishing.org/lookup/doi/10.1098/rsta).

Babarao, R. and Jiang, J. 2008. Exceptionally high CO_2 storage in covalent-organic frameworks: Atomistic simulation study. *Energy & Environmental Science*, **1**(1):139. Retrieved (http://xlink.rsc.org/?DOI=b805473h).

Babarao, R. and Jiang, J. 2008. Molecular screening of metal-organic frameworks for CO_2 storage. *Langmuir*, **24**(12):6270–6278.

Baschnagel, J., Binder, K., Doruker, P., Gusev, A. A. et al. 2000. Bridging the gap between atomistic and coarse-grained models of polymers: Status and perspectives. In: *Viscoelasticity, Atomistic Models, Statistical Chemistry*. *Advances in Polymer Science*, vol 152. Springer, Berlin, Heidelberg. Retrieved (http://link.springer.com/chapter/10.1007/3-540-46778-5_2).

Bezzu, C. G., Carta, M., Tonkins, A., Jansen, J. C. et al. 2012. A spirobifluorene-based polymer of intrinsic microporosity with improved performance for gas separation. *Advanced Materials*, **24**(44):5930–5933.

Bondi, A. A. 1916-. 1968. Physical properties of molecular crystals liquids, and glasses. Retrieved March 24, 2017 (http://agris.fao.org/agris-search/search.do?recordID=US201300598889#.WNUtapsn9k0.mendeley&title=Physical properties of molecular crystals liquids, and glasses&description=Looking for Agricultural Science and Technology Information?).

Bragg, W. H. 1912. X-rays and crystals [1]. *Nature*, **90**(2243).

Budd, P. M., Elabas, E. S., Ghanem, B. S., Makhseed, S. et al. 2004. Solution-processed, organophilic membrane derived from a polymer of intrinsic microporosity. *Advanced Materials*, **16**(5):456–459. Retrieved (http://dx.doi.org/10.1002/adma.200306053%5Cnhttp://doi.wiley.com/10.1002/adma.200306053).

Budd, P. M., Ghanem, B. S., Makhseed, S., McKeown, N. B. et al. 2004. Polymers of intrinsic microporosity (PIMs): Robust, solution-processable, organic nanoporous materials. *Chemical Communications*, **10**(2):230–231.

Budd, P. M., Msayib, K. J., Tattershall, C. E., Ghanem, B. S. et al. 2005. Gas separation membranes from polymers of intrinsic microporosity. *Journal of Membrane Science*, **251**(1–2):263–269.

Budd, P. M., McKeown, N. B., Ghanem, B. S., Msayib, K. J. et al. 2008. Gas permeation parameters and other physicochemical properties of a polymer of intrinsic microporosity: Polybenzodioxane PIM-1. *Journal of Membrane Science*, **325**(2):851–860.

Budd, P. M., McKeown, N. B., and Fritsch, D. 2005. Free volume and intrinsic microporosity in polymers. *Journal of Materials Chemistry*, **15**(20):1977–1986. Retrieved (http://dx.doi.org/10.1039/B417402J).

Budd, P. M., McKeown, N. B., and Fritsch, D. 2006. Polymers of intrinsic microporosity (PIMs): High free volume polymers for membrane applications. *Macromolecular Symposia*, **245–246**:403–405.

Bushell, A. F., Attfield, M. P., Mason, C. R., Budd, P. M. et al. 2013a. Gas permeation parameters of mixed matrix membranes based on the polymer of intrinsic microporosity PIM-1 and the zeolitic imidazolate framework ZIF-8. *Journal of Membrane Science*, **427**:48–62.

Bushell, A. F., Budd, P. M., Attfield, M. P., Jones, J. T. A. et al. 2013b. Nanoporous organic polymer/cage composite membranes. *Angewandte Chemie - International Edition*, **52**(4):1253–1256.

Carta, M., Malpass-Evans, R., Croad, M., Rogan, Y. et al. 2013. An efficient polymer molecular sieve for membrane gas separations. *Science*, **339**(6117):303–307. Retrieved (http://www.ncbi.nlm.nih.gov/pubmed/23329042%5Cnhttp://www.sciencemag.org/cgi/doi/10.1126/science.1228032).

Carta, M., Malpass-Evans, R., Croad, M., Rogan, Y. et al. 2014a. The synthesis of microporous polymers using Tröger's base formation. *Polymer Chemistry*, **5**(18):5267–5672. Retrieved (http://pubs.rsc.org/en/content/articlehtml/2014/py/c4py00609g%5Cnhttp://xlink.rsc.org/?DOI=C4PY00609G).

Carta, M., Croad, M., Malpass-Evans, R., Jansen, J. C. et al. 2014b. Triptycene induced enhancement of membrane gas selectivity for microporous Tröger's base polymers. *Advanced Materials*, **26**(21):3526–3531.

Chang, K.-S., Tung, K.-L., Lin, Y.-F., and Lin, H.-Y. 2013. Molecular modelling of polyimides with intrinsic microporosity: From structural characteristics to transport behaviour. *RSC Advances*, **3**(26):10403. Retrieved (http://xlink.rsc.org/?DOI=c3ra40196k).

Charpentier, J. C. 2016. What kind of modern 'green' chemical engineering is required for the design of the 'factory of future'? *Procedia Engineering*, **138**: 445–458.

Chen, Y.-R., Chena, L.-H., Changa, K.-S., Chen, T.-H. et al. 2016. Structural characteristics and transport behavior of triptycene-based PIMs membranes: A combination study using ab initio calculation and molecular simulations. *Journal of Membrane Science*, **514**:114–124.

Dirac, P. A. M. 1929. The basis of statistical quantum mechanics. *Mathematical Proceedings of the Cambridge Philosophical Society*, **25**(1):62.

Du, N., Robertson, G. P., Song, J., Pinnau, I. et al. 2008. Polymers of intrinsic microporosity containing trifluoromethyl and phenylsulfone groups as materials for membrane gas separation. *Macromolecules*, **41**(24):9656–9662.

Du, N., Park, H. B., Robertson, G. P., Dal-Cin, M. M. et al. 2011. Polymer nanosieve membranes for CO_2-capture applications. *Nature Materials*, **10**(5):372–375. Retrieved (http://dx.doi.org/10.1038/nmat2989).

Du, N., Robertson, G. P., Pinnau, I., and Guiver, M. D. 2010. Polymers of intrinsic microporosity with dinaphthyl and thianthrene segments. *Macromolecules*, **43**(20):8580–8587.

Du, N., Robertson, G. P., Song, J., Pinnau, I., and Guiver, M. D. 2009. High-performance carboxylated polymers of intrinsic microporosity (PIMs) with tunable gas transport properties. *Macromolecules*, **42**(16):6038–6043.

Düren, T., Millange, F., Férey, G., Walton, K. S., and Snurr, R. Q. 2007. Calculating geometric surface areas as a characterization tool for metal–organic frameworks. *Journal of Physical Chemistry C*, **111**(42):15350–15356.

Eddaoudi, M., Kim, J., Rosi, N., Vodak, D., et al. 2002. Systematic design of pore size and functionality in isoreticular MOFs and their application in methane storage. *Science*, **295**(5554): 469–472.

Emmler, T., Heinrich, K., Fritsch, D., Budd, P. M. et al. 2010. Free volume investigation of polymers of intrinsic microporosity (PIMs): PIM-1 and PIM1 copolymers incorporating ethanoanthracene units. *Macromolecules*, **43**(14):6075–6084.

ESF Programme. n.d. Challenges in molecular simulations: Bridging the length and time-scale gap (SIMU).

Fang, W., Zhang, L., and Jiang, J. 2010. Polymers of intrinsic microporosity for gas permeation: A molecular simulation study. *Molecular Simulation*, **36**(12):992–1003.

Fang, W., Zhang, L., and Jiang, J. 2011. Gas permeation and separation in functionalized polymers of intrinsic microporosity: A combination of molecular simulations and ab initio calculations. *Journal of Physical Chemistry C*, **115**(29):14123–14130.

Fang, W., Zhang, L., and Jiang, J. 2010. Polymers of intrinsic microporosity for gas permeation: A molecular simulation study. *Molecular Simulation*, **36**(12):992–1003.

Ferrese, A., Albertus, P., Christensen, J., and Newman, J. 2012. Lithium redistribution in lithium-metal batteries. *Journal of the Electrochemical Society*, **159**(10):A1615–A1623.

Freeman, B. D. 1999. Basis of permeability/selectivity tradeoff relations in polymeric gas separation membranes. *Macromolecules*, **32**(2):375–380. Retrieved (http://dx.doi.org/10.1021/ma9814548).

Frenkel, D. and Smit, B. 2001. *Understanding Molecular Simulation: From Algorithms to Applications*. Academic Press, Florida.

Frentrup, H., Hart, K. E., Colina, C. M., and Müller, E. A. 2015. In silico determination of gas permeabilities by non-equilibrium molecular dynamics: CO_2 and He through PIM-1. *Membranes*, **5**(1):99–119.

Friesner, R. A. 2005. Ab initio quantum chemistry: Methodology and applications. *Proceedings of the National Academy of Sciences of the United States of America*, **102**(19):6648–6653. Retrieved February 13, 2017 (http://www.pnas.org/cgi/content/short/102/19/6648).

Fuoco, A., Khdhayyer, M. R., Attfield, M. P., Esposito, E. et al. 2017. Synthesis and transport properties of novel MOF/PIM-1/MOF sandwich membranes for gas separation. *Membranes*, **7**(1):7. Retrieved (http://www.mdpi.com/2077-0375/7/1/7).

Gelb, L. D. and Gubbins, K. E. 1999. Pore size distributions in porous glasses: A computer simulation study. *Langmuir*, **15**(2):305–308. Retrieved (http://dx.doi.org/10.1021/la9808418%5Cnhttp://pubs.acs.org/doi/pdfplus/10.1021/la9808418).

Ghanem, B. S., McKeown, N. B., Budd, P. M., Al-Harbi, N. M. et al. 2009. Synthesis, characterization, and gas permeation properties of a novel group of polymers with intrinsic microporosity: PIM-polyimides. *Macromolecules*, **42**(20):7881–7888.

Ghanem, B. S., McKeown, N. B., Budd, P. M., and Fritsch, D. 2008a. Polymers of intrinsic microporosity derived from bis(phenazyl) monomers bader. *Macromolecules*, **41**(5):1640–1646.

Ghanem, B. S., McKeown, N. B., Budd, P. M., Selbie, J. D., and Fritsch, D. 2008b. High-performance membranes from polyimides with intrinsic microporosity. *Advanced Materials*, **20**(14):2766–2771.

Ghanem, B. S., Swaidan, R., Litwiller, E., and Pinnau, I. 2014. Ultra-microporous triptycene-based polyimide membranes for high-performance gas separation. *Advanced Materials*, **26**(22):3688–3692.

Gonciaruk, A., Althumayri, K., Harrison, W. J., Budd, P. M., and Siperstein, F. R. 2015. PIM-1/graphene composite: A combined experimental and molecular simulation study. *Microporous and Mesoporous Materials*, **209**:126–134.

Gooneie, A., Schuschnigg, S., and Holzer, C. 2017. A review of multiscale computational methods in polymeric materials. *Polymers*, **9**(1):16. Retrieved (http://www.mdpi.com/2073-4360/9/1/16).

Greenfield, M. L. and Theodorou, D. N. 2001. Coarse-grained molecular simulation of penetrant diffusion in a glassy polymer using reverse and kinetic Monte Carlo. *Macromolecules*, **34**(24):8541–8553.

Greenfield, M. L. and Theodorou, D. N. 1998. Molecular modeling of methane diffusion in glassy atactic polypropylene via multidimensional transition state theory. *Macromolecules*, **31**(20):7068–7090. Retrieved (http://dx.doi.org/10.1021/ma980750h).

Gubbins, K. E., Liu, Y.-C., Moore, J. D., and Palmer, J. C. 2011. The role of molecular modeling in confined systems: Impact and prospects. *Physical Chemistry Chemical Physics*, **13**(1):58–85.

Guiver, M. D. and Lee, Y. M. 2013. Polymer rigidity improves microporous membranes. *Science*, **339**(6117):284–285. Retrieved (http://www.ncbi.nlm.nih.gov/pubmed/23329040%5Cnhttp://www.sciencemag.org/cgi/doi/10.1126/science.1232714).

Gusev, A. A. and Suter, U. W. 1993. Dynamics of small molecules in dense polymers subject to thermal motion. *Journal of Chemical Physics*, 99(3). Retrieved (https://doi.org/10.1063/1.466198)

Gusev, A. A., Arizzi, S., Suter, U. W., and Moll, D. J. 1993. Dynamics of light gases in rigid matrices of dense polymers. *Journal of Chemical Physics*, **99**(3):2221–2227.

Hart, K. E., Abbott, L. J., and Colina, C. M. 2013. Analysis of force fields and BET theory for polymers of intrinsic microporosity. *Molecular Simulation*, **39**(5):397–404.

Hart, K. E., Abbott, L. J., McKeown, N. B., and Colina, C M. 2013a. Toward effective CO_2/CH_4 separations by sulfur-containing PIMs via predictive molecular simulations. *Macromolecules*, **46**(13):5371–5380.

Hart, K. E. and Colina, C. M. 2014. Estimating gas permeability and permselectivity of microporous polymers. *Journal of Membrane Science*, **468**:259–268. Retrieved (http://dx.doi.org/10.1016/j.memsci.2014.06.017).

Hart, K. E., Springmeier, J. M., McKeown, N. B., and Colina, C. M. 2013b. Simulated swelling during low-temperature N_2 adsorption in polymers of intrinsic microporosity. *Physical Chemistry Chemical Physics*, **15**(46):20161–20169. Retrieved (http://pubs.rsc.org/en/content/articlehtml/2013/cp/c3cp53402b).

Hastings, W. K. 1970. Monte Carlo sampling methods using Markov chains and their applications. *Biometrika*, **57**(1):97–109.

Head-Gordon, M. 1996. Quantum chemistry and molecular processes. *Journal of Physical Chemistry*, **100**(31):13213–13225.

Heuchel, M., Fritsch, D., Budd, P. M., McKeown, N. B., and Hofmann, D. 2008. Atomistic packing model and free volume distribution of a polymer with intrinsic microporosity (PIM-1). *Journal of Membrane Science*, **318**(1–2):84–99.

Heuchel, M., Hofmann, D., and Pullumbi, P. 2004. Molecular modeling of small-molecule permeation in polyimides and its correlation to free-volume distributions. *Macromolecules*, **37**(1):201–214.

Hofmann, D., Heuchel, M., Yampolskii, Y., Khotimskii, V., and Shantarovich, V. 2002. Free volume distributions in ultrahigh and lower free volume polymers: Comparison between molecular modeling and positron lifetime studies. *Macromolecules*, **35**(6):2129–2140.

Hölck, O. 2008. Gas sorption and swelling in glassy polymers: Combining experiment, phenomenological models and detailed atomistic molecular modeling. *Arbeit.* Doctoral Thesis, Technische Universität Berlin, Fakultät III - Prozesswissenschaften.

Hölck, O., Böhning, M., Heuchel, M., Siegert, M. R., and Hofmann, D. 2013. Gas sorption isotherms in swelling glassy polymers-detailed atomistic simulations. *Journal of Membrane Science*, **428**:523–532.

Ilinitch, O. M., Fenelonov, V. B., Lapkin, A. A., Okkel, L. G. et al. 1999. Intrinsic microporosity and gas transport in polyphenylene oxide polymers. *Microporous and Mesoporous Materials*, **31**(1):97–110.

Karger, J., Ruthven, D. M., and Theodorou, D. N. 2012. *Diffusion in Nanoporous Materials.* Wiley-VCH Verlag GMBH & Co. KGA.

Khan, M. M., Filiz, V., Bengtson, G., Shishatskiy, S. et al. 2013. Enhanced gas permeability by fabricating mixed matrix membranes of functionalized multiwalled carbon nanotubes and polymers of intrinsic microporosity (PIM). *Journal of Membrane Science*, **436**:109–120.

Kohn, W. 1999. Nobel lecture: Electronic structure of matter—wave functions and density functionals. *Reviews of Modern Physics*,

71(5):1253–1266. Retrieved (http://link.aps.org/doi/10.1103/RevModPhys.71.1253%5Cnhttp://journals.aps.org/rmp/pdf/10.1103/RevModPhys.71.1253).

Konnertz, N., Ding, Y., Harrison, W. J., Budd, P. M. et al. 2016. Molecular mobility of the high performance membrane polymer PIM-1 as investigated by dielectric spectroscopy. *ACS Macro Letters*, **5**(4):528–532.

Koros, W. J. and Zhang, C. 2017. Materials for next-generation molecularly selective synthetic membranes. *Nature Materials*, **16**: 289–297. Retrieved (http://www.nature.com/doifinder/10.1038/nmat4805).

Larsen, G. S., Hart, K. E., and Colina, C. M. 2014. Predictive simulations of the structural and adsorptive properties for PIM-1 variations. *Molecular Simulation*, **40**(7–9):599–609. Retrieved (http://dx.doi.org/10.1080/08927022.2013.829222).

Larsen, G. S., Lin, P., Hart, K. E., and Colina, C. M. 2011a. Molecular simulations of PIM-1-like polymers of intrinsic microporosity. *Macromolecules*, **44**(17):6944–6951.

Larsen, G. S., Lin, P., Siperstein, F. R., and Colina, C. M. 2011b. Methane adsorption in PIM-1. *Adsorption*, **17**(1):21–26.

Lau, C. H., Mulet, X., Konstas, K., Doherty, C. M. et al. 2016. Hypercrosslinked additives for ageless gas-separation membranes. *Angewandte Chemie International Edition*, **55**(6):1998–2001.

Law, S. 2010. Molecular modeling of matter : Impact and prospects in engineering. *Industrial & Engineering Chemistry Research*, **49**:3026–3046.

Lesch, D. A, Hoffman Estates, and Assistant Examinerianthony Shumate. 2010. UV-cross-linked membranes from polymers of intrinsic microporosity for Li Uid separations. US Patent 7758(12):751.

Li, F. Y., Xiao, Y., Chung, T. S., and Kawi, S. 2012a. High-performance thermally self-cross-linked polymer of intrinsic microporosity (PIM-1) membranes for energy development. *Macromolecules*, **45**(3):1427–1437.

Li, F. Y., Xiao, Y., Ong, Y. K., and Chung, T. S. 2012b. UV-rearranged PIM-1 polymeric membranes for advanced hydrogen purification and production. *Advanced Energy Materials*, **2**(12):1456–1466.

Liu, J., Xiao, Y., and Tai-Shung, N. C. 2017. Flexibly thermally treated 3D PIM-CD molecular sieve membranes exceeding the upper bound line for propylene/propane separation. *Journal of Materials Chemistry*

A, **5**: 4583–4595. Retrieved (http://pubs.rsc.org/en/Content/ArticleLanding/2017/TA/C6TA09751K).

Long, T. M. and Swager, T. M. 2003. Molecular design of free volume as a route to low-κ dielectric materials. *Journal of the American Chemical Society*, **125**(46):14113–14119.

Ma, L. K., Srivastava, R., Barpanda, D., Fowler, T. et al. 2013. An inverse approach to characterize anisotropic thermal conductivities of a dry fibrous preform composite. *Journal of Reinforced Plastics and Composites*, **32**(24):1916–1927.

Maginn, E. J. and Elliott, J. R. 2010. Historical perspective and current outlook for molecular dynamics as a chemical engineering tool. *Industrial & Engineering Chemistry Research*, **49**(7):3059–3078. Retrieved (http://pubs.acs.org/doi/pdf/10.1021/ie901898k%5Cnhttp://pubs.acs.org/doi/abs/10.1021/ie901898k).

Maier, G. 2013. Gas separation by polymer membranes: Beyond the border. *Angewandte Chemie International Edition*, **52**(19):4982–4984.

Mason, C. R., Maynard-Atem, L., Al-Harbi, N. M., Budd, P. M. et al. 2011. Polymer of intrinsic microporosity incorporating functionality: Preparation and gas transport properties. *Macromolecules*, **44**(16):6471–6479.

Mason, C. R., Maynard-Atem, L., Heard, K. W. J., Satilmis, B. et al. 2014. Enhancement of CO_2 affinity in a polymer of intrinsic microporosity by amine modification. *Macromolecules*, **47**(3):1021–1029.

Matteucci, S., Yampolskii, Y., Freeman, B. D., and Pinnau, I. 2006. Transport of gases and vapors in glassy and rubbery polymers. In: *Materials Science of Membranes for Gas and Vapor Separation*, Y. Yampolskii, I. Pinnau, and B. Freeman (eds.), John Wiley & Sons, Inc.

McDermott, A. G., Budd, P. M., McKeown, N. B., Colina, C. M., and Runt, J. 2014. Physical aging of polymers of intrinsic microporosity: A SAXS/WAXS study. *Journal of Materials Chemistry A*, **2**(30):11742–11752. Retrieved (http://dx.doi.org/10.1039/C4TA02165G).

McDermott, A. G., Larsen, G. S., Budd, P. M., Colina, C. M., and Runt, J. 2011. Structural characterization of a polymer of intrinsic microporosity: X-ray scattering with interpretation enhanced by molecular dynamics simulations. *Macromolecules*, **44**(1):14–16.

McKeown, N. B. 2012. Polymers of intrinsic microporosity. *ISRN Materials Science*, **2012**(7):1–16. Retrieved (http://www.hindawi.com/journals/isrn.materials.science/2012/513986/%5Cnpapers2://publication/doi/10.1021/la903483q).

McKeown, N. B. and Budd, P. M. 2002. Phthalocyanine-based nanoporous network polymers. *Chemical Communications*, **44**(0):2780–2781.

McKeown, N. B. and Budd, P. M. 2011. Polymers with inherent microporosity. In: *Porous Polymers*, M. S. Silverstein, N. R. Cameron, and M. A. Hillmyer (eds.), John Wiley & Sons, Inc.

McKeown, N. B., Hanif, S., Msayib, K., Tattershall, C. E., and Budd, P. M. 2002. Porphyrin-based nanoporous network polymers. *Chemical Communications*, **44**(23):2782–2783. Retrieved (http://xlink.rsc.org/?DOI=b208702m).

Merkel, T. C., Bondar, V., Nagai, K., and Freeman, B. D. 2000. Sorption and transport of hydrocarbon and perfluorocarbon gases in poly(1-trimethylsilyl-1-propyne). *Journal of Polymer Science, Part B: Polymer Physics*, **38**(2):273–296.

de Miranda, R. L., Kruse, J., Rätzke, K., Faupel, F. et al. 2007. Unusual temperature dependence of the positron lifetime in a polymer of intrinsic microporosity. *Physica Status Solidi Rapid Research Letters*, **1**(5):190–192.

Mitra, T., Bhavsar, R., Adams, D. J., Budd, P. M., and Cooper, A. 2016. PIM-1 mixed matrix membranes for gas separations using cost-effective hypercrosslinked nanoparticle fillers. *Chemical Communications*, **52**(32):5581–5584. Retrieved (http://dx.doi.org/10.1039/C6CC00261G).

Park, H. B., Jung, C. H., Lee, Y. M., Hill, A. J. et al. 2007. Polymers with cavities tuned for fast selective transport of small molecules and ions. *Science*, **318**(5848):254–258. Retrieved (http://science.sciencemag.org/content/318/5848/254.abstract).

Patel, H. A. and Yavuz, C. T. 2012. Noninvasive functionalization of polymers of intrinsic microporosity for enhanced CO_2 capture. *Chemical Communications*, **48**(80):9989–9991. Retrieved (http://dx.doi.org/10.1039/C2CC35392J).

Rajagopal, A. K. and Callaway, J. 1973. Inhomogeneous electron gas. *Physical Review B*, **7**(5):1912–1919.

Rallabandi, P. S. and Ford, D. M. 2000. Permeation of small molecules through polymers confined in mesoporous media. *Journal of Membrane Science*, **171**(2):239–252.

Robeson, L. M. 1991. Correlation of separation factor versus permeability for polymeric membranes. *Journal of Membrane Science*, **62**(2):165–185.

Robeson, L. M. 2008. The upper bound revisited. *Journal of Membrane Science*, **320**(1–2):390–400.

Robeson, L. M., Dose, M. E., Freeman, B. D., and Paul, D. R. 2016. Analysis of the transport properties of thermally rearranged (TR) polymers and polymers of intrinsic microporosity (PIM) relative to upper bound performance. *Journal of Membrane Science*, **525**: 18–24. Retrieved February 13, 2017 (http://www.sciencedirect.com/science/article/pii/S0376738816316878).

Robeson, L. M., Smith, Z. P., Freeman, B. D., and Paul, D. R. 2014. Contributions of diffusion and solubility selectivity to the upper bound analysis for glassy gas separation membranes. *Journal of Membrane Science*, **453**:71–83.

Rogan, Y., Malpass-Evans, R., Carta, M., Lee, M. et al. 2014. A highly permeable polyimide with enhanced selectivity for membrane gas separations. *Chemical Engineering Research Bulletin*, **2**(14):4874. Retrieved (http://xlink.rsc.org/?DOI=c4ta00564c%5Cnpapers2://publication/doi/10.1039/c4ta00564c).

Satilmis, B., Alnajrani, M. N., and Budd, P. M. 2015. Hydroxyalkylaminoalkylamide PIMs: Selective adsorption by ethanolamine- and diethanolamine-modified PIM-1. *Macromolecules*, **48**(16):5663–5669.

Sautet, P. and Van Santen, R. A. 2009. *Computational Methods in Catalysis and Materials Science: An Introduction for Scientists and Engineers*. Wiley-VCH Verlag GmbH & Co. KGaA.

Sing, K. S. W., Rouquerol, F., Rouquerol, J., and Llewellyn, P. 2013. Assessment of mesoporosity. In: *Adsorption by Powders and Porous Solids: Principles, Methodology and Applications*, 2nd edn, Academic Press, pp. 191–217.

Song, Q., Cao, S., Pritchard, R. H., Ghalei, B. et al. 2014. Controlled thermal oxidative crosslinking of polymers of intrinsic microporosity towards tunable molecular sieve membranes. *Nature Communications*, **5**:4813. Retrieved (http://www.nature.com/doifinder/10.1038/ncomms5813%5Cnhttp://www.scopus.com/inward/record.url?eid=2-s2.0-84923103246&partnerID=tZOtx3y1).

Staiger, C. L., Pas, S. J., Hill, A. J., and Cornelius, C. J. 2008. Gas separation, free volume distribution, and physical aging of a highly microporous spirobisindane polymer. *Chemistry of Materials*, **20**(8):2606–2608.

Sun, H. 1998. COMPASS: An ab initio force-field optimized for condensed-phase applications—overview with details on alkane and benzene compounds. *Journal of Physical Chemistry B*, **5647**(98):7338–7364.

Sun, H. and Rigby, D. 1997. Polysiloxanes: Ab initio force field and structural, conformational and thermophysical properties. *Spectrochimica Acta Part A: Molecular and Biomolecular Spectroscopy*, **53**(8):1301–1323. Retrieved (http://www.sciencedirect.com/science/article/pii/S1386142597000139).

Swaidan, R., Al-Saeedi, M., Ghanem, B., Litwiller, E., and Pinnau, I. 2014. Rational design of intrinsically ultramicroporous polyimides containing bridgehead-substituted triptycene for highly selective and permeable gas separation membranes. *Macromolecules*, **47**(15):5104–5114.

Swaidan, R., Ghanem, B., Litwiller, E., and Pinnau, I. 2015. Physical aging, plasticization and their effects on gas permeation in 'rigid' polymers of intrinsic microporosity. *Macromolecules*, **48**(18):6553–6561.

Tafipolsky, M., Amirjalayer, S., and Schmid, R. 2010. Atomistic theoretical models for nanoporous hybrid materials. *Microporous and Mesoporous Materials*, **129**(3):304–318.

Theodorou, D. N. 2005. Multiscale modeling of polymers. In: *Handbook of Materials Modeling*, S. Yip (ed.), Springer, Dordrecht, pp. 2757–2761.

Theodorou, D. N. 2010. Progress and outlook in Monte Carlo simulations. *Industrial and Engineering Chemistry Research*, **49**(7):3047–3058.

Tocci, E., Gugliuzza, A., Lorenzo, L. D., Macchione, M. et al. 2008. Transport properties of a co-poly(amide-12-B-ethylene oxide) membrane: A comparative study between experimental and molecular modelling results. *Journal of Membrane Science*, **323**(2):316–327.

Tocci, E., Lorenzo, L. D., Bernardo, P., Clarizia, G. et al. 2014. Molecular modeling and gas permeation properties of a polymer of intrinsic microporosity composed of ethanoanthracene and Tröger's base units. *Macromolecules*, **47**(22):7900–7916.

Tozawa, T., Jones, J. T. A., Swamy, S. I., Jiang, S. et al. 2009. Porous organic cages. *Nature Materials*, **8**(12):973–978. Retrieved (http://www.ncbi.nlm.nih.gov/pubmed/19855385).

van der Vegt, N. F. A. 2000. Temperature dependence of gas transport in polymer melts: Molecular dynamics simulations of CO_2 in polyethylene. *Macromolecules*, **33**(8):3153–3160. Retrieved (http://pubs.acs.org/doi/abs/10.1021/ma991737f).

Vogiatzis, K. D., Mavrandonakis, A., Klopper, W., and Froudakis, G. E. 2009. Ab initio study of the interactions between CO_2 and N-containing organic heterocycles. *ChemPhysChem*, **10**(2):374–383.

Wang, J., Zhu, J., Zhang, Y., Liu, J., and Van der Bruggen, B. 2017. Nanoscale tailor-made membranes for precise and rapid molecular sieve separation. *Nanoscale*, **9**: 2942–2957. Retrieved (http://pubs.rsc.org/en/Content/ArticleLanding/2017/NR/C6NR08417F).

Wang, R., Cao, C., and Chung, T. S. 2002. A critical review on diffusivity and the characterization of diffusivity of 6FDA-6FpDA polyimide membranes for gas separation. *Journal of Membrane Science*, **198**(2):259–271.

Wang, Y. C., Huang, S.-H., Hu, C.-C., Li, C.-L. et al. 2005. Sorption and transport properties of gases in aromatic polyimide membranes. *Journal of Membrane Science*, **248**(1–2):15–25.

Wang, Z. G., Liu, X., Wang, D., and Jin, J. 2014. Tröger's base-based copolymers with intrinsic microporosity for CO_2 separation and effect of Tröger's base on separation performance. *Polymer Chemistry*, **5**:2793. Retrieved (http://xlink.rsc.org/?DOI=c3py01608k).

Wang, Z., Wang, D., and Jin, J. 2014. Microporous polyimides with rationally designed chain structure achieving high performance for gas separation. *Macromolecules*, **47**(21):7477–7483.

Weng, X., Baez, J. E., Khiterer, M., Hoe, M. Y. et al. 2015. Chiral polymers of intrinsic microporosity: Selective membrane permeation of enantiomers. *Angewandte Chemie International Edition*, **54**(38):11214–11218. Retrieved February 14, 2017 (http://www.ncbi.nlm.nih.gov/pubmed/26352031).

Widom, B. 1963. Some topics in the theory of fluids. *Journal of Chemical Physics*, **39**(11):2808. Retrieved (http://scitation.aip.org/content/aip/journal/jcp/39/11/10.1063/1.1734110).

Wu, X., Tian, Z., Wang, S., Peng, D. et al. 2017. Mixed matrix membranes comprising polymers of intrinsic microporosity and covalent organic framework for gas separation. *Journal of Membrane Science*, **528**:273–283. Retrieved (http://linkinghub.elsevier.com/retrieve/pii/S0376738816320804).

Xiao, Y., Low, B. T., Hosseini, S. S., Chung, T. S., and Paul, D. R. 2009. The strategies of molecular architecture and modification of polyimide-based membranes for CO_2 removal from natural gas: A review. *Progress in Polymer Science*, **34**(6):561–580.

Yaghi, O. M., O'Keeffe, M., Ockwig, N. W., and Chae, H. K. 2003. Reticular synthesis and the design of new materials. *Nature*, **423**(6941):705–714. Retrieved (http://www.ncbi.nlm.nih.gov/pubmed/12802325%5Cnhttp://deepblue.lib.umich.edu/handle/2027.42/62718).

Yampolskii, Y., Pinnau, I., and Freeman, B. D. 2006. *Materials Science of Membranes for Gas and Vapor Separation*. John Wiley & Sons, Inc.

Zhang, J., Kang, H., Martin, J., Zhang, S. et al. 2016. The enhancement of chain rigidity and gas transport performance of polymers of intrinsic microporosity via intramolecular locking of the spiro-carbon. *Chemical Communications*, **52**(39):6553–6556. Retrieved (http://dx.doi.org/10.1039/C6CC02308H).

Zhang, L., Fang, W., and Jiang, J. 2011. Effects of residual solvent on membrane structure and gas permeation in a polymer of intrinsic microporosity: Insight from atomistic simulation. *Journal of Physical Chemistry C*, **115**(22):11233–11239.

Zhao, H., Xie, Q., Ding, X., Chen, J. et al. 2016. High performance post-modified polymers of intrinsic microporosity (PIM-1) membranes based on multivalent metal ions for gas separation. *Journal of Membrane Science*, **514**:305–312.

Zhao, L., Zhai, D., Liu, B., Liu, Z. et al. 2012. Grand canonical Monte Carlo simulations for energy gases on PIM-1 polymer and silicalite-1. *Chemical Engineering Science*, **68**(1):101–107.

Zhou, J., Zhu, X., Hu, J., Liu, H. et al. 2014. Mechanistic insight into highly efficient gas permeation and separation in a shape-persistent ladder polymer membrane. *Physical Chemistry Chemical Physics*, **16**(13):6075–6083. Retrieved (http://dx.doi.org/10.1039/C3CP55498H).

Zhuang, Y., Seong, J. G., Do, Y. S., Jo, H. J. et al. 2014. Intrinsically microporous soluble polyimides incorporating Trögers base for membrane gas separation. *Macromolecules*, **47**(10):3254–3262.

Chapter 3

Fabrication of Ordered Micro- and Nanoporous Membranes

Maria Luisa Perrotta and Annarosa Gugliuzza
Institute on Membrane Technology of the National Research Council of Italy (ITM-CNR), Via P. Bucci, Cubo 17C, 87036 Rende (CS), Italy
a.gugliuzza@itm.cnr.it

The field of nanostructures is part of one of the most attractive fields of nanoscience and nanotechnology. Currently, scientists from around the world are trying increasingly to give a contribution to the research of nanomaterials, whose structures are more and more close to those considered ideal. Hierarchical structures, well-controlled and ordered morphologies are the new goals in materials science. Manufacturing techniques are multifaceted and have been tested and evaluated over time, depending on the advantages and limitations that arise from their practice.

Here, reference is made, in particular, to the field of the fabrication of well-defined nanostructured membranes technology, which is turned particularly to the improvement in separation processes, in order to make more specific selectivity and increased productivity. Exceptional changes are envisaged toward a fine

Functional Nanostructured Membranes
Edited by Enrico Drioli, Lidietta Giorno, and Annarosa Gugliuzza

ISBN 978-981-4774-79-6 (Hardcover), 978-1-351-13511-5 (eBook)
www.panstanford.com

membrane technology where life science blends with the science of nanomaterials.

3.1 Introduction

Nanotechnologies involve the concept of membrane technology due to the fact that nanostructured films assembled in different configurations can work as permselective interfaces. Nowadays, a lot of research focuses on the preparation of micro- and nanostructured membranes having perfect arrangements of pores through bidimensional (2D) and tridimensional (3D) spaces. In the literature, there is a lot of effort in this direction, because different manufacturing techniques are proposed for many technological applications (Pitois and Francois, 1999b; Zhang et al., 2014), including chemical and (bio)separations (Bormashenko et al., 2012b), (bio)sensors (Male et al., 2016), textiles (Gugliuzza and Drioli, 2013), energy, electronics (Li et al., 2011b), and so on.

Key elements for making processes highly productive, in all of these fields, include a larger number of uniformly shaped pores that could provide a very high interfacial area taking the advantage of a larger and amplified mass transfer.

Referring to processes in the field of membrane technology, which requires use of porous membranes, i.e., microfiltration (MF), ultrafiltration (UF), and membrane contactors, structural features such as porosity, narrow pore distribution, pore size, and extensive surface become key concepts for successful operations. Currently, the market does not provide membranes whose structural features are close to those considered ideal, especially for processes such as membrane distillation, membrane crystallization (MCr), membrane condenser (MC), and membrane emulsification (ME) (Li et al., 2008). The availability of nanostructured membranes with highly defined pore shape, size, and distribution as well as controlled thickness and a high degree of free volume is considered a goal for improved correlations between morphology and transport properties. A large number of regular pores per surface area together with a high porosity is associated with ideal arrangements that possess an ordered structural pattern wherein every event can be reproduced point by point without additional resistance to transport. The major output is an optimization of the productivity–selectivity trade-off, thus resulting in highly performing membrane processes. The scientific research is always directed at the improvement of existing techniques, but today there is a great demand for the implementation of new greener, timesaving, and economically competitive technologies that can bring new advanced solutions and benefits in the membrane-manufacturing field. Highly defined and ordered nanostructures can open a way to new horizons in membrane preparation.

Herein, several kinds of manufacturing techniques are proposed for the development of nanostructured membranes with a high degree of geometric order through two-dimensional (2D) and three-dimensional (3D) space. Some manufacturing processes are described with a special focus on those that have followed and evolved over time:

- Lithographic techniques
- Phase separation and micro-molding combined technique
- Techniques based on colloidal templates
- Self-assembly of copolymers
- Breath figures

Each approach has advantages and limitations, even though over the years a lot of research has been done to improve their features and go beyond the state of the art. The major targets are not only under the costs profile, but also in terms of eco-sustainability and rapidity in realizing and producing nanostructures. In the latter case, the versatility of nanostructures in several applicative fields has led to the increasingly frequent search for techniques that are less time consuming but difficult to manage on scale. For example, compare between lithographic and breath figure (BF) techniques. In the first case, the implementation of nanostructured products is related to the use of "molds" forming nanostructures. Nanosized molds are quite expensive, and it is difficult to achieve good resolution on scale, thus certainly disadvantageous from the economics point of view. However, the lithography approach has a high percentage of success in getting extensive structural order. Instead, the BF technique mimics what happens in nature when fog condenses on cold surfaces so that water droplets, working as imprinting agents, give rise to the formation of ordered nanostructures in one step. In this case, water is largely available and no toxic agent is present, and the formation of nanostructure is economically competitive. However, it is fundamental to get perfect control of water dynamics in order to ensure extensive structural order on scale. In this chapter, the readers will find a thorough analysis of some attractive membrane-manufacturing techniques, which would enable them to compare and choose the most appropriate approach for tailoring nanostructured membranes with desired properties for applications of interest.

3.2 Lithographic Techniques

The lithographic technique frequently achieves polymeric surfaces characterized by a regular structure and a precise geometric order. Lithography is a mechanical technique of reproducing images. It originated in the fields of arts and writing, but over time it has shifted to science and specifically to nanotechnology. Mainly, nanolithography is used in the manufacture of semiconductors, advanced integrated circuits, or nanoelectromechanical systems. It refers to the manufacture of structures on the nanometer scale—at least one lateral dimension between the size of a single atom and that of about 100 nm. Generally, lithographic techniques allow imprinting patterns on micro- and nanoscale, thus achieving surfaces with extensive structural periodicity (Fig. 3.1). Figure 3.1 (Lee et al., 2008) shows how using a substrate on which the polymer or monomer solution (well known as the resist) is laid, a force is applied with a mold, forming a surface with a specific structure. Patterns are created by mechanical deformation of the imprint resist and subsequent processes (Schift and Kristensen, 2007).

The imprint resist is typically a monomer or polymer formulation that is cured by heat or UV light during the imprinting. Adhesion between the resist and the template is controlled to allow proper release. The molds are usually prepared from Si or SiO_2 by using electron beam lithography or dry etching, i.e., oxygen plasma (Guo, 2007).

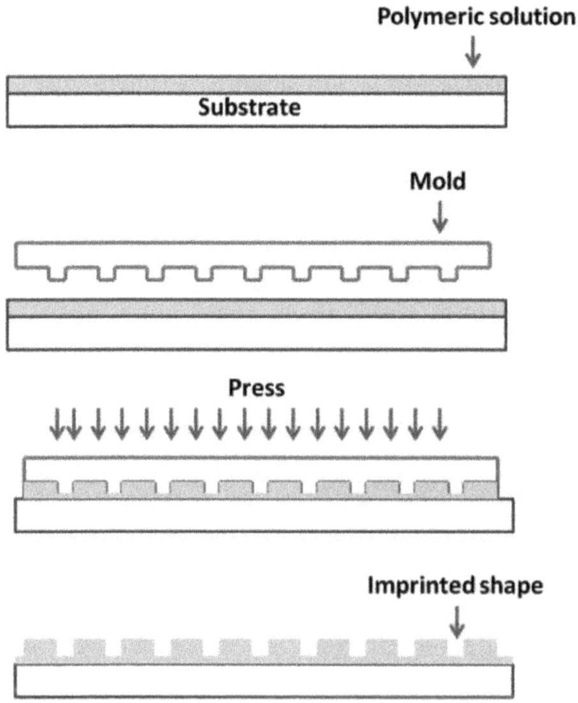

Figure 3.1 Schematic illustration of lithography.

3.2.1 Nanoimprint Lithography and Related Subcategories

Nanoimprint lithography can be classified into three subcategories:

1. Thermoplastic nanoimprint lithography
2. Photo nanoimprint lithography
3. Resist-free direct thermal nanoimprint lithography

Thermoplastic nanoimprint lithography is the first nanoimprint lithography technology to be developed (Chou et al., 1996). In this technology, a thin layer of a thermoplastic polymer is generally spin-coated on a substrate; then a shaped mold is placed in contact with the coated solution on the substrate and a force is applied at a precise pressure. The sample substrate is heated above the glass transition temperature (T_g) of the polymer so that the polymer solution becomes softer and the mold penetrates it better. After cooling, the lithographed surface is formed.

In *photo nanoimprint* lithography, a photo (UV) curable liquid resist is applied to the sample substrate and the mold is normally made of a transparent material such as fused silica or polydimethylsiloxane (PDMS). Afterward, both the mold and substrate are pressed together, while the resist is cured under UV light until it becomes a solid layer (Lan and Ding, 2010).

Unlike previous techniques, the *resist-free thermal nanoimprint lithography* does not require an etching step to transfer the patterned models from the imprinted resists to the device layer. In a typical process, photoresist patterns are first defined using photolithography. A PDMS-based elastomer is primarily shaped by pattern resists and then used as a mold. So a single-step process of nanoimprinting provides thin-film materials in the desired devices, with a specific geometry by using high pressures and high temperatures (Zou et al., 2014).

Lee et al. (2008) have studied the preparation of microfluidic channels and nanosized patterns through the use of a soft-lithographic technique, followed by UV and thermal curation of preceramic materials. The manufacture of microchannels, using cured inorganic polymers, has enormous potential in the field of microfluidics, where high-optical-transparency materials, thermal stability, and chemical inertness are in great demand. In this regard, the fabrication of micro-electromechanical systems (MEMS) and micro total analysis systems (MTAS) from ceramic materials has generated considerable attention, because they may be applicable for harsh environments that require high tolerance to temperatures, corrosion resistance, and also tribological property.

Despite that lithography plays a fundamental role in the field of microelectronics, because it allows reduction in the size of circuit paths and the consequent integration of an ever-increasing number of devices on a single chip. It has evolved over time, embracing experimentation of various types of materials to be used not only in microelectronics, but also in other fields of application, including membrane technology.

Schift et al. (2006) has developed a bi-layer process based on a polymeric sacrificial layer to generate freestanding polymeric membranes. Regular hole patterns have been generated in the upper layer using nanoimprint lithography and etched until the surface sacrificial layer using reactive ion etching. Large arrays with interconnected pores can be manufactured by using a wide range of polymeric materials reaching pore sizes of 200 nm. Below the imprinted functional layer, the so-called lift-off resist (LOR) was used (Chen et al., 2004; Carlberg et al., 2003).

Lift-off refers to the process of exposing a pattern into photoresist (or some other material), depositing a thin film (metals or dielectric) over the entire area, and then washing away the photoresist to leave behind the film only in the patterned area. This underlying LOR layer can be dissolved through the nanoimprint lithography pores in a selective solvent, thus creating an air gap. Several cases and different polymeric materials have to be distinguished for a regular array of holes with constant pitch and hole diameter (Schift et al., 2006), as shown in Fig. 3.2.

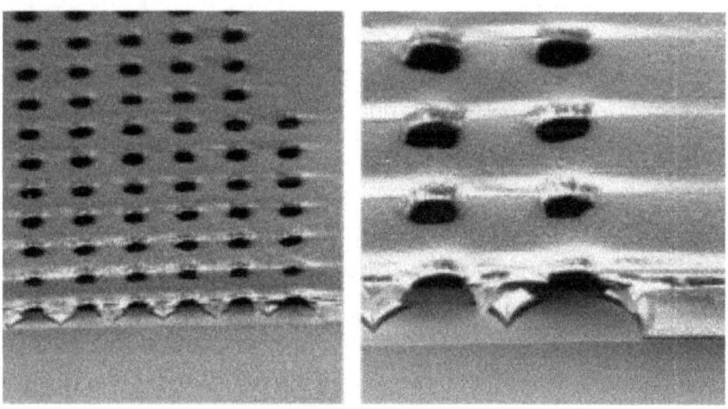

Figure 3.2 Nanostructured and regular array with ordered pores. Reprinted with permission from Schift et al. (2006). Copyright 2006, Elsevier B.V. All rights reserved.

Sainiemi et al. (2010) has demonstrated a simple technique in order to achieve nanoperforate, selective, and separating silicon single-crystalline membranes. They studied the membranes' capability to separate nanobeads of different sizes from the liquid suspension, exploiting the ability of high-performance membranes to separate gases from fluids. These membranes exhibit a thickness of 3 μm and pore size ranging from 250 to 470 nm.

For the first time, Choi et al. (2015) has created 10 nm pores in freestanding polymer membranes by a simple, cost-effective, high-throughput fabrication method such as nanoimprint lithography. Specifically, this approach has been used to build nanostructured SU-8 membranes for transient current measurements during the translocation of DNA molecules through the nanopores. SU-8 is a commonly used epoxy-based negative photoresist, so that the negative refers to a photoresist whereby the parts exposed to UV become crosslinked, while the remainder of the film remains soluble and can be washed away during development (Liu et al., 2004). The imprinted pore has a cone shape and reaches the underlying LOR layer in the freestanding polymer membrane. Conical nanopores have been used as sensors for translocation of single molecules such as λ-DNA.

Maruf et al. (2013) have studied the possibility to create well-defined, submicron surface patterns by means of nanoimprint lithography, on commercial membranes for ultrafiltration (UF), in order to demonstrate prevented flux decline during the separation of the colloidal species. Flux decline due to membrane fouling is, indeed, a drawback that takes place in almost all pressure-driven membrane processes. Here, the submicron patterns obtained by nanoimprint lithography are successfully used to reduce the deposition of silica colloidal particles on the membrane surface, leading to an increase in flux from 19 to 45%. Figure 3.3 shows SEM images collected on the top surface and cross section of virgin and patterned membranes, respectively (Maruf et al., 2013).

Furthermore, Maruf et al. (2015) have evaluated the mechanical properties of membranes manufactured by nanoimprint lithography and related performances in UF process, highlighting the nanoimprinting influence on structural and mechanical performance characteristics of a commercial polyethersulfone (PES) UF membrane. Yanagishita et al. (2007) have manufactured polymer through-hole membranes by nanoimprint lithography using a metal mold-like anodic porous alumina (Fig. 3.4).

Firstly, two polymeric layers are obtained; successively, the underlying layer is removed, leaving a structure with real ordered hole, resulting it interesting for a wide range of applications, including filtration, fuel cells, and bio-devices.

To summarize, nanoimprint lithography is envisaged as an innovative and promising technique for the fabrication of ordered and regular membranes, because it is a fast and generally an easy-to-handle technique. However, its current limitations are related mainly to the difficulty in designing molds under 10 nm, resulting in an expensive approach.

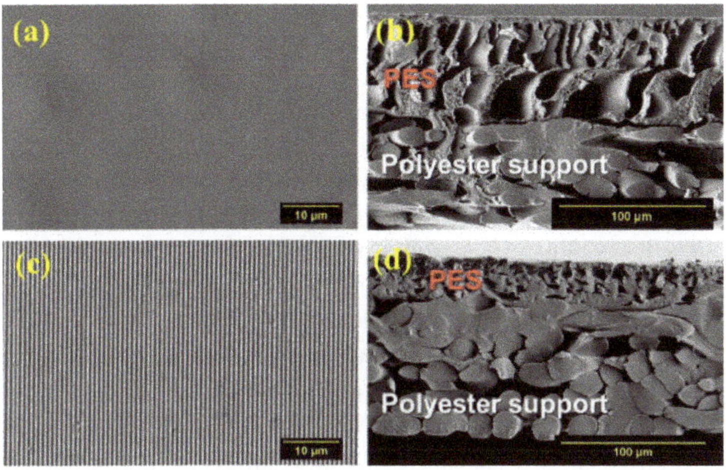

Figure 3.3 SEM images of top surface and cross section of virgin (a, b) and nanoimprinted membranes (c, d). Reprinted with permission from Maruf et al. (2013). Copyright 2013, Elsevier B.V. All rights reserved.

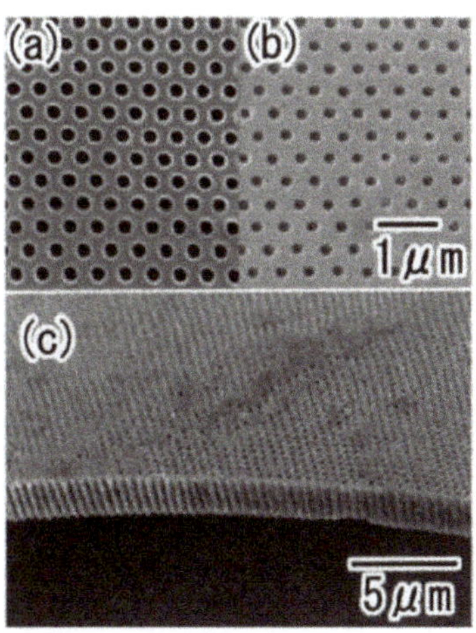

Figure 3.4 SEM images of top surface (a), bottom surface (b), and cross section (c) of final polymer through membrane hole structure. Reprinted with permission from Yanagishita et al. (2007), Copyright 2007, American Vacuum Society.

3.3 Phase Separation and Micro-Molding Combined Techniques

Micro-molding combined with phase separation is another attractive manufacturing approach for patterning nanostructures with hierarchical ordered morphology. It is envisaged as an evolution of traditional phase inversion (Baker, 2000). A polymer is precipitated from a solution assuming a desired configuration, including flat, cylindrical, or spherical shape according to the mold used. The solidification process is a traditional phase inversion and begins with a transition from a liquid to two liquid phases: one rich in polymer and the other rich in solvent. The first one solidifies to form a solid network; the second one generates the pores or voids in the matrix (Mulder, 1996). Conceptually, the phase separation can take place according to different types of phase separation:

- Precipitation by solvent evaporation where a polymer dissolved in a solvent is cast on a suitable support and then the solvent evaporates, leaving a homogeneous membrane.
- Precipitation from a vapor phase where the casting solution is placed in a vapor atmosphere, containing non-solvent saturated with the same solvent. The high solvent concentration in the vapor phase prevents solvent evaporation from the cast film, causing membrane formation due to the penetration of non-solvent into the polymer solution.
- Thermal precipitation where the polymeric solution is heated until a homogenous dope is obtained, and then the casting solution is cooled, inducing phase separation.

It is useful to stress that in traditional phase immersion, the polymer precipitation can be governed by two kinds of demixing: liquid–liquid or solid–liquid according to different kinetics and thermodynamics (Gugliuzza and Drioli, 2007), leading to different membrane morphologies, as shown in Fig. 3.5.

Although controlled pore distribution can be achieved and the technique is very feasible and easy to scale up, traditional phase inversion does not allow packing pores in perfect geometry in the membrane structure; rather, it yields different kinds of porous structures between perfectly regular pores obtained with the micro-molding approach (Johannes et al., 2009). Phase separation micro-molding is an innovative microfabrication technique to achieve micro- and nanostructures from a very broad range of polymers (Fig. 3.6).

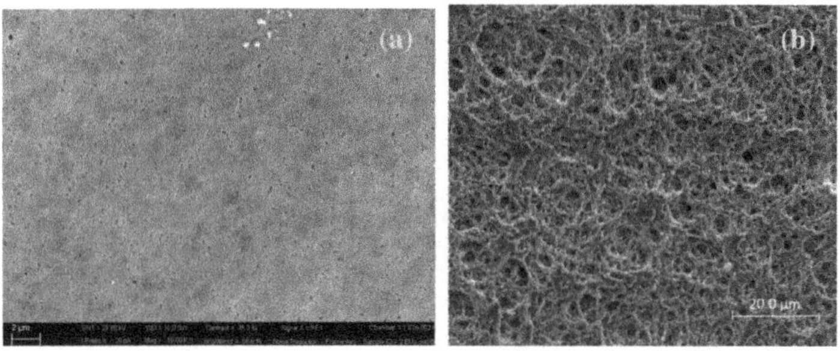

Figure 3.5 SEM images of two typical polyvinylidene difluoride (PVDF) membrane structure achieved with traditional phase separation technique by (a) liquid–liquid demixing and (b) solid–liquid demixing.

- Build-up of a mold having a corresponding predetermined micro- to nanosized structure at the surface.
- The contact between a casting fluid containing a casting material and the molded surface
- Coagulation of the casting solution into a non-solvent for the polymer
- Solidification and shrinkage of the polymeric film contacting the molded surface
- Removal of the patterned film from the molded surface

The replication process is combined with phase separation, while intrinsic shrinkage during the phase separation facilitates the release of the replica from the mold, which increases the reliability at small feature sizes, thin polymer films, or high aspect ratios. Under suitable circumstances, perforation of the polymer film can be obtained, resulting in completely open "through" microstructures. Furthermore, porosity can be introduced in a microstructure, which may result in important properties.

The advantage of combining this traditional technique with molding enables one to produce thousands, millions, even billions of parts with the same precision (Vogelaar and Wessling, 2002). However, the costs and the difficulty to produce molds with high precision at nanoscale remain somewhat high.

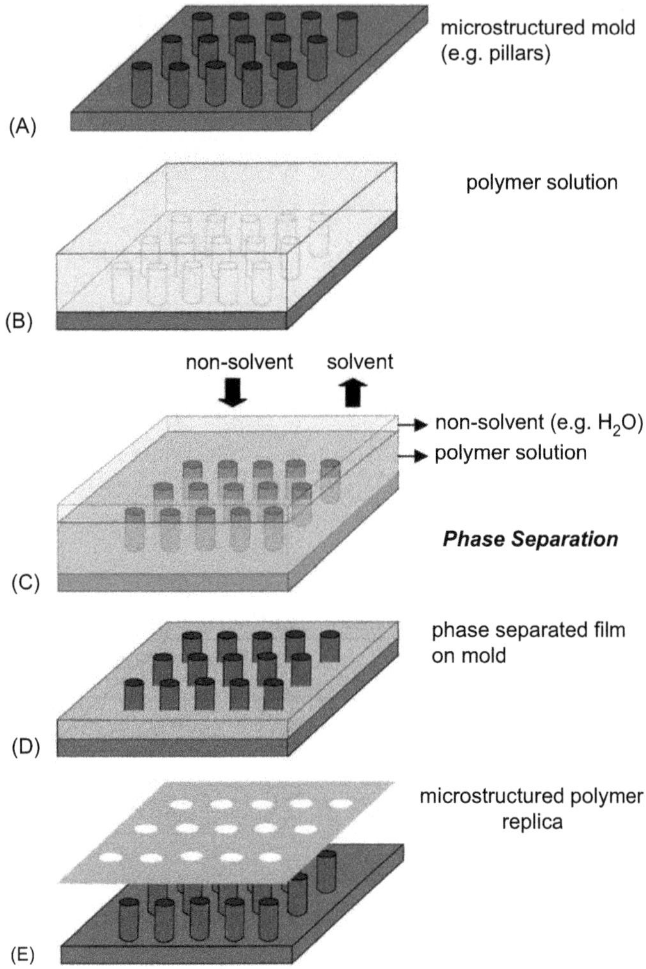

Figure 3.6 Schematic representation of liquid-induced phase separation micro-molding. (B) Casting a polymer film (B) on a structured mold (A). (C) Immersion into a non-solvent for the polymer and phase inversion or polymer coagulation. (D) Step where the polymer assimilates the mold structure and a polymer replica is obtained (E). This scheme represents the fabrication of polymeric microsieves. Reprinted with permission from Gironès et al. (2006). Copyright 2006, Elsevier B.V.

Attia and Alcock (2012) have reported an approach to produce metallic micro-scale components with complex hollow-like internal geometries at high replication accuracy. The approach is a variant of micro-powder injection molding (μPIM), a process for producing micro-parts. The process has demonstrated capabilities in the microfabrication of metals and ceramics, taking the advantage of reaching net-shape complexity and replication fidelity (Attia and Alcock, 2011). Results have shown that this manufacturing route is practical for fabricating closed microstructures for potential applications, including microfluidic channels and micro-scale chambers for energy applications, because 3D internal geometries can be obtained in miniaturized metallic structures.

Gironès et al. (2006) have realized perfectly structured membranes and accurate replicas of polyethersulfone (PES), copolymers of polyethersulfone and polyethylene oxide (PES–PEO), and blends of PES and hydrophilic additives by PSμM. The dimensions of the perforations have been tuned either by using molds with different pillar diameter or by thermal treatment of the polymer above its glass transition temperature. By the latter method, microsieves with initial pore sizes of about 5 or 2.5 μm have been reduced to 1.5 and 0.5 μm, respectively, whereas perforations down to 1.2 μm have been achieved by tuning the dimensions of the mold features (Fig. 3.7). These kinds of structures have been proposed for MF, because conventional microfiltration membranes are often compromised by the poor correlation between permeability and selectivity. In order to improve the properties of such membranes, microsieves have been introduced about a decade ago (Van Rijn, 2004; Kuiper et al., 1998). Microsieves differ from conventional membranes by their well-structured morphology and controlled porosity, which results in good separation behavior and high flow rates (Gironès et al., 2005).

PSμM membranes have also been used in cell culture systems, diagnostics, drug delivery, and tissue engineering. Talei Franzesi et al. (2006) have developed a simple method of micro-molding rapidly gelling, chemically crosslinkable hydrogels, which is based on the controlled release of the gelling agent from the mold. Hydrogels are a promising class of biomaterials since they can be easily tailored to produce desirable mechanical and chemical properties that resemble the native extracellular matrix and exhibit high permeability to oxygen, nutrients, and other water-soluble metabolites (Langer and Peppas, 2003; Rowley et al., 1999). A significant challenge to the micro-molding of chemical or pH-dependent crosslinking hydrogels is their rapid gelling, which occurs immediately upon contact with the gelling agent. In this way, it was possible to control the shape and size of microparticles and microstructures. The ability to engineer size, shape, and network density of the hydrogel particles can be used to control the release kinetics of the encapsulated molecules.

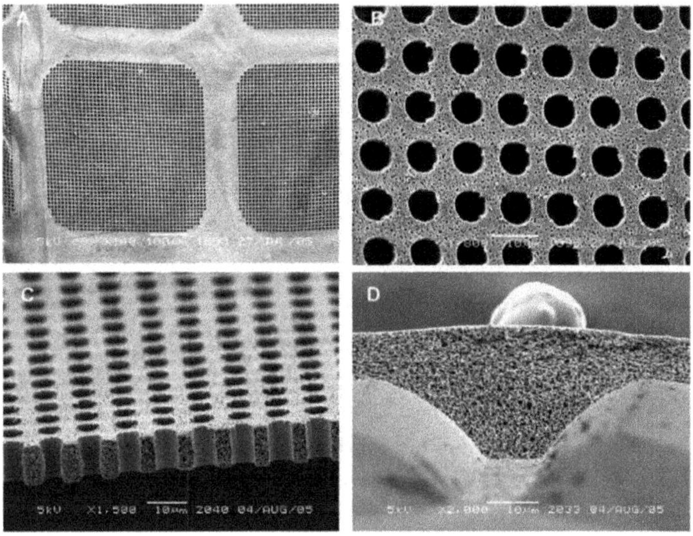

Figure 3.7 SEM images of PES–PEO microsieves: top view of a field (A), close-up of the pores (B), cross section of selective layer (C), and cross section of the support (D). Reprinted with permission from Gironès et al. (2006). Copyright 2006, Elsevier B.V. All rights reserved.

In many processes used for microfabrication of polymeric structures, there is a close, direct contact between the mold and the polymer imprinted (Wessling et al., 2002); obviously, this implies that it is very difficult to obtain a defects-free surface, especially in the mold release phase. In contrast, in PSμM an array of pillars, which are supposed to perforate a polymer film, forms according to the phase separation processes and the porous structure, but the solidified polymer film after perforation is achieved by an intrinsic shrinkage process. It is beneficial because clearly there is no contact between the pillars and the polymer layer, allowing a defect-free release, making this technology as a new and promising manufacturing approach in the field of hierarchical structures.

3.4 Micro- and Nanostructured Arrays by Colloidal Template Techniques

Membranes with morphological features such as high structural order at nanometric scale may be achieved according to the colloidal template method (Velev and Kaler, 2000). Colloid crystalline particles are 3D close-packed crystals of submicrometer spheres working as imprinting agents, whose long-ranged ordered structure is replicated in a solid matrix, thus yielding materials with ordered pores. Spheres work as templates to generate pores; however, their packing is relevant to homogenous material infusion and high degree of interconnection. In this way, the material solidifies without disrupting the order of the crystal lattice, and the spheres are removed subsequently, leaving imprinted in the material the long-ranged periodic structure of the crystal (Velev and Kaler, 2000) (Fig. 3.8).

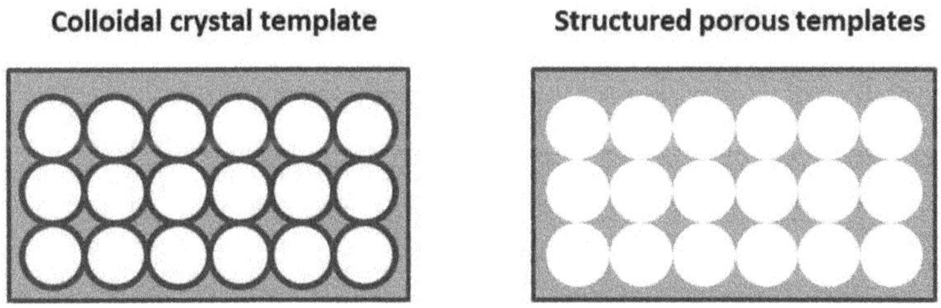

Figure 3.8 Schematic illustration of colloidal crystals replication and template in ordered porous material.

The general procedure that characterizes this templating method can be summarized as follows (Velev and Lenhoff, 2000):

- Assembly of particles to form a 3D crystal structure as a template
- Infusion of material into the cavity of the template crystal structure
- Solidification of the material
- Removal of templates by means of calcination or extraction with a solvent
- Formation of a solid skeleton containing a sorted array of 3D pores

Until today, the template technique showing simple and rapid processes with colloidal crystals could be the main method for preparing macroporous ordered materials (Fig. 3.9). The result shows not only narrow pore-size distribution of macroporous materials but also high void fraction.

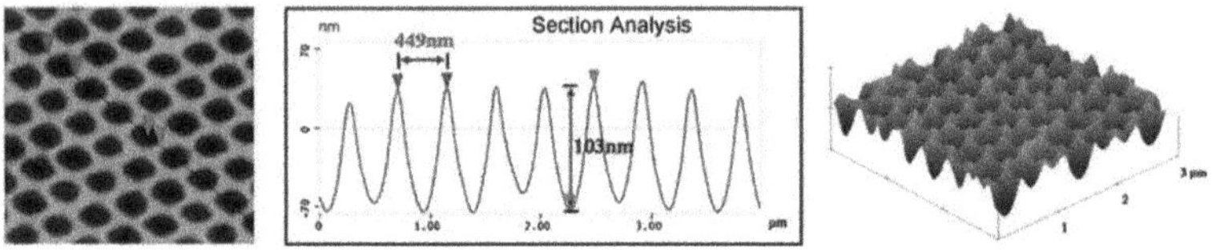

Figure 3.9 Example of morphological profile of ordered patterns realized by colloidal crystal template: AFM data of PDMS replica molds: (a) 2D image, (b) the cross-sectional line profile, and (c) 3D image. Reprinted with permission from Nam et al. (2008). Copyright 2008, Elsevier B.V.

Pores in the achieved structure are usually spherical in shape, because spherical particles are mostly used as templating agents (Hoa et al., 2006). However, this can be modified using different forms of particle models. The pore size and the periodicity of the porous structure can be adjusted by changing the size of colloidal particles (Velev et al., 1997, 1998; Hoa et al., 2006).

The first approach is to provide as a starting material a dispersion of latex particles to be filtered through a polycarbonate membrane. The particles so deposited and concentrated on the membrane surface crystallize as new particles. After washing, the cavities between the particles are mineralized for infusion with 0.5 M aqueous solution of $Si(OH)_4$. The polymerization is started with a surfactant, and finally the latex–silica composites are dried and calcined at 450°C. The organic components are burned during calcination, leaving behind silica flakes of low-density pores. An advantage of this approach is that the pore size can be expected depending on the size of the latex particles that act as templates. Pore size down to 150 nm can be achieved.

The choice of materials that act as templating agents depends not only on the morphological characteristics one wants to give to the final porous structure, but also on the ability to remove the templates without destroying the final pattern, the so-called "walls" of the microporous solid structure (Caruso, 2004). The formed polymeric network might get broken during the removal of the template, which represents a disadvantage of the technique.

Different organic and inorganic templating materials have been proposed, including oil emulsion droplets (Imhof and Pine, 1997) and silica microspheres (Johnson et al., 1999). Table 3.1 reports some examples.

Table 3.1 Typical materials and infusion techniques used to create colloidal particle crystals

Materials	Infusion scheme	References
TiO_2, ZrO_2, SiO_2	Sol–gel hydrolysis	(Imhof and Pine, 1997)
Polyurethane	UV-induced polymerization	(Park and Xia, 1998a; Park and Xia, 1998b; Gates et al., 1999)
Polymers- DVB, EDMA, PMMA, etc.	Heat or UV-induced polymerization	(Johnson et al., 1999)
Graphite, glassy carbon, diamond	Pyrolysis, chemical vapor deposition	(Zakhidov et al., 1998)
Semiconductor CdSe nanocrystals	Particle infusion	(Vlasov et al., 1999)

Colloidal crystal structures with ordered architecture are of great interest for tissue engineering, wherein the availability of arrays for cellular proliferation is requested to promote the optimum environment for good adhesion and consequent cells proliferation, as well as a regenerative activity through an adequate supply of oxygen and nutrients. João et al. (2016) have proposed a simple method for the construction of 3D scaffolds to hydroxyapatite (HAp), suggesting the inverted colloidal crystals (ICC) geometry for bone tissue engineering (Fig. 3.10). The fabrication of ceramic ICC implied a single-step procedure of sphere removal and sintering, composed of four temperature stages. In vitro tests have demonstrated the nontoxic nature of the produced HAp ICC scaffolds.

Concerning the membrane technology, this kind of nanostructures could have an interesting impact on traditional separation processes such as MF, UF, and membrane contactors. Wang (2010) has proposed a study on the fabrication of ICC membranes with uniform pore size, full pores interconnection, and high porosity for UF and MF separations, providing a new solution to overcoming problems related to low mechanical strength. A new ICC membrane formation method has been designed by using a vertical cell (VC) assembly, enabling one to produce colloidal crystal films (CCF) at low cost with surface areas up to 5 cm^2 and thickness up to 100 μm. Considering the difficulty in producing ICC membranes with pore size in the UF range relying on small SiO_2 particles, Wang (2010) described also a way to reduce the ICC membrane pore size by growing a uniform polyethylene-glycol-methacrylate (PPEGMA) nanolayer on membrane surface through initiated atom transfer radical polymerization, taking the further advantage of increasing the membrane surface hydrophilicity and providing good anti-fouling property.

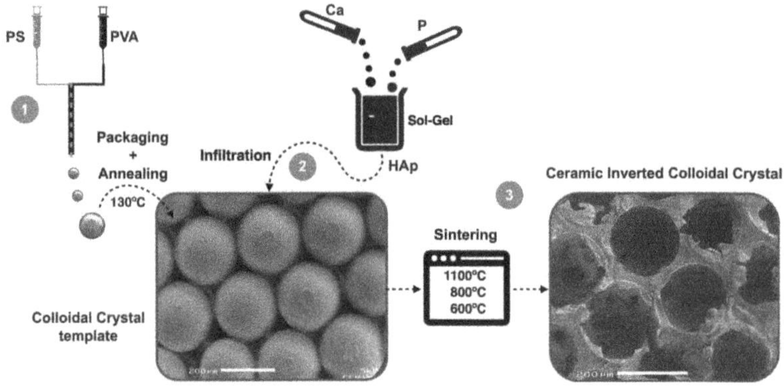

Figure 3.10 Schematic illustration of production of HAp ICC scaffolds. Reprinted with permission from João et al. (2016). Copyright 2016, Elsevier B.V.

Concerning the latest issue, titanium dioxide (TiO_2) has been proposed for generating ICC membranes due to its capability to oxidize numerous organic compounds under specific irradiation and provide cleaning effect (Fujishma and Honda, 1972; Nam et al., 2008). Zhang et al. (2012) have prepared ordered TiO_2 macroporous membranes using PBMA colloidal crystal as template. Other studies have evaluated the efficiency of TiO_2-based ICC membranes to contrast the flux decline due to fouling (Chen et al., 2006; Kim et al., 2005; Kudo, 2006). Casis et al. (2008) have developed ICC polymeric membranes by using silica particles as templating agents and polystyrene and other bio-inspired materials as polymers for water filtration. Bio-inspired polymers, similar to PD, have been chosen to introduce the photoreactivity of thymine into the polymeric system (Anastas and Warner, 1998; Grasshoff et al., 1998). Copolymers based on vinylbencyl thymine (VBT) have also been proposed for their ability to photo-crosslink upon irradiation with short UV wavelength (~280 nm) for causing photo-cyclodimerization reaction and making the material insoluble in water and more mechanically resistant (Cortalezzi et al., 2005; Reculusa et al., 2004).

High specificity of separation characteristics has been witnessed when specific structural parameters have been well addressed (Velev et al., 1997; Johnson et al., 1999). Park and Xia (1998b) have developed polyurethane membranes using polycrystalline lattice of 0.97 μm polystyrene beads as templates. The result has been a very hierarchical 3D structure with a controlled pore size and pore distribution and where each spherical hole is connected with the adjacent ones via small holes (Fig. 3.11).

Preparation of porous structures using colloidal crystal templating appears powerful and inexpensive, but difficult to manage a high degree of order upon scaling up. It would be instructive to make a distinction between the concepts of long-range order—typical of crystalline structures—and short-range order—referred to ordered arrangement of atoms, i.e., water, glass, and polymers (Félidj et al., 2004; Braun and Wiltzius, 2002; Kim, 2004). Long-range order implies extensive and periodical structural order throughout the material; short-range order is extended only to the nearest atoms. Periodic order is widely regarded as an important quality for colloidal arrays. However, there is some confusion in the matter of order since many systems have been described as having long-range order but, in reality, are composed of close-packed domains with local order. This is a current limitation of this kind of breakthrough techniques. An important aspect is, therefore, to obtain not only specific interactions between the various units making the pattern, and then the final array, but also the possibility to have precise control on interparticle distances (Xu et al., 2013). Another limitation is the delicacy of the final structure, which might be compromised during the elimination of templates.

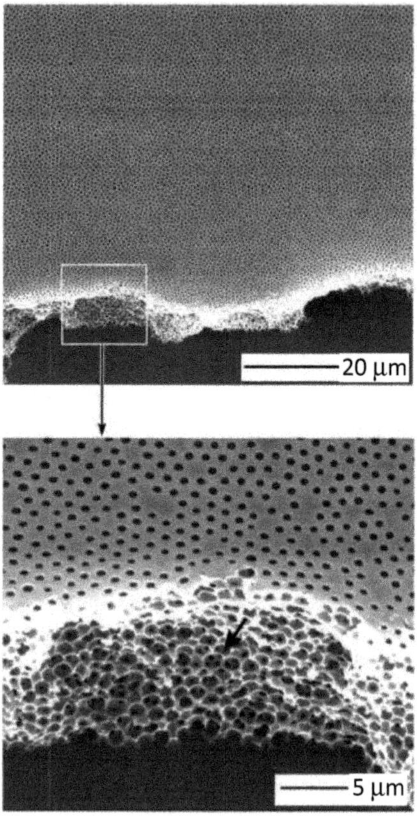

Figure 3.11 Cross-sectional SEMs of a polyurethane membrane that was fabricated by using a polycrystalline lattice of 0.97 μm polystyrene beads as template. Reprinted with permission from Park and Xia (1998a). Copyright 1998, American Chemical Society.

3.5 Self-Assembly of Nano-patterned Polymeric Architectures

Some materials have the capability to assemble spontaneously in complex nanostructures leading to sophisticated architectures. This is named as self-assembly (Whitesides, 1995; Singh et al., 1998). Single separate or connected components spontaneously form ordered aggregates (Whitesides and Boncheva, 2002) depending on the intrinsic structure and chemistry of the molecules involved

(Olenyuk et al., 1999; Lehn 1996): lipids, proteins, carbohydrates, and nucleic acids (Alberts et al., 1994); molecular crystals (Schwiebert et al., 1996); liquid crystals (Schmidt-Mende et al., 2001); semi-crystalline and phase-separated polymers (De Rosa et al., 2000). Specifically, success or failure of self-assembly processes depends on the following:

- Groups or segments forming the backbone of molecules and related capability to interact from less orderly state to a more orderly final state
- Attractive or repulsive weak and non-covalent chemical
- Environments that decide the degree of freedom of the assembling molecules
- Mass transport and agitation controlling for molecules mobility and contact points

High-frequency, controllable pores size and large surface area can be obtained providing new perspectives for patterning and structuring a large class of new materials, including membranes. Nanoporous membranes, obtained by using self-assembled block copolymers as precursors, represent a new class of high-performance materials for use in several processes of separation, purification, drug-delivery system, and sensors (Zhang et al., 2015a).

Ordered nanostructured membranes using self-assembled block copolymers can be obtained depending on the capability of these polymers to self-assemble on nanoscale (Hamley, 2009; Darling, 2007). It is important to yield permeable domains through the overall thickness of the structure, making both the feed and permeate sides accessible (Bates et al., 2012; Phillip et al., 2010b).

Two main strategies are used:

1. Using block polymers, which give rise to a cylindrical morphology with measures for making sure that the cylindrical domains are oriented perpendicularly to the thin dimension of the film (Zhang et al., 2015a).
2. Using block polymers with bi-continuous morphology (e.g., a gyroid-like phase) that obviates the need for alignment (Olson et al., 2008).

The most successful methodology for directing the assembly of nanostructured block polymer-based membranes has been the self-assembly and non-solvent-induced phase separation (SNIPS) procedure. This protocol combines the thermodynamically driven self-assembly of block polymers in solution with the soft-used membrane fabrication technique of NIPS (Mulvenna et al., 2014; Peinemann, 2007). In detail, a polymer solution is prepared by dissolving a block polymer in a solvent or mixture of solvents. The solution is then drawn into a thin film through simple casting techniques (e.g., using a doctor's blade). The solvent is allowed to evaporate from the thin film in a controlled manner for a predetermined period of time. Then the thin film is plunged into a non-solvent bath, thus causing the polymer to precipitate and kinetically trapping the nanostructure of the membrane (Zhang et al., 2015a).

Block copolymers have proven particularly advantageous for templating porous membranes. Some kinds of morphology adopted by block copolymers have nearly ideal attributes for liquid separations, making them attractive candidates for use as ultrafiltration (UF) membranes (Jackson and Hillmyer, 2010). The SNIPS process has been combined with standard wet spinning techniques to produce hollow-fiber membranes that possess a selective layer with a structure directed by a PS-P4VP template. Using a dry-jet wet spinning process, hollow-fiber membranes with pore sizes ranging from 20 to 40 nm in diameter have been produced by adjusting the ratio of THF to DMF in the casting solution, polymer molecular weight, and spinning conditions. This capability is particularly exciting given the trend toward the use of hollow-fiber modules for large-scale membrane separations (Hilke et al., 2013; Radjabian et al., 2014; Radjabian et al., 2013).

To have success in UF, four main criteria are typically needed: high selectivity, high permeability, mechanical integrity, and resistance to fouling (Phillip et al., 2010a). Nanoporous block-copolymer-based membranes seem to meet these requirements, because they are characterized by narrow pore-size distribution (high selectivity), high porosity (high permeability), and required chemical and physical properties.

Zhang et al. (2015a) reported a selectivity–permeability trade-off, i.e., Robeson Plot, between commercial membranes (Robeson, 1991) and block copolymer membranes. The high density of pores with a narrowly distributed pore size is resulted in more selective separations at higher permeability.

Phillip et al. (2010a) have fabricated nanoporous membranes (Fig. 3.12) containing monodisperse pores of 24 nm diameter, using poly(styrene-*b*-lactide) block copolymers to template the pore structure. In particular, 4 μm films have been cast onto a microporous membrane yielding good mechanical resistance. By casting the copolymer film from the appropriate solvents and controlling the solvent evaporation rate, a thin film greater than 100 cm^2 with polylactide cylinders oriented perpendicular to the thin dimension has been produced. Furthermore, the membranes have showed the ability to selectively reject dissolved solutes; in particular, a good rejection of poly(ethylene oxide) molecules of varying molecular weight has been reached. The rejection increased with the solute molecular weight, and rejection of the 100 kDa sample was nearly complete (>93%).

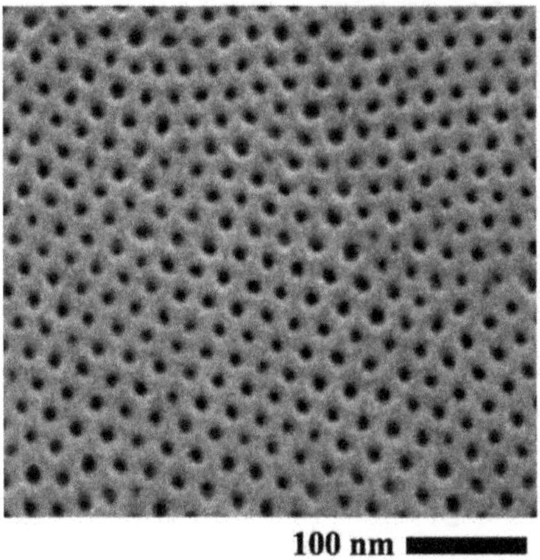

Figure 3.12 SEM image illustrating the structure surface of self-assembled block copolymer thin films, in this specific case achieved with a fast evaporation of the solvent. Reprinted with permission from Phillip et al. (2010a). Copyright 2010, American Chemical Society.

Hahn et al. (2014) have produced highly porous self-assembled integral asymmetric block copolymer membranes for using in the UF process for separation of biomolecules (Fig. 3.13). Four polystyrene-b-poly(4-vinylpyridine) (PS-b-P4VP) di-block copolymers have been investigated in order to prepare membranes with different pore sizes. When the molecular weight of polymer increased, the pore size increases as well in the range from 17 to 53 nm. Clean water fluxes and adsorption of lysozyme, myoglobin, hemoglobin, catalase, and ferritin on the membranes have been studied, evaluating diffusion rates and selectivity of the proteins. The membranes have showed promising results regarding the separation of different proteins in the range of around 2 to 12 nm, above all when compared with commercially available polycarbonate track-etched membranes. Figure 3.13 shows SEM micrographs referred to four types of membranes prepared using polystyrene-b-poly(4-vinylpyridine) (PS-b-P4VP) di-block copolymers as templates.

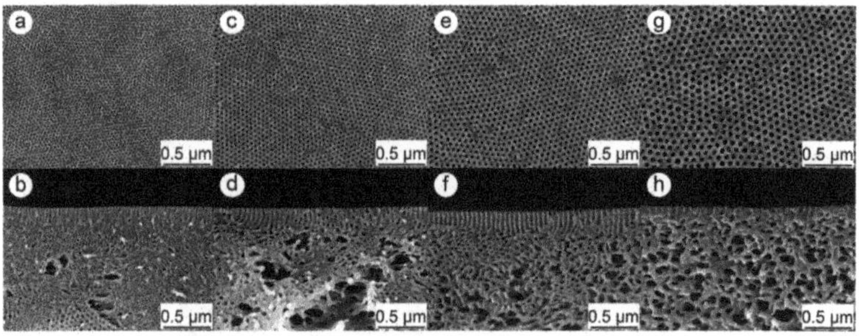

Figure 3.13 Surface and cross-sectional SEM micrographs of membranes achieved using four polystyrene-b-poly(4-vinylpyridine) (PS-b-P4VP) di-block copolymers: (a, b) $PS_{83}P4VP_{17}{}^{76}$; (c, d) $PS_{83.7}P4VP_{16.3}{}^{113}$; (e, f) $PS_{81}P4VP_{19}{}^{160}$; (g, h) $PS_{80}P4VP_{20}{}^{252}$. Reprinted from Hahn et al. (2014) under Creative Commons Attribution Licence, Copyright 2014, The Royal Society of Chemistry.

3.6 Breath Figure for Bio-inspired High-Defined Membranes

Breath figures assembly is one of the most innovative and bio-inspired methods for achieving microporous membranes with ordered pore geometry. BF is a commonly observed phenomenon in daily life, for instance fog, which forms when water vapor contacts a cold surface. During this event, condensing water droplets tend to rearrange in ordered geometry resembling honeycomb patterns (Speranza et al., 2010). First, Aitken (1895, 1911) and later Raylegh (1911) investigated the BF phenomena, revealing the self-assembly of water droplets on cold surfaces. Indeed, the BF method mimics exactly what happens in nature when water vapor comes in contact with a cold surface, thus allowing the formation of droplets rearranged in well-defined geometries. This capability can be exploited to shape well-defined pores in polymeric films under well-established thermodynamics, phenomena of capillarity, and Marangoni effects (Xu and Luo, 2007).

The simplicity and inexpensiveness of the approach makes this manufacturing method a powerful and practical route for the fabrication of high-definition membranes. It is essentially based on the condensation of water droplets on the surface of low-viscous polymer solutions, containing immiscible or partially miscible solvents. This makes the recovery of the solvent easier on scale. In addition, water is a nontoxic and cheap template, so the overall approach can be regarded as a next-generation green manufacturing technology.

3.6.1 Potential Controlling Mechanisms

In spite of handle manufacturing approach, the mechanism controlling the BF geometry formation can be complex and not unique (Gugliuzza et al., 2006). This is due to the affinity of polymeric materials and solvents used, and also uncontrolled changes in the surrounding experimental conditions.

Srinivasarao et al. (2001) have proposed an interesting mechanism to understand what happens in the BF structure. When moist air is blown and induced on the surface of a polymeric solution, there is a decrease in the temperature related to the evaporation cooling. This causes nucleation and growth of water droplets on the surface. After a certain period of time, rafts of water droplets are formed; in the absence of local perturbation, droplet islands are prevented from coalescence phenomena, leading to hexagonal structures. Each single array sinks into the polymeric solution, forming initially a single layer and leaving, at the same time, a balance between the formation of the polymer network and the evaporation of the solvent. Depending on the solvent density, the rafts of water droplets leave an imprint on the polymeric matrix, moving from the 2D monolayer, as the solvent is heavier than water, to 3D polymeric networks, as the solvent is lighter than water. Conversely, Speranza et al. (2010) have demonstrated the possibility to realize also 3D structures using halogenated solvents at controlled viscosity.

Srinivasarao et al. (2001) have assert that the non-coalescence of the droplets is essentially due to the combination of thermo-capillarity and Marangoni effect, which leaves the bubble array to levitate at the surface and form an initial crystalline BF bubbles array that precipitates in solution. In the case of bubbles-array formation, the solvent-saturated air/solvent boundary takes over the role of the second liquid. In such systems, droplets of identical material composition are immiscible. Another important effect is the Marangoni convection, in which two droplets of the same liquid do not coalesce if a sufficiently large temperature gradient exists (Schatz and Neitzel, 2001) (Fig. 3.14).

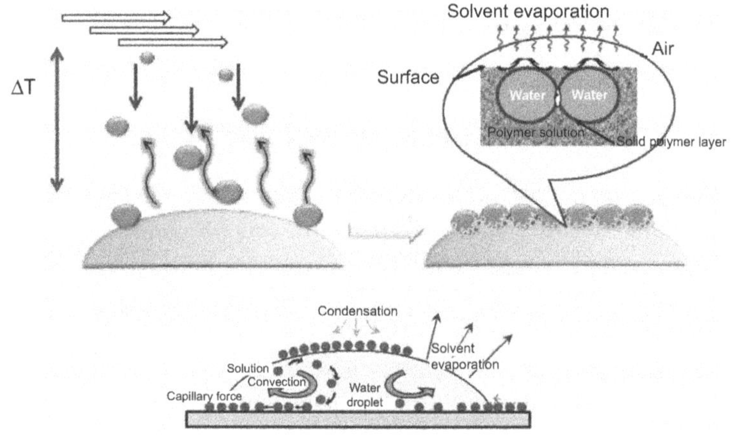

Figure 3.14 Schematic representation of the BF method.

Another hypothesis has been suggested by Pitois and Francois (1999a): when water vapor droplets come in contact with the surface of the polymeric solution, they touch the same surface and a polymeric layer formed at the interface prevents droplet coalescence (Maruyama, 1998).

Karthaus (2000) has proposed another mechanism, which is different from that suggested by Francois and Srinivasarao. The initial water droplets do not levitate, and they are not ordered on the surface, but thermocapillary and Marangoni forces cause the water droplets to submerge into the organic solution, causing a hexagonal order (Stenzel et al., 2006). Speranza et al. (2010) have demonstrated evidences of mixed mechanisms, confuting the hypothesis that 3D air bubble networks can be formed only when solvents lighter than water are used (Fig. 3.15).

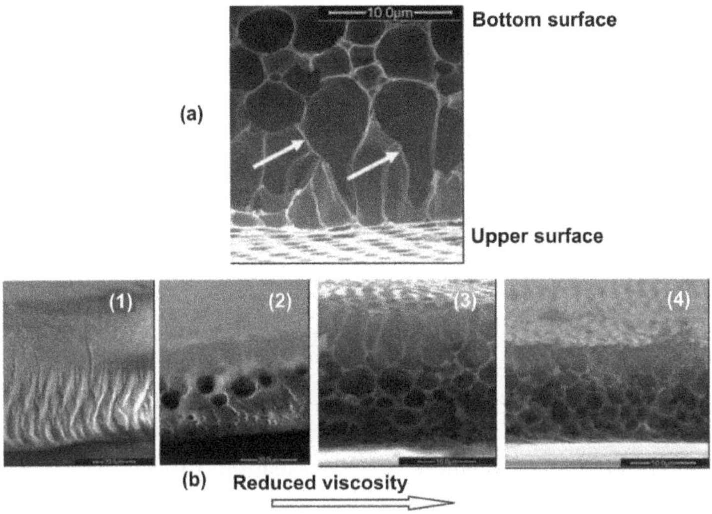

Figure 3.15 (a) SEM image illustrating the interconnection of pores in the upper layer of the polymer array. (b) SEM images of cross section related to polymer arrays prepared by solutions having decreasing viscosity. Reprinted with permission from Speranza et al. (2010). Copyright 2010, American Chemical Society.

It is also relevant to highlight how a high degree of interconnection can be achieved. Due to the fact that the polymer layer embedding the droplets is very thin and weak between the touching spherical templates, smaller pores are formed around the walls of the pores leading to a high degree of interconnection when all solvents are completely evaporated. This is a critical point for promoting a large degree of interconnection and, consequently, reduced resistance to mass diffusion.

3.6.2 Polymers Chemistry

The simplicity of BF pattern formation has led to the preparation of ordered porous films using a wide variety of polymers and different macromolecular architectures, including linear, star, graft, dendritic, hyperbranched polymers, and coil–coil or rod–coil di-block copolymers (Kim et al., 2009).

The literature refers that polystyrene (PS) is used as a component in many polymers that form BF architectures. It is known that the architecture of the polymer used as well as the presence of surfactants or terminals groups favors the formation of porous microstructures, where the holes are strongly interconnected. The fluidity of the solution is a key issue for the formation of ordered structures via BF (Zhang et al., 2015b). Other elements contribute to the optimization of the structure and to the success of the final process (Stenzel et al., 2006):

- **Molecular weight**: In general, an increase in molecular weight results in the formation of bigger pores. However, molecular weight is not crucial to the size of pores when the molar concentration of the solution is maintained constant. Generally, the molar concentration decreases when the molecular weight increases, and this results in higher precipitation of water droplets, leading to the formation of bigger pores.
- **Number of arms of star polymers**: A highly branched polymer tends to stabilize the water–organic interface more. This has been demonstrated experimentally by increasing the number of arms on a star polymer, while keeping the molecular weight constant (Stenzel et al., 2006). The result is an increase in segment density and also in a much stronger end-group effect. This means that with more branches, pores smaller in size are generated. This is why terminal groups of the polymer structure are of considerable importance in stabilizing the interface.
- **Polymer architecture and polymer structure**: An effective empirical relationship between the structure of the polymer and the pore size has not been established. Rather, polymer structures tend to influence the stabilization of droplets, the point of precipitation, and the viscosity of the solution.

The most recent research has further expanded the portfolio of usable polymers, such as oligomers (Sun et al., 2009a), complexes of DNA (Ke et al., 2010a), and glycol-polymers (Ke et al., 2010b; Munoz-Bonilla et al., 2010). Often additives are used to promote the formation of BF and to introduce functionality to the porous polymeric film (Gugliuzza et al., 2008; Speranza et al., 2010). Among these, surfactant films (Gugliuzza et al., 2016), perfluoropolyethers (Pilati et al., 2007), various kinds of nanoparticles (Sun et al., 2009b; Jiang et al., 2010), or carbon nanotubes (Lee et al., 2009) are frequently used.

Speranza et al. (2010) have proposed self-supporting membranes patterned from PEEK-WC-NO_2 with different substitution degrees (SDs) in order to evaluate the influence of the nitro group on solution properties, including viscosity. An increase in the degree of chemical substitution of the polymer chains has been effectively demonstrated to cause a reduction in the polymer solution viscosity due to the steric nitro portion that limited the degree of interlacing of the chains. On this basis, viscosity has also been hypothesized to be a driving force parameter for the construction of well-defined surface structures. In order to establish the role of polymer viscosity in the rapidity of rearrangement and crystallization of the droplets, equal amounts of PEEK-WC and PEEK-WC-NO_2 at increasing SDs have been dissolved in the same chlorinated solvent, yielding polymer dopes at decreasing viscosities. As a result, it has been observed that condensed droplets can freely move through the media at lower viscosity, thus forming highly ordered lattices in a shorter time. These experimental findings strengthen the role of polymer chain rearrangements in the kinetics controlling the formation of BF membranes.

3.6.3 Nature of Solvents

The most used solvents are highly volatile and water-immiscible. This is because solvents play a key role, since evaporation cools the surface of the solution and density affects flotation. Surface free tension and immiscibility with water also decide the final shape of droplets and, hence, the pore shape and size imprinted in the polymeric pattern. Some of the many solvents reported in the literature are carbon disulfide, benzene, toluene, chloroform, dichloromethane, 1,2-dichloroethane, Freon, and so on.

Speranza et al. (2010) have showed that a slower droplet condensation and a more extended growth rate are expected when a solvent with a higher boiling point, such as chloroform (T_{bp}= 61°C, Mw = 119.38 Da), is used, thus yielding enlarged pores. Differently, solvents with higher volatility, such as dichloromethane (T_{bp}= 39.9°C, Mw = 84.93 Da), yield quicker coverage and self-assembly of submerged droplets, leaving smaller air bubbles as the effect of an enhanced difference in temperature. Other authors (Tian et al., 2006) have speculated that the size of the droplets is decided by the evaporation velocity of the solvent and, therefore, by the difference in the temperature between the liquid surface and humid air.

3.6.4 Temperature Effect

The temperature of the solution is another key element in the formation of ordered arrays using BFs, because the difference in temperature between moist air and the surface of the polymeric solution allows the spontaneous condensation of water vapor as the first step of the process. As an effect of solvent evaporation, the surface of the polymeric solution becomes cold, reaching a temperature lower than the dew point. This causes water droplets to condense and grow through the solution. So the difference in temperature can be regarded as the real driving force of BF formation. Given the crucial importance of temperature, Bormashenko et al. (2012a) have suggested pre-cooling of the solution. It has been shown that all the solvents, including acetone, dichloromethane, chloroform, carbon tetrachloride, tetrahydrofuran, toluene, xylenes, carbon disulfide, and N,N-dimethylformamide, used with the polymers polystyrene and polycarbonate gave rise to microstructured surfaces, precisely honeycomb arrays, when exposed to humidity and pre-cooled sufficiently.

3.6.5 Surface Free Tension

The major hindrance to commercially producing large-scale films is the lack of control on the structural order at the long range through the surface area of the films realized (Fig. 3.16).

In order to make the BF process scalable, it is imperative to suppress any event that can affect seriously the extensive degree of structural order over the entire surface. This imposes the formation of local chemical microenvironments through suitable hydrophilic/hydrophobic balance, which enables one to direct the rearrangement of droplets in ordered lattices (Gugliuzza et al., 2016). With this regard, it has been demonstrated that combining a nonionic surfactant such as Tween 20 with alcohols at different hydrocarbon chain length, it is possible to get the droplets in extensively ordered lattices, yielding modulation in pore size and switching from self-assembly to assisted self-assembly (Fig. 3.17).

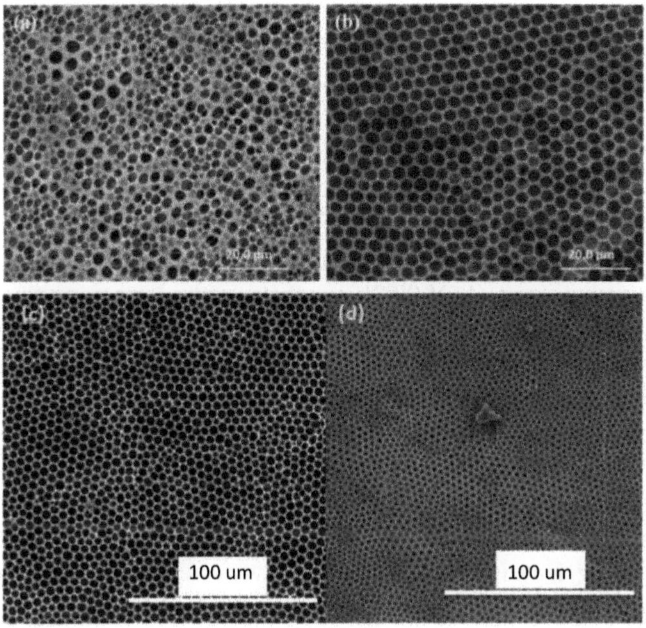

Figure 3.16 SEM images of typical honeycomb membrane: (a) without surfactant (self-assembly); (b–d) with amphiphylic additive (assisted self-assembly).

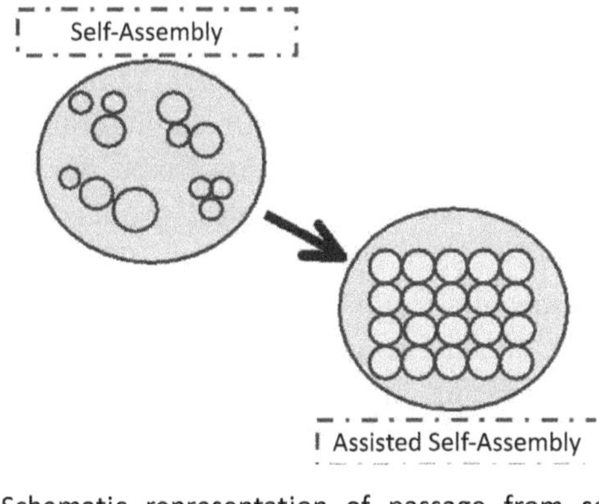

Figure 3.17 Schematic representation of passage from self-assembly to assisted self-assembly.

3.6.6 Potential Applications of Honeycomb Structures

Perfectly ordered structures have advantages in many fields of application due to the benefits arising from the peculiarities of specific structural elements (Lee et al., 2016; Wienk et al., 1996; Trotta et al., 2001). The size of the periodic microstructures in breath figure arrays (BFAs) falls within the range of hundreds of nanometers to tens of micrometers, matching the requirements of optical and optoelectronic devices, including microlens arrays (MLAs), micropatterned light-emitting diodes (LEDs), antireflective (AR) coatings, solar cells, photonic crystals, and optical sensors (Kurono et al., 2002; Ghannam et al., 2007). Li et al. (2011a) prepared honeycomb porous lanthanum strontium manganite/yttria-stabilized zirconia (LSM/YSZ) composite cathodes having the ideal size lowering resistance polarization. Honeycomb polymeric films can also represent a chemical power source; lithium ion battery is one of the most important energy-storage devices and the porous structure of the BFA film anode can promote mass transfer in the electrolyte and provide sufficient space to realize volume expansion in the charge/discharge process (Tsai et al., 2011).

In the last years, BF structures have been envisaged to have interesting features to be used in advanced fields of membrane science as well, including biomedicine, food, and water treatment. For example, surface topography might have a vital role in the cellular response, working as a support to direct the growth of cells within the porous structure to form functional tissue (Lanza et al., 2000; Agrawal and Ray, 2001). It has been found that the behavior of adhesion, growth, and proliferation of different types of cells on scaffolds depends on surface characteristics such as wettability, chemistry, charge, rigidity, and topography (Yamamoto et al., 2007; Dalby et al., 2003). Advances in tissue engineering have revealed that porosity architectures such as pore shape, pore size, and pore interconnectivity are key morphological properties for scaffolds (Karageorgiou and Kaplan, 2005; Hollister, 2005). The honeycomb structure has showed a significant effect on the adhesion, growth, and proliferation of cells. In addition, cell signaling has been more effectively activated on a honeycomb-structured microporous polymer film than on a flat counterpart (Tanaka, 2011). Yamamoto et al. (2007) have investigated the mechanism behind cellular response to surface topography, preparing honeycomb-patterned films from polycaprolactone (PCL) with micropatterned surface topography and evaluating the response of porcine aortic endothelial cells (PAECs) to adsorbed Fn molecules onto them. Bormashenko et al. (2012b) have studied the possibility to use honeycomb structures deposited on stainless steel gauzes in water/oil separation, with surfaces that are simultaneously superhydrophobic and superoleophilic (Feng et al., 2004; Pan et al., 2008). For the first time, membranes with a perfectly regular structure have been used in membrane contactors (Gugliuzza et al., 2008), in particular in membrane distillation for water desalination, giving performances superior to what is commercially available (Perrotta et al., 2017).

New types of structured membranes have been targeted for developing advanced membrane distillation operations, which are part of membrane integrated systems dedicated to the production of fresh water. A lot of research has been addressed at the fabrication of new membranes with such structural and chemical features to overcome some current shortcomings, including productivity and thermal efficiency. A perfluorinated polymer, Hyflon® AD, has been used to generate unprecedented suspended honeycomb nanofilm onto the top surface of the PES honeycomb membrane (Fig. 3.18) in order to split the area of the biggest pores in smaller pores, preserve high surface interfacial area, and reduce the pathway through which water vapor can diffuse while crystals are retained. The establishment of fine materials structure–transport relationships appears a practical route for obtaining extraordinary productivity–efficiency trade-offs during water desalination via thermally driven membrane distillation. These novel nanostructured membranes are promising to catalyze the accomplishment of new concepts of water desalination processes.

The goals of the future are to explore the potential of this type of honeycomb membranes not only in the field of membrane desalination but also in other areas of water treatment such as ultrafiltration and microfiltration (UF and MF) for the recovery of specific compounds or water/oil separation and also gas separation. Liu et al. (2011) have developed new honeycomb-structured and gas-tight La0.6Sr0.4Co0.2Fe0.8O3-a (LSCF) perovskite hollow-fiber membranes with an ultra-thin dense layer using phase inversion/

sintering technique with an EtOH–NMP mixture as the internal coagulant (Fig. 3.19).

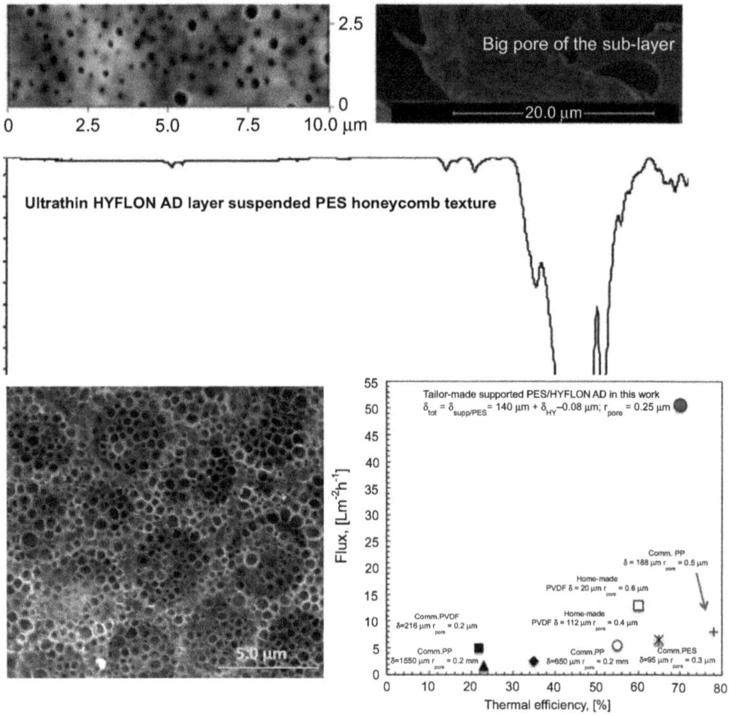

Figure 3.18 SEM images of top surface of PES honeycomb membranes with Hyflon-suspended nanofilm. Reprinted with permission from Perrotta et al. (2017). Copyright 2017, Elsevier B.V.

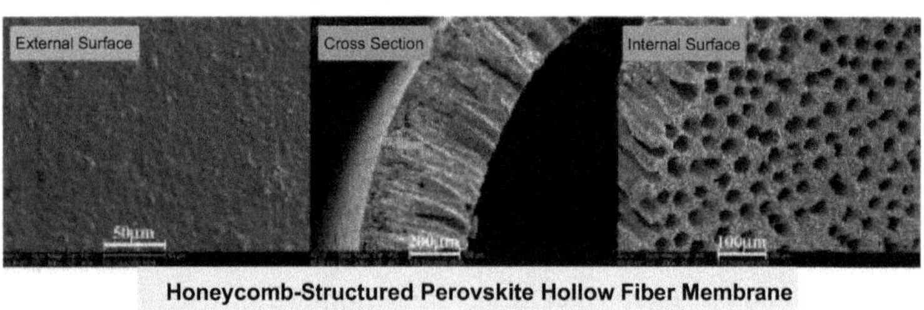

Figure 3.19 SEM images of honeycomb-structured perovskite hollow-fiber membrane, in order: external surface, cross section, and internal surface. Reprinted with permission from Liu et al. (2011). Copyright 2011, Elsevier B.V.

These membranes have provided significantly higher O_2 fluxes than conventional types, with an improvement factor up to 15 at 800°C, while maintaining membrane quality and mechanical stability (Fig. 3.20). The oxygen flux for honeycomb-structured LSCF hollow fibers is higher than that obtained with conventional LSCF hollow fibers. This enhanced flux can be contributed to two factors: (1) thin effective membrane thickness with a reduced bulk diffusion resistance; (2) porous downstream surface with a much larger membrane area for the surface exchange reactions, as was explained previously.

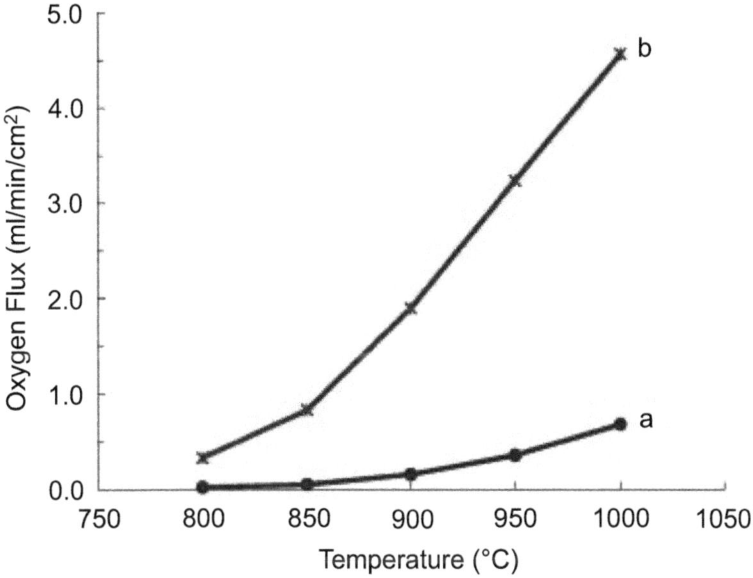

Figure 3.20 Oxygen fluxes through the conventional LSCF hollow fibers (a) and the honeycomb-structured LSCF hollow fibers (b) (measured at air feed flow rate of 250 mL/min in the shell side and He sweep gas rate of 147 mL/min in the hollow-fiber lumen). Reprinted with permission from Liu et al. (2011). Copyright 2011, Elsevier B.V.

Therefore, this type of mixed conducting ceramic membranes with new honeycomb structure could find application not only for gas separation, but also as catalytic membrane reactors for light hydrocarbon oxidation.

3.7 Future Prospects and Concluding Remarks

Reviewing some of the attractive techniques used in manufacturing nanostructured and hierarchically organized membranes, it is possible to arrive at the following conclusions:

- By defining the geometry of the pores of membranes, membrane separation performance can be controlled in different fields.
- One of the main goals is to obtain a more homogeneous and narrow pore distribution, in order to predict the performance depending on the specific morphology and ordered structure of the interface used.
- The fields of application are varied, ranging from molecular biology and tissue engineering to photonics as well as water and gas purification, with a major ambition to get very high productivity–selectivity trade-offs.
- Critical aspects are still to be considered, including manufacturing costs and working time, as well as the ease of managing and scaling up products to industrial scale. However, the fine structure–property relationships make these new-generation nanostructured membranes extraordinarily appealing due to low resistance to transport, high efficiency energy, and separation capability.

Appendix

Abbreviations

AFM	microscopic atomic force
ALD	atomic layer deposition
AR	antireflective
ATRP	atom transfer radical polymerization
BF	breath figures
CCF	colloidal crystal film
DMF	dimethyl formamide
DNA	deoxy-ribonucleic acid
DRIE	deep reactive ion etching
DVB	divinylbenzene

EDMA	ethylene dimethacrylate
HAp ICC	hydroxyapatite inverted colloidal crystals
ICC	inverted colloidal crystals
LEDs	light emitting diodes
LIPS	liquid-induced phase separation
LOR	lift-off resist
LRO	long range order
MC	membrane condenser
MCr	membrane crystallization
MD	membrane distillation
ME	membrane emulsification
MEMS	micro-electromechanical systems
MF	microfiltration
MLAs	microlens arrays
m-TAS	micro total analysis system
NEMS	nanoelectromechanical system
NIL	nanoimprinting lithography
NMP	n-methyl-pirrolidone
PCL	polycaprolactone
PDMS	polydimethylsiloxane
PEEK-WC	poly(ether ether ketone) with a cardo group
PEEK-WC-NO_2	nitrated poly(ether ether ketone) with a cardo group
PEGMA	poly-ethylene-glycol-methacrylate
PES	poly-ethersulfone
PES–PEO	polyethylene oxide
PMMA	poly-methyl-methacrylate
PS	polystyrene
PS-b-P4VP	polystyrene-b-poly(4-vinylpyridine)
PSµM	phase separation micro molding
SDs	substitution degrees
SEM	microscopic electronic system
SNIPS	non-solvent-induced phase separation
SOI	silicon-on- insulator
SRO	short range order
TB	triblock copolymer
THF	tetrahydrofuran
UF	ultrafiltration
UV	ultraviolet

VBT	vinyl bencyl thymine
VC	vertical cell
VIPS	vapor-induced phase separation
μPIM	micro-powder injection molding

Notation

M	molar (mol/L)
MW	molecular weight (Da)
T	temperature (°C)
T_{bp}	boiling point (°C)
T_g	transition temperature (°C)

Symbols

α	separation factor
μ	micron

References

Agrawal, C. M. and Ray, R. B. (2001). Biodegradable polymeric scaffolds for musculoskeletal tissue engineering, *J. Biomed. Mater. Res.*, **55**, pp. 141–150.

Aitken, J. (1895). Breath figures, *Proc. R. Soc. Edinburgh*, **20**, pp. 94–97.

Aitken, J. (1911). Breath figures, *Nature*, **86**, pp. 516–517.

Alberts, B., Bray, D., Lewis, J., Raff, M., Roberts, K., and Watson, J. D. (1994). *Molecular Biology of the Cell*, 3rd edn, Garland, New York.

Anastas, P. and Warner, J. (1998). *Green Chemistry: Theory and Practice*, Oxford University Press, New York.

Attia, U. M. and Alcock, J. R. (2011). A review of micro-powder injection moulding as a microfabrication technique, *J. Micromech. Microeng.*, **21**, 043001, pp. 1–22.

Attia, U. M. and Alcock, J. R. (2012). Fabrication of hollow, 3D, micro-scale metallic structures by micro-powder injection moulding, *J. Mater. Proc. Tech.*, **212**, pp. 2148–2153.

Baker, R. W. (2000). *Membrane Technology and Applications*, McGraw-Hill.

Bates, F. S., Hillmyer, M. A., Lodge, T. P., Bates, C. M., Delaney, K. T., and Fredrickson, G. H. (2012). Multiblock polymers: Panacea or Pandora's Box? *Science*, **336**, pp. 434–440.

Bormashenko, E., Balter, S., and Aurbach, D. (2012a). On the nature of the breath figures self-assembly in evaporated polymer solutions: Revisiting physical factors governing the patterning, *Macromol. Chem. Phys.*, **213**, pp. 1742–1747.

Bormashenko, E., Balter, S., Bormashenko, Y., and Aurbach, D. (2012b). Honeycomb structures obtained with breath figures self-assembly allow water/oil separation, *Colloids Surf. A*, **415**, pp. 394–398.

Braun, P. B. and Wiltzius, P. (2002). Macroporous materials: Electrochemically grown photonic crystals, *Curr. Opin. Colloid Interface Sci.*, **7**, pp. 116–123.

Carlberg, P., Graczyk, M., Sawe, E.-L., Maximov, I., Beck, M., and Montelius, L. (2003). Lift-off process for nanoimprint lithography, *Microelectron. Eng.*, **67–68**, pp. 203–207.

Caruso, F. (2004). *Colloid and Colloid Assemblies: Synthesis, Modification, Organization and Utilization of Colloid Particles*, Wiley-VCH, pp. 468.

Casis, N., Fidalgo de Cortalezzi, M. M., Estenoz, D., Ravaine, S., and Meira, G. (2008). Polymeric membranes from colloidal crystal templates, Conamet/SAM Congress Chile, October 28–31.

Chen, D., Weavers, L. K., and Walker, H. W. (2006). Ultrasonic control of ceramic membrane fouling: Effect of particles characteristics, *Water Res.*, **40**, pp. 840–850.

Chen, Y., Peng, K., and Cui, Z. (2004). A lift-off process for high resolution patterns using PMMA/LOR resist stack, *Microelectron. Eng.*, **73–74**, pp. 278–281.

Choi, J. and Park, S. (2015). Fabrication of nanopores in polymer membranes using nanoimprint lithography for DNA analysis, *Proc. 19th International Conference on Miniaturized Systems for Chemistry and Life Sciences*, Gyeongju, Korea.

Chou, S. Y., Krauss, P. R., and Renstrom, P. J. (1996). Imprint lithography with 25-nanometer resolution, *Science*, **272**, pp. 85–87.

Cortalezzi, M. M., Colvin, V. L., and Wiesner, M. R. (2005). Controlling submicron-particle template morphology: Effect of solvent chemistry, *J. Colloid Interface Sci.*, **283**, pp. 366–372.

Dalby, M. J., Riehle, M. O., Yarwood, S. J., Wilkinson, C. D., and Curtis, A. S. (2003). Nucleus alignment and cell signaling in fibroblasts: Response to a micro-grooved topography, *Exp. Cell Res.*, **284**, pp. 274–284.

Darling, S. B. (2007). Directing the self-assembly of block copolymers, *Prog. Polym. Sci.*, **32**, pp. 1152–1204.

De Rosa, C., Park, C., Thomas, E. L., and Lotz, B. (2000). Microdomain patterns from directional eutectic solidification and epitaxy, *Nature*, **405**, pp. 433–437.

Félidj, N., Lau Truong, S., Aubard, J., Lévi, G., Krenn, J. R., Hohenau, A., Leitner, A., and Aussenegg, F. R. (2004). Gold particle interaction in regular arrays probed by surface enhanced Raman scattering, *J. Chem. Phys.*, **120**, pp. 7141–7146.

Feng, L., Zhang, Z., Ma, Y., Liu, B., Jiang, L., and Zhu, D. (2004). A super-hydrophobic and superpleophilic coating mesh film for the separation of oil and water, *Angew. Chem. Int. Ed.*, **43**, pp. 2012–2014.

Fujishima, A. and Honda, K. (1972). Electrochemical photolysis of water at a semiconductor electrode, *Nature*, **238**, pp. 37–38.

Gates, B., Yin, Y., and Xia, Y. N. (1999). Fabrication and characterization of porous membranes with highly ordered three-dimensional periodic structures, *Chem. Mater.*, **11**, pp. 2827–2836.

Ghannam, L., Garay, H., Francois, J., and Billon, L. (2007). New polymeric materials with interferential optical properties, *Macromol. Chem. Phys.*, **208**, pp. 1469–1479.

Gironès, M., Lammertink, R. G. H., Borneman, Z., and Wessling, M. (2005). The role of wetting on the water flux performance of microsieve membranes, J. Membr. Sci., 259, pp. 55–64.

Gironès, M., Akbarsyah, I. J., Nijdam, W., Van Rijn, C. J. M., Jansen, V., Lammertink, R. G. H., and Wessling, M. (2006). Polymeric microsieves produced by phase separation micromolding, *J. Membr. Sci.*, **283**, pp. 411–424.

Grasshoff, J., Taylor, L., and Warner, J. (1998). Copolymers having pendant functional thymine groups, U.S. Patent 5, 708, 106.

Gugliuzza, A. and Drioli, E. (2007). PVDF and HYFLON AD membranes: Ideal interfaces for contactor applications, *J. Membr. Sci.*, **300**, pp. 51–62.

Gugliuzza, A. and Drioli, E. (2013). A review on membrane engineering for innovation in wearable fabrics and protective textiles, *J. Membr. Sci.*, **446**, pp. 350–375.

Gugliuzza, A., Aceto, M. C., Macedonio, F., and Drioli, E. (2008). Water droplets as template for next generation self-assembled poly-(etheretherketone) with Cardo membranes, *J. Phys. Chem. B*, **112**, pp. 10483–10496.

Gugliuzza, A., Perrotta, M. L., and Drioli, E. (2016a). Controlled bulk properties of composite polymeric solutions for extensive structural order of honeycomb polysulfone membranes, *Membranes*, **6**, pp. 27–38.

Gugliuzza, A., Perrotta, M. L., Saielli, G., Casella, G., Macedonio, F., and Drioli, E. (2016b). Fabrication of hierarchically structured honeycomb PES/HYFLON AD membranes and potential use in water desalination, Permea & Melpro, May 15–19, Prague, Czech Republic, *Poster presentation.*

Gugliuzza, A., Siciliano, C., Costabile, R., Liguori, A., and Drioli, E. (2008). Novel porous functional membranes, *IV Proceedings, Convegno Congiunto delle Sezioni Calabria e Sicilia 2008*, SCI, December 1–2, Rende, Italy.

Guo, L. J. (2007). Nanoimprint lithography: Methods and material requirements, *Adv. Mater.*, **19**, pp. 495–513.

Hahn, J., Clodt, J. I., Filiza, V., and Abetz, V. (2014). Protein separation performance of self-assembled block copolymer membranes, *RSC Adv.*, **4**, pp. 10252–10260.

Hamley, I. W. (2009). Ordering in thin films of block copolymers: Fundamentals to potential applications, *Prog. Polym. Sci.*, **34**, pp. 1161–1210.

Hilke, R., Pradeep, N., Madhavan, P., Vainio, U., Behzad, A. R., Sougrat, R., Nunes, S. P., and Peinemann, K.-V. (2013). Block copolymer hollow fiber membranes with catalytic activity and pH-response, *ACS Appl. Mater. Interfaces*, **5**, pp. 7001–7006.

Hoa, M. L. K., Lu, M., and Zhang, Y. (2006). Preparation of porous materials with ordered hole structure, *Adv. Colloid Interface Sci.*, **121**, pp. 9–23.

Hollister, S. J. (2005). Porous scaffold design for tissue engineering, *Nat. Mater*, **4**, pp. 518–524.

Imhof, A. and Pine, D. J. (1997). Ordered macroporous material by emulsion templating, *Nature*, **389**, pp. 948–951.

Jackson, E. A. and Hillmyer, M. A. (2010). Nanoporous membranes derived from block copolymers: From drug delivery to water filtration, *ACS Nano*, **4**, pp. 3548–3553.

Jiang, X., Zhou, X., Zhang, Y., Zhang, T., Guo, Z., and Gu, N. (2010). Interfacial effects of in situ-synthesized Ag nanoparticles on breath figures, *Langmuir*, **26**, pp. 2477–2483.

João, C., Almeida, R., Silva, J., and Borges, J. (2016). A simple sol-gel route to the construction of hydroxyapatite inverted colloidal crystals for bone tissue engineering, *Mater. Lett.*, **185**, pp. 407–410.

Johannes, C., Van Rijn, M., Vogelaar, L., Nijdam, W., Barsema, J. N., and Wessling, M. (2009). Method of making a product with a micro- or nano-sized structure and product. Patent No. US7531120 B2.

Johnson, S. A., Ollivier, P. J., and Mallouk, T. E. (1999). Ordered mesoporous polymers of tunable pore size from colloidal silica templates, *Science*, **283**, pp. 963–965.

Karageorgiou, V. and Kaplan, D. (2005). Porosity of 3D biomaterial scaffolds and osteogenesis, *Biomaterials*, **26**, pp. 5474–5491.

Karthaus, O., Maruyama, N., Cieren, X., Shimomura, M., Hasegawa, H., and Hashimoto, T. (2000). Water-assisted formation of micrometer-size honeycomb patterns of polymers, *Langmuir*, **16**(15), pp. 6071–6076.

Ke, B.-B., Wan, L.-S., and Xu, Z.-K. (2010a). Controllable construction of carbohydrate microarrays by site-directed grafting on self-organized porous films, *Langmuir*, **26**, pp. 8946–8952.

Ke, B.-B., Wan, L.-S., Zhang, W.-X., and Xu, Z.-K. (2010b). Controlled synthesis of linear and comb-like glycopolymers for preparation of honeycomb-patterned films, *Polymer*, **51**, pp. 2168–2176.

Kim, B., Carignano, M. A., Tripp, S. L., and Wei, A. (2004). Cluster size analysis of two-dimensional order in colloidal gold nanoparticle arrays, *Langmuir*, **20**, pp. 9360–9365.

Kim, J. H., Seo, M., and Kim, S. Y. (2009). Lithographically patterned breath figure of photoresponsive small molecules: Dual-patterned honeycomb lines from a combination of bottom-up and top-down lithography, *Adv. Mater.*, **21**, pp. 4130–4133.

Kim, T. K., Lee, M. N., Lee, S. H., Lee, S. H., Park, Y. C., Jung, C. K., and Boo, J. H. (2005). Development of surface coating technology of TiO_2 powder and improvement of photocatalytic activity by surface modification, *Thin Solid Films*, **75**, pp. 171–177.

Kudo, A. (2006). Development of photocatalyst materials for water splitting, *Int. J. Hydrogen En.*, **2**, pp. 197–202.

Kuiper, S., Van Rijn, C. J. M., Nijdam, W., and Elwenspoek, M. C. (1998). Development and applications of very high flux microfiltration membranes, J. Membr. Sci., 150, pp. 1–8.

Kurono, N., Shimada, R., Ishihara, T., and Shimomura, M. (2002). Fabrication and optical property of self-organized honeycomb patterned films, *Mol. Cryst. Liq. Cryst.*, **377**, pp. 285–288.

Lan, H. and Ding, Y. (2010). Nanoimprint lithography, In: Lithography, M. Wang (Ed.). InTech, Available from: http://www.intechopen.com/books/lithography/nanoimprintlithography.

Langer, R. and Peppas, N. A. (2003). Advances in biomaterials, drug delivery, and bionanotechnology, *AIChE J.*, **49**, pp. 2990–3006.

Lanza, R. P., Langer, R., and Vacanti, J. P. (2000). *Principles of Tissue Engineering*, Academic Press, New York, pp. 251–252.

Lee, H. J., Yoon, T. H., Park, J. H., Perumal, J., and Kim, D. P. (2008). Characterization and fabrication of polyvinylsilazane glass microfluidic channels via soft lithographic technique, *J. Ind. Eng. Chem.*, **14**, pp. 45–51.

Lee, J.-Y., Tan, W. S., An, J., Chua, C. K., Tang, C. Y., Fane, A. G., and Chong, T. H. (2016). The potential to enhance membrane module design with 3D printing technology, *J. Membr. Sci.*, **499**, pp. 480–490.

Lee, S. H., Park, J. S., Lim, B. K., Mo, C. B., Lee, W. J., Lee, J. M., Hong, S. H., and Kim, S. O. (2009). Highly entangled carbon nanotube scaffolds by self-organized aqueous droplets, *Soft Matter*, **5**, pp. 2343–2346.

Lehn, J. M. (1996). Supramolecular chemistry and chemical synthesis, *NATO ASI Ser. Ser.*, **320**, pp. 511–524.

Li, J., Zhang, N., Ni, D., and Sun, K. (2011a). Preparation of honeycomb porous solid oxide fuel cell cathodes by breath figures method, *Int. J. Hydrogen Energy*, **36**, pp. 7641–7648.

Li, L., Schulte, L., Clausen, L. D., Hansen, K. M., Jonsson, G. E., and Ndoni, S. (2011b). Gyroid nanoporous membranes with tunable permeability, *ACS Nano*, **5**(10), pp. 7754–7766.

Li, N., Fane, A. G., Winston Ho, W., and Matsura, T. (2008). *Advanced Membrane Technology and Application*, John Wiley & Sons, Inc.

Liu, J., Cai, B., Zhu, J., Ding, G., Zhao, X., Yang, C., and Chen, D. (2004). Process research of high aspect ratio microstructure using SU-8 resist, *Microsyst. Technol.*, **10**, pp, 265–268.

Liu, N., Tan, X., Meng, B., and Liu, S. (2011). Honeycomb-structured perovskite hollow fibre membranes with ultra-thin densified layer for oxygen separation, *Sep. Purif. Technol.*, **80**, pp. 396–401.

Male, U., Jo, E. J., Park, J. Y., and Huh, D. S. (2016). Surface functionalization of honeycomb-patterned porous poly(ε-caprolactone) films by interfacial polymerization of aniline, *Polymer*, **99**, pp. 623–632.

Maruf, S. H., Li, Z., Yoshimura, J. A., Xiao, J., Greenberg, A. R., and Ding, Y. (2015). Influence of nanoimprint lithography on membrane structure and performance, *Polymer*, **69**, pp. 129–137.

Maruf, S. H., Wang, L., Greenberg, A. R., Pellegrino, J., and Ding, Y. (2013). Use of nanoimprinted surface patterns to mitigate colloidal deposition on ultrafiltration membranes, *J. Membr. Sci.*, **428**, pp. 598–607.

Maruyama, N., Koito, T., Nishida, J., Sawadaishi, T., Cieren, X., Kjiro, K., Karthaus, O., and Shimomura, M. (1998). Mesoscopic patterns of molecular aggregates on solid substrates, *Thin Solid Films*, **327**, pp. 854–856.

Mulder, M. (1996). *Basic Principles of Membrane Technology*, 2nd edn, Springer Netherlands.

Mulvenna, R. A., Weidman, J. L., Jing, B., Pople, J. A., Zhu, Y., and Boudouris, B. W. (2014). Tunable nanoporous membranes with chemically-tailored pore walls from triblock polymer templates, *J. Membr. Sci.*, **470**, pp. 246–256.

Munoz-Bonilla, A., Ibarboure, E., Bordege, V., Fernandez-Garcia, M., and Rodriguez-Hernandez, J. (2010). Fabrication of honeycomb-structured porous surfaces decorated with glycopolymers, *Langmuir*, **26**, pp. 8552–8558.

Nam, H. J., Kim, J. H., Jung, D. Y., Park, J. B., and Lee, H. S. (2008). Two-dimensional nanopatterning by PDMS relief structures of polymeric colloidal crystals, *App. Surf. Sci.*, **254**, pp. 5134–5140.

Olenyuk, B., Whiteford, J. A., Fechtenkotter, A., and Stang, P. J. (1999). Self-assembly of nanoscale cuboctahedra by coordination chemistry, *Nature*, **398**, pp. 796–799.

Olson, D. A., Chen, L., and Hillmyer, M. A. (2008). Templating nanoporous polymers with ordered block copolymers, *Chem. Mater.*, **20**, pp. 869–890.

Pan, Q., Wang, M., and Wang, H. (2008). Separating small amount of water and hydrophobic solvents by novel superhydrophobic copper meshes, *Appl. Surf. Sci.*, **254**, pp. 6002–6006.

Park, S. H. and Xia, Y. N. (1998a). Fabrication of three-dimensional macroporous membranes with assemblies of microspheres as templates, *Chem. Mater.*, **10**, pp. 1745–1747.

Park, S. H. and Xia, N. Y. (1998b). Macroporous membranes with highly ordered and three-dimensionally interconnected spherical pores, *Adv. Mater.*, **10**, pp. 1045–1048.

Peinemann, K.-V., Abetz, V., and Simon, P. F. W. (2007). Asymmetric superstructure formed in a block copolymer via phase separation, *Nat. Mater.*, 6, pp. 992–996.

Perrotta, M. L., Saielli, G., Casella, G., Macedonio, F., Giorno, L., Drioli, E., and Gugliuzza, A. (2017). An ultrathin suspended hydrophobic porous membrane for high-efficiency water desalination, *Appl. Mater. Today*, **9**, pp. 1–9.

Phillip, W. A., O'Neill, B., Rodwogin, M., Hillmyer, M. A., and Cussler, E. L. (2010a). Self-assembled block copolymer thin films as water filtration membranes, *ACS Appl. Mater. Interface*, **2**, pp. 847–853.

Phillip, W. A., Hillmyer, M. A., and Cussler, E. L. (2010b). Cylinder orientation mechanism in block copolymer thin films upon solvent evaporation, *Macromolecules*, **43**, pp. 7763–7770.

Pilati, F., Montecchi, M., Fabbri, P., Synytska, A., Messori, M., Toselli, M., Grundke, K., and Pospiech, D. (2007). Design of surface properties of PET films: Effect of fluorinated block copolymers, *J. Colloid Interface Sci.*, **315**, pp. 210–222.

Pitois, O. and Francois, B. (1999a). Crystallization of condensation droplets on a liquid surface, *Colloid Polym. Sci.*, **277**, pp. 574–578.

Pitois, O. and Francois, B. (1999b). Formation of ordered micro-porous membranes, *Eur. Phys. J. B*, **8**, pp. 225–231.

RadjabianKollBuhrVainioAbetzHandgeAbetzRadjabian, M., Koll, J., Buhr, K., Handge, U. A., and Abetz, V. (2013). Hollow fiber spinning of block copolymers: Influence of spinning conditions on morphological properties, *Polymer*, **54**, pp. 1803–1812.

Radjabian, M., Koll, J., Buhr, K., Vainio, U., Abetz, C., Handge, U. A., and Abetz, V. (2014). Tailoring the morphology of self-assembled block copolymer hollow fiber membranes, *Polymer*, **55**, pp. 2986–2997.

Rayleigh, L. (1911). Breath figures, *Nature*, **86**, pp. 416–417.

Reculusa, S., Massé, P., and Ravaine, S. (2004). Three-dimensional colloidal crystals with a well-defined architecture, *J. Colloid Interface Sci.*, **279**, pp. 471–478.

Robeson, L. M. (1991). Correlation of separation factor versus permeability for polymeric membranes, *J. Membr. Sci.*, **62**, pp. 165–185.

Rowley, J. A., Madlambayan, G., and Mooney, D. J. (1999). Alginate hydrogels as synthetic extracellular matrix materials, *Biomaterials*, **20**, pp. 45–53.

Sainiemi, L., Viheriala, J., Sikanen, T., Laukkanen, J., and Niemi, T. (2010). Nanoperforated silicon membranes fabricated by UV-nanoimprint lithography, deep reactive ion etching and atomic layer deposition, *J. Micromech. Microeng.*, **20**, pp. 8.

Schatz, M. F. and Neitzel, G. P. (2001). Experiments on thermocapillary flow, *Annu. Rev. Fluid Mech.*, **33**, pp. 93–127.

Schift, H. and Kristensen, A. (2007). *Springer Handbook of Nanotechnology*, Springer Berlin Heidelberg, pp. 239–278.

Schift, H., Bellini, S., and Gobrecht, J. (2006). Perforated polymer membranes fabricated by nanoimprint, *Microelectron. Eng.*, **83**, pp. 873–875.

Schmidt-Mende, L., Fechtenkotter, A., Mullen, K., Moons, E., Friend, R. H., and MacKenzie, J. D. (2001). Self-organized discotic liquid crystals for high-efficiency organic photovoltaics, *Science*, **293**, pp. 1119–1122.

Schwiebert, K. E., Chin, D. N., MacDonald, J. C, and Whitesides, G. M. (1996). Engineering the solid state with 2-benzimidazolones, *J. Am. Chem. Soc.*, **118**, pp. 4018–4029.

Singh, R., Maru, V. M., and Moharir, P. S. (1998). Complex chaotic systems and emergent phenomena, *J. Nonlinear Sci.*, **8**, pp. 235–259.

Speranza, V., Trotta, F., Drioli, E., and Gugliuzza, A. (2010). High-definition polymeric membranes: Construction of 3D lithographed channel arrays through control of natural building blocks dynamics, *App. Mater. Inter.*, **2**, pp. 459–466.

Srinivasarao, M., Collings, D., Philips, A., and Patel, S. (2001). Three-dimensionally ordered array of air bubbles in a polymer film, *Science*, **292**, pp. 79–83.

Stenzel, M. H., Barner-Kowollik, C., and Davis, T. P. (2006). Formation of honeycomb structured, porous films via breath figures with different polymer architectures, *J. Polym. Sci. A*, **44**, pp. 2363–2375.

Sun, H., Li, H., and Wu, L. (2009a). Micro-patterned polystyrene surfaces directed by surfactant-encapsulated polyoxometalate complex via breath figures, *Polymer*, **50**, pp. 2113–2112.

Sun, H., Li, W., and Wu, L. X. (2009b). Honeycomb-patterned films fabricated by self-organization of DNA: Surfactant complexes, *Langmuir*, **25**, pp. 10466–10472.

Talei Franzesi, G., Ni, B., Ling, Y., and Khademhosseini, A. (2006). A controlled-release strategy for the generation of cross-linked hydrogel microstructures, *J. Am. Chem. Soc.*, **128**, pp. 15064–15065.

Tanaka, M. (2011). Design of novel 2D and 3D biointerfaces using self-organization to control cell behavior, *Biochimi. Biophys. Acta*, **1810**, pp. 251–258.

Tian, Y., Jiao, Q., Ding, H., Shi, Y., and Liu, B. (2006). The formation of honeycomb structure in polyphenylene oxide films, *Polymer*, **47**, pp. 3866–3873.

Trotta, F., Drioli, E., and Gordano, A. (2001). Nitro derivatives of PEEK-WC, *J. Appl. Polym. Sci.*, **80**, pp. 1037–1045.

Tsai, H. H., Xu, Z. H., Pai, R. K., Wang, L. Y., Dattelbaum, A. M., Shreve, A. P., Wang, H. L., and Cotlet, M. (2011). Structural dynamics and charge transfer via complexation with fullerene in large area conjugate, *Chem. Mater*, **23**, pp. 759–761.

van Rijn, C. J. M. (2004). *Nano and Micro Engineered Membrane Technology*, Elsevier, Amsterdam.

Velev, O. D., Jede, T. A., Lobo, R. F., and Lenhoff, A. M. (1997). Porous silica via colloidal crystallization, *Nature*, **389**, pp. 447–448.

Velev, O. D., Jede, T. A., Lobo, R. F., and Lenhoff, A. M. (1998). Microstructured porous silica obtained via colloidal crystal templates, *Chem. Mater*, **10**, pp. 3597–3602.

Velev, O. D. and Kaler, E. W. (2000). Structured porous materials via colloidal crystal templating: From inorganic oxides to metals, *Adv. Mater.*, **12**, pp. 531–534.

Velev, O. D. and Lenhoff, A. M. (2000). Colloidal crystals as templates for porous materials, *Curr. Opin. Colloid Interface Sci.*, **5**, pp. 56–63.

Vlasov, Y. A., Yao, N., and Norris, D. J. (1999). Synthesis of photonic crystals for optical wavelengths from semiconductor quantum dots, *Adv. Mater.*, **11**, pp. 165–169.

Vogelaar, L. and Wessling, M. (2002). Phase separation micro moulding, In: *OSPT Process Technology*, Akse, H. A. & Stegehuis-de Vegte, M. G. (Eds.), Graduate School on Process Technology, Enschede, p. 329, ISBN 9789036517751.

Wang, X. (2010). Inverse colloidal crystal membranes: Formation, surface modification and applications, Department of Chemical and Biological Engineering, For the Degree of Doctor of Philosophy Colorado State University Fort Collins, Colorado.

Whitesides, G. M. (1995). Self-assembling materials, *Sci. Am.*, **273**, pp. 146–149.

Whitesides, G. M. and Boncheva, M. (2002). Beyond molecules: Self-assembly of mesoscopic and macroscopic components, *Proc. Natl Acad. Sci. USA*, **99**, pp. 4769–4774.

Wienk, I. M., Boom, R. M., Beerlage, M. A. M., Bulte, A. M. W., Smolders, C. A., and Strathmann, H. (1996). Recent advances in the formation of phase inversion membranes made from amorphous or semi-crystalline polymers, *J. Membr. Sci.*, **113**, pp. 361–371.

Xu, L., Ma, W., Wang, L., Xu, C., Kuang, H., and Kotov, N. A. (2013). Nanoparticle assemblies: Dimensional transformation of nanomaterials and scalability, *Chem. Soc. Rev.*, **42**, pp. 3114–3126.

Xu, X. and Luo, J. (2007). Marangoni flow in an evaporating water droplet, *Appl. Phys. Lett.*, **124102**, pp. 1–3.

Yamamoto, S., Tanaka, M., Sunami, H., Ito, E., Yamashita, S., Morita, Y., and Shimomura, M. (2007). Effect of honeycomb-patterned surface topography on the adhesion and signal transduction of porcine aortic endothelial cells, *Langmuir*, **23**, pp. 8114–8120.

Yanagishita, T., Nishio, K., and Masuda, H. (2007). Polymer through-hole membrane fabricated by nanoimprinting using metal molds with high aspect ratios, *J. Vac. Sci. Technol. B*, **25**, pp. 35–38.

Zakhidov, A. A., Baughman, R. H., Iqbal, Z., Cui, C. X., Khayrullin, I., Dantas, S. O., Marti, I., and Ralchenko, V. G. (1998). Carbon structures with three-dimensional periodicity at optical wavelengths, *Science*, **282**, pp. 897–901.

Zhang, H., Dong, F., Zhai, S., Kang, X., and Fang, S. (2012). Preparation of ordered TiO_2 macroporous membrane using PBMA colloid crystal as template, *Adv. Mater. Res.*, **399–401**, pp. 677–682.

Zhang, Y., Sargent, J. L., Boudouris, B. W., and Phillip, W. A. (2015a). Nanoporous membranes generated from self-assembled block polymer precursors: Quo Vadis? *J. Appl. Pol. Sci.*, **132**, 41683, pp. 1–17.

Zhang, A., Bai, H., and Li, L. (2015b). Breath figure: A nature-inspired preparation method for ordered porous films, *Chem. Rev.*, **115**, pp. 9801–9868.

Zou, Y., Moreel, L., Lin, H., Zhou, J., Li, L., Danto, S., Musgraves, J. D., Koontz, E., Richardson, K., Dobson, K. D., Birkmire, R., and Hu, J. (2014). Solution processing and resist-free nanoimprint fabrication of thin film chalcogenide glass devices: Inorganic–organic hybrid photonic integration, *Adv. Opt. Mater.*, **2**, pp. 759–764.

Chapter 4

Nanostructured Ion-Exchange Membranes

Enrica Fontananova and Enrico Drioli
Institute on Membrane Technology of the National Research Council of Italy (ITM-CNR), Via P. Bucci, Cubo 17C, 87036 Rende (CS), Italy
e.fontananova@itm.cnr.it

Ion-exchange membranes have today consolidated applications in several strategic fields (from water desalination to energy conversion), and more progresses can be anticipated in the near future. In this perspective, nanoscience and nanotechnology are playing a key role in bringing performance advances and industrial scale-up opportunities. The development of advanced ion-exchange membranes includes (i) the realization of membranes with nanostructured ion channels produced by microphase separation between hydrophobic and hydrophilic regions of a block copolymer; (ii) the production of mixed matrix membranes containing nanostructured functional additives. This work gives an overview of the recent progress and the existing limitations of nanostructured ion-exchange membranes, and opens a window to future perspectives.

Functional Nanostructured Membranes
Edited by Enrico Drioli, Lidietta Giorno, and Annarosa Gugliuzza

ISBN 978-981-4774-79-6 (Hardcover), 978-1-351-13511-5 (eBook)
www.panstanford.com

4.1 Introduction to Ion-Exchange Membranes

An ion-exchange membrane (IEM) is a charge-selective membrane characterized by the presence of charged groups covalently linked to a polymeric chain. IEMs are ideally impermeable to ions having the same electrical charge as the fixed charged groups (or co-ions) by Donnan exclusion (Strathmann, 2004). On the contrary, ions with opposite charge (or counter-ions) can be transported through the IEM under the effect of an electrochemical potential.

Considering their chemical composition, IEMs can be classified as cation-exchange membranes (CEMs) containing negatively charged fixed groups and anion-exchange membranes (AEMs) containing positively charged fixed groups (Fig. 4.1). Moreover, a third type of IEM is bipolar membrane (BPM), which is composed of an anion-exchange and a cation-exchange layer laminated together. BPMs are used in specific applications such as water splitting reaction for producing acids and bases (Vermass et al., 2015).

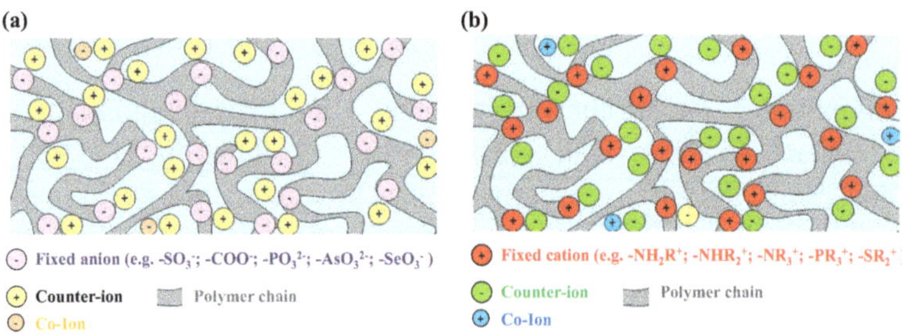

Figure 4.1 Schematic of a homogeneous cation-exchange membrane (a) and a homogeneous anion-exchange membrane (b).

Based on the structure, IEMs can be divided into homogenous and heterogeneous membranes (Strathmann, 2010). In the first, the fixed charged groups are evenly distributed over the entire membrane polymer matrix (Fig. 4.1). In the second, distinct macroscopic domains of the ion-exchange material and the non-conductive matrix used as a binder are present (Fig. 4.2).

A high membrane permselectivity (i.e., the selective permeability for anions or cations) combined with an elevated ionic conductivity are demanding requirements for IEMs. These properties are basically determined by the type, concentration, and distribution of fixed ionic moieties, as well as by the composition of the main polymer chain. A trade-off relationship between permselectivity and conductivity is usually observed: Membranes with higher permselectivity generally have lower conductivity (Fontananova et al., 2017).

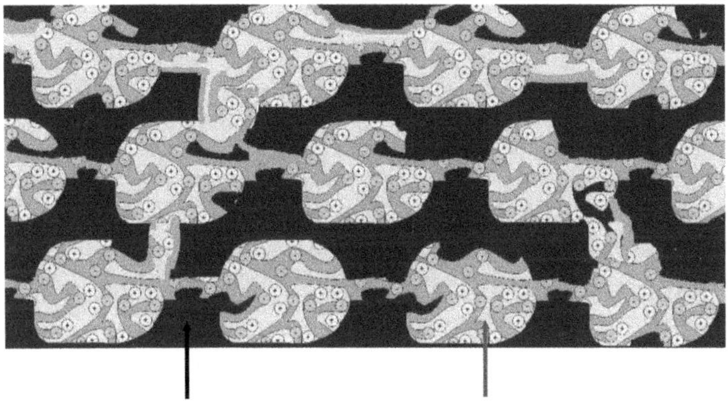

Figure 4.2 Schematic illustration of a heterogeneous cation-exchange membrane.

Homogeneous IEMs usually have higher permselectivity and lower electrical resistance in comparison with heterogeneous IEMs (Strathmann, 2004). However, heterogeneous membranes usually have lower costs than homogeneous membranes. Homogeneous IEMs are often prepared by casting and solvent evaporation of a solution of an ion-exchange polymer. Another method to produce this type of membranes is grafting of ion-exchange groups on the already been formed film. For the first strategy, the polymers can be prepared by polymerization of monomers that contain a moiety, which either is, or can be made, anionic or cationic. Alternatively, the anionic or cationic moieties are introduced into a dissolved polymer by a chemical reaction.

Homogeneous IEMs can also be prepared through melting and extrusion of ion-exchange resins. The preparation can be carried out by mixing an ion-exchange powder with a dry binder polymer followed by extrusion of sheets under the appropriate conditions of pressure and temperature. Alternatively, casting and solvent evaporation of a dispersion of ion-exchange particles in the solution of a film-forming binder polymer can be carried out.

Table 4.1 Properties and main applications of some commercial ion-exchange membranes

Membrane	Company	Type	Thickness (mm)	IEC[a] (mmol/g)	AMR[b] ($\Omega\ cm^2$)	Application
Fumasep® FKB	FuMA-TechGmbH, Germany	CEM	0.10–0.13	1.2–1.3	4–6	Desalination
Fumasep® FKD		CEM	0.08–0.089	1.2–1.4	4–7	Base recovery
Fumasep® FAB		AEM	0.10–0.13	1.0–1.1	4	Desalination
Fumasep® FAD		AEM	0.08–0.09	1.5–1.7	1.0–1.2	Acid recovery
Fumasep® FBM		BPM	0.18–0.2	—	0.4–0.8	Production of acid and base
NEOSEPTA® CMX	ASTOM Co., Japan	CEM	0.17	—	3.0	Desalination
NEOSEPTA® CMB		CEM	0.21	—	4.5	Base recovery
NEOSEPTA® AMX		AEM	0.14	—	2.4	Desalination
NEOSEPTA® AFN		AEM	0.16	—	0.5	Acid recovery
NEOSEPTA® BP-1		BPM	0.22	—		Production of acid and base
Fuji-CEM-0150	Fujifilm Manufacturing Europe B.V.	CEM	0.170	1.6	2.6	Reverse electrodialysis
Fuji-CEM-80050		CEM	0.114	1.1	3.0	Reverse electrodialysis
Fuji-CEM-90026		CEM	0.113	1.0	1.6	Reverse electrodialysis
Fuji-AEM-0150		AEM	0.166	1.1	1.6	Reverse electrodialysis
Fuji-AEM-80045		AEM	0.129	1.4	1.6	Reverse electrodialysis
Fuji-AEM-90025		AEM	0.109	1.6	1.1	Reverse electrodialysis

Membrane	Company	Type	Thickness (mm)	IEC[a] (mmol/g)	AMR[b] (Ω cm²)	Application
Nafion™ N115	DuPont Co.,USA	CEM	0.127	0.95–1.01	—	Fuel cell
Nafion™ N117		CEM	0.183	0.95–1.01	—	Fuel cell
Nafion™ N1110		CEM	0.254	0.95–1.01	—	Fuel cell
Chemjoy® CJMCED	Chemjoy Co., Ltd., China	CEM	0.1–0.16	0.9–1.5	<2	Desalination
Chemjoy® CJMAED		AEM	0.14–0.16	0.9–1.2	<4	Desalination
Chemjoy® CJBPM		BPM	0.2–0.25	—	—	Production of acid and base
JCM-II	Tingrun Co., Ltd., China	CEM	0.16–0.23	1.8–2.9	1–7	Desalination
JAM-II		AEM	0.16–0.23	1.6–2.2	4–10	Desalination
BPM-I		BMP	0.16–0.23	—	—	Production of acid and base
CMI-7000	Membrane International, USA	CEM	0.45	1.6	<30	Electrodeionization
AMI-7001		AEM	0.45–0.60	1.3	<40	Electrocoating Electrolysis Electrodialysis Electroplating

[a]IEC: Ion-exchange capacity

[b]AMR: Areal membrane resistance (measured in 0.5 mol/L NaCl, 20–25°C)

Note Data from Fontananova et al. (2014), Ran et al. (2017), and Membranes International Inc. Reprinted with permission from Fontananova et al. (2014) and Ran et al. (2017). Copyright 2014 and 2017, Elsevier.

Today, IEMs have consolidated uses in several electro-membrane processes (i.e., membrane processes coupling mass transport with electrical current transport through ion-selective membranes). The application areas are mainly three: (i) de-ionization of salt solutions (e.g., electrodialysis), (ii) electrochemical synthesis of inorganic and organic compounds (e.g., chlorine-alkaline process), and (iii) conversion of chemical into electrical energy (e.g., fuel cells and reverse electrodialysis) (Vermaas et al., 2015). Table 4.1 lists some examples of commercialized IEMs and their main applications.

4.2 Role of Physical and Chemical Microenvironment on IEM Transport Properties

The conductivity and permselectivity of IEMs are not constant for a given membrane at a certain temperature, but they strongly depend on the liquid or gaseous phase contacting the membrane (e.g., a gaseous atmosphere with a given relative humidity (RH) or a liquid electrolyte solution). The surrounding environment influences the water volume fraction in the membrane phase, which, in turn, influences the membrane microstructure (Fontananova et al., 2014; Fontananova et al., 2017). A typical example is the increase in the ionic conductivity of CEMs used for polymer electrolyte membrane fuel cells (PEMFCs) with the increase in the humidification level of the gaseous streams fed to the PEMFCs (Kreuer, 2001; Marechal et al., 2007). In the presence of water, the ionic fixed groups of IEMs can interconnect by a network of hydrogen bonds forming swelled hydrophilic domains, which represent the pathway for the transport of ions by a vehicular or Grotthus mechanism (Kreuer, 1996; 2000). In vehicular transport, protons are linked to a vehicle, water in the specific case (e.g., as H_3O^+ and $H_5O_2^+$ ions), and they diffuse together under a gradient of electrochemical potential. In the Grotthus mechanism, the vehicle molecules are stationary (i.e., the fixed charged groups) and the transport involves the structural intermolecular reorganization of hydrogen bonds with the concomitant reorientation of the vehicle molecules (protons hopping).

Currently, the most implemented CEM for applications in PEMFCs is Nafion membrane (from Du Point). Nafion is a perfluorosulfonic acid (PFSA) membrane characterized by elevated proton conductivity and high chemical, thermal, and mechanical stability. PFSA includes long-side-chain (LSC) polymers, such as Nafion (Fig. 4.3a), Aciplex (Asahi Chemical), Flemion (Asahi Glass), and Gore-Select (Gore and Associates) (Kreuer, 2001), and short-side-chain (SSC) polymers initially proposed in the 1980s by the Dow Company, under the trade name Dow Ionomer, and more recently by Solvay Solexis, under the trade name Hyflon Ion and now as Aquivion (Fig. 4.3b) (Arcella et al., 2005). These SSC ionomers are characterized by shorter pendent group carrying the ionic functionality, higher crystallinity, and higher glass transition (T_g) than LSC ionomers at given equivalent weight (Ghielmi et al., 2005).

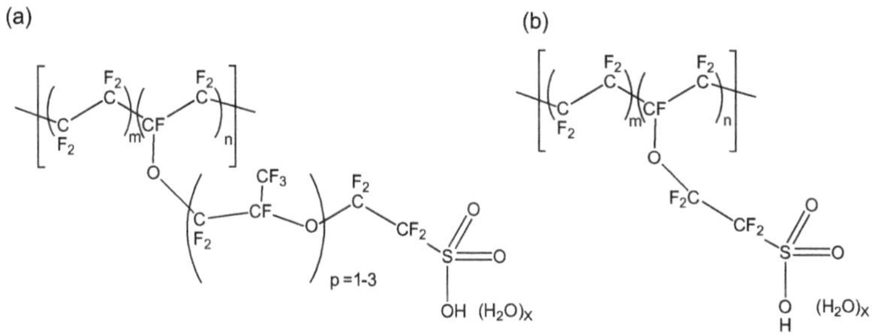

Figure 4.3 Chemical formula of (a) Nafion and (b) Hyflon Ion or Aquivion.

However, Nafion and, in general, PFSA polymers have high costs (Neburchilov et al., 2007), which limit their practical applications. Moreover, these membranes have an elevated fuel crossover (H_2 and methanol, in H_2-PEMFC and methanol-PEMFC, respectively), which reduces process performances. In addition, the high hydraulic permeability of PFSA membranes can have severe consequences in PEMFC applications because of easy membrane dehydration at medium-high temperature, with successive decrease in conductivity, as well as cathode flooding, with the consequent restricted oxygen transport through the pores of the gas diffusion electrode (Neburchilov et al., 2007).

Considering these limitations, a huge number of less expensive partially fluorinated, non-fluorinated hydrocarbon, non-fluorinated aromatic and acid–base blend membranes have been proposed (Hickner et al., 2004; Roziere and Jones, 2003; Smitha et al., 2005; Thiam et al., 2011; Tripathi and Shahi, 2011).

Among the various non-fluorinated polymers investigated as possible alternatives to PFSA, interesting results have been obtained with sulfonated aromatic polymers (SAPs), such as the sulfonated derivate of polyetherketone (SPEK), poly(etheretherketone) (SPEEK, Fig. 4.4a), poly(etherketoneketone) (SPEKK), poly(etheretherketoneketone) (SPEEKK), and polyetheretherketones (SPEEK) (Li et al., 2003; Mikhailenko et al., 2004; Smitha et al., 2005).

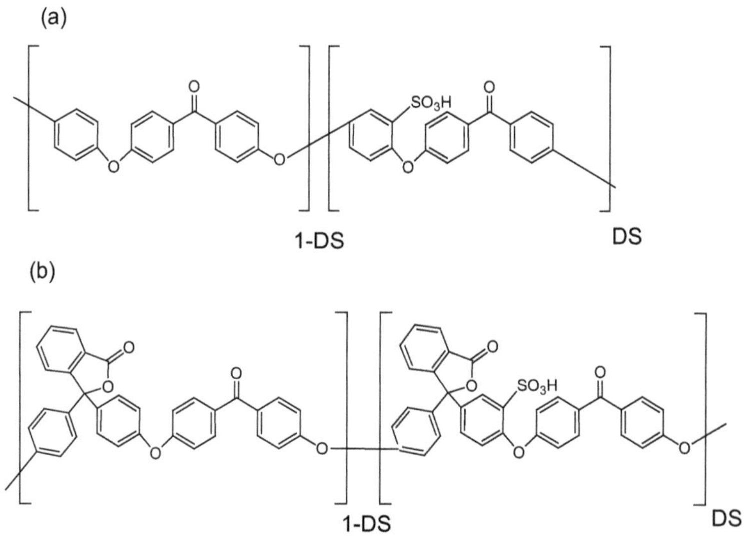

Figure 4.4 Chemical formula of (a) SPEEK and (b) SPEEK-WC (B) polymers (DS is the degree of sulfonation, which is assumed to be ≤1).

The SPEEK polymer is prepared by sulfonation of the semicrystalline thermoplastic polymer polyetheretherketone (PEEK). PEEK is not soluble in common organic solvents under mild conditions and, therefore, hardly suitable for the preparation of membranes by the casting technique (Beck, 1992). On the contrary, the sulfonated PEEK is amorphous and soluble in several organic solvents. However, the low solubility of the starting material limits the sulfonation process, which is initially heterogeneous and, as a consequence, produces different fractions of polymer with a variable degree of sulfonation (DS) (Zaidi et al., 2000).

Unlike the classical PEEK, a modified PEEK, known as PEEK-WC (Fig. 4.4b), is amorphous and soluble in organic solvents with medium polarity (e.g., chloroform, dimethylsulfoxide, dimethylacetamide, dimethylformamide). PEEK-WC is characterized by the presence of a lactonic group, called Cardo group (WC in the polymer name means With Cardo), which prevents the crystalline organization of the polymer chain occurring with the PEEK. Thanks to its amorphous properties, PEEK-WC can be used as membrane material by casting technique (Fontananova et al., 2003). PEEK-WC is characterized by an elevated glass transition temperature (T_g = 222°C) and excellent chemical, thermal, and mechanical stability (Drioli and Zhang, 1989). Membranes made of PEEK-WC have been prepared by casting and phase separation techniques, and they have been applied with good results in gaseous and liquid separations (Jansen et al., 2005; Tasselli et al., 2004).

PEEK-WC polymer can be functionalized by reaction with chlorosulfonic acid at room temperature, introducing a sulfonic group on the aromatic ring obtaining the SPEEK-WC with ion-exchange capacity (IEC) (Trotta et al., 1998). The DS of SPEEK-WC can be modulated by changing the reaction time and/or the reaction temperature. Obviously, the DS influences the IEC and the proton conductivity of the system.

Sulfonated PEEK-WC is a promising non-fluorinated ion-exchange material for application in PEMFCs (Drioli et al., 2004). SPEEK-WC membranes have a higher resistance to water and methanol vapor transport than commercial Nafion 117 membrane, as well as a lower permeability to H_2 and O_2 (Fontananova et al., 2010). As a consequence, minor crossover and electrode flooding problems may occur during PEMFC operations using SPEEK-WC membranes compared with Nafion. The lower permeabilities of the SPEEK-WC membranes are related to their lower diffusion coefficients, caused by the higher stiffness of the aromatic polymer chains of SPEEK-WC with respect to Nafion (Fontananova et al., 2010).

SAP membranes have in general lower water and methanol permeabilities compared to PFSA membranes because of the specific differences in their microstructures (Kreuer, 2001). In a PFSA membrane, extremely hydrophobic perfluorinated regions are combined with highly hydrophilic sulfonic groups. In the presence of water, the hydrophilic portions of the polymer aggregate and form hydrophilic nano-domains. In the cluster-network model, used to rationalize the transport mechanism in Nafion membranes (Mauritz and Moore, 2004), hydrophilic pockets of about 4 nm in diameter are formed by sulfonic acid groups, counter-ions, and absorbed water. These pockets are interconnected by a network of short and narrow channels of approximately 1 nm in diameter and separated from the hydrophobic region formed by the fluorinated backbone of the polymer (Fig. 4.5).

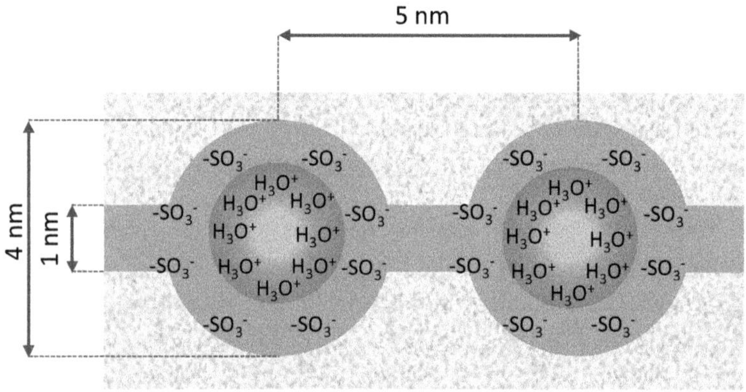

Figure 4.5 Cluster-network model used to rationalize the transport mechanism in Nafion membranes. Adapted with permission from Mauritz and Moore (2004). Copyright 2004, American Chemical Society.

The transport of proton, water, and methanol occurs through these hydrophilic regions. The hydrophobic domains are responsible for the morphological stability of the system. For example, the mean dimension of the hydrophilic pockets in Nafion, with equivalent weight (EW) 1200, shrinks from 4 to 2.44 nm when the RH is changed from 100 to 34% (Hsu and Gierke, 1982). As a consequence, the transport properties vary with the level of humidification.

In the SAP membranes, the hydrophilic nanochannels are narrower compared to Nafion, less separated from the hydrophobic regions, and more branched because the polymer backbone is less hydrophobic and the sulfonic groups are less acidic than those of Nafion (Fig. 4.6) (Kreuer, 2001; Xue and Yin, 2006). Numerous authors confirmed these different microstructures by small angle

X-ray scattering (SAXS) experiments carried out on Nafion and various SAP (Kreuer, 2001; Xue and Yin, 2006).

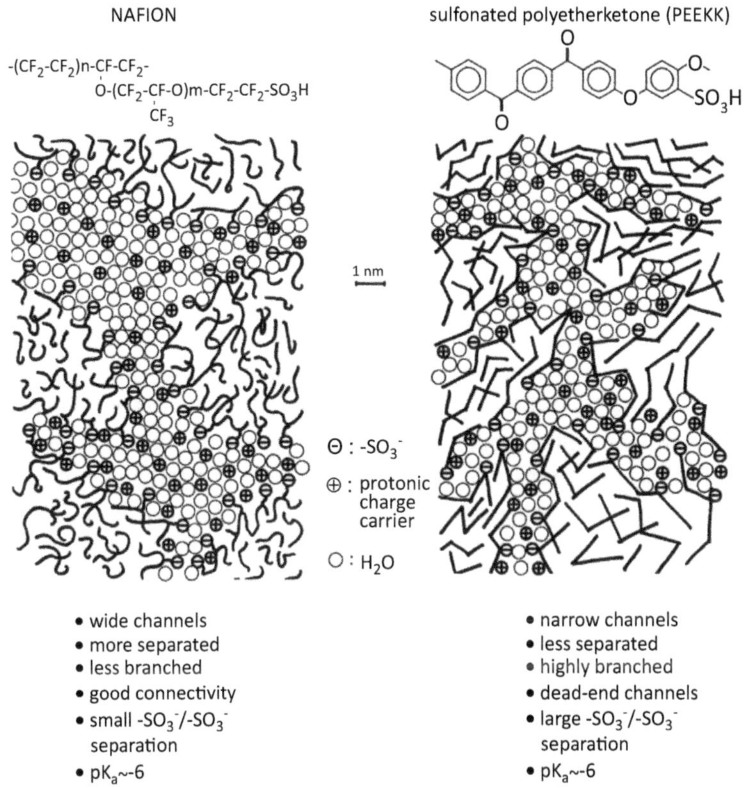

Figure 4.6 Model of Nafion and SPEEKK microstructure. Reprinted with permission from Kreuer (2001). Copyright 2001, Elsevier.

As a consequence, the proton, water, and methanol diffusion coefficients of this family of polymers are typically lower compared to those of Nafion (Li et al., 2003; Yang and Manthiram, 2006). Moreover, the presence of the fluorinated backbone induces a higher acidity of the sulfonic group in the PFSA membranes and, in general, a higher proton conductivity in comparison with non-fluorinated membranes (Kreuer, 2001).

Further discussion is necessary when the IEMs are in contact with a liquid electrolyte solution, such as in reverse electrodialysis (RED) applications (Logan and Elimelech, 2012; Tufa et al., 2015).

The ionic strength and composition of the external liquid solution have a relevant impact on the membrane microstructure and transport properties by influencing water and mobile ionic species uptake.

Recent studies (Fontananova et al., 2017) demonstrated that on increasing the ionic strength of solution, the conductivity of IEMs (homogeneous reinforced non-fluorinated AEM and CEM indicated as AEM-80045 and CEM-80050, Table 4.1) decreased because of a lower water uptake and the consequent shrinking of the hydrophilic channels through which ionic transport occurs. On decreasing the water content in the membrane, the hydrophilic channels of the membrane shrunk and the ionic transport resistance increased (Kreuer, 2001). Moreover, the concentration of the fixed charged groups increased with the reduction in water uptake, and consequently, ion migration through the membrane was hindered more because of the stronger interactions with the fixed charged groups that can form isolated ionic domains not well interconnected with each other. In general, the membrane with a lower concentration of fixed charged group resulted in more sensitivity to the ionic strength of external solution (Fontananova et al., 2014). Moreover, the ionic concentration profiles inside the membrane and at its interfaces can be influenced by the counter-ion-exchange phenomena with multi-component electrolyte solutions, as well as co-ion diffusion in the case of non-ideally permselective membranes.

The effect of multivalent ions on membrane electrochemical properties was investigated in the perspective of using seawater as green feed solution for salinity-gradient power (SGP) harvesting by RED (Fontananova et al., 2014). The study focused on the role of the most common multivalent ions present in seawater (Husain et al., 2010).

The results indicated a critical role of Mg^{2+}, the third most abundant ion in seawater after sodium and chloride, in the ionic conductivity of CEM-800050. Switching the feed from 0.5M NaCl to mixed electrolyte solutions containing NaCl and $MgCl_2$ but with similar ionic strength of the previous ones, the CEM-80050 conductivity decreased in a relevant way. The Mg^{2+} ions have a higher affinity for sulfonic groups (i.e., the fixed groups of CEM-80050) than Na^+ ions have (Ersöz et al., 2001; Strathmann, 2010). This means that the membrane initially in the Na^+ form (the membranes were activated in a 0.5 M NaCl solution) tends to exchange Na^+ (i.e., the mobile ion) with Mg^{2+}. The reduction in conductivity was due to the lower mobility of the bivalent Mg^{2+} ion with respect to the monovalent Na^+ caused by the higher hydrated radius of the first one (Saracco, 1997). The presence of multivalent ions can also alter the membrane microstructure working as a bridge between two different fixed charged groups and reduce the effective dimension of the hydrophilic nanochannels of the IEM. This blocking effect of the multivalent ions contributes to the reduction in the membrane ionic conductivity.

On the contrary, the conductivity of AEM-80045 was poorly sensitive to the presence of this bivalent cation. These results are expected considering that AEM-80045 tends to repel Mg^{2+}, and cations in general, because of the presence of covalently linked quaternary ammonium groups (i.e., positively charged groups) on its polymer chains.

Permselectivity is also influenced by the external solution's ionic strength and compositions. On increasing the ionic strength, the permselectivity tends to decrease because of the higher difference in ion activity between the electrolyte solution and the membrane phase, resulting in a diffusive transport of the co-ion (Strathmann, 2010).

When the membranes were tested in mixed electrolyte solutions containing Mg^{2+}, the permselectivities of both membranes were found to be decreased and the effect was more relevant for CEM-80050. The permselectivities of CEM-80050 dropped because of the shielding effect of this bivalent cation on the fixed charged groups, reducing the capacity of the membrane to exclude co-ions (Geise et al., 2014; Jörissen et al., 2003).

However, Mg^{2+} influences also the permselectivity of AEM-80045. This effect could be related to a decrease in the dielectric exclusion of the non-charged portion of AEM-800 (Yaroshchuk, 2000) induced by the Mg^{2+} ion having a higher charge density than Na^+. Consequently, Mg^{2+} could induce a partial polarization of the polymer with an increase in the co-ion uptake in AEM-80045.

It is evident that the proprieties of IEMs depend on the chemical and physical environment of ion nanochannels. The development of new nanostructured IEMs with tailored properties has received much attention in the last decades. Nanostructured membranes are systems in which the building blocks and/or their periodic alignment are at the nanometer scale (He et al., 2015). Nanostructured IEMs can be divided into two main groups: nanostructured polymeric IEMs and nanostructured mixed matrix IEMs. Nanostructured polymer IEMs are made of block copolymer ionomers containing hydrophilic and hydrophobic polymer moieties, which give self-organized nanochannels. Nanostructured mixed matrix IEMs combine, in a synergic way, a polymeric component with a nanostructured filler material.

4.3 Nanostructured Polymeric IEMs

Nanoscience and nanotechnology hold highly promising prospects for the design and development of IEMs with tailored nanochannels for ion transport. In the first generation of IEMs, the ion-exchange groups were directly and randomly attached to the polymer backbone, principally by sulfonation in the case of CEMs (Fig. 4.7a) or by chloromethylation followed by an amination with a tertiary amine in the case of AEMs (Fig. 4.8a).

The synthesis of polymers with more complex topological architectures than the first-generation ones has attracted a lot of interest thanks to the improved performance that can be achieved (Fig. 4.6b–g and Fig. 4.7b–i). An overview of the different polymer architectures developed for CEMs and AEMs was recently reported (Ran et al., 2017).

Block copolymers formed by two or more segments having different composition and polarity (e.g., one functionalized with ion-exchange groups and the second one un-functionalized) have the capability of phase separation into an ordered morphology at the nanometric scale (Li et al., 2011; Ran et al., 2017; Shin et al., 2017). A transmission electron microscopy (TEM) image of a membrane made of a block copolymer of styrene and sodium styrene sulfonate (PS-g-PSSA; Fig. 4.9a) shows dark regions corresponding to ionic domains and lighter regions formed by hydrophobic polystyrene domains (Ding et al., 2001). No similar ordered morphology was observed in the TEM image of the membrane prepared from a random copolymer of styrene and sodium styrene sulfonate (PS-r-SSA; Fig. 4.9b) (Ding et al., 2001).

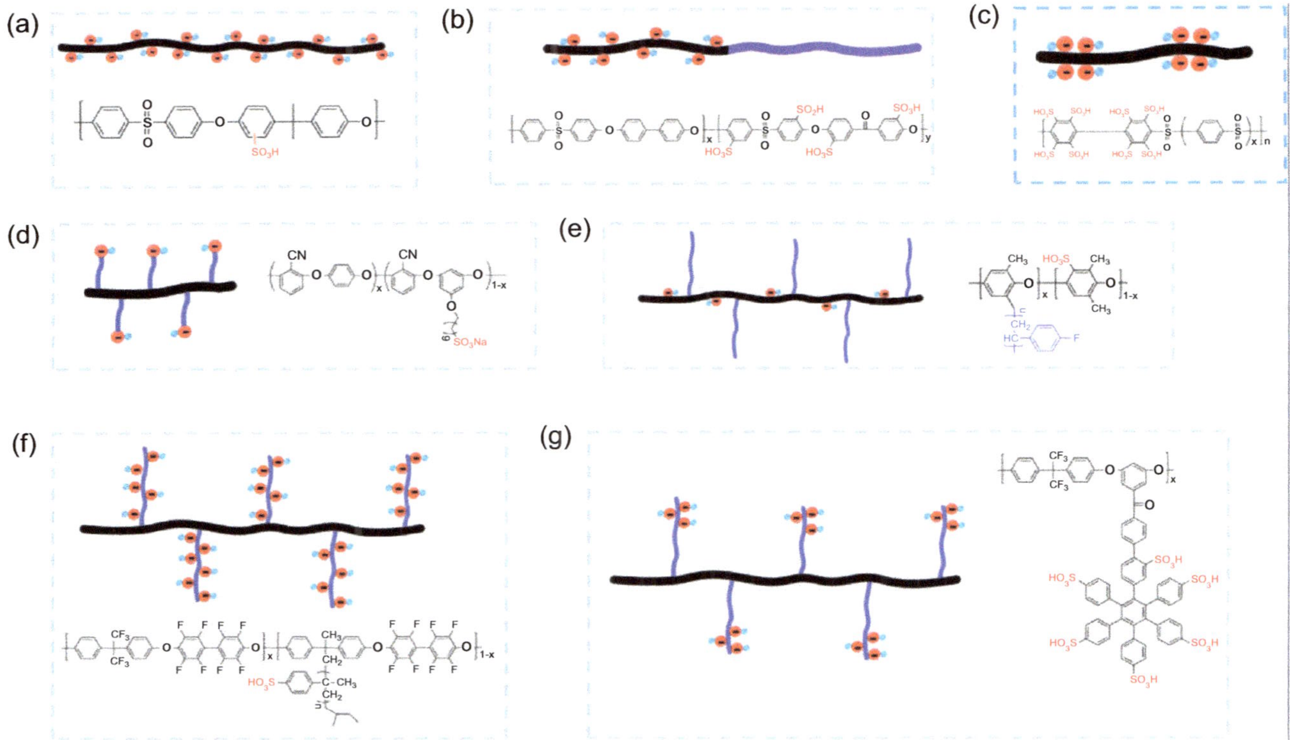

Figure 4.7 Different types of polymer architecture of cation-exchange membranes. Reprinted with permission from Ran et al. (2017). Copyright 2017, Elsevier.

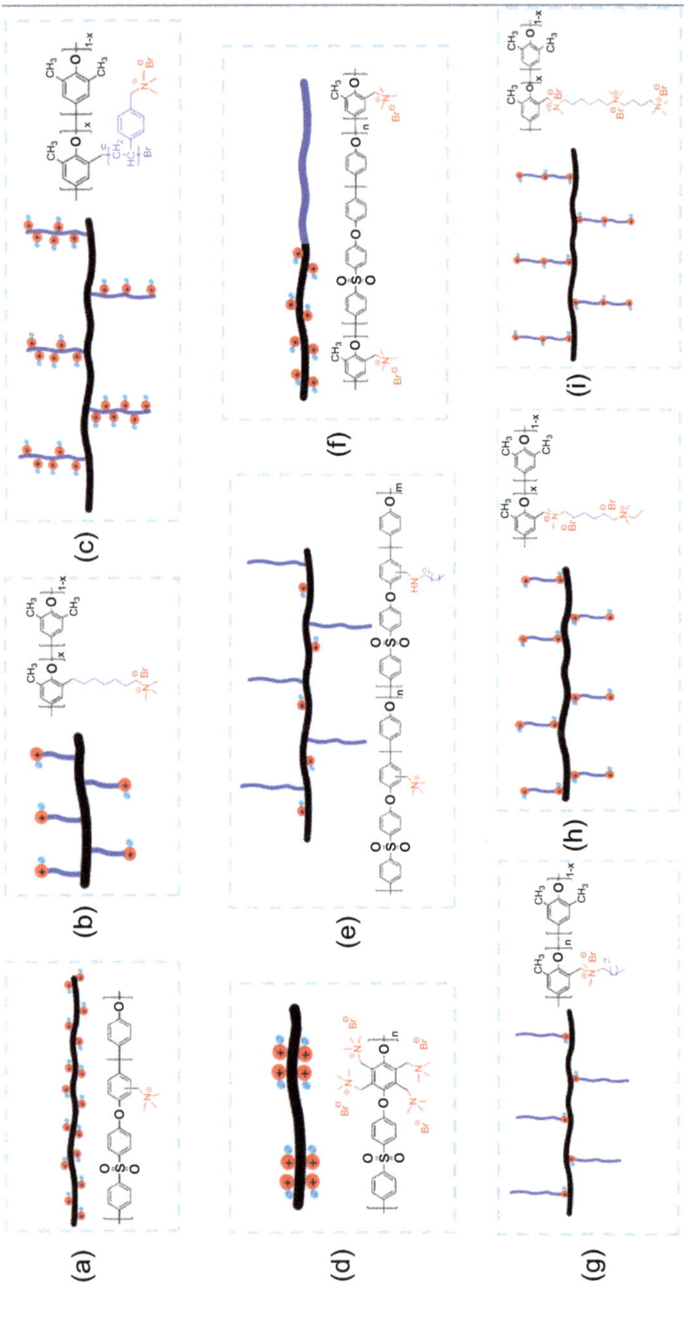

Figure 4.8 Different types of polymer architecture of anion-exchange membranes. Reprinted with permission from Ran et al. (2017). Copyright 2017, Elsevier.

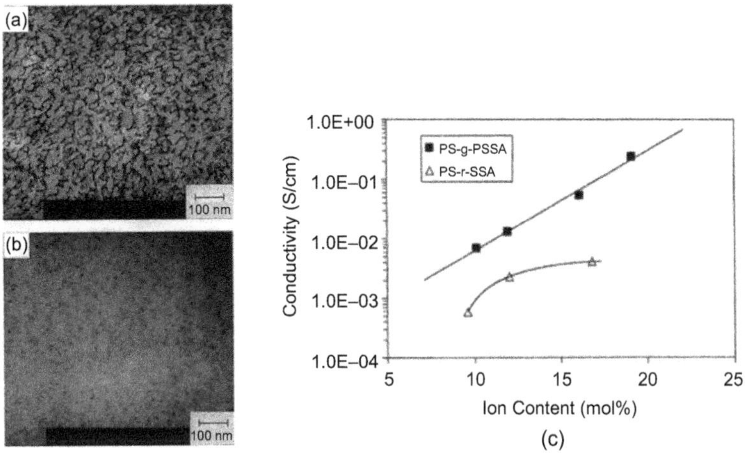

Figure 4.9 TEM images of (a) a block copolymer of styrene and sodium styrene sulfonate (PS-g-PSSA; 11.9% SSA); (b) a random copolymer of styrene and sodium styrene sulfonate (PS-r-SSA; 12.0% SSA), and (c) their proton conductivity at various contents of SSA. Reprinted with permission from Ding et al. (2001). Copyright 2001, American Chemical Society.

The factors influencing the physical nanostructures of copolymers, in terms of tortuosity and size, include the following: distribution of ionic and non-ionic segments within the polymer structure; polarity and mobility differences between them; and membrane processing conditions (e.g., solvent type, thermal annealing conditions) (Li et al., 2011; He et al., 2015). The chemical environment in these nanostructured copolymers depends on the type, concentration, and distribution of ion-exchange groups (Li et al., 2011; He et al., 2015). The IEMs based on block copolymers have typically higher conductivity than graft or random copolymers due to the presence of well-interconnected ionic nanochannels (Ding et al., 2001) (Fig. 4.9c), and in some cases, they also have higher dimensional stability upon hydration, thanks to the entanglement of polymer chains (Ding et al., 2001; Kinsinger et al., 2015).

Sulfonated poly(styrene-b-isobutylene-b-styrene) (S-SIBS) triblock copolymer membranes showed proton conductivity about one order of magnitude higher than those of sulfonated polystyrene (S-PS) at similar IEC. The values reported are also higher than those of Nafion 117 (Fig. 4.10) (Elabd et al., 2006).

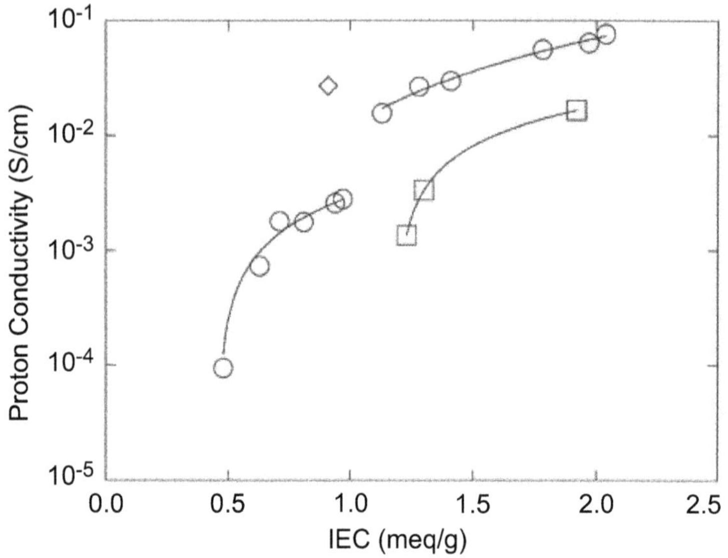

Figure 4.10 Proton conductivity as a function of ion-exchange capacity (IEC) for Nafion 117 (◊), S-SIBS (○), and S-PS (□) membranes. Reprinted with permission from Elabd et al. (2006). Copyright 2006, American Chemical Society.

4.4 Nanostructured Mixed Matrix IEMs

Nanostructured fillers can interact with the polymeric matrix to form tailored nanostructures for ion transport. Numerous nanofillers have been used for modifying IEMs, including metal oxides (TiO_2, ZrO_2, WO_3), silicon oxide (SiO_2), zirconium phosphates, heteropolyacids (HPAs), and nanostructured carbon materials. Nanostructured mixed matrix IEMs can be prepared by the following methods: in situ synthesis of nanoparticles inside the polymeric matrix; dispersion of nanoparticles in the polymeric solution and successive casting; and decoration of the membrane with nanoparticles by chemical methods (Mikhailenko et al., 2004).

Nanosized metal and silica oxides were synthesized by the in situ sol–gel method inside Nafion membranes with the goal of increasing the proton conductivity and water retention at high temperatures and low relative humilities, as well as to improve thermomechanical properties (Jalani et al., 2005). The Nafion membrane was converted to the sodium form by immersion in an NaOH solution to ensure good mechanical strength; it was swelled in an ethanol–water solution (10:1) and then immersed in a precursor solution (zirconium (IV) propoxide, titanium (IV) tertbutoxide or tetraethyl orthosilicate); finally, it was thermally treated. The authors reported a lower size and a better dispersion of the nanoparticles for the in situ sol–gel method than for the dispersion of the already formed metal oxide and successive casting. The mixed matrix membranes of Nafion nanocomposites exhibited better thermomechanical properties than unmodified Nafion (Jalani et al., 2005). In particular, Nafion–ZrO_2 membranes showed higher water uptake and conductivity than unmodified membranes at 120°C and 40% RH. This was attributed to the higher acidity of zirconia-based membranes in comparison with other nanocomposites containing Ti and Si oxides.

The in situ sol–gel technique was also used for the preparation of nanocomposite AEMs. Silica/poly(2,6-dimethyl-1,4-phenylene oxide) (PPO) AEMs were prepared by PPO bromination, hydroxylation, quaternization, and sol–gel reaction with monophenyl triethoxysilane or/and tetraethoxysilane, followed by heat treatment at 120–140°C for different times (Wu et al., 2009). The results indicated a relatively high alkaline resistance and thermal stabilities of these membranes, comparable to those of AEMs based on partially fluorinated polymers. This result is particularly relevant considering that the existing limitation of using AEMs in alkaline fuel cells is their poor alkaline stability due to the displacement of the ammonium group by the OH^- anions (Neagu et al., 2000).

Mixed matrix Nafion IEMs containing phosphonic acid-functionalized nano-silica were prepared by the sonication of the already formed nanoparticles inside a Nafion dispersion, casting, and solvent evaporation (Wang et al., 2008). The functionalized silica nanoparticles were well dispersed into the Nafion membrane (Fig. 4.11).

The phosphonic acid-functionalized silica additive enhanced the proton conductivity and water retention at high temperature by introducing both acidic groups and porous silica. The highest conductivity was obtained with a membrane containing 10% of additive. This membrane had a proton conductivity of 0.026 S/Cm at 85°C and 50% RH (24% higher than unmodified Nafion membranes). The fuel cell performance of the composite membrane was also higher than that of pristine Nafion (60 mV higher at 1 A/cm^2 and 95°C–35% RH; 80 mV higher at 0.8 A/cm^2 and 120°C–35% RH) (Wang et al., 2008).

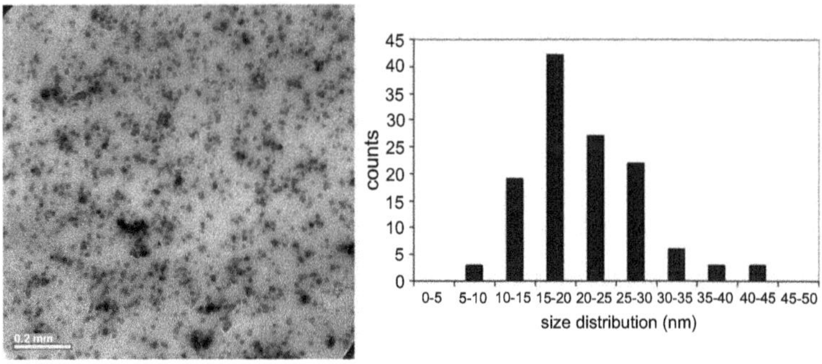

Figure 4.11 TEM image of a Nafion mixed matrix membrane containing phosphonic acid-functionalized silica nanoparticles. Reprinted with permission from Wang et al. (2008). Copyright 2008, Elsevier.

Nanostructured additives were immobilized in SPEEK-WC membranes modifying successfully their properties for applications in PEMFC. Mixed matrix SPEEK-WC membranes were prepared by dispersing amorphous zirconium phosphate sulfophenylenphosphonate [$Zr(HPO_4)(O_3PC_6H_4SO_3H$, hereafter Zr(SPP))] (Regina et al., 2006) in the polymeric casting solution. The membranes obtained were characterized by a good distribution of the inorganic filler. The SPEEK-WC membranes containing Zr(SPP) had a significantly lower swelling degree in water and methanol compared to the polymeric membranes prepared without the filler. HPAs were also used as functional additive in SPEK-WC- mixed matrix membranes prepared by dispersion of the nanofillers in the polymeric solution and successive casting (Fontananova et al., 2010). Several HPAs were tested, but the best results in term of proton conductivity were obtained for silicotungstic acid ($H_4SiW_{12}O_{40}\bullet nH_2O$). The inorganic filler was uniformly distributed on nanometric scale in SPEEK-WC membranes, improving the interconnection of the ionic cluster of the polymeric matrix and providing a better pathway for protons hopping (Fontananova et al., 2010). The SPEEK-WC mixed matrix membranes were also characterized by lower H_2, O_2, water, and methanol permeability than the polymeric ones (lower crossover and flooding problems in PEMFCs).

Nanostructured carbon materials, such as one-dimensional carbon nanotubes (CNTs) and two-dimensional graphene oxide (GO), are being developed as materials for energy-storage and -conversion devices such as supercapacitors and batteries (Lade et al., 2017; Thiam et al., 2011). The loading of graphitic carbon nanomaterials in membrane is usually lower (≤2 wt.%) in comparison to three-dimensional inorganic nanofillers such as TiO_2 and ZrO_2 (usually blended at loading ≥5 wt.%, up to 60 wt.%). This is because of the high specific surface, elevated aspect ratio, and the intrinsic properties of graphitized nanostructure (Ismail et al., 2009). In addition, the relatively easy functionalization of their surface makes them ideal candidate for tailoring the polymer/nanofiller interface.

Multiwalled carbon nanotubes (MWCNTs) were incorporated into a sulfonated polyethersulfone (SPES) membrane enhancing the proton conductivity at low RH% with respect to the unmodified SPES membrane (Sriram et al., 2016).

Sulfonic acid-functionalized MWCNTs were dispersed in Nafion membranes improving the proton conductivity due to the enhancement of ionic cluster domain size (Kannan et al., 2009). These nanocomposite membranes showed improved performance in PEMFC application with a higher maximum power density than those of Nafion 115 and recast Nafion 115 (380, 250, and 230 mW/cm^2, respectively) (Kannan et al., 2009).

GO was also frequently used as nanofillers for IEMs with the following aims: to improve water retention ability, to reduce fuel crossover, to enhance proton conductivity, as well as mechanical, thermal, and chemical stability (Pandey et al., 2017). GO IEMs were realized in the form of free-standing system by filtration of GO dispersion through a cellulose acetate filter (Ji et al., 2017). Pristine GO nanosheets are negatively charged and have cation-exchange properties. Positively charged GO with anion-exchange capacity was obtained by conversion of the carboxylic acid groups of the GO to alkyl-substituted imidazolium (p-GO) by a two-step chemical reaction. The scanning electron microscopy (SEM) image of the self-standing GO and p-GO showed a compact lamellar structure (Fig. 4.12).

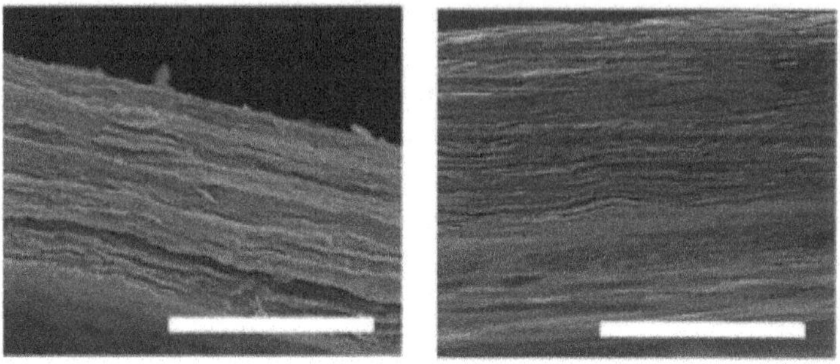

(a) (b)

Figure 4.12 SEM images of (a) self-standing GO and (b) positively charged GO membranes (scale bar 10 μm). Reprinted with permission from Ji et al. (2017). Copyright 2016, John Wiley and Sons.

These membranes were used in a nanofluidic RED system (Ji et al., 2017) based on a two-dimensional material. The power density achieved using 0.5 M and 0.01 M solutions as concentrate and dilute feed, respectively, was 0.77 W/m^2. The power output was 54% higher than that obtained using commercial IEMs (CMI and AMI, Table 4.1).

Mixed matrix Nafion membranes containing sulfonated GO were obtained by dispersing the GO platelets within a Nafion dispersion and successive casting (Nicotera et al., 2014). A reduction in methanol diffusion and an increase in proton mobility and water retention at high temperatures were observed for the mixed matrix membrane in comparison with unmodified Nafion. Moreover, the presence of GO in the polymeric matrix reduced the ohmic losses at high temperatures in direct methanol-PEMFC applications (Nicotera et al., 2014).

4.5 Conclusion

Huge progresses have been made in the past decades for the realization of new materials with tailored properties, as well as suitable preparation methods of nanostructured IEMs characterized by superior properties than the first-generation IEMs. In addition to the development of nanostructured membrane materials, advanced membrane preparation methods such as electrospinning also offer unique opportunities to obtain nanostructured IEMs.

Block copolymers with self-assembling properties can give ordered nanostructures for fast ionic transport over a wide range of conditions. Nanostructured mixed matrix membranes represent also a viable way to improve the mechanical and transport properties of polymeric IEMs.

However, some open questions still linger regarding the long-term use of nanostructured IEMs without decline in performance, fine control of nanostructures and interfaces, and understanding of the depth of transport mechanisms and factors influencing them. Moreover, efforts should be intensified to produce environmentally sustainable membranes by reducing the consumption of solvents and favoring the use of nontoxic materials in membrane fabrication. Molecular dynamic simulations and modelling studies can significantly contribute to achieve relevant progress in the development of IEMs using nanotechnology, relating the performance of IEMs with their nanostructures. Also the science and technology of AEMs is less advanced than that of CEMs and requires much attention with the aim to improve their alkaline stability and performance.

Abbreviations

AEM	anion-exchange membranes
BPM	bipolar membranes
CEM	cation-exchange membranes
CNT	carbon nanotubes
DS	degree of sulfonation
EW	equivalent weight
GO	graphene oxide
HPA	heteropolyacids
IEC	ion-exchange capacity
IEM	ion-exchange membrane
LSC	long-side-chain
MWCNT	multiwalled carbon nanotubes

PEEK	polyetheretherketone
PEEK-WC	polyetheretherketone with cardo
PEMFC	polymer electrolyte membrane fuel cell
PFSA	perfluorosulfonic acid
PPO	poly(2,6-dimethyl-1,4-phenylene oxide)
PS-g-PSSA	block copolymer of styrene and sodium styrene sulfonate
PS-r-SSA	random copolymer of styrene and sodium styrene sulfonate
RED	reverse electrodialysis
RH	relative humidity
SAP	sulfonated aromatic polymers
SAXS	small angle X-ray scattering
SEM	scanning electron microscopy
SGP	salinity-gradient power
SPEEK	sulfonated poly(etheretherketone)
SPEEKK	sulfonated poly(etheretherketoneketone)
SPEEK-WC	sulfonated polyetheretherketone with cardo
SPEK	sulfonated polyetherketone
SPEKK	sulfonated poly(etherketoneketone)
SPES	sulfonated polyethersulfone membrane
S-PS	sulfonated polystyrene
SSC	short-side-chain
S-SIBS	triblock copolymer of sulfonated poly(styrene-b-isobutylene-b-styrene)
TEM	transmission electron microscopy
Tg	glass transition temperature
Zr(SPP)	zirconium phosphate sulfophenylenphosphonate

References

Arcella, V., Troglia, C., and Ghielmi, A. 2005. Hyflon ion membranes for fuel cells. *Ind. Eng. Chem. Res.*, **44**: 7645–7651.

Beck, H. N. 1992. Solubility characteristics of poly(etheretherketone) and poly(phenylene sulphide). *J. Appl. Polym. Sci.*, **45**: 36–40.

Ding, J., Chuy, C., and Holdcroft, S. 2001. A self-organized network of nanochannels enhances ion conductivity through polymer films. *Chem. Mater.*, **13**: 2231–2233.

Drioli, E. and Zhang, H.-C. 1989. A study of polyetheretherketone and polyarylsulphone ultra-filtration membranes. *Chimica Oggi*, **11**: 59–63.

Drioli, E., Regina, A., Casciola, M., Oliveti, A., Trotta, F., and Massari, T. 2004. Sulfonated PEEK-WC membranes for possible fuel cell applications. *J. Membr. Sci.*, **228**: 139–148.

Elabd, Y. A., Napadensky, E., Walker, C. W., and Winey, K. I. 2006. Transport properties of sulfonated poly(styrene-b-isobutylene-b-styrene) triblock copolymers at high ion-exchange capacities. *Macromolecules*, **39**: 399–407.

Ersöz, M., Güğül, İ. H., Çimen, A., Leylek, B., and Yildiz, S. 2001. The sorption of metals on polysulfone cation exchange membranes. *Turkish J. Chem.*, **25**: 39–48.

Fontananova, E., Basile, A., Cassano, A., and Drioli, E. 2003. Preparation of polymeric membranes entrapping β-cyclodextrins and their molecular recognition of naringin. *J. Inclusion Phenom. Macrocyclic Chem.*, **47**: 33–37.

Fontananova, E., Messana, D., Tufa, R. A., Nicotera, I., Kosma, V., Curcio, E., van Baak, W., Drioli, E., and Di Profio, G. 2017. Effect of solution concentration and composition on the electrochemical properties of ion exchange membranes for energy conversion. *J. Power Sources*, **340**: 282–293.

Fontananova, E., Trotta, F., Jansen, J. C., and Drioli, E. 2010. Preparation and characterization of new non-fluorinated polymeric and composite membranes for PEMFCs. *J. Membr. Sci.*, **348**: 326–336.

Fontananova, E., Zhang, W., Nicotera, I., Simari, C., van Baak, W., Di Profio, G., Curcio, E., and Drioli, E. 2014. Probing membrane and interface properties in concentrated electrolyte solutions. *J. Membr. Sci.*, **459**: 177–189.

Geise, G. M., Cassady, H. J., Paul, D. R., Logan, B. E., and Hickner, M. A. 2014. Specific ion effects on membrane potential and the permselectivity of ion exchange membranes. *Phys. Chem. Chem. Phys.*, **16**: 21673–21681.

Ghielmi, A., Vaccarono, P., Troglia, C., and Arcella, C. 2005. Proton exchange membranes based on the short-side-chain perfluorinated ionomer. *J. Power Sources*, **145**: 108–115.

He, G., Li, Z., Zhao, J., Wang, S., Wu, H., Guiver, M. D., and Jiang, Z. 2015. Nanostructured ion-exchange membranes for fuel cells: Recent advances and perspectives. *Adv. Mater.*, **27**: 5280–5295.

Hickner, M. A., Ghassemi, H., Kim, Y. S., Einsla, B. R., and McGrath, J. E. 2004. Alternative polymer systems for proton exchange membranes (PEMs). *Chem. Rev.*, **104**: 4587–4612.

Hsu, W. Y. and Gierke, T. D. 1982. Elastic theory for ionic clustering in perfluorinated ionomers. *Macromolecules*, **15**: 101–105.

Husain, A., El Nashar, A., Al Radif, A., and Bushara, M. 2010. Properties of natural waters, in physical, chemical and biological aspects of water. In: *Encyclopedia of Desalination and Water Resources* (*DESWARE*), pp. 210–222.

Ismail, A. F., Goh, P. S., Sanip, S. M., and Aziz, M. 2009. Transport and separation properties of carbon nanotube-mixed matrix membrane. *Sep. Purif. Technol.*, **70**: 12–26.

Jalani, N. H., Dunn, K., and Datta, R. 2005. Synthesis and characterization of Nafione MO_2 (M = Zr, Si, Ti) nanocomposite membranes for higher temperature PEM fuel cells. *Electrochimica Acta*, **51**: 553–560.

Jansen, J. C., Macchione, M., and Drioli, E. 2005. High flux asymmetric gas separation membranes of modified poly(etheretherketone) prepared by the dry phase inversion technique. *J. Membr. Sci.*, **255**: 167–180.

Ji, J., Kang, Q., Zhou, Y., Feng, Y., Chen, X., Yuan, J., Guo, W., Wei, Y., and Jiang, L. 2017. Osmotic power generation with positively and negatively charged 2D nanofluidic membrane pairs. *Adv. Funct. Mater.*, **27**: 1603623–1603623.

Jörissen, J., Breiter, S. M., and Funk, C. 2003. Ion transport in anion exchange membranes in presence of multivalent anions like sulfate or phosphate. *J. Membr. Sci.*, **213**: 247–261.

Kannan, R., Parthasarathy, M., Maraveedu, S. U., Kurungot, S., and Pillai, V. K. 2009. Domain size manipulation of perflouorinated polymer electrolytes by sulfonic acid-functionalized MWCNTs to enhance fuel cell performance. *Langmuir*, **25**: 8299–8305.

Kinsinger, C. L., Liu, Y., Liu, F., Yang, Y., Seifert, S., Knauss, D. M., Herring, A. M., and Maupi, C. M. 2015. Random and block sulfonated polyaramides as advanced proton exchange membranes. *J. Phys. Chem. C*, **119**(44): 24724–24732.

Kreuer, K. D. 1996. Proton conductivity: Materials and applications. *Chem. Mater.*, **8**: 610–641.

Kreuer, K. D. 2000. On the complexity of proton conduction phenomena, *Solid State Ionics*, **136–137**: 149–160.

Kreuer, K. D. 2001. On the development of proton conducting polymer membranes for hydrogen and methanol fuel cells. *J. Membr. Sci.*, **185**: 29–39.

Lade, H., Kumar, V., Arthanareeswaran, G., and Ismail, A. F. 2017. Sulfonated poly(arylene ether sulfone) nanocomposite electrolyte membrane for fuel cell applications: A review. *Int. J. Hydrogen Energy*, **42**: 1063–1074.

Li, L., Zhang, J., and Wang, Y. 2003. Sulfonated poly(etheretherketone) membranes for direct methanol fuel cell. *J. Membr. Sci.*, **226**: 159–167.

Li, N., Wang, C., Lee, S. Y., Park, C. H., Lee, Y. M., and Guiver, M. D. 2011. Enhancement of proton transport by nanochannels in comb-shaped copoly(arylene ether sulfone)s. *Angew. Chem. Int. Ed.*, **50**: 9158–9161.

Logan, B. E. and Elimelech, M. 2012. Membrane-based processes for sustainable power generation using water. *Nature*, **488**: 313–319.

Marechal, M., Souquet, J.-L., Guindet, J., and Sanchez, J.-Y. 2007. Solvation of sulphonic acid groups in Nafion® membranes from accurate conductivity measurements. *Electrochem. Commun.*, **9**: 1023–1028.

Mauritz, K. A. and Moore, R. B. 2004. State of understanding of Nafion. *Chem. Rev.*, **104**: 4535–4585.

Membranes International Inc. Retrieved from https://www.membranesinternational.com (accessed on June 23, 2017).

Mikhailenko, S. D., Wang, K., Kaliaguine, S., Xing, P., Robertson, G. P., and Guiver, M. D. 2004. Proton conducting membranes based on cross-linked sulfonated poly(etheretherketone) (SPEEK). *J. Membr. Sci.*, **233**: 93–99.

Neagu, V., Bunia, I., and Plesca, I. 2000. Ionic polymers VI. Chemical stability of strong base anion exchangers in aggressive media. *Polym. Degrad. Stabil.*, **70**: 463.

Neburchilov, V., Martin, J., Wang, H., and Zhang, J. 2007. A review of polymer electrolyte membranes for direct methanol fuel cells. *J. Power Sources*, **169**: 221–238.

Nicotera, I., Simari, C., Coppola, L., Zygouri, P., Gournis, D., Brutti, S., Minuto, D., Aricò, A. S., Sebastian, D., and Baglio, V. 2014. Sulfonated graphene oxide platelets in Nafion nanocomposite membrane: Advantages for application in direct methanol fuel cells. *J. Phys. Chem. C*, **118**: 24357–24368.

Pandey, R. P., Shukla, G., Manohar, M., and Shahi, V. K. 2017. Graphene oxide based nanohybrid proton exchange membranes for fuel cell applications: An overview. *Adv. Colloid Interface Sci.*, 240: 15–30.

Ran, J., Wu, L., He, Y., Yang, Z., Wang, Y., Jiang, C., Ge, L., Bakangura, E., and Xu, T. 2017. Ion exchange membranes: New developments and applications. *J. Membr. Sci.*, **522**: 267–291.

Regina, A., Fontananova, E., Drioli, E., Casciola, M., Sganappa, M., and Trotta, F. 2006. Preparation and characterization of sulfonated PEEK-WC membranes for fuel cell applications. A comparison between polymeric and composite membranes. *J. Power Sources*, **160**: 139–147.

Roziere, J. and Jones, D. J. 2003. Non-fluorinated polymer materials for proton exchange membrane fuel cells. *Annu. Rev. Mater. Res.*, **33**: 503–555.

Saracco, G. 1997. Transport properties of monovalent-ion-permselective membranes. *Chem. Eng. Sci.*, **52**: 3019–3031.

Shin, D. W., Guiver, M. D., and Lee, Y. M. 2017. Hydrocarbon-based polymer electrolyte membranes: Importance of morphology on ion transport and membrane stability. *Chem. Rev.*, **117**: 4759–4805.

Smitha, S., Sridhar, S., and Khan, A. 2005. Solid polymer electrolyte membranes for fuel cell applications: A review. *J. Membr. Sci.*, **259**: 10–26.

Sriram, K., Arthanareeswaran, G., Ismail, A. F., and Paul, D. 2016. Effects of special nanoparticles on fuel cell properties of sulfonated polyethersulfone membrane. *Int. J. Polym. Mater. Polym. Biomat.*, **65**: 294–301.

Strathmann, H. 2004. *Ion Exchange Membrane Separation Processes*. Elsevier.

Strathmann, H. 2010. Electromembrane processes: Basic aspects and applications. In: *Comprehensive Membrane Science and Engineering*, 1st edn. Oxford: Elsevier.

Tasselli, F., Jansen, J. C., and Drioli, E. 2004. PEEK-WC ultrafiltration hollow fibre membranes: Preparation, morphology and transport properties. *J. Appl. Polym. Sci.*, **91**: 841–853.

Thiam, H. S., Daud, W. R. W., Kamarudin, S. K., Mohammad, A. B., Kadhum, A. A. H., Loh, K. S., and Majlan, E. H. 2011. Overview on nanostructured membrane in fuel cell applications. *Int. J. Hydrogen Energy*, **36**: 3187–3205.

Tripathi, B. P. and Shahi, V. K. 2011. Organic–inorganic nanocomposite polymer electrolyte membranes for fuel cell applications. *Prog. Polym. Sci.*, **36**: 945–979.

Trotta, F., Drioli, E., Moraglio, G., and Baima Poma, E. 1998. Sulfonation of polyetheretherketone by chlorosulfuric acid. *J. Appl. Polym. Sci.*, **70**: 477–482.

Tufa, R. A., Curcio, E., Brauns, E., Van Baak, W., Fontananova, E., and Di Profio, G. 2015. Membrane distillation and reverse electrodialysis for near-zero liquid discharge and low energy seawater desalination. *J. Membr. Sci.*, **496**: 325–333.

Vermaas, D. A., Sassenburg, M., and Smith, W. A. 2015. Photo-assisted water splitting with BPM induced pH gradients for practical solar fuel devices. *J. Mater. Chem. A*, **3**: 19556–19562.

Wang, K., McDermid, S., Li, J., Kremliakova, N., Kozak, P., Song, C., Tang, Y., Zhang, J., and Zhang, J. 2008. Preparation and performance of nano silica/Nafion composite membrane for proton exchange membrane fuel cells. *J. Power Sources*, **184**: 99–103.

Wu, Y., Wu, C., Xu, T., Lin, X., and Fu, Y. 2009. Novel silica/poly(2,6-dimethyl-1,4-phenylene oxide) hybrid anion-exchange membranes for alkaline fuel cells: Effect of heat treatment. *J. Membr. Sci.*, **338**: 51–60.

Xue, S. and Yin, G. 2006. Methanol permeability in sulfonated poly(ethretherketone) membranes: A comparison with Nafion membranes. *Eur. Pol. J.*, **42**: 776–785.

Yang, B. and Manthiram, A. J. 2006. Comparison of the small angle X-ray scattering study of sulfonated poly(etheretherketone) and Nafion membranes for direct methanol fuel cells. *Power Sources*, **153**: 29–35.

Yaroshchuk, A. E. 2000. Dielectric exclusion of ions from membranes. *Adv. Colloid Interface Sci.*, **85**: 193–230.

Zaidi, S. M. J., Mikhailenko, S. D., Robertson, G. P., Guiver, M. D., and Kaliaguine, S. 2000. Proton conducting composite membranes from polyether ether ketone and heteropolyacids for fuel cell applications. *J. Membr. Sci.*, **173**: 17–34.

Chapter 5

Nanostructured Zeolite Membranes: Synthesis and Applications

Catia Algieri, Angelo Garofalo, and Enrico Drioli
Institute on Membrane Technology of the National Research Council of Italy (ITM-CNR), Via P. Bucci, Cubo 17C, 87036 Rende (CS), Italy
c.algieri@itm.cnr.it

Zeolite membranes, for their high thermal and chemical stability (high resistance to chlorine, oxidants, and solvents), can be used when the polymeric membranes cannot operate. However, the industrial application of zeolite membranes is limited by the costs and low reproducibility during the synthesis process. To date, their use at the industrial level is only represented by LTA zeolite membranes for organic solvent dehydration. Different methods for their preparation were developed; but the most used is the secondary growth for its different advantages such as easier operation, higher controllability in crystal orientation, microstructure and film thickness, and better reproducibility with respect to the in situ method. However, further improvements based on the preparation of reproducible defect-free zeolite membranes with larger area and suitable configuration (capillary or multichannel supports) are necessary. Another

Functional Nanostructured Membranes
Edited by Enrico Drioli, Lidietta Giorno, and Annarosa Gugliuzza

ISBN 978-981-4774-79-6 (Hardcover), 978-1-351-13511-5 (eBook)
www.panstanford.com

important aspect to consider is the reduction in their manufacturing cost to facilitate their introduction in the industry.

This chapter gives an overview of the methods for synthesizing nanostructured zeolite membranes. Besides, the chapter also illustrates the application of membranes in water desalination and membrane reactors.

5.1 Introduction

Zeolites are hydrated crystalline aluminosilicate, having SiO_4^{4-} and AlO_4^{5-} tetrahedral groups as structural unit (Fig. 5.1). These units are connected to each other through shared oxygen atoms, giving a wide range of structures.

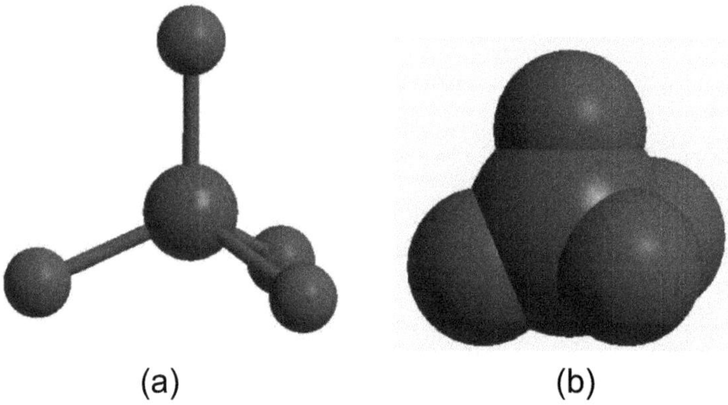

Figure 5.1 Tetrahedral structure: (a) 3D balls and (b) space filling.

The silica frameworks are electrically neutral in a three-dimensional lattice, while substitution of Si atoms with Al leads to negative charges; these can be balanced by extra-structural organic or inorganic cations, which can be exchanged by other cations. This type of framework has a large number of small cavities, which may be connected by pores or channels of molecular dimensions comparable to many chemical substances (Breck, 1974; Kaduk and Faber, 1995; Weller, 1994). The pore size varies between 0.3 and 2 nm, so zeolites present a high internal surface area, which allows to adsorb a wide range of chemical species. For this reason, zeolites act as molecular sieves and can separate molecules with different shape, size, and polarity. More than 200 zeolite structures are currently known (Ismail et al., 2015).

The Si–O bond in the zeolite structure is more stable than the Al–O bond, so the chemical and thermal stability increases with the Si/Al ratio. However, zeolite properties can change significantly by varying this value. In fact, crystals are hydrophilic at low ratio and are hydrophobic at high ratio. Furthermore, the ionic amount increases when the Si/Al ratio decreases, because the ion exchange depends linearly to the aluminum content.

Therefore, zeolites present unique properties: uniform pore size; ionic exchange; ability to have internal acidity; high thermal, chemical, and mechanical stability; absorption, separation, and catalytic properties. These characteristics offer the opportunity to obtain zeolite frameworks and use them in a wide range of applications (Kosinov, 2016; Maghsoudi, 2016). In fact, in the last decades, great attention was given to the use of nanostructured zeolite membranes as separators, chemical sensors, catalysts, and protection or insulation layer. This growing interest was observed for both industrial and academic community (Caro and Noack, 2008).

For their high thermal and chemical stability (high resistance to chlorine, oxidants, and solvents), zeolite membranes can be used where the polymeric membranes cannot operate. However, the industrial application of zeolite membranes is limited by the costs and low reproducibility during the synthesis process. To date, their use at the industrial level is only represented by LTA zeolite membranes for organic solvent dehydration by means of pervaporation and vapor permeation processes (Kondo and Kita, 2010).

Zeolite membranes are commonly prepared on porous supports because self-standing zeolite layers are difficult to synthesize and these structures are very fragile (Sano et al., 1991). These supports offer mechanical strength and structural stability and do not influence permeability and selectivity. For this purpose, porous materials such as ceramics, metals, glass, carbon, and clay are used as supports (Caro et al., 2000); however, the most frequent ones are alumina and stainless steel.

In particular, alumina supports (α-Al_2O_3 and γ-Al_2O_3) present a smooth surface, allowing the preparation of defect-free zeolite layers. On the other hand, these materials are very expensive, so it is essential to find cheaper ones for obtaining more economic membranes.

5.2 Preparation of Nanostructured Zeolite Membranes

The most common methods to prepare nanostructured zeolite membranes are the in situ or one-step and the secondary growth methods (Kazemimoghadam and Mohammadi, 2007). The first one uses a porous support in contact with a zeolite precursor solution under hydrothermal treatment to achieve both nucleation and growth of crystals. For example, a crystallization temperature of about 180°C is necessary to prepare MFI-type membranes, while the synthesis of FAU or LTA membrane requires 90–100°C. In situ hydrothermal crystallization is a simple method even though the preparation of reproducible and thin zeolite membranes by this method is not easy (Huang and Yang, 2008). Usually, this procedure needs long crystallization times (from hours to days), causing the formation of impure zeolites (Xu et al., 2004). Moreover, nucleation on the support surface is not simultaneous and non-homogeneous heating leads to the formation of zeolite crystals with different sizes.

However, the main negative aspect of the in situ method is poor reproducibility. To overcome these problems, the secondary growth method was developed (Lovallo and Tsapatsis, 1996). This procedure presents two different steps, seeding and growth, which permit control of both nucleation and crystal growth and so they can be optimized independently. The most critical step is seeding, since it influences the membrane quality. Therefore, the secondary growth requires an additional step compared to the in situ method, but it offers several advantages such as better reproducibility, higher crystal orientation control, and the possibility to obtain very thin membranes.

About the seeding step, several procedures are used to cover the porous support with zeolite crystals, but the most common are rubbing (Hasegawa et al., 2002a; Kusakabe et al., 1998) and dip-coating (Bernal et al., 2001; Lovallo and Tsapatsis, 1996; Lovallo et al., 1998).

Rubbing allows the seeding of inner and outer tubular supports by means of small brushes (Fig. 5.2). This process must be repeated several times to obtain a high-density seeding on the support; however, it is not a very reproducible procedure and is more appropriate for the outer tubular surface.

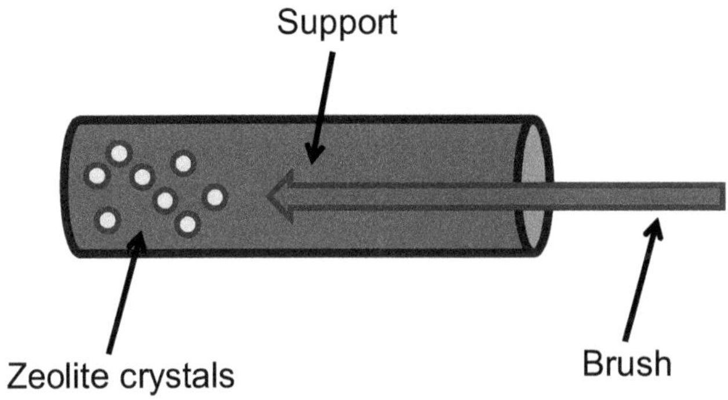

Figure 5.2 Simplified scheme of rubbing.

The dip-coating method is commonly used to obtain thin films on a support. It can be described by three steps (Fig. 5.3): (a) immersion of the support in a zeolite suspension at a constant rate; (b) permanence of the support in contact with the suspension to allow the deposition of crystals; and (c) extraction of the support, with the same immersion rate. Also in this case, the coating must be repeated a number of times to achieve a uniform zeolite layer, considering the negative action of the gravitational force on the zeolite crystals.

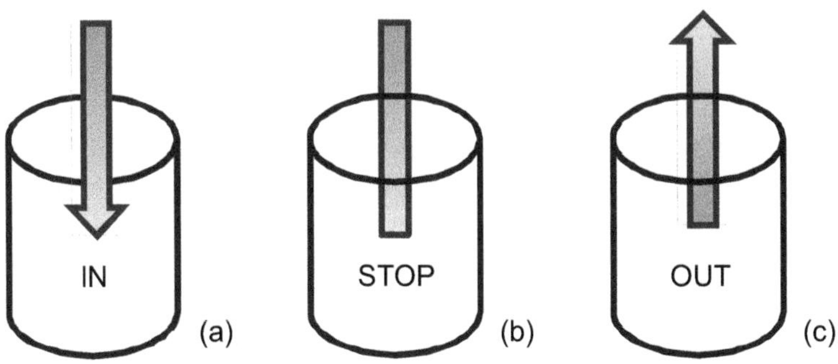

Figure 5.3 Simplified scheme of dip-coating.

Other seeding procedures used in the literature are spin coating (Mintova and Bein, 2001) and use of cationic polymers (Hedlund et al., 1999; Weh et al., 2002).

Spin coating is employed, in particular, for flat supports. Cationic polymers are used to adjust different zeta potentials between zeolite crystals and support, in order to cover the latter by electrostatic force.

However, more controllable seeding procedures involve the filtration of a zeolite suspension through a porous support. The filtration process may be described by two different methods: dead-end (Huang et al., 2004; Takata et al., 2002) and cross-flow (Pera-Titus et al., 2005) (Fig. 5.4).

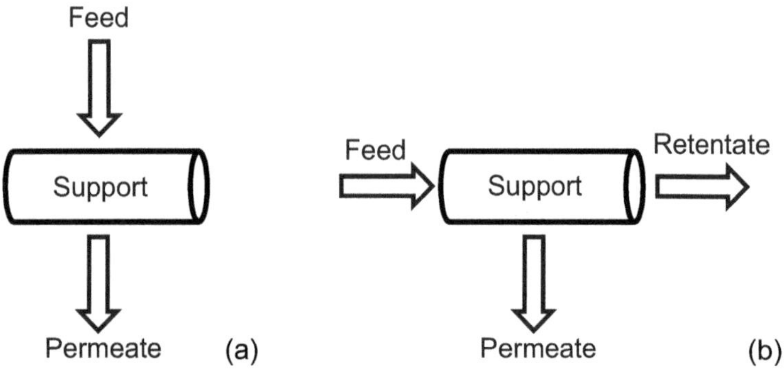

Figure 5.4 Schematic representation of (a) dead-end and (b) cross-flow filtration.

In dead-end filtration, the suspension flows perpendicularly to the support, inducing the formation of a cluster of crystals and so an excessive crystal accumulation; instead, in the cross-flow procedure, the suspension flows tangentially along the support surface, obtaining a more uniform, compact, and thin zeolite layer.

In cross-flow filtration, a high flow rate is commonly used to enhance the sweeping action through the membrane. On the contrary, for the seeding of nanocrystals, a very low filtration rate is required to avoid the shearing forces on the zeolite layer and to obtain a uniform support coverage.

At the Institute for Membrane Technology (ITM-CNR), a great number of nanostructured zeolite membranes with different topology were prepared, using both the in situ and the secondary growth methods. For example, Algieri et al. (2001; 2003) synthesized tubular supported MFI membranes by a modified one-step method. In particular, the α-Al_2O_3 supports were put in contact with two different hydrothermal solutions. The first was used to promote nucleation on the support surface, directly, skipping the seeding of the nanocrystals, while the second one allowed the growth of the crystals present on the support. This approach permitted synthesis of nanostructured membranes with the same gas selectivity, but with a better permeability than those obtained using the traditional in situ procedure. However, this is not a very reproducible method and with the formation of an inhomogeneous zeolite layer. Therefore, in the next studies, the secondary growth method was performed using α-Al_2O_3 tube as support. To cover the inner support surface with zeolite seeds, the suspension was pushed inside the support by means of a peristaltic pump. However, poor-quality membranes were obtained in terms of the zeolite layer homogeneity using tubular supports. In fact, during the filtration process, the support is fixed in the horizontal position and only the lower part of the support was totally covered with the seeds, due to the action of the gravitational force. Therefore, the deposition of crystals on the inner tubular surface is not completely uniform. To overcome the coverage uniformity problems, a novel seeding procedure was developed and then patented at the ITM-CNR (Algieri et al., 2011; Bernardo et al., 2006; Bernardo et al., 2008). This new method combines at the same time the support tilting and rotation with the cross-flow filtration of a suspension of zeolite seeds (Fig. 5.5). The tilting is performed with respect to the horizontal plane, and it is useful to eliminate air bubbles that may be formed into the support during the filtration. Besides, the support rotation allows a uniform coverage and a more homogeneous deposition of zeolite crystals.

Figure 5.5 Apparatus used for the cross-flow seeding procedure.

The newly developed seeding procedure can be exploited for different support and zeolite types. Garofalo et al. (2014) demonstrated, for the first time, the possibility of synthesizing tubular MFI (silicalite) membranes (10 cm long) using the innovative cross-flow seeding system. The authors investigated the influence of different seeding parameters on the membrane quality, as the support pore size and the zeolite slurry concentration. This study showed that the secondary growth method coupled with the cross-flow seeding procedure, using α-Al_2O_3 support of 70 nm and a zeolite slurry concentration of 0.032 wt%, is a potential route to preparing almost defect-free MFI zeolite membranes.

Other developments in terms of zeolite membrane synthesis involve the use of microwave heating for the aging of the growth solution and for the secondary growth (Li et al., 2010a; Liu et al., 2012). This methodology permits to obtain membranes with small crystals and high crystallinity in very short times (Cundy, 1998; Huang and Yang, 2007; Xu et al., 2004). Microwave synthesis can be performed in the same way as in situ crystallization or after substrate coating with the zeolite precursor. Moreover, membranes with different characteristics were obtained in a few hours by changing the synthesis parameters such as temperature and duration of the microwave treatment (Cheng et al., 2002; Xu et al., 2000). For instance, it is possible to control the membrane thickness using microwave heating after the seeding step. However, long heating times at high temperature can induce breaking of the membrane layer.

Some zeolite membranes, such as the MFI-type (silicalite, ZSM5), require the use of an organic compound as a structure-directing agent (SDA) for the formation of the zeolite layer. The organic template orients the crystal growth, and its removal is essential to have free pores and to activate the membrane framework. For this purpose, calcination process at high temperature (about 550°C) is achieved. Unfortunately, in many cases this treatment leads to the formation of cracks in the membrane structure, most probably due to the different physical properties of support material and zeolite. For example, it was observed that on increasing the temperature, the alumina support expands while the zeolite layer shrinks. Furthermore, alumina has an isotropic thermal expansion, whereas several zeolite crystals show an anisotropic thermal expansion:

the combination of these two different behaviors could induce the formation of defects in the membranes (den Exter et al., 1997; Dong et al., 2000; Gues and van Bekkum, 1995).

Template removal from zeolite membranes starts at 260–350°C and ends at 500°C. Recently, an innovative procedure for removing defects in the zeolite membrane layer was developed by Choi et al. (2009). This method, called "rapid thermal processing," consists of a quick membrane treatment at 700°C (0.5 min) with a very fast heating rate (1 min), after both the hydrothermal process and standard calcination. The authors found that this methodology allowed better separation performance than those obtained for the classic calcined membranes. A possible explanation is the condensation of the Si–OH groups between adjacent crystals that occurs during the rapid thermal processing, leading to the formation of a uniform and compact zeolite layer.

Many research groups investigated the opportunity to reduce the use of organic template by means of a seed-assisted synthesis procedure (Rangnekar et al., 2015). In this approach, zeolite crystals were prepared using the structural agent, while membrane synthesis was carried out without the template in the solution growth (Davis, 2014; Itabashi et al., 2012; Iyoki et al., 2014; Oleksiak and Rimer, 2014). In this case, the calcination step is eliminated, and it is possible to avoid crack formation in the zeolite layer. However, the microstructure negatively influenced the membrane separation performance (Caro and Noack, 2008).

5.3 Characterization of Nanostructured Zeolite Membranes

To evaluate the morphological characteristics of the nanostructured zeolite membranes, several techniques can be exploited. For example, scanning electron microscopy (SEM) is used to investigate the size and shape of zeolite crystals. The study of surface morphology also allows detection of defects in the zeolite layer. Besides, SEM analyses are performed to explore the membrane cross section in order to determine membrane thickness. Figure 5.6 reports the top view and cross section of a tubular MFI zeolite membrane prepared with the secondary growth method and using the cross-flow seeding procedure.

(a) (b)

Figure 5.6 (a) Top view and (b) cross section of silicalite membrane on tubular α-Al_2O_3.

SEM analyses showed a homogeneous and compact zeolite layer on the alumina support, with a membrane thickness of about 5 μm.

The elemental composition of zeolite membranes is determined by means of energy-dispersive X-ray (EDX) analysis. Instead, X-ray diffraction is employed to investigate membrane topology. The X-ray spectra allow identification of different species and observation of the characteristic peaks of both zeolites and support materials by comparison with the literature data.

The presence of defects in the zeolite layer can be detected by means of single gas permeation tests at room temperature. In particular, for zeolite membranes that require template removal (as MFI topology), this test is performed before and after the calcination process, according to the pressure drop method (PDM) (Li et al., 2004a). The bubble soap flowmeter is a recommended tool for estimating the gas flux that passes through the membrane in steady state. In the PDM, one end of the membrane module is closed and the pressure at the feed side is controlled in order to set the transmembrane pressure difference without a sweep gas. Obviously, if the single gas does not pass through the membrane before calcination, the membrane is gas-tight. This indicates the absence of defects in the zeolite layer.

5.4 Scale-Up of MFI Zeolite Membranes

Optimization, reproducibility, and scale-up are three of the most critical components of the development process of nanostructured zeolite membranes. In fact, design and implementation of membrane preparation on a large scale follow an elaborate pathway, exclusive for each material.

Until now, much of the current research on zeolite membranes was restricted to laboratory testing and just a few works are present in the literature regarding the scale-up of zeolite membranes. However, even if laboratory tests showed the potential of zeolite membranes in many fields, scaling-up of membranes is essential to study how the technology can get as close as possible to the real application conditions. Unfortunately, the problem of developing easy, cheap, and reproducible synthesis procedures is complicated to overcome.

As previously reported, LTA membranes were the only ones to be synthesized and used at the industrial level (Hoof et al., 2006; Kondo et al., 1997; Li et al., 2009). At the lab scale, different research groups have tried to prepare membranes with different topology and larger area (Li et al., 2010b; Sato et al., 2008; van den Bergh et al., 2010). However, other zeolite membrane types are also strongly attractive for different industrial applications. In this context, particular attention was recently given to the preparation and reproducibility of scaled-up MFI zeolite membranes. The MFI zeolite topology is highly suitable in many membrane separation processes, owing to small pore size (about 0.55 nm).

Drobek et al. (2012a) prepared tubular MFI zeolite membranes 5 cm long, using the secondary growth method. The researchers synthesized MFI nanocrystals by microwave-assisted hydrothermal treatment. Subsequently, the zeolite seeds were deposited on the α-Al_2O_3 surface by dip-coating. Finally, the secondary growth was carried out by means of hydrothermal treatment with conventional heating. Characterization by gas permeation led to good results, so the same synthesis protocol was applied to α-Al_2O_3 support 15.5 cm long, in order to prepare scaled-up MFI membranes. The authors demonstrated the reproducibility of the synthesis procedure, obtaining MFI membranes 15.5 cm long with separation properties comparable with those obtained for shorter membranes.

A further improvement in the scale-up of MFI zeolite membranes was achieved by Garofalo et al. (2016). In fact, supported MFI zeolite membranes 30 cm long were prepared, for the first time, using the secondary growth method coupled with the cross-flow seeding procedure, optimized in their previous work (Garofalo et al., 2014).

To obtain high-quality membranes, the authors investigated the influence of rotation time during the seeding step, keeping the other operating parameters constant. Obviously, an increase in the seeding time leads to an increase in the amount of nanocrystals present in the inner support surface. Table 5.1 reports a comparison among the best parameters used in the previous work (Garofalo et al., 2014) to prepare 10 cm long MFI membranes and these new seeding conditions.

Table 5.1 Comparison among the seeding parameters for MFI membranes 10 cm and 30 cm long

Sample	α-Al_2O_3 support length (cm)	Seeding time (min)	Crystals amount (g)	$mass_{crystals}/area_{membrane}$ (gm^{-2})	Ref.
ZM1	10	36	0.09	54.6	Garofalo et al. (2014)
ZM2		36	0.09	15.2	
ZM3	30	72	0.18	30.5	Garofalo et al. (2016)
ZM4		126	0.32	54.2	

Characterization by single gas permeation test before calcination showed that all the membranes prepared with the same operating conditions of the ZM3 sample were gas-tight. A small concentration of defects was found for those prepared as ZM2 membrane. On the contrary, ZM4 was very unstable due to the very thick zeolite layer. In fact, the authors observed a weak interaction between the support and the zeolite nanocrystals. Consequently, the layer collapsed during the growth treatment.

During the membrane synthesis, a new type of autoclave was used, in which the Teflon tank was locked in a stainless steel chamber (Fig. 5.7). This new autoclave was made because on using the conventional autoclave, a deformation of the Teflon tank was observed during the hydrothermal treatment of the scaled-up membranes. As a result, a partial leak of the growth solution was detected. This behavior was most probably due to the combined action of both pressure and temperature on the autoclave.

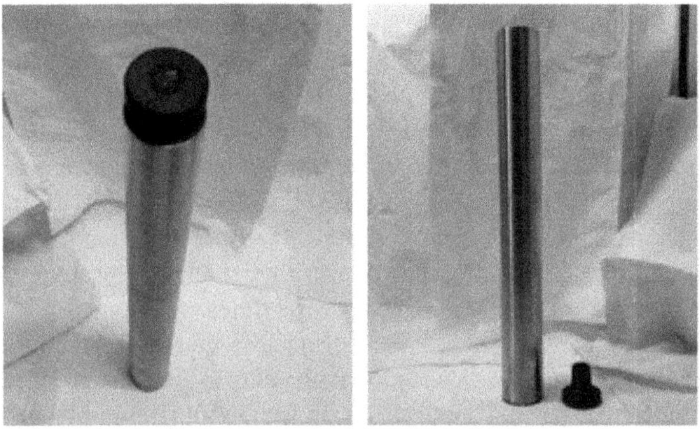

Figure 5.7 Teflon-lined autoclave in a stainless steel chamber.

In conclusion, this work demonstrated the possibility to upscale the synthesis of nanostructured zeolite membranes. In fact, uncalcined defect-free MFI membranes 30 cm long were prepared by means of a reproducible and simple protocol. However, many efforts must be made in the membrane preparation field to improve the manufacturing methodology and to lower production costs.

5.5 MFI Zeolite Membranes: Application in Desalination Process

The growing population and global warming have increased the demand for fresh water (Huges and Gale, 2010). To meet this demand, desalination of sea and brackish water to make sweet water in arid regions has gained growing interest. Desalination removes salts and other minerals to obtain pure water for human consumption and domestic/industrial use (Bremere et al., 2001). The application of membrane technology in the reuse of wastewater effluent largely increased over the past decade. Considering the different membrane processes, membrane distillation (MD) is a promising thermally driven desalination technology better than reverse osmosis (RO) plants (Criscuoli and Drioli, 1999). In fact, the last one requires electricity to pump feed water above the osmotic pressure, while MD requires only heat, reducing the electrical energy demand, with a consequent reduction in energy and desalination costs (Meindersma et al., 2006). MD is performed with hydrophobic microporous membranes that are not selective and act as inert barrier to prevent the passage of liquid water inside pores. However, polymeric membranes present different problems: fouling (Hsu et al., 2002; Wang et al., 2008) and degradation, when in contact with acid or alkaline environments (Farahbakhsh et al., 2004). Considering these drawbacks, different research groups have directed their attention toward inorganic membranes due to their high chemical stability and the possibility to reuse them several times.

Zeolite membranes are nanostructured inorganic membranes with well-defined pore size in the range of 0.3–1 nm and so suitable for molecular sieving applications. In the desalination process, the MFI zeolite topology is the most appropriate, having a pore size (of about 0.55 nm) lower than the kinetic diameters of different hydrated ions (Kazemimoghadam and Mohammadi, 2007; Liu et al., 2007) and larger than the kinetic diameter of water. Different research groups have studied the possibility of using them in desalination (Drobek et al., 2012b; Duke et al., 2009; Garofalo et al., 2014; Li et al., 2004a; Li et al., 2004b; Li et al., 2007; Zhu et al., 2014).

For the first time, Li et al. (2004a) used zeolite membranes with MFI topology for water desalination through the RO process. The rejections of Na^+, K^+, NH_4^+, Ca^{2+}, and Mg^{2+} reached 58.1%, 62.6%, 79.9%, 80.7%, and 88.4%, respectively. In another work, the same authors (Li et al., 2004b) established that different mechanisms control the transport of ions and water through the zeolite pores and intercrystalline spaces. Li et al. (2007) evaluated the effect of the aluminum content on the performance of MFI membranes in the desalination process by RO. In particular, the rejection and flux increased with the aluminum content. This behavior is due to an increase in the surface charge and water affinity of the membrane structure.

The same result was found by Duke et al. (2009). In this work, the rejection increased from 97% to 99% moving from silicalite to ZSM5 zeolite membrane. The desalination performance of silicalite membranes by RO, prepared by the in situ crystallization method, was evaluated by Zhu et al. (2014). A rejection value higher than 93% was obtained for major seawater ions (Ca^{2+}, Mg^{2+}, and Na^{+}) at an operating pressure of 700 kPa and room temperature. Increasing the temperature increased the permeation flux and decreased the rejection. Temperature probably determined acceleration of ion diffusion, alterations in the zeolite structure, and hydration of the ion diameter. All these effects caused an increase in ion penetration through the membrane. These experimental data indicated that the rejection of ions using an MFI-type zeolite membrane is achievable at low temperature in RO.

In membrane processes, use of membranes with tubular or hollow-fiber configuration is preferred for optimizing the membrane area–volume ratio. The first example of tubular MFI membranes (area 11 cm^2) applied in desalination process was presented by Drobek et al. (2012b). In this work, silicalite and ZSM5 membranes were prepared by the secondary growth method and were used in desalination; the tests were carried out using a pervaporation setup. In this work, ZSM5 membranes presented high fluxes with respect to the silicalite ones. However, both silicalite and ZSM5 membranes were severely affected by the salt feed concentration and temperature cycling. However, the long-term tests suggested that silicalite membranes exhibited better performance than the ZSM5 ones. Nevertheless, on extending further the thermal test (>560 h), the zeolite layers on the supports were destroyed for both membranes. In 2014, the laboratory of ITM (Garofalo et al., 2014) prepared MFI tubular membranes with a larger area (18 cm^2) by the secondary growth method. The prepared membranes were tested for the first time in vacuum MD for water desalination using two different salt concentrations as feed: 1 wt.% and 3.5 wt.%. The less concentrated salty solution showed a very high flux (7 kg/m^2h) and rejection (99%). On the other hand, increasing further the salt concentration of the feed decreased both the permeate water flux (5.2 kg/m^2h) and the rejection (96%).

To date, MFI membranes of various shapes (flat, tubular mono- and multi-channels) and sizes/lengths have been developed and tested in desalination, with ion rejections in the range of 80–99% for pervaporation (PV) and 60–95% for RO (Drobek et al., 2012b; Duke et al., 2009; Li et al., 2004a; Li et al., 2004b; Li et al., 2007;

Zhu et al., 2014). However, the synthesized membranes presented a small size, usually 2–2.5 cm in diameter (flat configuration) and up to 10 cm in length (tubular configuration). Although laboratory testing showed interesting results about the possibility to use zeolite membranes for water desalination, two problems are still present: synthesis reproducibility and membrane scale-up (Gascon et al., 2012; Maloncy et al., 2005). Supported zeolite membranes having MFI topology were scaled up by Garofalo et al. (2016). The membranes were synthesized by the secondary growth method using the cross-flow procedure during the seeding step, coupled with the rotation of the tubular support as previously reported. In particular, the influence of rotation time (36, 72, 136 min) during the seeding step was studied to optimize the membrane quality. The silicalite membranes were characterized and tested in VMD using NaCl solutions at different concentrations. First, the preparation method is reproducible, and the membrane prepared using a rotation time of 72 min showed the best performance. The experimental data indicated that on increasing the salt concentration from 1.0 wt.% to 5.0 wt.%, the fluxes decreased and the rejection value remained almost constant at very high values (>99.5%). Using the brine solution as feed (7.5 wt.%), a decrease in both flux and rejection was detected. This behavior is probably due to the salt trapped in the zeolite structure as also observed by Zhu et al. (2014).

Table 5.2 summarizes the results obtained by different research groups. The results reported by ITM (Garofalo et al., 2016) are very promising since one major drawback in the zeolite membrane field is reproducibility, especially on a large scale.

5.6 Zeolite Membranes in Catalytic Process

Membrane reactors (MRs) are mainly used to carry out reactions limited by the equilibrium conversion. Compared with a traditional reactor, an MR improves the reaction conversion and products selectivity. In fact, the thermodynamic equilibrium restrictions can be overcome by means of the shift effect.

In catalytic processes, membranes resistant to the temperature with high chemical stability are required. In this context, zeolite membranes meet these requirements. Keeping in mind the

Table 5.2 Separation results of MFI zeolite membranes present in the open literature

Zeolite-type	Configuration	Process	Diameter (cm)	Effective length (cm)	Amb (cm^2)	Feed (gl^{-1})	J ($kgm^{-2}\ h^{-1}$)	R (%)	Ref.
Silicalite	Flat	RO	2.8		6.15	5.8	0.2	21.6	Li et al., 2004b
Silicalite	Flat	PV	2.0	—	3.14	38.0	0.5	97.0	Duke et al., 2009
Silicalite	Tubular	PV	—	4	8.79	<35.0	4.0	>99.0	Drobek et al., 2012b
						35.0	2.5	98.5	
						75.0	1.5	80.0	
Silicalite	Tubular	VMD		27	59.38	10.0	13.8	99.8	Garofalo et al., 2016
						35.0	8.4	99.7	
						50.0	8.2	99.6	
						75.0	3.7	94.6	

properties of the zeolite membranes, it is possible to distinguish two types of catalytic membrane processes; in one case, the membrane is inert, and in the second case, the membrane presents intrinsic catalytic property.

Zeolite membranes were extensively studied for the removal of hydrogen in dehydrogenation reactions, in particular, for the conversion of alkanes in olefins. In 2001, isobutane dehydrogenation using an MR combining the PtIn/zeolite catalyst with a tubular MFI membrane was investigated (Ciavarella et al., 2001). A comparison between the performances of traditional reactor and MRs showed an isobutene yield four times higher with MR.

Another important reaction is the production of styrene from ethylbenzene. Styrene is one of the most important monomers used for the production of synthetic rubbers and plastics. Kong et al. (2007) used MFI zeolite membrane for the production of styrene. In this work, conversion with the MR was 7% higher than the fixed-bed reactor above 600°C without decrease in styrene selectivity. A recent theoretical study (Choi et al., 2015) on propane dehydrogenation (PD) showed that small zeolite pores (<0.5 nm) are required for improving alkane conversion (>55%). Besides, the membranes must exhibit a H_2 permeance in the range of $(2–6) \times 10^{-7}$ mol/m^{2}sPa at 600°C. However, in most cases, the nanoporous zeolite membranes present two drawbacks: low permeance values [$(0.1–2.0) \times 10^{-7}$ mol/m^{2}sPa] and low stability at high temperature. Kim et al. (2016) demonstrated the possibility of enhancing conversion and selectivity at high temperature for the PD reaction using a nanostructured zeolite membrane (SAPO-34, pore size of 0.38 nm). The researchers prepared membranes with reduced thickness (about 1 μm), larger area (18 cm^2), and good performance in terms of H_2 selectivity. The selectivity was greater than 80% with high propane conversion (65–75%). These interesting results are due to the efficient removal of hydrogen from the reaction medium with the zeolite membranes. These experimental data indicated that this type of zeolite membrane is a good candidate for the PD process.

Today, dimethyl ester (DME) is regarded as a clean, high-efficiency alternative to diesel fuel and liquefied petroleum gas (Semelsberger et al., 2006). The main method for producing this compound is methanol dehydration:

$$2CH_3OH \rightleftharpoons CH_3OCH_3 + H_2O$$

This process is usually carried out in a fixed-bed reactor at 250–400°C and with a conversion of about 80% on pure γ-alumina or γ-alumina modified with phosphates or titanates (Kong et al., 2009). Recently, catalytic dehydration of methanol in a reactor using a NaA zeolite membrane with γ-alumina as catalyst was studied (Fedosov et al., 2015). The separation factor of the water/methanol mixture in the temperature range of 150–250°C varied from 22 to 5 with a permeate flux of 0.6 kg/m^{2}h. The influence of temperature on methanol conversion is reported in Fig. 5.8. At low temperatures (150–200°C), catalyst activity was very low. An increase in methanol conversion was observed only above 200°C. The effect of the zeolite membrane on alcohol conversion was more pronounced at high temperature. In fact, at 250°C with the MR, the conversion was equal to 85%, while with the traditional one, the conversion was only 61%. This difference revealed the positive effect of water removal (by means of the hydrophilic zeolite membrane) on DME production.

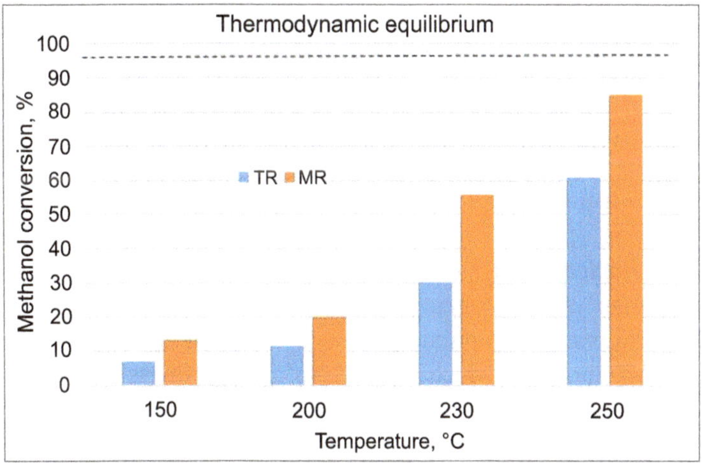

Figure 5.8 Effect of temperature on methanol conversion. (Elaborated from Fedosov et al., 2015).

Hydrogen is one of the most promising energy carriers for the future as it is a high-efficiency and zero-emission fuel due to the production of only water on being burned. Hydrogen is not a primary fuel; in fact, it is mainly produced at high temperature by the steam methane reforming (SMR) process. This catalytic process involves a reaction between natural gas (or other light hydrocarbons) and steam in the temperature range of 700–1000°C. A second reaction, called water gas shift (WGS), is subsequently used for decreasing the amount of CO and producing more hydrogen. The use of a catalytic MR with an H_2 permselective membrane can improve the WGS reaction conversion and the separation of hydrogen from carbon dioxide, simultaneously. Dong et al. (2015) prepared tubular ZSM5/silicalite bilayer membrane composed of a 3 μm ZSM5 layer, an 8 μm silicalite layer, and a 2 μm YSZ barrier layer supported on an α-alumina substrate. Experimental study coupled with modelling analysis identified the best operating conditions (550°C, feed pressure of 20 atm, H_2O/CO ratio of 2.0, and GHSV of 60,000 h^{-1}) to have both high CO conversion (>95%) and high H_2 recovery (>90%).

Until now, the different examples reported the use of catalytically inert zeolite membranes coupled with a conventional fixed bed of catalysts put on one of the membrane sides. In other studies, the zeolite membranes exhibited both catalytic activity and permselectivity; some examples are as follows. In 2006, researchers synthesized two-layered H-ZSM5-mordenite membranes for merging both the catalytic activity of H-ZSM5 and the water separation of mordenite membrane (de la Iglesia et al., 2005). The prepared membrane was used in acetic acid esterification to yield ethyl acetate and water. This work demonstrated that an H-ZSM5 membrane achieved a conversion of only 7.5%. On the contrary, using the two-layered membrane increased the conversion to 22.8%.

Another very important process is m-xylene isomerization because it results in obtaining p-xylene. It is used for the production of terephtalic acid, which is used for the synthesis of polyester resin and fibers (Xomeritakis et al., 2001). The effect of the zeolite MR on this reaction was evaluated by Yeong et al. (2010). For this purpose, acid-modified silicalite membranes were used. In particular, propylsulfonic (PS) acid sites and arenesulfonic (AS) acid sites were provided to the silicalite membranes by means of post-synthetic modifications. Due to higher acid density, AS silicalite membrane exhibited better catalytic activity; for example, at 450°C, the m-xylene conversion was more than 11% compared to the other membrane type. Besides, both catalytic membranes showed good structural stability after a long time of catalytic tests (120 h).

A catalytic H-MFI membrane, prepared by means of an ion-exchange treatment, was also used for m-xylene isomerization (Zhang et al., 2012). The authors obtained very high p-xylene selectivity (92.1%) and a reagent conversion of 6.5% at 270°C. These experimental data were due to the effective removal of p-xylene from the reaction medium. However, the conversion was lower than those reported in the literature due to the limited active sites on the disk-shaped membrane.

Zeolite membranes can be only used as catalytic contactor. In this case, the catalytic particles are dispersed in a very thin zeolite layer and this ensures good contact between reactants and the active site of the catalysts. In this way, the bypass problems present in the packed bed reactor and the pressure drop are reduced (Bernardo et al., 2008).

The carbon monoxide selective oxidation (Selox) was found to be the most effective method for removing the trace amount of CO from the hydrogen produced by reforming processes or available in petrochemical plants (e.g., in ethylene plants) (Avgouropoulos and Ioannides, 2003). In this case, the hydrogen could be used in other operations as ammonia synthesis, hydrogenations, and fuel cell applications (proton-exchange membrane fuel cells) (Song, 2002). It is well known that systems based on palladium, ruthenium, and gold are effective catalysts for this process. Most of the Selox catalysts were studied over fixed-bed or monolithic catalysts; the use of catalytic MRs was investigated only in a few papers where γ-alumina and nanostructured FAU zeolite membranes are used (Hasegawa et al., 2001; Hasegawa et al., 2002a; Hasegawa et al., 2002b; Sotowa et al., 2002). The Pt/γ-Al_2O_3 membrane presented a maximum CO conversion of ca. 50% with a 20% selectivity at 230°C (Hasegawa et al., 2002a). The other papers (Hasegawa et al., 2002a; Hasegawa et al., 2002b; Sotowa et al., 2002) did not report data in terms of CO conversion and selectivity. At the ITM laboratory (Bernardo et al., 2006; Bernardo et al., 2008), researchers deeply studied the Selox by using catalytic zeolite membranes (Pt/NaY). Particularly in this work, active catalytic Pt/NaY membranes were prepared from NaY membranes by means of ion exchange followed by air treatment. The membranes with a low concentration of defects (membranes Q and H) exhibited complete CO removal with a selectivity around 50%. The pressure effect is negligible for the less permeable membranes (H and Q), while it is positive for more permeable membrane (U). The authors explained the influence of pressure on the CO conversion considering the effect of two different paths on the zeolite membrane: into the zeolite channels and through the defects. In the first, considering its very small size, the probability of interaction between the catalyst and molecules is very high. On the contrary, the defects have a large volume owing to the pore size of the support, and the interaction of the reagents with the catalyst depends on the pressure.

A comparison of these data with the literature is given in Fig. 5.9. The Pt/NaY catalytic membrane prepared at the ITM expands the CO conversion–O_2 selectivity region of the traditional reactors. In addition, these results confirm the good potentiality of catalytic membranes for deep purification of H_2-rich streams.

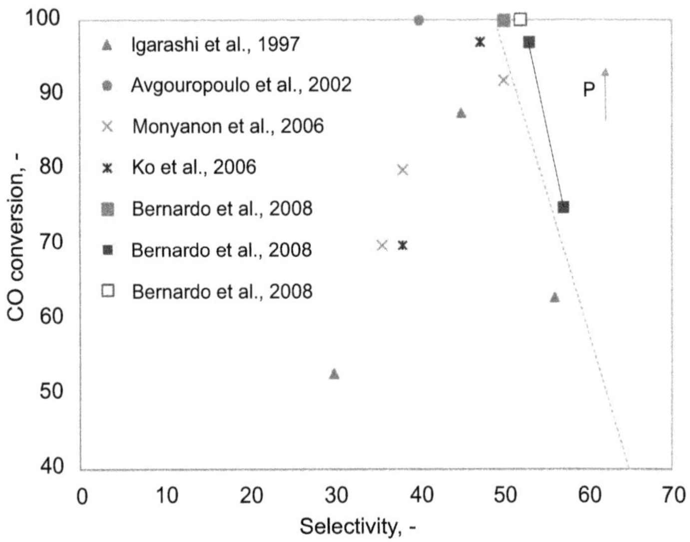

Figure 5.9 CO conversion versus selectivity at 200°C. H membrane: green square; Q membrane: white square; U membrane: brown square.

5.7 Future Perspective and Conclusion

In the last 20 years, significant progress has been achieved in the synthesis of thin nanostructured zeolite layer, with different topology, high flux, and good separation performance. The variety of zeolite membranes with a definite nanostructure, shape, and separation performance makes them appropriate for solving different problems. In particular, they have been extensively studied for enhancing conversion and selectivity of different catalytic reactions and gas and liquid separations. Recently, they have also been used in the water desalination process by RO and MD. However, problems related to reproducibility during the upscaling of synthesis and long-term stability hindered their industrial application. In fact, the use of zeolite membranes on a large scale is confined to NaA and T-type for the dehydration of different organic solvents. Therefore, further improvements are required based on the preparation of reproducible defect-free zeolite membranes with a larger area and suitable configuration (capillary or multichannel supports). Another important aspect to consider is the reduction in their manufacturing cost to facilitate their introduction in the industry.

Abbreviations

AS	arenesulfonic
DME	dimethyl ester
EDX	energy dispersive X-ray
FAU	faujasite
GHSV	gas hourly space velocity
LTA	Lynde Type A
MD	membrane distillation
MFI	mordenite framework inverted
MR	membrane reactor
NaA	sodium A
NaY	sodium Y
PD	propane dehydrogenation
PDM	pressure drop method
PS	propylsulfonic
PV	pervaporation
RO	reverse osmosis
SAPO-34	silicoaluminum phosphate-34
SDA	structure-directing agent
SELOX	selective oxidation
SEM	scanning electron microscopy

SMR	steam methane reforming
TR	traditional reactor
WGS	water gas shift
YSZ	yttria-stabilized zirconia
ZM	zeolite membrane
ZSM5	Zeolite Socony Mobil–5

References

Algieri, C., Golemme, G., Kallus, S., and Ramsay, J. D. F. (2001). Preparation of thin supported MFI membranes by in situ nucleation and secondary growth, *Microporous Mesoporous Mater.,* **54**, pp. 257–268.

Algieri, C., Bernardo, P., Golemme, G., Barbieri, G., and Drioli, E. (2003). Permeation properties of a thin silicalite-1 (MFI) membrane, *J. Membr. Sci.*, **222**, pp. 181–190.

Algieri, C., Bernardo, P., Barbieri, G., and Drioli, E. (2009). A novel seeding procedure for preparing tubular NaY zeolite membranes, *Microporous Mesoporous Mater.,* **119**, pp. 129–136.

Algieri, C., Bernardo, P., Barbieri, G., and Drioli, E. (2011). Procedimento per la produzione di membrane zeolitiche su supporti porosi tubolari, RM 0001388750.

Avgouropoulos, G., Ioannides, T., Papadopoulou, C., Batista, J., Hocevar, S., and Matralis, H. K. (2002). A comparative study of Pt/Al_2O_3, Au/Fe_2O_3 and CuO-CeO_2 catalysts for the selective oxidation of carbon monoxide in excess hydrogen, *Catal. Today*, **75**, pp. 157–167.

Avgouropoulos, G. and Ioannides, T. (2003). Selective CO oxidation over CuO-CeO_2 catalysts prepared via the urea-nitrate combustion method, *Appl. Catal. A*, **224** (1), pp. 155–167.

Bernal, M. P., Xomeritakis, G., and Tsapatsis, M. (2001). Tubular MFI zeolite membranes made by secondary (seeded) growth, *Catal. Today*, **67** (1–3), pp. 101–107.

Bernardo, P., Algieri, C., Barbieri, G., and Drioli, E. (2006). Catalytic (Pt-Y) membranes for the purification of H_2-rich streams, *Catal. Today*, **118**, pp. 90–97.

Bernardo, P., Algieri, C., Barbieri, G., and Drioli, E. (2008). Hydrogen purification from carbon monoxide by means of selective oxidation using zeolite catalytic membranes, *Sep. Purif. Technol.*, **62**, pp. 629–635.

Breck, D. W. (1974) *Zeolite Molecular Sieves: Structure, Chemistry, and Use* (John Wiley & Sons, Inc., New York, USA).

Bremere, I., Kennedy, M., Stikker, A., and Schipper, J. (2001). How water scarcity will affect the growth in the desalination market in the coming 25 years, *Desalination*, **138**, pp. 7–9.

Caro, J., Noack, M., Kolsch, P., and Schafer, R. (2000). Zeolite membranes-state of their development and perspective, *Microporous Mesoporous Mater.*, **38**, pp. 3–24.

Caro, J. and Noack, M. (2008). Zeolite membranes: Recent development and progress, *Microporous Mesoporous Mater.*, **115**, pp. 215–233.

Cheng, Z. L., Chao, Z. S., and Wan, H. L. (2002). Research of A type zeolite as well as zeolite membrane by microwave heating, *Chin. J. Inorg. Chem.*, **18**, pp. 528–532.

Choi, J., Jeong, H. K., Snyder, M. A., Stoeger, J. A., Masel, R. I., and Tsapatsis, M. (2009). Grain boundary defect elimination in a zeolite membrane by rapid thermal processing, *Science*, **325**, pp. 590–593.

Choi, S. W., Jones, C. W., Nair, S., Sholl, D. S., Moore, J. S., Liu, Y. J., Dixit, R. S., and Pendergast, J. G. (2015). Material properties and operating configurations of membrane reactors for propane dehydrogenation, *AIChE J.*, **61**, pp. 922–935.

Ciavarella, P., Casanave, D., Moueddeb, H., Miachon, S., Fiaty, K., and Dalmon, J. A. (2001). Isobutane dehydrogenation in a membrane reactor influence of the operating conditions on the performance, *Catal. Today*, **67**, pp. 177–184.

Criscuoli, A. and Drioli, E. (1999). Energetic and exergetic analysis of an integrated membrane desalination system, *Desalination*, **124**, pp. 243–249.

Cundy, C. S. (1998). Microwave techniques in the synthesis and modification of zeolite catalysts. A review, *Collect. Czech. Chem. Commun.*, **63**, pp. 1699–1723.

Davis, M. E. (2014). Zeolites from a materials chemistry perspective, *Chem. Mater.*, **26**, pp. 239–245.

de la Iglesia, O., Irusta, S., Mallada, R., Menendez, M., Coronas, J., and Santamarıa, J. (2006). Preparation and characterization of two-layered mordenite-ZSM-5 bi-functional membranes, *Microporous Mesoporous Mater.*, **93**, pp. 318–324.

den Exter, M. J., van Bekkum, H., Rijn, C. J. M., Kapteijn, F., Moulijn, J. A., Schellevis, H., and Beenakker, C. I. N. (1997). Stability of oriented silicalite-1 films in view of zeolite membrane preparation, *Zeolites*, **19**, pp. 13–20.

Dong, J. H., Lin, Y. S., Hu, M. Z. C., Peascoe, R. A., and Payzant, E. A. (2000). Template removal-associated microstructural development of porous-ceramic-supported MFI zeolite membranes, *Microporous Mesoporous Mater.*, **34** (3), pp. 241–253.

Dong, X., Wang, H., Rui, Z., and Lin, Y. S. (2015). Tubular dual-layer MFI zeolite membrane reactor for hydrogen production via the WGS reaction: Experimental and modeling studies, *Chem. Eng. J.*, **268**, pp. 219–229.

Drobek, M., Motuzas, J., van Loon, M., Dirrix, R. W. J., and Terpstra, R. A. (2012a). Coupling microwave assisted and classical heating methods for scaling-up MFI zeolite membrane synthesis, *J. Membr. Sci.*, **401–402**, pp. 144–151.

Drobek, M., Yacou, C., Motuzas, J., Julbe, A., Ding, L., and Diniz da Costa, J. C. (2012b). Long term pervaporation desalination of tubular MFI zeolite membranes, *J. Membr. Sci.*, **415–416**, pp. 816–823.

Duke, M. C., O'brien-Abraham, J., Milne, N., Zhu, B., Lin, J. Y. S., and da Costa, J. C. (2009). Seawater desalination performance of MFI type membranes made by secondary growth, *Sep. Purif. Technol.*, **68**, pp. 343–350.

Farahbakhsh, K., Svrcek, C., Guest, F. K., and Smith, D. W. (2004). A review of the impact of chemical pretreatment on low-pressure water treatment membranes, *J. Environ. Eng. Sci.*, **3**, pp. 237–253.

Fedosov, D. A., Smirnov, A. V., Shkirskiy, V. V., Voskoboynikov, T. V., and Ivanova, I. I. (2015). Methanol dehydration in NaA zeolite membrane reactor, *J. Membr. Sci.*, **486**, pp. 189–194.

Garofalo, A., Donato, L., Drioli, E., Criscuoli, A., Carnevale, M. C., Alharbi, O., Aljlil, S. A., and Algieri, C. (2014). Supported MFI zeolite membranes by cross flow filtration for water treatment, *Sep. Purif. Technol.*, **137**, pp. 28–35.

Garofalo, A., Carnevale, M. C., Donato, L., Drioli, E., Alharbi, O., Aljlil, S. A., Criscuoli, A., and Algieri, C. (2016). Scale-up of MFI zeolite membranes for desalination by vacuum membrane distillation, *Desalination*, **397**, pp. 205–212.

Gascon, J., Kapteijn, F., Zornoza, B., Sebastian, V., Casaedo, C., and Coronas, J. (2012). Practical approach to zeolitic membranes and coatings: State of the art, opportunities, barriers and future perspectives, *Chem. Mater.*, **24**, pp. 2829–2844.

Gues, E. R. and van Bekkum, H. (1995). Calcination of large MFI-type single crystals, Part 2: Crack formation and thermomechanical properties in view of the preparation of zeolite membranes, *Zeolites*, **15**, pp. 333–341.

Hasegawa, Y., Kusakabe, K., and Morooka, S. (2001). Selective oxidation of carbon monoxide in hydrogen-rich mixtures by permeation through a platinum-loaded Y-type zeolite membrane, *J. Membr. Sci.*, **190**, pp. 1–8.

Hasegawa, Y., Sotowa, K. I., Kusakabe, K., and Morooka, S. (2002a). The influence of feed composition on CO oxidation using zeolite membranes loaded with metal catalysts, *Microporous Mesoporous Mater.*, **53**, pp. 37–43.

Hasegawa, Y., Ueda, A., Kusakabe, K., and Morooka, S. (2002b). Oxidation of CO in hydrogen-rich gas using a novel membrane combined with a microporous SiO_2 layer and a metal-loaded Al_2O_3 layer, *Appl. Catal. A: General*, **225** (1–2), pp. 109–115.

Hedlund, J., Noack, M., Kolsch, P., Creaser, D., Caro, J., and Sterte, J. (1999). ZSM-5 membranes synthesized without organic templates using a seeding technique, *J. Membr. Sci.*, **159**, pp. 263–273.

Hoof, V. V., Dotremont, C., and Buekenhoudt, A. (2006). Performance of Mitsui NaA type zeolite membranes for the dehydration of organic solvents in comparison with commercial polymeric pervaporation membranes, *Sep. Purif. Technol.*, **48**, pp. 304–309.

Hsu, S. T., Cheng, K. T., and Chiou, J. S. (2002). Seawater desalination by direct contact membrane distillation, *Desalination*, **143**, pp. 279–287.

Huang, A., Lin, Y. S., and Yang, W. (2004). Synthesis and properties of A-type zeolite membranes by secondary growth method with vacuum seeding, *J. Membr. Sci.*, **245**, pp. 41–51.

Huang, A. and Yang, W. (2007). Hydrothermal synthesis of NaA zeolite membrane together with microwave heating and conventional heating, *Mater. Lett.*, **61**, pp. 5129–5132.

Huang, A. and Yang, W. (2008). Enhancement of NaA zeolite membrane properties through organic cation addition, *Sep. Purif. Technol.*, **61**, pp. 175–181.

Hughes, Z. E. and Gale, J. D. (2010). A computational investigation of the properties of a reverse osmosis membrane, *J. Mater. Chem.*, **20**, pp. 7788–7799.

Igarashi, H., Uchida, H., Suzuki, M., Sasaki, Y., and Watanabe, M. (1997). Removal of carbon monoxide from hydrogen-rich fuels by selective oxidation over platinum catalyst supported on zeolite, *Appl. Catal. A: Gen.*, **159**, pp. 159–169.

Ismail, A. F., Khulbe, K., and Matsuura, T. (2015). *Gas Separation Membranes: Polymeric and Inorganic* (Springer International Publishing, Switzerland).

Itabashi, K., Kamimura, Y., Iyoki, K., Shimojima, A., and Okubo, T. (2012). A working hypothesis for broadening framework types of zeolites in seed-assisted synthesis without organic structure-directing agent, *J. Am. Chem. Soc.*, **134**, pp. 11542–11549.

Iyoki, K., Itabashi, K., and Okubo, T. (2014). Progress in seed-assisted synthesis of zeolites without using organic structure-directing agents, *Microporous Mesoporous Mater.*, **189**, pp. 22–30.

Kaduk, J. A. and Faber, J. (1995). Crystal structure of zeolite Y as a function of ion exchange, *Rigaku J.*, **12** (2), pp. 14–34.

Kazemimoghadam, M. and Mohammadi, T. (2007). Synthesis of MFI zeolite membranes for water desalination, *Desalination*, **206**, pp. 547–553.

Kim, S. J., Liu, Y., Moore, J. S., Dixit, R. S., Pendergast, J. G., Sholl, D., Jones, C. W., and Nair, S. (2016). Thin hydrogen-selective SAPO-34 zeolite membranes for enhanced conversion and selectivity in propane dehydrogenation membrane reactors, *Chem. Mater.*, **28**, pp. 4397–4402.

Ko, E. Y., Park, E. D., Seo, K. W., Lee, H. C., Lee, D., and Kim, S. (2006). Selective CO oxidation in the presence of hydrogen over supported Pt catalysts promoted with transition metals, *Korean J. Chem. Eng.*, **2**, pp. 183–187.

Kondo, M., Komoriki, M., Kita, H., and Okamoto, K. I. (1997). Tubular-type pervaporation module with zeolite NaA membrane, *J. Membr. Sci.*, **133**, pp. 133–141.

Kondo, M. and Kita, H. (2010). Permeation mechanism through zeolite NaA and T-type membranes for practical dehydration of organic solvents, *J. Membr. Sci.*, **361**, pp. 223–231.

Kong, C., Lu, J., Yang, J., and Wang, J. (2007). Catalytic dehydrogenation of ethylbenzene to styrene in a zeolite silicalite-1 membrane reactor, *J. Membr. Sci.*, **306**, pp. 29–35.

Kong, W., Dai, W., Li, N., Guan, N., and Xiang, S. (2009). A one-step route to SAPO-46 using H_3PO_3 containing gel and its application as the catalyst for methanol dehydration, *J. Mol. Catal. A: Chem.*, **308**, pp. 127–133.

Kosinov, N., Gascon, J., Kapteijn, F., and Hensen, E. J. M. (2016). Recent developments in zeolite membranes for gas separation, *J. Membr. Sci.*, **499**, pp. 65–79.

Kusakabe, K., Kuroda, T., and Mooroka, S. (1998). Separation of carbon dioxide from nitrogen using ion-exchanged faujasite-type zeolite membranes formed on porous support tubes, *J. Membr. Sci.*, **148**, pp. 13–23.

Li, L. X., Dong, J. H., Nenoff, T. M., and Lee, R. (2004a). Desalination by reverse osmosis using MFI zeolite membranes, *J. Membr. Sci.*, **243**, pp. 401–404.

Li, L. X., Dong, J. H., Nenoff, T. M., and Lee, R. (2004b). Reverse osmosis of ionic aqueous solutions on a MFI zeolite membrane, *Desalination*, **170**, pp. 309–316.

Li, L., Liu, N., McPherson, B., and Lee, R. (2007). Enhanced water permeation of reverse osmosis through MFI type zeolite membranes with high aluminum contents, *Ind. Eng. Chem. Res.*, **46**, pp. 1584–1589.

Li, X., Kita, H., Zhu, H., Zhang, Z., and Tanaka, K. (2009). Synthesis of long-term acid-stable zeolite membranes and their potential application to esterification reactions, *J. Membr. Sci.*, **339**, pp. 224–232.

Li, X., Yan, Y., and Wang, Z. (2010a). Continuity control of b-oriented MFI zeolite films by microwave synthesis, *Ind. Eng. Chem. Res.*, **49**, pp. 5933–5938.

Li, S., Carreon, M. A., Zhang, Y., Funke, H. H., Noble, R. D., and Falconer, J. L. (2010b). Scale-up of SAPO-34 membranes for CO_2/CH_4 separation, *J. Membr. Sci.*, **352**, pp. 7–13.

Liu, Y., Li, Y., Cai, R., and Yang, W. (2012). Suppression of twins in b-oriented MFI molecular sieve films under microwave irradiation, *Chem. Commun.*, **48**, pp. 6782–6784.

Lovallo, M. C. and Tsapatsis, M. (1996). Preferentially oriented submicron silicalite membranes, *AICHE J.*, **42**, pp. 3020–3029.

Lovallo, M. C., Gouzinis, A., and Tsapatsis, M. (1998). Synthesis and characterization of oriented MFl membranes prepared by secondary growth, *AICHE J.*, **44**, pp. 1903–1913.

Maghsoudi, H. (2016). Defects of zeolite membranes: Characterization, modification and post-treatment techniques, *Sep. Purif. Rev.*, **45**, pp. 169–192.

Maloncy, M. L., Maschmeyer, T., and Jansen, J. C. (2005). Technical and economical evaluation of a zeolite membrane based heptane hydroisomerization process, *Chem. Eng. J.*, **106**, pp. 187–195.

Meindersma, G. W., Guijt, C. M., and de Haan, A. B. (2006). Desalination and water recycling by air gap membrane distillation, *Desalination*, **187**, pp. 291–301.

Mintova, S. and Bein, T. (2001). Microporous films prepared by spin-coating stable colloidal suspensions of zeolites, *Adv. Mater.*, **13** (24), pp. 1880–1883.

Monyanon, S., Pongstabode, S., and Luengnaruemitchai, A. (2006). Catalytic activity of Pt–Au/CeO_2 catalyst for the preferential oxidation of CO in H_2-rich stream, *J. Power Sources*, **163**, pp. 547–554.

Oleksiak, M. D. and Rimer, J. D. (2014). Synthesis of zeolites in the absence of organic structure-directing agents: Factors governing crystal selection and polymorphism, *Rev. Chem. Eng.*, **30**, pp. 1–49.

Pera-Titus, M., Llorens, J., Cunill, F., Mallada, R., and Santamarıa, J. (2005). Preparation of zeolite NaA membranes on the inner side of tubular supports by means of a controlled seeding technique, *Catal. Today*, **104**, pp. 281–287.

Rangnekar, N., Mittal, N., Elyassi, B., Caro, J., and Tsapatsis, M. (2015). Zeolite membranes: A review and comparison with MOFs, *Chem. Soc. Rev.*, **44**, pp. 7128–7154.

Sano, T., Kiyozumi, Y., Kawamura, M., Mizukami, F., Takaya, H., Mouri, T., Inaoka, W., Watanabe, M., and Toyoda, K. (1991). Preparation and characterization of ZSM-5 zeolite film, *Zeolites*, **11**, pp. 842–845.

Sato, K., Sugimoto, K., and Nakane, T. (2008). Mass production of tubular NaY zeolite membranes for industrial purpose and their application to ethanol dehydration by vapor permeation, *J. Membr. Sci.*, **319**, pp. 244–255.

Semelsberger, T. A., Borup, R. L., and Greene, H. L. (2006). Dimethylether (DME) as an alternative fuel, *J. Power Sources*, **156**, pp. 497–511.

Song, C. (2002). Fuel processing for low-temperature and high-temperature fuel cells: Challenges and opportunities for sustainable development in the 21st century, *Catal. Today*, **77**(2), pp. 17–49.

Sotowa, K. I., Hasegawa, Y., Kusakabe, K., and Morooka, S. (2002). Enhancement of CO oxidation by use of H_2-selective membranes impregnated with noble-metal catalysts, *Int. J. Hydrog. Energy*, **27**, pp. 339–346.

Takata, Y., Tsuru, T., Yoshioka, T., and Asaeda, M. (2002). Gas permeation properties of MFI zeolite membranes prepared by the secondary growth of colloidal silicalite and application to the methylation of toluene, *Microporous Mesoporous Mater.*, **54**, pp. 257–268.

van den Bergh, J., Tihaya, A., and Kapteijn, F. (2010). High temperature permeation and separation characteristics of an all-silica DDR zeolite membrane, *Microporous Mesoporous Mater.*, **132**, pp. 137–147.

Wang, J., Qu, D., Tie, M., Ren, Peng, X., and Luan, Z. (2008). Effect of coagulation pretreatment on membrane distillation process for desalination of recirculating cooling water, *Sep. Purif. Technol.*, **64**, pp. 108–115.

Weh, K., Noack, M., Sieber, I., and Caro, J. (2002). Permeation of single gases and gas mixtures through faujasite type molecular sieve membranes, *Microporous Mesoporous Mater.*, **54** (1–2), pp. 27–36.

Weller, M. T. (1994). *Inorganic Materials* (Oxford University Press, Inc., New York, USA).

Xomeritakis, G., Lai, Z., and Tsapatsis, M. (2001). Separation of xylene isomer vapors with oriented MFI membranes made by seeded growth, *Ind. Eng. Chem. Res.*, **40** (2), pp. 544–552.

Xu, X. C., Yang, W. S., Liu, J., and Lin, L. W. (2000). Synthesis of a high-performance NaA zeolite membrane by microwave heating, *Adv. Mater.*, **12**, pp. 195–198.

Xu, X., Bao, Y., Song, C., Yang, W., Liu, J., and Lin, L. (2004). Microwave-assisted hydrothermal synthesis of hydroxy-sodalite zeolite membrane, *Microporous Mesoporous Mater.*, **75**, pp. 173–181.

Yeong, Y. F., Abdullah, A. Z., Ahmad, A. L., and Bhatia, S. (2010). Synthesis, characterization and reactive separation activity of acid-functionalized silicalite-1 catalytic membrane in m-xylene isomerization, *J. Membr. Sci.*, **360**, pp. 109–122.

Zhang, C., Hong, Z., Chen, J., Gu, X., Wanqin, J., and Xu, N. (2012). Catalytic MFI zeolite membranes supported on α-Al_2O_3 substrates for m-xylene isomerization, *J. Membr. Sci.*, **389**, pp. 451–458.

Zhu, B., Hong, Z., Milne, N., Doherty, C. M., Zou, L., Lin, Y. S., Hill, A. J., Gu, X., and Duke, M. (2014). Desalination of seawater ion complexes by MFI-type zeolite membranes: Temperature and long term stability, *J. Membr. Sci.*, **453**, pp. 126–135.

Chapter 6

Two-Dimensional Materials for Membranes

Annarosa Gugliuzza,[a] Antonio Politano,[b] and Enrico Drioli[a]
[a]*Institute on Membrane Technology of the National Research Council of Italy (ITM-CNR), Via P. Bucci, Cubo 17C, 87036 Rende (CS), Italy*
[b]*Fondazione Istituto Italiano di Tecnologia, Graphene Labs, Via Morego 30, 16163, Genova, Italy*
a.gugliuzza@itm.cnr.it

This chapter deals with two-dimensional (2D) materials and related potential in membrane science. Advanced methods for exfoliation, properties, and perspectives of 2D materials in advanced applicative fields are described. An excursus through the state of the art is proposed together with future prospects in advanced branches of membrane technology, including ion filtering, gas separation, and energy power and storage. The expectancy from using membranes of 2D materials is discussed in relation to intrinsic material features but also the need to overcome some current limitations for membrane operations, which are expected to provide key solutions for several urgent requirements from scientific communities and the society.

Functional Nanostructured Membranes
Edited by Enrico Drioli, Lidietta Giorno, and Annarosa Gugliuzza

ISBN 978-981-4774-79-6 (Hardcover), 978-1-351-13511-5 (eBook)
www.panstanford.com

6.1 Introduction

Two-dimensional (2D) materials are characterized by strong in-plane chemical bonds and weak interlayer bonds (Gupta et al., 2015). By means of micromechanical (Jakob, 2015; Lin et al., 2013) or chemical (Coleman, 2012; Coleman et al., 2011) exfoliation techniques, it is possible to isolate a few single freestanding nanosheets exhibiting unusual and different physicochemical properties with respect to their three-dimensional (3D) counterpart. The confinement of charge and heat transport into a 2D space is regarded as a key issue to enable novel phenomena and devices.

The progenitor of 2D materials is graphene, a single-layer sheet of graphite isolated by Novoselov et al. (2004). In graphene, each carbon atom is bonded with three neighboring carbon atoms by planar σ orbitals, which are separated by 120° with each other, while π orbitals are delocalized in the perpendicular direction with respect to the graphene sheet. Electrons in linearly dispersing π and π^* bands, forming the so-called Dirac cone (Kumar et al., 2016), behave as massless particles moving at quasi-relativistic group velocity ($c/300$ with c being the light velocity) (Novoselov et al., 2005; Qian et al., 2015). Furthermore, graphene is characterized by high stiffness (Booth et al., 2008) together with outstanding elastic properties (Lee et al., 2008). All these features make graphene particularly promising for applications in several fields (Ferrari et al., 2015).

Other 2D materials "beyond graphene" have currently engaged the scientific community for their huge application capabilities (Bhimanapati et al., 2015; Butler et al., 2013), often complementary with respect to the case of graphene (Castellanos-Gomez, 2016; Novoselov, 2007). Different classes of 2D materials "beyond graphene" have recently emerged: black phosphorus (Castellanos-Gomez, 2015), hexagonal boron nitride (Pakdel et al., 2014), transition-metal dichalcogenides (Mak et al., 2016), 2D carbides (Anasori et al., 2015), and 2D metal–organic framework (MOF) (Koitz et al., 2016).

The family of 2D materials is envisaged to bring innovation in the fields of nanoelectronics, nanophotonics, and nanomechanics (Ferrari et al., 2015). Moreover, a great commitment has been recently concerted on the manipulation of these featuring materials for potential applications in other technological areas, including membrane separation, catalysis, and energy storage. In particular, 2D materials can be used as membranes or nanofillers in polymeric membranes to engineer novel functional membranes (Gugliuzza et al., 2015; Gugliuzza et al., 2017). Indeed, 2D materials like nanosheets or membranes are predicted to work as ion-filter in water desalination and molecular sieves for gas separation, but also as sources for catalytic production of green energy. With this regard, it should be pointed out that the opening of uniformly sized pores or interlayer gaps in the sheets is necessary in order to enable selective molecular passage (Drioli et al., 2016). The ultrathin thickness together with a well-controlled distribution of pores serving as pathways represents critical issues for selective membrane operations (Perrotta et al., 2017). In addition, both electrical and thermal conduction are key factors in different membrane processes, including membrane distillation and catalytic membrane reactors. The design and massive fabrication of 2D membranes, which should work under realistic operating conditions, represent another open challenge (Wang et al., 2017). At the state of the art, a few contributions are reported in the literature on the design and fabrication of 2D material membranes for liquid and gas separation. Theoretical predictions are more numerous compared with pioneer experimental works. Frequently, systems studied by theory are characterized by nonrealistic structure–property relationships, as a consequence of the complexity of this topic for both experimentalists and theoreticians.

Herein, an overview on the types of membranes based on 2D materials, including both synthesis approaches and applications in membrane science and technology, is presented.

6.2 Methods for Exfoliation of 2D Materials

Devices based on 2D materials can be fabricated by top-down or bottom-up approaches. In some cases, the two approaches can be followed with similar final results, as for the case of graphene nanoribbons (Cai et al., 2010; Narita et al., 2014; Tour, 2013) or graphene quantum dots (Shen et al., 2012) (Fig. 6.1).

The top-down approach is suitable for large arrays of devices on a chip by means of common patterning tools (Tour, 2013).

Nevertheless, the edges of the flakes are compromised in the nanofabrication process, which has not yet atomic precision. The bottom-up approach allows a good control of edges, but it is often impossible to locate the structures with precision. Moreover, the achievement of sufficiently long device structures with the bottom-up approach is complicated.

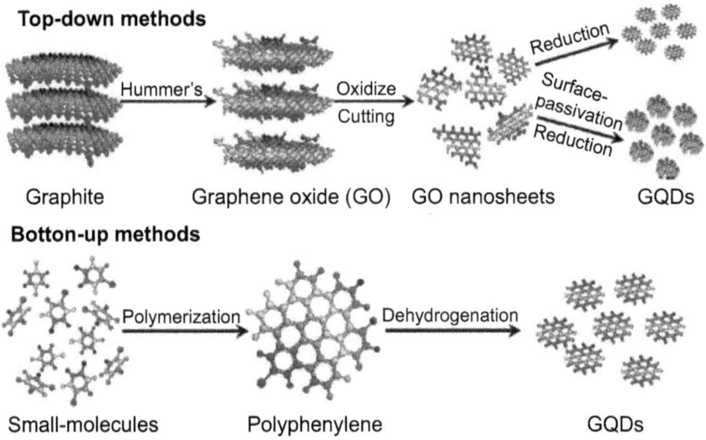

Figure 6.1 Schematic diagram of the top-down and bottom-up methods for synthesizing graphene quantum dots (GQDs). Republished with permission of Royal Society of Chemistry (Great Britain), from Shen et al. (2012); permission conveyed through Copyright Clearance Center, Inc.

The liquid-phase exfoliation procedure to isolate sheets of 2D materials usually comprises three stages: (1) dispersion of the bulk material in a liquid medium, (2) exfoliation via ultrasonication, and (3) purification.

The dispersion of layered materials in common solvents is essential for applications ranging from electronics to energy storage (Bonaccorso et al., 2014). Shear forces and cavitation produced due to the propagation of high-amplitude sonication waves act on the surface of bulk materials inducing its exfoliation. To stabilize the isolated 2D flakes, minimization of the interfacial tension with the liquid medium is crucial.

By combination with suspensions of polymer solutions and/or other nanomaterials, it is possible to achieve hybrid dispersions or composites, which can be cast into films (Coleman et al., 2011).

Nevertheless, processing of 2D materials is currently in its infancy. Reference properties are currently established by isolating flakes with micromechanical cleavage and by comparative studies on films grown with chemical vapor deposition or molecular beam epitaxy, which are more appropriate for large-scale production. However, in order to exploit films of 2D materials in technological processes, the growth conditions yielding high-quality films should be recognized and optimized. Moreover, suitable n- and p-type dopants should be identified (Fiori et al., 2014).

6.3 The Graphene Era: Status and Prospect

Considering its potential for many applications, including capacitors, displays, data storage, thermal management, solar cells, imaging, and sensing, graphene is estimated to cover a wide size of the global research market with a growth rate of $1.5m in 2015 and $2.1b in 2025 (Gagliardi et al., 2011).

Membrane processes could represent proficient industrial applications of graphene. However, graphene is completely impermeable to molecular compounds, due to the localization of electrons within aromatic rings that form a perfect barrier to every molecule, including the smallest ones. Its potential use as a membrane implicates, therefore, the drilling of the nanosheet in order to generate sub-nanometer pores through which the smallest molecules can pass whereas the biggest ones are blocked (Fig. 6.2). Hence, the enhancement of the permeability/selectivity trade-off to achieve ultrahigh flux and outstanding rejection remains a major challenge.

Various approaches, including electron beam and energetic ions bombardment (Celebi et al., 2014), ultraviolet-induced oxidative etching (Koenig et al., 2012), and oxygen–plasma etching (Surwade et al., 2015), as well as molecular building blocks combined with nanoimprinting lithography, doping, and chemical vapor deposition (Chandra et al., 2012; Srinivas et al., 2012), have been proposed to produce sub-nanoporous graphene sheets (Yuan et al., 2014). However, whenever sub-nanometer pores are formed, some troublesome events can take place, including shrinkage and expansion of graphene nanopores and self-healing. Specifically, when using electron beam at temperature in the 400–1200°C range, pores of 10 nanometer can be completely closed. This process could be stopped if the electron beam is properly blocked (Lu et al., 2012). On the other hand, interactions with metal impurities catalyze self-healing (Zan et al., 2012). Recently, Robertson et al. (2015) have presented the first image with atomic resolution of sub-nanometer pores in graphene by keeping graphene at ~500°C in order to prevent self-healing of nanopores with mobile carbon species. The electron-beam irradiation using a transmission electron microscope (TEM) remains, however, too expensive for mass production. The use of scanning electron microscope (SEM) with the assistance of nitrogen ions and an electron beam less than 10 keV has been proposed by Fox et al. (2011) to generate nanopore patterns in graphene layers.

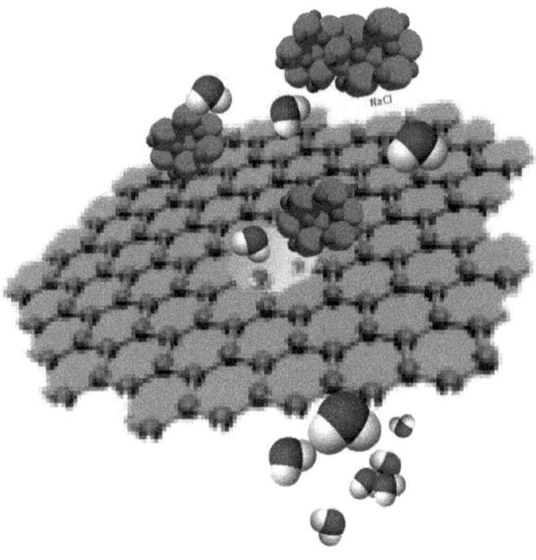

Figure 6.2 Representative scheme for selective transport through graphene nanopores.

In addition, an electrical pulse fabrication method based on a simple fluidic cell has been proposed for the production of single nanopores down to sub-nanometer sizes. Short voltage pulses have been applied across the membrane in electrolyte solution to create or enlarge nanopores in order to remove the minimum amount of material and achieve a higher control of pore size (Kuan et al., 2015).

Induced oxidative etching approaches, commonly used for preparing polymeric membranes, together with bombardment with gallium ions, have also been used for creating pores in graphene sheets (O'Hern et al., 2014). Similarly, Koenig et al. (2012) exposed graphene, mechanically exfoliated and pressurized with H_2, to ultraviolet light under ambient conditions to create nanopores.

Moreover, block copolymers have been proposed as building blocks for generating pores in graphene according to lithography principles. A layer of porous block copolymer has been annealed on a graphene sheet and, then, covered with an in situ formed SiO_x-based nanopore hard mask. Therefore, O_2–plasma etching has been used to drill the graphene and obtain regular pores through the sheet once the SiO_x mask is lifted (Bai et al., 2010; Kim et al., 2010).

Chemical activation by KOH and combined thermal shock or microwave irradiation have also been employed to produce GO with a specific surface area (SSA) up to 1900 m^2/g and pore volume up to 1.65 cm^3/g (Srinivas et al., 2012).

A major concern for the most part of these manufacturing techniques is the difficulty to control pore size and pore distribution at the atomic scale, causing serious restriction for any future applications in membrane science, due to the loss of selectivity. Recently, thermal activation of aerogel graphene has been used to get pores in graphene sheets in a controlled and scalable way (Maiti et al., 2015). The reduction of GO and the exploitation of different thermal resistance of two amorphous and disordered domains and highly crystalline domains have been suggested as a practical route for the mass production of nanoporous graphene with uniform pore size.

6.4 2D Materials "Beyond Graphene"

Despite the huge application capabilities of graphene, the absence of a band gap makes it unsuitable for different applications. The energy gap between the highest occupied and lowest unoccupied electronic levels represents a crucial parameter in the determination of the electronic and optical properties of a material (Novoselov, 2007). In applications for which a switching of the conductivity is required, a finite band gap prevents undesired effects, such as the occurrence of large dark currents, usually observed in graphene-based photodetectors (Koppens et al., 2014). In fact, field-effect transistors necessitate a large value of the ON/OFF ratio (Zhou et al., 2015), which is defined as the ratio of currents in the on and off states of the nanodevice. Furthermore, applications requiring flexibility would benefit from a reduced value of Young's modulus compared to that of graphene (~1 TPa) (Lee et al., 2008).

For these reasons, recently the vast community working on 2D materials is going "beyond graphene" (Bhimanapati et al., 2015; Chen et al., 2015). An abundant number of layered materials "beyond graphene" (semimetals, semiconductors, insulators) (Miro et al., 2014) exists in nature. The synthesis of novel 2D materials, such as transition-metal dichalcogenides (TMDC), 2D carbides, IV-VI compounds or atomically thin elemental materials (silicene, germanene, phosphorene, borophene) could open new pathways, since these materials show exotic physicochemical properties, which have never been accessed so far with 3D materials. In particular, the combination of flexibility and transparency with an existing electronic band gap enables novel device applications. One should consider that silicene, germanene, or borophene does not have a three-dimensional bulk counterpart. As a matter of fact, they can be grown only on top of single-crystal surfaces with a subsequent limitation of their potential applications.

For the sake of clarity, it is relevant to remind that one has to consider as 2D a material that is characterized by the following properties: (i) strong in-plane bonds, (ii) highly crystalline atomic planes, and (iii) van der Waals interactions between atomically thin planes in the normal direction (i.e., missing dangling bonds). The absence of dangling bonds between the planes and the weak van der Waals interaction enable the isolation of single-unit-cell-thickness flakes by mechanical or chemical exfoliation of a parental bulk crystal.

Two-dimensional materials are essential building blocks for novel optoelectronic devices (Fig. 6.3). Ultrasensitive, broadband, and ultrafast photodetectors have been fabricated with 2D materials (Echtermeyer et al., 2015; Fang et al., 2012) and their heterostructures (Viti et al., 2016) and integrated with existing silicon photonic technology (Koppens et al., 2014). Recently, graphene has also entered the fabrication of smartphone touch panels.

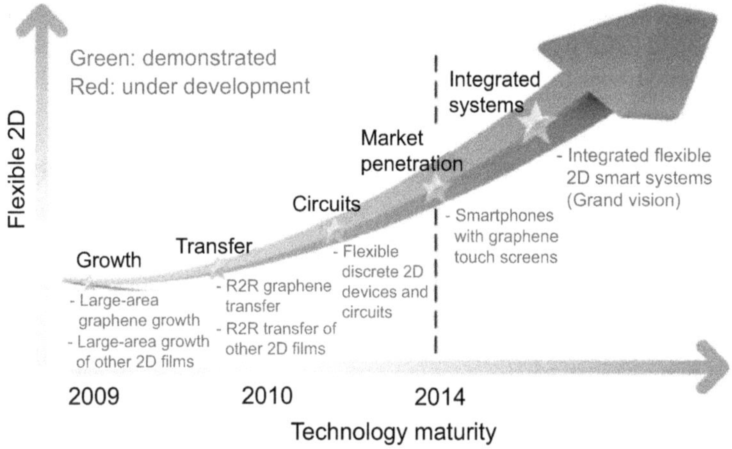

Figure 6.3 Perspective of technological maturity for flexible 2D nanotechnology. Reprinted with permission from Akinwande et al. (2014). Copyright 2014, Springer Nature.

Flexible electronics represent another promising prospect for nanoelectronics. Flexible transistors based on graphene (Akinwande et al., 2014), phosphorene (Zhu et al., 2016; Zhu et al., 2015), and TMDC (Chang et al., 2013) have performances already superior compared to organic or metal oxide thin-film transistors. The high mobility of charge carriers of phosphorene and its flexibility could enable flexible high-performance electronics (Zhu et al., 2016; Zhu et al., 2015).

The rise of novel 2D materials is also stimulating studies on mass transport even if a number of investigations on materials "beyond graphene" are still limited only to TMDC. Among the various TMDC nanosheets, thin layers of MoS_2 have been used for water desalination as well (Sun et al., 2013). The hydrophilicity of single-unit-cell MoS_2 is due to the S termination, which implies high affinity to water and certain gases, such as CO_2 and NO_2 (Zheng et al., 2016).

The chemical stability of MoS_2, arising from the S termination, makes membrane reusable. Nanochannels in MoS_2 membranes remain stable even under pressures as high as 1.0 MPa (Sun et al., 2013), thanks to high Young's modulus (370 GPa) (Bertolazzi et al., 2011), comparable to that of steel.

Concerning other 2D materials, it should be mentioned that Yi et al. (2014) found that thin layers of boron nitride are impermeable to oxygen atoms.

6.5 The Advent of 2D Materials in Membrane Technology

The availability of membranes is auspicious for achieving highly productive and efficient membrane operations, especially for ion filtering, gas separation, and energy storage. Despite preliminary studies suggesting these 2D materials as promising precursors of membranes, some concerns must be considered. Scalable approaches to produce nanosized and uniform pores to provide high permeability–selectivity trade-off, reduced energy consumption together with controlled (bio)fouling and scaling represent matter on which there is a great claim for in depth evaluations.

6.5.1 Ion Filtering

Ion filtering represents one of the most promising applications of graphene membranes (Fig. 6.4). Molecular dynamics simulations have showed that water transport through a porous graphene membrane is more efficient in comparison with the case of carbon nanotubes and other polymeric membranes (Celebi et al., 2014; Cohen-Tanugi et al., 2012). The salt rejection–water permeability trade-off has been found to be ~100 for hydrogenated and hydroxylated graphene. Advantageously, the atomic thickness requires less pressure and, accordingly, less energy to push water through it.

Despite that the fabrication of porous graphene with uniformly distributed nanosized pores is still challenging, porous graphene is nowadays proposed for the removal of ions from water by reverse osmosis (O'Hern et al., 2015; Surwade et al., 2015). Nanopores, implanted in graphene membranes through ion bombardment and oxidative etching, have a crucial role in salt rejection (O'Hern et al., 2014). In particular, short oxidation times make the pores cation-selective, due to electrostatic repulsions from negatively charged functional groups terminating the pore edges. Conversely, a longer oxidation affects the capability to reject salts against organic compounds at higher molecular weight, resulting in steric size exclusion. This corroborates the difficulty to get uniformly controlled and suitably sized pores in graphene without losing ion selectivity. Nevertheless, the use of porous graphene has been demonstrated to be rather efficient in electrodialysis under electrical field (He et al., 2013; Sint et al., 2008; Zhao et al., 2013). Rollings et al. (2016) have recently demonstrated that graphene nanopores act as voltage-activated channels for removing monovalent cations such as K^+ over Cl^- with a selectivity ratio over 100 and a monovalent cations mobility five times higher than bivalent ones.

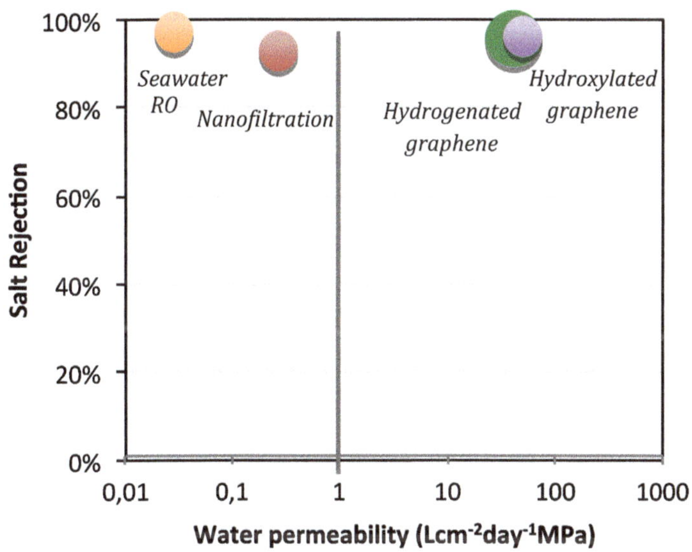

Figure 6.4 Salt rejection versus water permeability for traditional polymeric and graphene-based membranes.

Stacked graphene oxide layer membranes have also been proposed to let water molecules pass through nanochannels between each graphene oxide layer, while hydrated salt ions are blocked (You et al., 2016). In this regard, a combined experimental and molecular dynamics simulation study has envisaged in the length of nanochannels the key issue to regulate the salt selectivity through-channel as a concentration-gradient-driven diffusion is applied to separate water from ions (Sun et al., 2016). Instead, the role of the applied pressure has been recognized as a negative factor for filtrations, because of poor salt rejections. On the other hand, thicker graphene oxide lamellar membranes have been identified as a promising solution for desalination processes, according to forward diffusion-driven osmosis (Fig. 6.5).

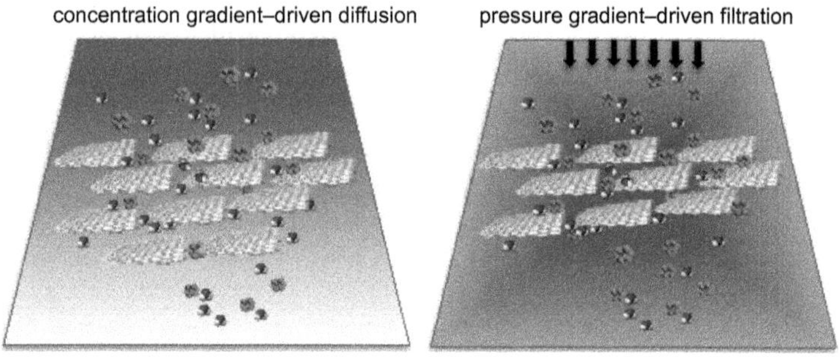

Figure 6.5 Compared ion filtering through graphene membranes: (a) concentration gradient–driven diffusion and (b) pressure gradient–driven filtration.

Realistically, the bottleneck still remaining is the univocal identification of structure–transport relationships, which regulate a separation mechanism rather than another (Fig. 6.6). Only a few attempts have been made to investigate structural and chemical effects on the diffusion of differently sized molecules through modified graphene.

Sun et al. (2014b) have recently investigated the possible mechanisms involving ion transport through GO membranes (Fig. 6.6). It has been proved that the formation of nanocapillary interlayer spacing between GO sheets and the establishment of non-covalent π interactions enable guest molecules to be accommodated or quickly pass (Sun et al., 2012). The presence of hydrophilic groups favors an increase in the interlayer stacking up to 1.4 nm (Raidongia et al., 2012).

Depending on the interlayer spacing, small solutes can be separated from species with larger hydrated radii according to size sieving mechanisms. However, hydroxylated chemical moieties locally change the hybridization of carbon atoms from sp^2 to sp^3, thus modifying the interaction between guest molecules and matrix. Water is a suitable medium where ions can be dissolved and can migrate through. An interlayer hexagonal ice bilayer (Boukhvalov et al., 2013) has been envisaged to form and melt at the sheet edges, allowing monovalent cations and strong acids or bases to diffuse more quickly than bivalent cations or weak acids and bases. On the other hand, sp^2 regions within sp^3 systems are expected to promote selective π–π^* interactions (Ma et al., 1997; Mahadevi et al., 2012) between transition metals and functional groups attached to GO sheets, thus conferring to the membrane a real ion-selectivity character. The predominance of regions with sp^2 hybridization allows, however, the membrane to discriminate among cations without *d* electrons. More recently, a density functional theory (DFT) study has been proposed to quantify the interactions between alkali and alkaline earth cations with counter Cl^- anions and nanosized and microsized graphene membranes within a range of 10 to 40°C (Sun et al., 2014b). It has been confirmed that cations without *d* electrons interact with the sp^2 clusters via cation–π interactions, while transition-metal ions coordinate to sp^3-configured sites of the membrane.

Graphene oxide has also been studied as a hydrophilic filler in PVDF membrane for desalination via direct contact membrane distillation (Bhadra et al., 2016). Graphene oxide flakes immobilized in the polymer matrix have produced fluxes up to 97 kg/m^2h at 80°C, suggesting mixed mechanisms, including selective sorption, nanocapillary effect, and reduced temperature polarization. In this regard, it is relevant to observe that the direct contact membrane distillation is a relatively standardized membrane operation used to remove salts from water by using membranes as physical interfaces between two aqueous solutions. Briefly, the mechanism controlling the selective transport includes three steps: the water is (a) evaporated on the feed side; (b) pushed through the membrane pores under a difference of temperature, osmotic concentration, or vacuum; and (c) condensed on the permeate side, while the salt is retained (Fig. 6.7). Thus, the membrane for this kind of operation would have some ideal features, including symmetric structure with very narrow pore distribution (less than 0.1 to 0.2 μm), excellent and stable hydrophobic properties, super porosity, outstanding mechanical strength, excellent capability to prevent polarization gradient in membrane and scaling, but also minimized thermal conductivity.

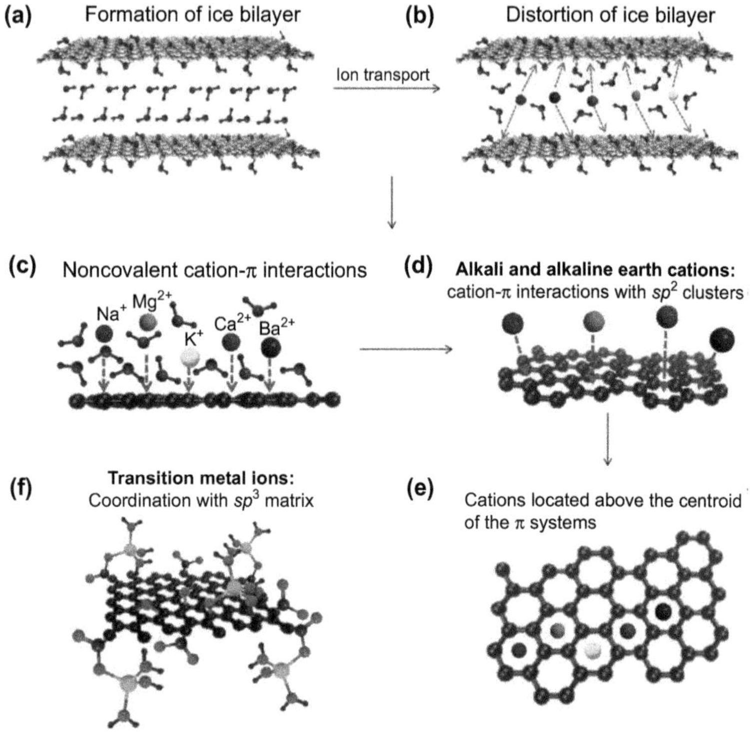

Figure 6.6 Proposed mechanisms of transport through GO laminar membranes: (a) ice bilayer within interlayered GO flakes; (b) distorted structure of ice bilayer due to water–cations interactions; (c,d,e) non-covalent cation–π interactions of alkali and alkaline earth cations with sp^2 clusters of GO sheet; (f) coordination of transition-metal ions with sp^3 matrix of GO sheets. Reprinted with permission from Sun et al., (2014b). Copyright 2014, American Chemical Society.

In this respect, it is well known that the thermal conductivity of graphene is ~5000 W/mK at 300 K. This may be envisaged as a strong limitation for thermally driven membrane distillation operations due to the heat lost by conduction. However, the implementation of highly productive and thermally efficient membrane desalination is mandatory for upscaling. For this reason, other classes of atomically thick materials, including TMDC and layered metal oxides, with competitive thermal conductivity are currently considered an alternative to graphene. The possibility of using 2D materials beyond graphene, with thermal conductivity less than graphene and, in some cases, traditional polymers, opens new scenarios for desalination membrane processes (Table 6.1).

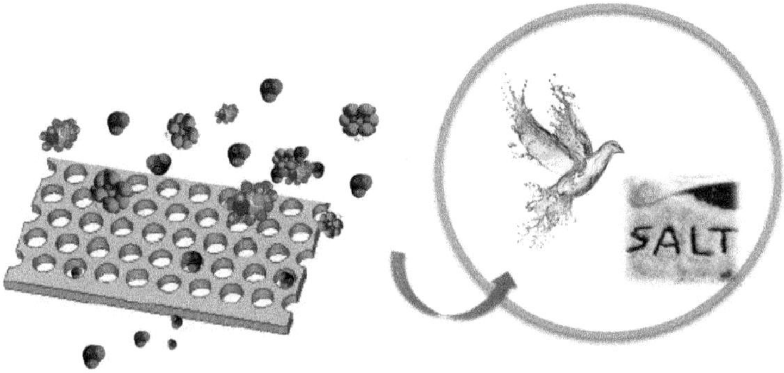

Figure 6.7 Scheme of ideal membrane for water desalination.

Table 6.1 Thermal conductivity of 2D materials, as compared with polymers used in membrane distillation

Material	Thermal conductivity at $T = 300$ K (W/m K)	Reference
Monolayer graphene	~5000	Balandin et al. (2008)
Monolayer SnSe	0.46–0.68 (anisotropic)	Zhao et al. (2014)
Sb_2Te_3	0.23–0.50	Chen et al. (2010)
Bi_2Se_3	0.4 (few layers)	Jana et al. (2013)
MoS_2	108 (monolayer)	Li et al. (2013)
WSe_2	0.05 (monolayer) 1.5 (bulk)	Chiritescu et al. (2007)
Phosphorene	Zigzag direction: 110 armchair direction: 36	Jain et al. (2015)
PVDF	0.19	
Polytetrafluoroethylene	0.24–0.35	
Polypropylene	0.10–0.22	
Cellulose acetate	0.16–0.36	

Liu et al. (2016) have illustrated in a recent review the potential of 2D materials in advanced membrane separations. They refer to chemically exfoliated monolayer of molybdenum disulfide (MoS_2) (Sun et al., 2013; Wang et al., 2015), tungsten disulfide (WS_2) (Sun et al., 2014a), and MXene (Ren et al., 2015) to be used as membranes. Membranes formed from MoS_2 are regarded as cheaper and higher performing than graphene for what concerns the capability to transport water and reject salts (Fig. 6.9). Molecular dynamics simulations have suggested an interesting capability of MoS_2 membranes—pores of 20–60 Å—to retain more than 88% of ions. Water flux is two to five orders of magnitude higher than traditional nanoporous membranes and 70% greater than porous graphene (Heiranian et al., 2015).

Water fluxes and ion rejection have been simulated through three pore edge types for MoS_2 (Mo only, S only, and Mo+S mixed) (Fig 6.8). These three types of pores have an accessible area comparable to graphene (mixed, A = 55.45 Å^2; Mo only, A = 56.42 Å^2; S only, A = 57.38 Å^2; and graphene, A = 59.67 Å^2). However, the water flux through the mixed pore is intermediary between Mo-only and S-only nanopores. With respect to graphene, the highest water flux through MoS_2 nanopore is due to smaller applied pressure. Flat MoS_2 membranes have also been produced with 3 nm pores, showing water flux three to five times higher than graphene oxide membranes and exhibiting a half-life of 1 week, due to the absence of hydrophilic moieties (Sun et al., 2013). Tunable pores in MoS_2 monolayer have been simulated, and mechanical stretching (up to 12%) has been envisaged as an "open–closed" pore switcher. As a result, the passage of water can be stimulated up to two orders of magnitude higher than the commercial RO membranes (Li et al., 2016). It is found that both steric and electrostatic effects provide to the membrane a different filtering capability toward monovalent ions. Sodium is solvated by water molecule forming a shell much tighter than chloride ions. This theoretical study demonstrates that the free energy for extracting bounded water molecules is 9.90 kJ/mol against the 4.36 kJ/mol necessary for hydrated chloride anions. This implies a water exchange time of 27.22 ± 0.26 ps for Na^+, while 8.58 ± 0.03 ps for Cl^-. Layers of $Ti_3C_2T_x$, which is part of the MXene family, have also been proposed for monovalent cations. Due its hydrophilic nature, water passage through the interspaces is facilitated with a flux of 37.4 L/h m^2 bar (Ren et al., 2015).

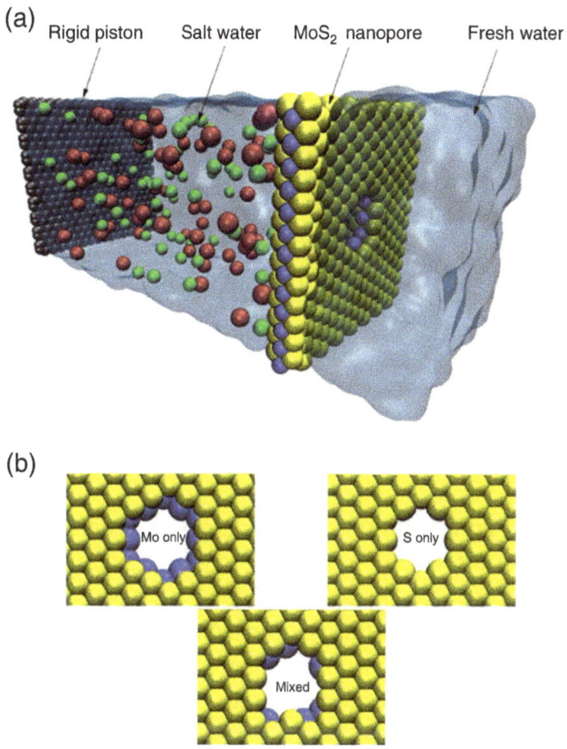

Figure 6.8 Molecular dynamics simulation of different MoS_2 pore architectures. (a) Schematic of the simulation box consisting of a MoS_2 sheet (molybdenum in blue and sulfur in yellow), water (transparent blue), ions (red and green), and a graphene sheet (gray). (b) Left: Mo-only pore type. Right: S-only pore type. Bottom: mixed pore type. Reprinted from Heiranian et al. (2015) under Creative Commons Attribution Licence. Copyright 2015, Nature Publishing Group.

Despite MD processes requiring the use of materials at low thermal conductivity, there are many other membrane operations that need, on the contrary, membranes with specific area, thickness, and high thermal conductivity. This concerns membrane contactors used as membrane condensers for heat recovery (Yan et al., 2015). It regards heat exchanger to recover the latent heat of existing vapor from the desorber, thus resulting in higher energy saving. Generally, membranes with high heat transfer coefficient are required and 2D materials such as graphene could provide new propulsion to the design of new advanced membrane condenser and dryers (Drioli et al., 2014; Macedonio et al., 2013).

6.5.2 Gas Separation

As though ion filtering, graphene and 2D materials beyond graphene have been envisaged to work as membranes for gas separation as well (Tsetseris et al., 2014). However, pristine graphene is regarded as the thinnest barrier against gases due to weak binding energy with various molecules. Thus, the major concern is to increase the binding by forming nanoholes in the Å range through perforation, implanted defects at edge and boundary, or chemical doping. In every case, the critical issue is to reach a chemical activation enough to let the molecules pass through the layer, thereby preventing complete molecular decomposition. It should be mentioned that the range of binding energy has been estimated from 10 to 40 kJ/mol (Gadipelli et al., 2015). Considering that the gas transport through membranes is usually modelled according to the solution-diffusion mechanism, a great ambition is to enhance the current selectivity–permeability trade-off for specified couple of gases. This requires the combination of well-controlled and uniform nanoholes with selective physisorption. Although the transport mechanism through 2D membranes is not yet fully understood, local defects together with wrinkling and insertion of functional groups into the pore edges or between stacked layers have been proposed to achieve gas-selective layers according to sieve mechanisms. Nevertheless, impurities and molecular residues can give rise to obstructions to gas diffusion. Also physisorption related to surface characteristics of both the adsorbent membrane and adsorbed molecules together with polarizability, magnetic susceptibility, permanent dipole moment, and quadrupole moment should be considered for improving the sorption selectivity (Gadipelli et al., 2015). In this regard, ionic metal atoms such as Li^+, Na^+, Mg^{2+}, and Al^{3+} have also been demonstrated to increase the binding energy of H_2 (12–340) kJ/mol, due to the polarization of H_2 molecules by an electric field and/or hybridization of σ or σ^* orbitals of H_2 with transition-metal *d* orbitals (Bodrenko et al., 2012; Kubas, 2001).

Stable adsorption of CO_2 has been demonstrated in graphene with pores of 0.5 nm at low pressure (Terzyk et al., 2009), while CH_4 has been adsorbed (27.2 kJ/mol) more favorably in Li-doped nanostructured graphene at a pressure of 35 bar (Chen et al., 2012). Binding energy, interlayer spacing, and metal doping have also been envisaged to make graphene nanostructures somewhat discriminate between H_2S–CH_4 and CO_2–CH_4 gas couples, resulting in a higher adsorption for the first gas pair. In Li-doped graphene, CH_4 seems to be adsorbed by nonbonding interactions, whereas CO_2 molecules dissociate (Gadipelli et al., 2015). The chemical functionalization of graphene with polar groups, including –COOH, $-NH_2$, $-NO_2$, and $-H_2PO_3$, has also been demonstrated to increase significantly the CO_2 and CH_4 adsorption energy (Kandagal et al., 2012; Wood et al., 2012). DFT and MD simulations have confirmed high selectivity for H_2 over other gases with the energy barrier of 0.12, 0.26, 0.25, and 0.82 eV for H_2, N_2, CO, and CH_4 (Tao et al., 2014), respectively, while a H_2/CH_4 selectivity of 10^8 has been estimated through nitrogen-doped graphene sheets. Selectivity up to 10^{23} has been found in graphene with all pores saturated by hydrogen. A CO_2/N_2 selectivity of 11 has been estimated through nanoporous graphene membranes with all pores saturated with nitrogen, due to enhanced electrostatic interactions (Shan et al., 2012). Larger uptake and diffusion have been found for CO_2 when moving from OH- to CH_3-functionalized graphene membranes, whereas reduced dispersive interaction for CO_2 leads to higher SO_2/N_2 and SO_2/CO_2 selectivity (Schrier, 2011). GO membranes deposited on polyethersulfone films have been proved to get high CO_2 permeation under humidification, leading to high selectivity over other gases, according to the sequence $CO_2 > H_2 > He > CH_4 > O_2 > N_2$. Indeed, the highest CO_2 permeation is ascribed to the favorable interaction with polar moieties in GO, which work as preferential sites for adsorption. The presence of water enhances the permeation behavior (Kim et al., 2013).

6.5.3 Energy Power and Storage

The use of graphene in fuel cells (Li et al., 2015), batteries (Chang et al., 2011), supercapacitors (Raccichini et al., 2015), photovoltaic modules (Guo et al., 2010), and devices for hydrogen generation provides chances to tackle challenges driven by the increasing global energy demand (Bonaccorso et al., 2015).

The current technology of batteries and supercapacitors is not able to ensure bendable devices, since cell damage and electrolyte leakage usually occur upon bending.

The superb surface-to-mass ratio of graphene (~2600 m^2/g), together with its outstanding electrical conductivity and flexibility, enables the storage of electric charge, ions, or hydrogen. Moreover, other 2D materials, such as TMDC and transition-metal oxides, also represent suitable candidates for energy devices. Their 2D nature allows bendability and the possibility to tailor their properties by means of layer-by-layer assembly.

Innovative energy-storage devices have been fabricated with novel features, but the low-cost mass production of graphene with controlled microstructure and low residual oxygen content remains challenging (El-Kady et al., 2016). Water splitting and energy production represent other possible applications.

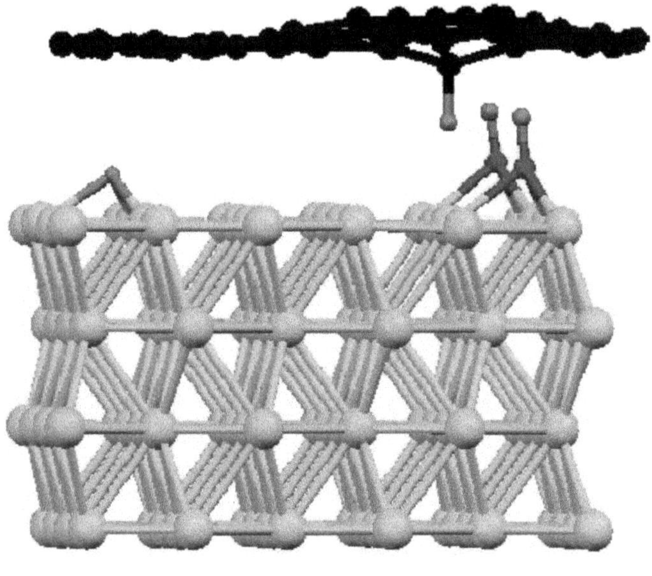

Figure 6.9 Hydrogen storage in graphene supported by a nickel substrate. Adapted from Politano et al. (2016).

For a diffuse usage, energy devices must be compact, combining both fuel production/storage and conversion of primary energy sources into the energy service. A smart solution could be thus represented by the use of graphene, since it is active for both the production and the storage of hydrogen (Fig. 6.9). As a matter of fact, recently, epitaxial graphene has been found to be an unexpectedly efficient catalyst for the production of hydrogen from water dosed at room temperature (Politano et al., 2016). Enhanced hydrogen production has also been achieved by combining graphene with titanium dioxide (Wang et al., 2016). The graphene sheet also acts as a storage medium for hydrogen, with performances already competitive with state-of-the-art materials (Bonaccorso et al., 2015; Tozzini et al., 2013). This challenging capability could be exploited for developing membrane-based self-powered devices and processes.

6.6 Conclusion and Outlook

Two-dimensional materials represent a huge opportunity for many research and market areas. The use of these novel materials has strongly influenced the fabrication of ultrafast, ultralight, and low-powered wearable electronics, as well as sensors and energy-storage devices. However, the potential of 2D materials still remains nearly unexplored in the field of membrane science and technology. Membrane processes involving 2D materials enable novel functional applications as desalination ion-filters and gas separation with potential use also in the energy, automotive, and biomedical fields. Two-dimensional materials can be used as an atomically thin membrane or as a nanofiller in polymeric nanocomposites. Undoubtedly, the attractive possibility of using these atom-thick layers in some areas of membrane science originates from the goal to get reduced molecular pathway, high discrimination factor, and in some cases relatively controlled thermal conductivity. The design of nanofilters should benefit from the stability of the inorganic nanomaterials, also in consideration of the prospect of strain-based control. However, the production costs are still unaffordable compared with the state-of-the-art polymeric membranes. New manufacturing techniques are expected to make this new class of materials marketable, while combined theoretical and experimental studies are necessary to identify accurate structure–property relationships and unquestionable mechanisms controlling for selective transport through these membrane types. The accomplishment of these targets would open new horizons toward the design of new competitive functional devices and processes.

References

Akinwande, D., Petrone, N., and Hone, J. (2014). Two-dimensional flexible nanoelectronics, *Nat. Commun.*, **5**, pp. 5678.

Anasori, B., Xie, Y., Beidaghi, M., Lu, J., Hosler, B. C., Hultman, L., Kent, P. R. C., Gogotsi, Y., and Barsoum, M. W. (2015). Two-dimensional, ordered, double transition metals carbides (MXenes), *ACS Nano*, **9**, pp. 9507–9516.

Bai, J., Zhong, X., Jiang, S., Huang, Y., and Duan, X. (2010). Graphene nanomesh, *Nat. Nanotechnol.*, **5**, pp. 190–194.

Balandin, A. A., Ghosh, S., Bao, W., Calizo, I., Teweldebrhan, D., Miao, F., and Lau, C. N. (2008). Superior thermal conductivity of single-layer graphene, *Nano Lett.*, **8**, pp. 902–907.

Bertolazzi, S., Brivio, J., and Kis, A. (2011). Stretching and breaking of ultrathin MoS_2, *ACS Nano*, **5**, pp. 9703–9709.

Bhadra, M., Roy, S., and Mitra, S. (2016). Desalination across a graphene oxide membrane via direct contact membrane distillation, *Desalination*, **378**, pp. 37–43.

Bhimanapati, G. R., Lin, Z., Meunier, V., Jung, Y., Cha, J., Das, S., Xiao, D., Son, Y., Strano, M. S., Cooper, V. R., Liang, L., Louie, S. G., Ringe, E., Zhou, W., Kim, S. S., Naik, R. R., Sumpter, B. G., Terrones, H., Xia, F., Wang, Y., Zhu, J., Akinwande, D., Alem, N., Schuller, J. A., Schaak, R. E., Terrones, M., and Robinson, J. A. (2015). Recent advances in two-dimensional materials beyond graphene, *ACS Nano*, **9**, pp. 11509–11539.

Bodrenko, I. V., Avdeenkov, A. V., Bessarabov, D. G., Bibikov, A. V., Nikolaev, A. V., Taran, M. D., and Tkalya, E. V. (2012). Hydrogen storage in aromatic carbon ring based molecular materials decorated with alkali or alkali-earth metals, *J. Phys. Chem. C*, **116**, pp. 25286–25292.

Bonaccorso, F., Colombo, L., Yu, G., Stoller, M., Tozzini, V., Ferrari, A. C., Ruoff, R. S., and Pellegrini, V. (2015). Graphene, related two-dimensional crystals, and hybrid systems for energy conversion and storage, *Science*, **347**, pp. 1246501.

Bonaccorso, F. and Sun, Z. (2014). Solution processing of graphene, topological insulators and other 2d crystals for ultrafast photonics, *Opt. Mater. Express*, **4**, pp. 63–78.

Booth, T. J., Blake, P., Nair, R. R., Jiang, D., Hill, E. W., Bangert, U., Bleloch, A., Gass, M., Novoselov, K. S., Katsnelson, M. I., and Geim, A. K. (2008). Macroscopic graphene membranes and their extraordinary stiffness, *Nano Lett.*, **8**, pp. 2442–2446.

Boukhvalov, D. W., Katsnelson, M. I., and Son, Y.-W. (2013). Origin of anomalous water permeation through graphene oxide membrane, *Nano Lett.,* **13**, pp. 3930–3935.

Butler, S. Z., Hollen, S. M., Cao, L., Cui, Y., Gupta, J. A., Gutiérrez, H. R., Heinz, T. F., Hong, S. S., Huang, J., Ismach, A. F., Johnston-Halperin, E., Kuno, M., Plashnitsa, V. V., Robinson, R. D., Ruoff, R. S., Salahuddin, S., Shan, J., Shi, L., Spencer, M. G., Terrones, M., Windl, W., and Goldberger, J. E. (2013). Progress, challenges, and opportunities in two-dimensional materials beyond graphene, *ACS Nano,* **7**, pp. 2898–2926.

Cai, J. M., Ruffieux, P., Jaafar, R., Bieri, M., Braun, T., Blankenburg, S., Muoth, M., Seitsonen, A. P., Saleh, M., Feng, X. L., Mullen, K., and Fasel, R. (2010). Atomically precise bottom-up fabrication of graphene nanoribbons, *Nature,* **466**, pp. 470–473.

Castellanos-Gomez, A. (2015). Black phosphorus: Narrow gap, wide applications, *J. Phys. Chem. Lett.,* **6**, pp. 4280–4291.

Castellanos-Gomez, A. (2016). Why all the fuss about 2D semiconductors? *Nat. Photon.,* **10**, pp. 202–204.

Celebi, K., Buchheim, J., Wyss, R. M., Droudian, A., Gasser, P., Shorubalko, I., Kye, J.-I., Lee, C., and Park, H. G. (2014). Ultimate permeation across atomically thin porous graphene, *Science,* **344**, pp. 289–292.

Chandra, V., Yu, S. U., Kim, S. H., Yoon, Y. S., Kim, D. Y., Kwon, A. H., Meyyappan, M., and Kim, K. S. (2012). Highly selective CO_2 capture on N-doped carbon produced by chemical activation of polypyrrole functionalized graphene sheets, *Chem. Commun.,* **48**, pp. 735–737.

Chang, H.-Y., Yang, S., Lee, J., Tao, L., Hwang, W.-S., Jena, D., Lu, N., and Akinwande, D. (2013). High-performance, highly bendable MoS_2 transistors with high-K dielectrics for flexible low-power systems, *ACS Nano,* **7**, pp. 5446–5452.

Chang, K. and Chen, W. (2011). L-Cysteine-assisted synthesis of layered MoS_2/graphene composites with excellent electrochemical performances for lithium ion batteries, *ACS Nano,* **5**, pp. 4720–4728.

Chen, J.-J., Li, W.-W., Li, X.-L., and Yu, H.-Q. (2012). Improving biogas separation and methane storage with multilayer graphene nanostructure via layer spacing optimization and lithium doping: A molecular simulation investigation, *Environ. Sci. Technol.,* **46**, pp. 10341–10348.

Chen, J., Sun, T., Sim, D., Peng, H., Wang, H., Fan, S., Hng, H. H., Ma, J., Boey, F. Y. C., Li, S., Samani, M. K., Chen, G. C. K., Chen, X., Wu, T., and Yan, Q. (2010). Sb_2Te_3 nanoparticles with enhanced Seebeck coefficient and low thermal conductivity, *Chem. Mater.,* **22**, pp. 3086–3092.

Chen, K., Song, S., and Xue, D. (2015). Beyond graphene: Materials chemistry toward high performance inorganic functional materials, *J. Mater. Chem. A,* **3**, pp. 2441–2453.

Chiritescu, C., Cahill, D. G., Nguyen, N., Johnson, D., Bodapati, A., Keblinski, P., and Zschack, P. (2007). Ultralow thermal conductivity in disordered, layered WSe_2 crystals, *Science,* **315**, pp. 351–353.

Cohen-Tanugi, D. and Grossman, J. C. (2012). Water desalination across nanoporous graphene, *Nano Lett.,* **12**, pp. 3602–3608.

Coleman, J. N. (2012). Liquid exfoliation of defect-free graphene, *Acc. Chem. Res.,* **46**, pp. 14–22.

Coleman, J. N., Lotya, M., O'Neill, A., Bergin, S. D., King, P. J., Khan, U., Young, K., Gaucher, A., De, S., Smith, R. J., Shvets, I. V., Arora, S. K., Stanton, G., Kim, H.-Y., Lee, K., Kim, G. T., Duesberg, G. S., Hallam, T., Boland, J. J., Wang, J. J., Donegan, J. F., Grunlan, J. C., Moriarty, G., Shmeliov, A., Nicholls, R. J., Perkins, J. M., Grieveson, E. M., Theuwissen, K., McComb, D. W., Nellist, P. D., and Nicolosi, V. (2011). Two-dimensional nanosheets produced by liquid exfoliation of layered materials, *Science,* **331**, pp. 568–571.

Drioli, E., Carnevale, M. C., Figoli, A., and Criscuoli, A. (2014). Vacuum membrane dryer (VMDr) for the recovery of solid microparticles from aqueous solutions, *J. Membr. Sci.,* **472**, pp. 67–76.

Drioli, E., Gugliuzza, A., and Politano, A. (2016). Composite graphene and beyond graphene membranes, in *GRAPCHINA 2016*, Qingdao, China.

Echtermeyer, T. J., Milana, S., Sassi, U., Eiden, A., Wu, M., Lidorikis, E., and Ferrari, A. C. (2015). Surface plasmon polariton graphene photodetectors, *Nano Lett.,* **16**, pp. 8–20.

El-Kady, M. F., Shao, Y., and Kaner, R. B. (2016). Graphene for batteries, supercapacitors and beyond, *Nat. Rev. Mater.,* pp. 16033.

Fang, Z., Liu, Z., Wang, Y., Ajayan, P. M., Nordlander, P., and Halas, N. J. (2012). Graphene-antenna sandwich photodetector, *Nano Lett.,* **12**, pp. 3808–3813.

Ferrari, A. C., Bonaccorso, F., Fal'ko, V., Novoselov, K. S., Roche, S., Boggild, P., Borini, S., Koppens, F. H. L., Palermo, V., Pugno, N., Garrido, J. A., Sordan, R., Bianco, A., Ballerini, L., Prato, M., Lidorikis, E., Kivioja, J., Marinelli, C., Ryhanen, T., Morpurgo, A., Coleman, J. N., Nicolosi, V., Colombo, L., Fert, A., Garcia-Hernandez, M., Bachtold, A., Schneider, G. F., Guinea, F., Dekker, C., Barbone, M., Sun, Z., Galiotis, C., Grigorenko, A. N., Konstantatos, G., Kis, A., Katsnelson, M., Vandersypen, L., Loiseau, A., Morandi, V., Neumaier, D., Treossi, E., Pellegrini, V., Polini, M., Tredicucci, A., Williams, G. M., Hee Hong, B., Ahn, J.-H., Min Kim,

J., Zirath, H., van Wees, B. J., van der Zant, H., Occhipinti, L., Di Matteo, A., Kinloch, I. A., Seyller, T., Quesnel, E., Feng, X., Teo, K., Rupesinghe, N., Hakonen, P., Neil, S. R. T., Tannock, Q., Lofwander, T., and Kinaret, J. (2015). Science and technology roadmap for graphene, related two-dimensional crystals, and hybrid systems, *Nanoscale,* **7**, pp. 4598–4810.

Fiori, G., Bonaccorso, F., Iannaccone, G., Palacios, T., Neumaier, D., Seabaugh, A., Banerjee, S. K., and Colombo, L. (2014). Electronics based on two-dimensional materials, *Nat. Nanotech,* **9**, pp. 768–779.

Fox, D., O'Neill, A., Zhou, D., Boese, M., Coleman, J., and Zhang, H. (2011). Nitrogen assisted etching of graphene layers in a scanning electron microscope, *Appl. Phys. Lett.,* **98**, pp. 243117.

Gadipelli, S. and Guo, Z. X. (2015). Graphene-based materials: Synthesis and gas sorption, storage and separation, *Prog. Mater. Sci.,* **69**, pp. 1–60.

Gagliardi, M., McWilliams, A., and Moran, R. (2011). *Graphene: Technologies, Applications, and Markets,* BCC Research Market Forecasting, Wellesley, MA.

Gugliuzza, A., Giorno, L., and Drioli, E. (2015). Graphene membranes, in *Encyclopedia of Membranes,* Drioli, E. and Giorno, L. (eds.), Springer, Berlin.

Gugliuzza, A., Politano, A., and Drioli, E. (2017). The advent of graphene and other two-dimensional materials in membrane science and technology, *Curr. Opin. Chem. Eng.,* **16**, pp. 78–85.

Guo, C. X., Yang, H. B., Sheng, Z. M., Lu, Z. S., Song, Q. L., and Li, C. M. (2010). Layered graphene/quantum dots for photovoltaic devices, *Angew. Chem.,* **49**, pp. 3014–3017.

Gupta, A., Sakthivel, T., and Seal, S. (2015). Recent development in 2D materials beyond graphene, *Prog. Mater. Sci.,* **73**, pp. 44–126.

He, Z., Zhou, J., Lu, X., and Corry, B. (2013). Bioinspired graphene nanopores with voltage-tunable ion selectivity for Na^+ and K^+, *ACS Nano,* **7**, pp. 10148–10157.

Heiranian, M., Farimani, A. B., and Aluru, N. R. (2015). Water desalination with a single-layer MoS_2 nanopore, *Nat. Commun.,* **6**, pp. 8616.

Jain, A. and McGaughey, A. J. H. (2015). Strongly anisotropic in-plane thermal transport in single-layer black phosphorene, *Sci. Rep.,* **5**, pp. 8501.

Jakob, B. (2015). Adhesive tape exfoliation: Why it works for graphene, *Europhys. Lett.,* **109**, pp. 58004.

Jana, M. K., Biswas, K., and Rao, C. N. R. (2013). Ionothermal synthesis of few-layer nanostructures of Bi_2Se_3 and related materials, *Chem. Eur. J.*, **19**, pp. 9110–9113.

Kandagal, V. S., Pathak, A., Ayappa, K., and Punnathanam, S. N. (2012). Adsorption on edge-functionalized bilayer graphene nanoribbons: Assessing the role of functional groups in methane uptake, *J. Phys. Chem. C*, **116**, pp. 23394–23403.

Kim, B. H., Kim, J. Y., Jeong, S.-J., Hwang, J. O., Lee, D. H., Shin, D. O., Choi, S.-Y., and Kim, S. O. (2010). Surface energy modification by spin-cast, large-area graphene film for block copolymer lithography, *ACS Nano*, **4**, pp. 5464–5470.

Kim, H. W., Yoon, H. W., Yoon, S.-M., Yoo, B. M., Ahn, B. K., Cho, Y. H., Shin, H. J., Yang, H., Paik, U., Kwon, S., Choi, J.-Y., and Park, H. B. (2013). Selective gas transport through few-layered graphene and graphene oxide membranes, *Science*, **342**, pp. 91–95.

Koenig, S. P., Wang, L., Pellegrino, J., and Bunch, J. S. (2012). Selective molecular sieving through porous graphene, *Nat. Nanotech*, **7**, pp. 728–732.

Koitz, R., Hutter, J., and Iannuzzi, M. (2016). Formation and properties of a terpyridine-based 2D MOF on the surface of water, *2D Mater.*, **3**, pp. 025026.

Koppens, F. H. L., Mueller, T., Avouris, P., Ferrari, A. C., Vitiello, M. S., and Polini, M. (2014). Photodetectors based on graphene, other two-dimensional materials and hybrid systems, *Nat. Nanotechnol.*, **9**, pp. 780–793.

Kuan, A. T., Lu, B., Xie, P., Szalay, T., and Golovchenko, J. A. (2015). Electrical pulse fabrication of graphene nanopores in electrolyte solution, *Appl. Phys. Lett.*, **106**, pp. 203109.

Kubas, G. J. (2001). Metal-dihydrogen and σ-bond coordination: The consummate extension of the Dewar–Chatt–Duncanson model for metal–olefin π bonding, *J. Organomet. Chem.*, **635**, pp. 37–68.

Kumar, C., Kuiri, M., Jung, J., Das, T., and Das, A. (2016). Tunability of 1/f noise at multiple Dirac cones in hBN encapsulated graphene devices, *Nano Lett.*, **16**, pp. 1042–1049.

Lee, C., Wei, X., Kysar, J. W., and Hone, J. (2008). Measurement of the elastic properties and intrinsic strength of monolayer graphene, *Science*, **321**, pp. 385–388.

Li, W., Carrete, J., and Mingo, N. (2013). Thermal conductivity and phonon linewidths of monolayer MoS_2 from first principles, *Appl. Phys. Lett.*, **103**, pp. 253103.

Li, W., Yang, Y., Weber, J. K., Zhang, G., and Zhou, R. (2016). Tunable, strain-controlled nanoporous MoS_2 filter for water desalination, *ACS Nano*, **10**, pp. 1829–1835.

Li, Y., Song, J., and Yang, J. (2015). Graphene models and nano-scale characterization technologies for fuel cell vehicle electrodes, *Renewable Sustainable Energy Rev.*, **42**, pp. 66–77.

Lin, T., Chen, J., Bi, H., Wan, D., Huang, F., Xie, X., and Jiang, M. (2013). Facile and economical exfoliation of graphite for mass production of high-quality graphene sheets, *J. Mater. Chem. A*, **1**, pp. 500–504.

Liu, G., Jin, W., and Xu, N. (2016). Two-dimensional-material membranes: A new family of high-performance separation membranes, *Angew. Chem.*, **55**, pp. 13384–13397.

Lu, N., Wang, J., Floresca, H. C., and Kim, M. J. (2012). In situ studies on the shrinkage and expansion of graphene nanopores under electron beam irradiation at temperatures in the range of 400–1200°C, *Carbon*, **50**, pp. 2961–2965.

Ma, J. C. and Dougherty, D. A. (1997). The cation–π interaction, *Chem. Rev.*, **97**, pp. 1303–1324.

Macedonio, F., Brunetti, A., Barbieri, G., and Drioli, E. (2013). Membrane condenser as a new technology for water recovery from humidified "waste" gaseous streams, *Ind. Eng. Chem. Res.*, **52**, pp. 1160–1167.

Mahadevi, A. S. and Sastry, G. N. (2012). Cation–π interaction: Its role and relevance in chemistry, biology, and material science, *Chem. Rev.*, **113**, pp. 2100–2138.

Maiti, U. N., Thapa, R., Lim, J., Li, D. J., Kim, K. H., and Kim, S. O. (2015). Self-size-limiting nanoscale perforation of graphene for dense heteroatom doping, *ACS Appl. Mater. Interfaces*, **7**, pp. 25898–25905.

Mak, K. F. and Shan, J. (2016). Photonics and optoelectronics of 2D semiconductor transition metal dichalcogenides, *Nat. Photon.*, **10**, pp. 216–226.

Miro, P., Audiffred, M., and Heine, T. (2014). An atlas of two-dimensional materials, *Chem. Soc. Rev.*, **43**, pp. 6537–6554.

Narita, A., Verzhbitskiy, I. A., Frederickx, W., Mali, K. S., Jensen, S. A., Hansen, M. R., Bonn, M., De Feyter, S., Casiraghi, C., Feng, X., and Müllen, K. (2014). Bottom-up synthesis of liquid-phase-processable graphene nanoribbons with near-infrared absorption, *ACS Nano,* **8**, pp. 11622–11630.

Novoselov, K. (2007). Graphene: Mind the gap, *Nat. Mater.,* **6**, pp. 720–721.

Novoselov, K. S., Geim, A. K., Morozov, S. V., Jiang, D., Katsnelson, M. I., Grigorieva, I. V., Dubonos, S. V., and Firsov, A. A. (2005). Two-dimensional gas of massless Dirac fermions in graphene, *Nature,* **438**, pp. 197–200.

Novoselov, K. S., Geim, A. K., Morozov, S. V., Jiang, D., Zhang, Y., Dubonos, S. V., Grigorieva, I. V., and Firsov, A. A. (2004). Electric field effect in atomically thin carbon films, *Science,* **306**, pp. 666–669.

O'Hern, S. C., Boutilier, M. S., Idrobo, J.-C., Song, Y., Kong, J., Laoui, T., Atieh, M., and Karnik, R. (2014). Selective ionic transport through tunable subnanometer pores in single-layer graphene membranes, *Nano Lett.,* **14**, pp. 1234–1241.

O'Hern, S. C., Jang, D., Bose, S., Idrobo, J.-C., Song, Y., Laoui, T., Kong, J., and Karnik, R. (2015). Nanofiltration across defect-sealed nanoporous monolayer graphene, *Nano Lett.,* **15**, pp. 3254–3260.

Pakdel, A., Bando, Y., and Golberg, D. (2014). Nano boron nitride flatland, *Chem. Soc. Rev.,* **43**, pp. 934–959.

Perrotta, M., Saielli, G., Casella, G., Macedonio, F., Giorno, L., Drioli, E., and Gugliuzza, A. (2017). An ultrathin suspended hydrophobic porous membrane for high-efficiency water desalination, *Appl. Mater. Today,* **9**, pp. 1–9.

Politano, A., Cattelan, M., Boukhvalov, D. W., Campi, D., Cupolillo, A., Agnoli, S., Apostol, N. G., Lacovig, P., Lizzit, S., Farías, D., Chiarello, G., Granozzi, G., and Larciprete, R. (2016). Unveiling the mechanisms leading to H_2 production promoted by water decomposition on epitaxial graphene at room temperature, *ACS Nano,* **10**, pp. 4543–4549.

Qian, Z., Liu, F., Hui, Y., Kar, S., and Rinaldi, M. (2015). Graphene as a massless electrode for ultrahigh-frequency piezoelectric nanoelectromechanical systems, *Nano Lett.,* **15**, pp. 4599–4604.

Raccichini, R., Varzi, A., Passerini, S., and Scrosati, B. (2015). The role of graphene for electrochemical energy storage, *Nat. Mater.,* **14**, pp. 271–279.

Raidongia, K. and Huang, J. (2012). Nanofluidic ion transport through reconstructed layered materials, *J. Am. Chem. Soc.,* **134**, pp. 16528–16531.

Ren, C. E., Hatzell, K. B., Alhabeb, M., Ling, Z., Mahmoud, K. A., and Gogotsi, Y. (2015). Charge- and size-selective ion sieving through $Ti_3C_2T_x$ MXene membranes, *J. Phys. Chem. Lett.,* **6**, pp. 4026–4031.

Robertson, A. W., Lee, G.-D., He, K., Gong, C., Chen, Q., Yoon, E., Kirkland, A. I., and Warner, J. H. (2015). Atomic structure of graphene subnanometer pores, *ACS Nano,* **9**, pp. 11599–11607.

Rollings, R. C., Kuan, A. T., and Golovchenko, J. A. (2016). Ion selectivity of graphene nanopores, *Nat. Commun.,* **7**, pp. 11408.

Schrier, J. (2011). Fluorinated and nanoporous graphene materials as sorbents for gas separations, *ACS Appl. Mater. Interfaces,* **3**, pp. 4451–4458.

Shan, M., Xue, Q., Jing, N., Ling, C., Zhang, T., Yan, Z., and Zheng, J. (2012). Influence of chemical functionalization on the CO_2/N_2 separation performance of porous graphene membranes, *Nanoscale,* **4**, pp. 5477–5482.

Shen, J., Zhu, Y., Yang, X., and Li, C. (2012). Graphene quantum dots: Emergent nanolights for bioimaging, sensors, catalysis and photovoltaic devices, *Chem. Commun.,* **48**, pp. 3686–3699.

Sint, K., Wang, B., and Král, P. (2008). Selective ion passage through functionalized graphene nanopores, *J. Am. Chem. Soc.,* **130**, pp. 16448–16449.

Srinivas, G., Burress, J., and Yildirim, T. (2012). Graphene oxide derived carbons (GODCs): Synthesis and gas adsorption properties, *Energy Environ. Sci.,* **5**, pp. 6453–6459.

Sun, L., Huang, H., and Peng, X. (2013). Laminar MoS_2 membranes for molecule separation, *Chem. Commun.,* **49**, pp. 10718–10720.

Sun, L., Ying, Y., Huang, H., Song, Z., Mao, Y., Xu, Z., and Peng, X. (2014a). Ultrafast molecule separation through layered WS_2 nanosheet membranes, *ACS Nano,* **8**, pp. 6304–6311.

Sun, P., Ma, R., Deng, H., Song, Z., Zhen, Z., Wang, K., Sasaki, T., Xu, Z., and Zhu, H. (2016). Intrinsic high water/ion selectivity of graphene oxide lamellar membranes in concentration gradient-driven diffusion, *Chem. Sci.,* **7**, pp. 6988–6994.

Sun, P., Zheng, F., Zhu, M., Song, Z., Wang, K., Zhong, M., Wu, D., Little, R. B., Xu, Z., and Zhu, H. (2014b). Selective trans-membrane transport of alkali and alkaline earth cations through graphene oxide membranes based on cation– π interactions, *ACS Nano,* **8**, pp. 850–859.

Sun, P., Zhu, M., Wang, K., Zhong, M., Wei, J., Wu, D., Xu, Z., and Zhu, H. (2012). Selective ion penetration of graphene oxide membranes, *ACS Nano,* **7**, pp. 428–437.

Surwade, S. P., Smirnov, S. N., Vlassiouk, I. V., Unocic, R. R., Veith, G. M., Dai, S., and Mahurin, S. M. (2015). Water desalination using nanoporous single-layer graphene, *Nat. Nanotechnol.,* **10**, pp. 459–464.

Tao, Y., Xue, Q., Liu, Z., Shan, M., Ling, C., Wu, T., and Li, X. (2014). Tunable hydrogen separation in porous graphene membrane: First-principle and molecular dynamic simulation, *ACS Appl. Mater. Interfaces,* **6**, pp. 8048–8058.

Terzyk, A. P., Furmaniak, S., Gauden, P. A., and Kowalczyk, P. (2009). Fullerene-intercalated graphene nano-containers: Mechanism of argon adsorption and high-pressure CH_4 and CO_2 storage capacities, *Adsorpt. Sci. Technol.,* **27**, pp. 281–296.

Tour, J. M. (2013). Top-down versus bottom-up fabrication of graphene-based electronics, *Chem. Mater.,* **26**, pp. 163–171.

Tozzini, V. and Pellegrini, V. (2013). Prospects for hydrogen storage in graphene, *Phys. Chem. Chem. Phys.,* **15**, pp. 80–89.

Tsetseris, L. and Pantelides, S. (2014). Graphene: An impermeable or selectively permeable membrane for atomic species? *Carbon,* **67**, pp. 58–63.

Viti, L., Hu, J., Coquillat, D., Politano, A., Consejo, C., Knap, W., and Vitiello, M. S. (2016). Heterostructured hBN-BP-hBN nanodetectors at terahertz frequencies, *Adv. Mater.,* **28**, pp. 7390–7396.

Wang, D., Wang, Z., Wang, L., Hu, L., and Jin, J. (2015). Ultrathin membranes of single-layered MoS_2 nanosheets for high-permeance hydrogen separation, *Nanoscale,* **7**, pp. 17649–17652.

Wang, L., Boutilier, M. S., Kidambi, P. R., Jang, D., Hadjiconstantinou, N. G., and Karnik, R. (2017). Fundamental transport mechanisms, fabrication and potential applications of nanoporous atomically thin membranes, *Nat. Nanotechnol.,* **12**, pp. 509–522.

Wang, Z., Yin, Y., Williams, T., Wang, H., Sun, C., and Zhang, X. (2016). Metal link: A strategy to combine graphene and titanium dioxide for enhanced hydrogen production, *Int. J. Hydrogen Energy,* **41**, pp. 22034–22042.

Wood, B. C., Bhide, S. Y., Dutta, D., Kandagal, V. S., Pathak, A. D., Punnathanam, S. N., Ayappa, K. G., and Narasimhan, S. (2012). Methane and carbon dioxide adsorption on edge-functionalized graphene: A comparative DFT study, *J. Chem. Phys.,* **137**, pp. 054702.

Yan, S., Zhao, S., Wardhaugh, L., and Feron, P. H. (2015). Innovative use of membrane contactor as condenser for heat recovery in carbon capture, *Environ. Sci. Technol.,* **49**, pp. 2532–2540.

Yi, M., Shen, Z., Zhao, X., Liang, S., and Liu, L. (2014). Boron nitride nanosheets as oxygen-atom corrosion protective coatings, *Appl. Phys. Lett.,* **104**, pp. 143101.

You, Y., Sahajwalla, V., Yoshimura, M., and Joshi, R. K. (2016). Graphene and graphene oxide for desalination, *Nanoscale,* **8**, pp. 117–119.

Yuan, W., Chen, J., and Shi, G. (2014). Nanoporous graphene materials, *Mater. Today,* **17**, pp. 77–85.

Zan, R., Ramasse, Q. M., Bangert, U., and Novoselov, K. S. (2012). Graphene reknits its holes, *Nano Lett.,* **12**, pp. 3936–3940.

Zhao, L.-D., Lo, S.-H., Zhang, Y., Sun, H., Tan, G., Uher, C., Wolverton, C., Dravid, V. P., and Kanatzidis, M. G. (2014). Ultralow thermal conductivity and high thermoelectric figure of merit in SnSe crystals, *Nature,* **508**, pp. 373–377.

Zhao, S., Xue, J., and Kang, W. (2013). Ion selection of charge-modified large nanopores in a graphene sheet, *J. Chem. Phys.,* **139**, pp. 114702.

Zheng, Z., Grünker, R., and Feng, X. (2016). Synthetic two-dimensional materials: A new paradigm of membranes for ultimate separation, *Adv. Mater.,* **28**, pp. 6529–6545.

Zhou, C., Wang, X., Raju, S., Lin, Z., Villaroman, D., Huang, B., Chan, H. L.-W., Chan, M., and Chai, Y. (2015). Low voltage and high ON/OFF ratio field-effect transistors based on CVD MoS_2 and ultra high-k gate dielectric PZT, *Nanoscale,* **7**, pp. 8695–8700.

Zhu, W., Park, S., Yogeesh, M. N., McNicholas, K. M., Bank, S. R., and Akinwande, D. (2016). Black phosphorus flexible thin film transistors at gigahertz frequencies, *Nano Lett.,* **16**, pp. 2301–2306.

Zhu, W., Yogeesh, M. N., Yang, S., Aldave, S. H., Kim, J.-S., Sonde, S., Tao, L., Lu, N., and Akinwande, D. (2015). Flexible black phosphorus ambipolar transistors, circuits and AM demodulator, *Nano Lett.,* **15**, pp. 1883–1890.

Chapter 7

Nanostructured Material Membranes for Gas Separation

Paola Bernardo
Institute on Membrane Technology of the National Research Council of Italy (ITM-CNR), Via P. Bucci, Cubo 17C, 87036 Rende (CS), Italy
p.bernardo@itm.cnr.it

Nanostructured materials such as microporous polymers, 2D materials like graphene, and crystalline metal–organic frameworks (MOFs) experienced fast development in the last few years. Such materials are of interest for different applications, such as gas storage and heterogeneous catalysis. In addition, membranes based on nanostructured materials displayed outstanding properties when considered for molecular separations. This chapter focuses on microporous polymers with intrinsic microporosity (PIMs), which are gaining increasing interest in developing membranes for large-scale gas separations in commercial and environmental applications. The chapter presents the strategies pursued in order to tailor the transport properties of these ultrapermeable materials.

Functional Nanostructured Membranes
Edited by Enrico Drioli, Lidietta Giorno, and Annarosa Gugliuzza

ISBN 978-981-4774-79-6 (Hardcover), 978-1-351-13511-5 (eBook)
www.panstanford.com

7.1 Introduction

Membrane gas separation rapidly became a competitive separation technology since the serial production of polymeric membranes in late 1970. In general, membrane processes give several benefits than many conventional separation techniques (e.g., distillation, extraction, absorption, and adsorption). Membrane systems do not require a phase change, thus resulting in a low energy requirement. They do not involve expensive adsorbents and/or solvents. Membrane systems are fully automated, modular, with small footprint and weight, and thus, very attractive for offshore gas processing platforms. The absence of moving parts makes gas separation systems particularly suited for remote locations where reliability is critical. All these features contribute to implementing the logic of Process Intensification in various industrial contexts.

Considering gas and vapor separations, today there are different industrial applications for membrane plants, ranging from hydrogen recovery from off-gases in refinery operations, to nitrogen generation from air and to the upgrading of natural gas and biogas (i.e., predominantly removing CO_2 from CH_4) (Bernardo et al., 2009). Engineering and material science both contributed to the development of these applications. The commercially available membranes are based on a few polymers. Indeed, polymeric materials are cheap and can be easily processed in thin films required for cost-effective applications. However, for many potential large-scale applications, such as the capture of CO_2 from power station flue gases, there is a need for new materials that offer high permeability combined with good selectivity.

The performance of a membrane material depends on productivity and separation efficiency, which can be expressed in terms of a permeability coefficient and a selectivity, respectively. Typically, in polymeric materials, the permeability P is the product of a diffusion coefficient D and a solubility coefficient S, i.e., $P = DS$. The ideal selectivity for two gases (i and j) is obtained as the ratio of the individual permeability, $\alpha_{i/j} = P_i/P_j$.

In 1991, Robeson demonstrated a limitation of polymeric membranes for gas separation applications as a trade-off between permeability and selectivity (Robeson, 1991). Polymers recently developed allowed a shift in the upper bounds toward more interesting performance (Robeson, 2008). According to Freeman's theoretical analysis of the permeability–selectivity trade-off for polymeric membranes (Freeman, 1999), higher gas permeability can be achieved by increasing the interchain separation and, hence, the polymer free volume, whereas selectivity can be improved by enhancing polymer rigidity. These considerations were the basis for the development of very permeable and also selective polymers.

Typically, intermolecular interactions in the polymer macromolecules allow their bending and twisting, leading to an efficient packing. However, every polymer matrix has an unoccupied void space referred to as free volume. In the glassy state, the fraction of free volume that remains upon cooling is about 2.5% (Budd et al., 2004a). However, glassy polymers having a rigid molecular structure (e.g., polyimides or polysulfones) can present additional free volume (up to 20%) "frozen-in" by rapid cooling to below the glass-transition temperature or rapid removal of a solvent (Budd et al., 2004b). The free volume theory correlates the free volume with the diffusion coefficient: A large free volume amount corresponds to a large D and hence P. The size distribution of the free volume elements can have a great influence on the selectivity. Indeed, the size and connectivity of free volume elements can strongly affect the solubility of the penetrant in the polymer. However, the differences in solubility are reduced with respect to those in the diffusion coefficient when analyzing different polymers. Therefore, the packing and flexibility of polymer chains affect more the gas transport than the solubility.

In recent years, different research institutions have been actively working on the synthesis of polymers with high free volumes and tailored free volume distributions. The recognized strategies to achieve polymeric membrane materials with superior separation performance are the design of nanostructured materials, the thermal treatment of suited polymers, as well as the combination of existing commercial polymers with organic or inorganic additives.

Above a certain amount of free volume, the voids in a polymer matrix would be interconnected. Accordingly, the polymer performs as a conventional microporous material (McKeown et al., 2005). Representative "ultra-high free volume" polymers are the polyacetylene derivative poly(1-trimethylsilyl-1-propyne) (PTMSP) and the perfluoropolymer Teflon AF 2400 (copolymer of 4,5-difluoro-2,2-bis(trifluoromethyl)-1,3-dioxole and tetrafluoroethylene)

(Alentiev et al., 1997). A decade ago, the synthesis of a new class of polymers, termed polymers of intrinsic microporosity (PIMs), provided a step change in the field of polymer membranes for gas separation (Budd et al., 2004a; Budd et al., 2008; Budd et al., 2010). PIMs, being solution processable, were also considered for application as membrane-forming materials. They have attracted attention for their high gas permeability, which is due to the high free volume and can be two to three orders of magnitude higher than that displayed by conventional high free volume polymers, combined with moderate permselectivity. The exceptionally high gas permeability is an indication that their fractional free volume (~30%) is interconnected, and thus these polymers have been classified as microporous (Srinivasan et al., 1994).

The addition of an inorganic, metal–organic, or organic filler to a polymer to form a mixed matrix membrane (MMM) (Chung et al., 2007) is a strategy that can give synergistic improvements in permeation properties and enhance the mechanical performance of neat polymers. This route was recently exploited also for high free volume polymers, together with their thermal rearrangement. An interesting additional consequence is the possibility to control plasticization and aging effects.

This chapter addresses the progresses in the field of membranes for gas separation based on nanostructured materials. Indeed, the last years brought a fast development in the development of microporous polymers, 2D materials like graphene, and crystalline metal–organic frameworks (MOFs). Such nanostructured materials, which are of interest for different applications such as gas storage and heterogeneous catalysis, displayed outstanding properties when considered for membrane materials. The focus is on the novel PIMs that have gained importance as materials for molecular separation and, particularly, gas separation (Carta et al., 2008; Carta et al., 2009; Carta et al., 2014a; Carta et al., 2014b; Du et al., 2008; Du et al., 2010; Emmler et al., 2010; Fritsch et al., 2011; Ghanem et al., 2008a; Ghanem et al., 2008b; Ghanem et al., 2009; Ma et al., 2013; McKeown, 2012; Ritter et al., 2009; Song et al., 2008). The chapter presents the approaches pursued for their modification, including their molecular design, thermal rearrangement, as well as the incorporation of fillers, and highlights the achievable benefits. It also presents their peculiarities, outlining the role of thermomechanical history in the achievable performance and transport properties.

7.2 Nanostructured Materials

The synthesis of materials with novel properties by controlled manipulation of their microstructure on the atomic scale has become an emerging interdisciplinary field involving solid-state physics, chemistry, and materials science. Membrane science, being multidisciplinary, is also involved in the production of novel nanostructures with tailored transport properties.

The physical blending of different materials is an approach to customize the final properties of a membrane. Different additives can be considered in order to modify the transport properties of polymers. An interesting example is the blending of low-molecular-weight polyethylene glycol (PEG) or PEG dimethyl ether (PEG-DME) with a commercial block copolymer of polyether (PE) and polyamide (PA), referred to as Pebax® (Car et al., 2008; Yave et al., 2010). This approach could double the already high CO_2 permeability of Pebax®, increasing the CO_2/H_2 selectivity (Car et al., 2008). Indeed, for polymeric membrane-based selective CO_2 separation, ethylene oxide units have been identified as favorable chemical groups, since the ether oxygen interacts with CO_2, resulting in high CO_2 solubility selectivity (Lin and Freeman, 2005). Fractional free volume increased upon adding PEG-DME, while the glass-transition temperature decreased from 220 to 195 K (Car et al., 2008). In this nanostructured blend, the PEG-DME chains might be located between the chains of polyethylene oxide (PEO) and PA, thus inhibiting the packaging of PEO and PA chains. Therefore, crystallinity in PEO and PA phases is decreased, while the amorphous phase and, thus, the fractional free volume are increased. Accordingly, an increase in the diffusivity of gases was observed.

Interesting results were obtained by blending ionic liquids (ILs) with barrier polymers such as poly(vinylidene fluoride-co-hexafluoropropylene) (p(VDF-HFP)), producing gels (Friess et al., 2012). The presence of [EMIM][TFSI] increases significantly the gas permeability, particularly for CO_2. A correlation of the transport properties with Young's modulus reveals a transition from a diffusion-controlled transport to a solubility-controlled transport with increasing IL content in the membrane. The permeation parameters for water, alcohols, linear and cyclic hydrocarbons measured in a membrane containing 80% of IL were reported to have different trends. Polar compounds (linear C1–C4 alcohols, water) are more permeable and more sorbing than the corresponding hydrocarbons. The two groups show opposite trends in permeability as a function of the critical volume, with an increase for alkanes and a decrease for alcohols. The same trends are observed for solubility as a function of the critical temperature, evidencing that at such high IL concentrations, the mass transport is controlled by solubility. Instead, diffusion is mainly correlated with the molecular size, regardless of the chemical nature of permeants. Significant differences in the transport of toluene and cyclohexane suggest that these membranes are also suitable for the separation of alkanes and aromatics.

The isolation of graphene, a single-atom-thick sheet of sp^2-hybridized carbon atoms arranged in honeycomb lattice, in 2004 led to an increasing interest in 2D materials (Novoselov, 2004). Graphene, having an atomic thickness, demonstrated superior properties such as high thermal conductivity and high mechanical strength (strongest nanomaterial). Indeed, graphene's p-orbitals create a dense, delocalized cloud that rejects atoms and molecules trying to pass through these rings (Berry, 2013; Liu et al., 2016). Consequently, graphene is impermeable, even to the smallest gas molecules such as hydrogen and helium (Bunch et al., 2008). Graphene and its derivatives, graphene oxide (GO) or reduced graphene oxide, are explored as promising membrane materials for gas and liquid separation (Yoon et al., 2016). However, this requires the creation of holes at a nanometric or subnanometric scale, thus producing nanoporous graphene membranes (Yoon et al., 2016). Kim et al. demonstrated instead that few- and several-layered graphene and GO sheets can be engineered to exhibit the desired gas separation characteristics (Kim et al., 2013). Selective gas diffusion can be achieved by controlling the nanochannels formed between GO sheets via different stacking methods. This approach, minimizing transport resistance, boosts the flux, allowing ultrafast molecular separation also in water desalination. Interesting prospects are offered by the combination of 2D materials with some selected polymers in nanocomposite membranes.

The subsequent part of the chapter deals with microporous organic polymers with interconnected voids in the porous structure and a quite large accessible surface area (hundreds of square meters per gram), as measured by gas adsorption. Representative examples are the "thermally rearranged" (TR) polymers, microporous organic polymers tuned to optimize the cavity sizes and distribution for difficult separation applications (Han et al., 2012; Park et al., 2007). TR membranes have been fabricated by the conversion of soluble aromatic polyimides with ortho-positioned functional groups into polybenzoxazoles (PBOs), which takes place at high temperature (typically above 400°C) (Park et al., 2007). The process changes the molecular connectivity and conformation, thus altering chain packing of the polymeric macromolecules and resulting in a narrow free volume distribution and hourglass shaped cavities (Park et al., 2007). The fully aromatic PBOs present exceptional thermal and chemical resistance coupled with high separation performance that exceeds the trade-off relationships for many gas pairs. The thermal treatment makes these materials insoluble, but at the same time, they become resistant to CO_2-induced plasticization (Park et al., 2007).

7.2.1 Polymers of Intrinsic Microporosity

PIMs are glassy and solution-processable polymers with relatively inflexible and randomly twisted backbones. These polymers demonstrate microporosity but do not possess a network of covalent bonds (Budd et al., 2004b). The shape of the molecules is randomly contorted, and there are no single bonds in the backbone about which rotation can occur. This conformation hinders the conformational changes that instead happen in conventional polymers allowing their rearrangement and space filling. Accordingly, PIMs showed a high free volume and large gas and vapor permeability that result from a poor packing of the constituent macromolecules (Budd et al., 2004b).

Originally, PIMs presented fused ring sequences interrupted by spirocenters, as in PIM-1 (Budd et al., 2004), a fluorescent, yellow polymer, which is the archetypal membrane-forming PIM (Fig. 7.1).

PIM-1, with an excess free volume FFV = 0.26, shows a very attractive combination of permeability and selectivity for gas separation. Indeed, the performance of PIM-1 surpasses the 1991 Robeson's upper bound lines (Robeson, 1991) for important gas pairs such as O_2/N_2 and CO_2/CH_4, contributing to define the 2008 upper bound lines (Robeson, 2008).

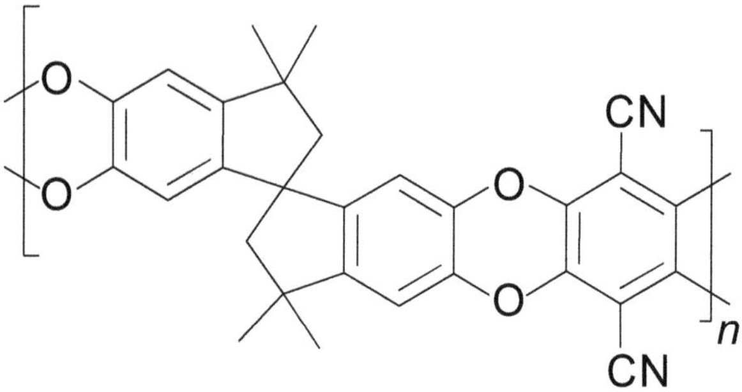

Figure 7.1 Repeat unit of PIM-1.

Detailed studies of gas permeation, thermodynamic properties, and free volume are available for the solution-processable PIM-1. A distinguishing feature of this polymer is the large solubility coefficients among all the polymers so far studied. This is particularly true for CO_2, since the electronegative nitrogen atom in the polar nitrile groups preferentially interacts with the electrophilic carbon atom in CO_2. The gas permeability of PIM-1 decreases in the following order: $CO_2 \gg H_2 \gg O_2 > He \gg N_2$. The predominant role of solubility in PIM-1 produced the so-called "reverse selectivity": CO_2 is more permeable than other smaller gases such as H_2. Rubbery materials typically present this behavior (Lau et al., 2013). Instead, more rigid and glassy polymers usually display a permeability that is larger for smaller gases (Koschine et al., 2015).

Table 7.1 reports a comparison of the permeability and selectivity for the CO_2/H_2 separation of different polymers and PIM-1. Despite its super glassy state, PIM-1 displays larger permeability values with respect to other conventional glassy polymers that can be further enhanced by a treatment in an alcohol that partially cancels the previous history of PIM-1 membranes by removing residual solvent and relaxing accumulated stresses.

Table 7.1 Permeability and selectivity for the H_2/CO_2 gas pair in different polymers

Polymer	Polymer state	Permeability (Barrer)		Selectivity CO_2/H_2	Reference
		H_2	CO_2		
Polysulfone (PSf)	Glassy	14.0	5.6	0.40	Aitken et al. (1992)
Poly(2,6-dimethylphenylene oxide) (PPO)	Glassy	69	35	0.51	Zhuang et al. (2015)
Cellulose acetate (CA)	Glassy	12	4.8	0.40	Puleo et al. (1989)
Polyimide (Matrimid)	Glassy	17.5	7.3	0.42	Zhang et al. (2008)
Amorphous polyethylene oxide (PEO)	Rubbery	21	143	6.8	Lin and Freeman (2004)
Pebax® 1657	Rubbery	8.0	73	9.1	Car et al. (2008)
Polydimethyl siloxane (PDMS)	Rubbery	890	3800	4.3	Merkel et al. (2000)
PIM-1, as cast	Glassy	1630	4390	2.7	Bushell et al. (2013a)
PIM-1, after treatment in etOH	Glassy	3300	12600	3.8	Bushell et al. (2013a)

Note: 1 Barrer = 10^{-10} cm^3 cm cm^{-2} s^{-1} $cmHg^{-1}$

The transport properties of highly rigid glassy polymers depend on the amount and distribution of free volume within the material (Moe et al., 1988). Glasses are in a non-equilibrium state, and many factors, such as the film formation procedure, may influence the packing of the macromolecules and, thus, the free volume distribution. The gas permeability in PIM-1 is very sensitive to various film formation and treatment protocols, including the type of casting solvent, drying conditions, and post-treatments (Budd et al., 2008). Contact with water results in relatively low gas permeability ($P(CO_2)$ = 1550 Barrer), while soaking in methanol leads to a marked increase in gas permeability ($P(CO_2)$ = 12,600

Barrer) (Budd et al., 2008). Treatment of PIM-1 membranes with methanol or ethanol is used to remove prior membrane formation history. The gas permeability of PIM-1 after a soaking in methanol increases approximately twofold. Indeed, these treatments affect the free volume amount and size distribution, resulting in a strong increase in the diffusion coefficient for permanent gases and a small increase in solubility.

PIM-1 demonstrated a gas permeability somewhat lower than PTMSP, but comparable hydrocarbon/hydrogen selectivity (Thomas et al., 2009). Condensable components in the feed mixture were reported to induce a swelling dilation of the PIM-1 polymer matrix. Accordingly, n-butane permeability increases with increasing concentration of n-butane in the feed mixture, leading to a large increase in n-butane/hydrogen selectivity. The blocking behavior with a reduction in permanent gas permeability in the presence of large condensable gases that was observed in PTMSP is also found in PIM-1 (Thomas et al., 2009). Therefore, PIM-1 is a novel highly hydrocarbon/hydrogen-selective microporous glassy polymer that could find use as membrane material in petrochemical applications.

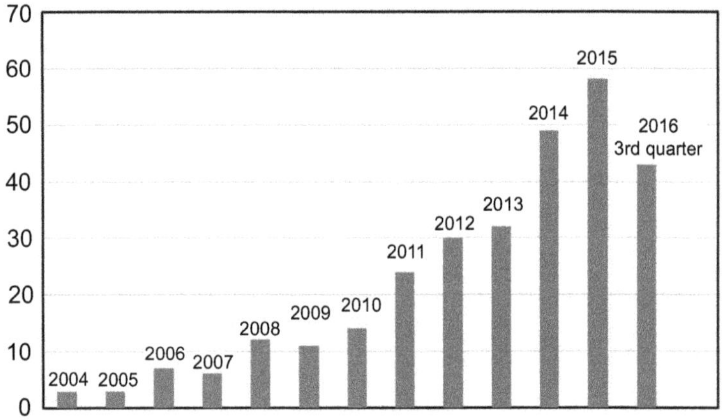

Figure 7.2 Number of yearly published scientific papers on PIMs since 2004. *Source*: Scopus, Title-Abstract-Keywords containing the words "polymer," "intrinsic," and "microporosity."

Owing to the interesting characteristics, PIM-1 was the starting point for a growing polymer family. An increasing number of scientific publications on these materials can be found (Fig. 7.2). In particular, two strategies were pursued for the production of different PIMs: the design of novel backbone structures or the chemical modification of PIM-1. The progresses achieved are presented in the following sections:

7.2.1.1 Chemical modification of PIM-1

Different research groups succeeded in the chemical modification of PIM-1. This strategy involves the conversion of nitrile groups to other functionalities (Fig. 7.3), leaving the same monomer structure (Du et al., 2009; Du et al., 2011; Mason et al., 2011; Mason et al., 2014; Patel and Yavuz, 2012; Swaidan et al., 2011; Weber et al., 2011).

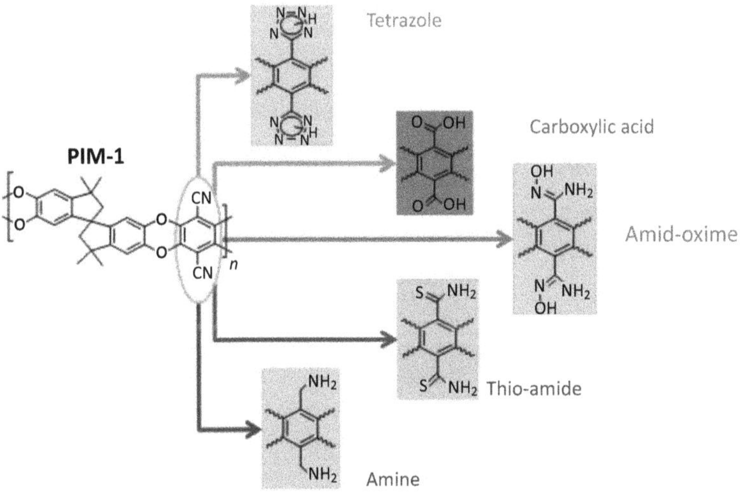

Figure 7.3 Structure of the repeat unit of PIM-1 and of some functional groups considered for its modification.

The chemical modification distinctly affected the amount and distribution of free volume within the polymer, particularly when groups capable of strong interactions such as H-bonding are inserted. As a result, a strong effect on the transport properties can be achieved. Typically, the chemical modification of PIM-1 induces a compaction of the polymer matrix, as evidenced by the decrease in the surface area of the polymer (Table 7.2). Accordingly, the chemical modification influences both gas permeability and selectivity.

Table 7.2 Effect of the chemical modification of PIM-1

Polymer	BET surface area (m^2/g)	Effect of the modification
PIM-1	770–850	
Carboxylic acid PIM-1	400	Reduced apparent BET surface area by increasing the carboxylation degree, due to strong interchain H-bonding (Patel and Yavuz, 1012)
Amidoxime PIM-1	480	Rigidification of the polymeric matrix; finer ultra-microporosity, by an extensive network of intermolecular H-bonding facilitated by ample electronegative and electropositive centers (Swaidan et al., 2011)
Thioamide PIM-1 50% conversion	544	Reduced apparent BET surface area and polymer's free volume • The thioamide is bulkier than the nitrile; • Groups capable of H-bonding → intermolecular interactions between neighboring chains (Mason et al., 2011).
80% conversion	263	

In particular, for the O_2/N_2 gas pair, amine-PIMs followed the usual selectivity/permeability trade-off, with a reduction in permeability associated with a higher selectivity (Mason et al, 2014). Instead, for mixtures involving CO_2 (e.g., CO_2/CH_4 or CO_2/N_2), amine-PIM-1 exhibited anomalous results. Gravimetric sorption proved that amine modification enhances the CO_2 capture capacity (Mason et al., 2014). However, this interaction with CO_2 is sufficiently strong that diffusion and, thus, permeability are substantially reduced. Conversely, a high permeability is obtained for small, non-interacting gases. Accordingly, while the H_2/CO_2 permselectivity is smaller than 1 in PIM-1, there is a permselectivity toward H_2 in amine-PIM-1. The creation of very small pores (ultramicropores, <0.7 nm), inaccessible to the slightly larger N_2 molecule, could be the reason of a strong adsorption of

CO_2. Another explanation could be the strong absorption by specific interaction of CO_2 with functional groups of the polymer that makes the desorption of CO_2 from polymer difficult, thus depressing the gas mobility. The net effect of increased polymer–penetrant interaction is a delicate balance between increased sorption, favoring transport, and possible "immobilization," penalizing diffusion.

Crosslinking of polymers is an effective strategy for reducing plasticization, but it is limited to polymers with crosslinkable organic functionalities and often reduce membrane permeability. The chemical crosslinking carried out on PIM-1 films resulted in enhanced C_3H_6/C_3H_8, CO_2/N_2, and CO_2/CH_4 gas separation, with adequate gas permeability, even if reduced from the ultra-high values of the pristine polymer (Khan et al., 2013a).

7.2.1.2 Novel PIM structures

The design of polymers with performance that surpasses the current Robeson's upper bounds for important gas pairs is a challenge. Over the past few years, different studies focused on the structure–property relationships of PIMs in order to optimize their performance based on Freeman's theoretical analysis (Freeman, 1999).

Novel structures were obtained by exchanging the spirobisindane (single tetrahedral carbon atom shared by two rings), which was the original contortion center in the ladder PIMs, with highly rigid units like the ethanoanthracene (EA) (Bezzu et al., 2012; Carta et al., 2013; Carta et al., 2014a; Carta et al., 2014b; Carta et al., 2014c; Ghanem et al., 2014; Rogan et al., 2014; Rose et al., 2015; Swaidan et al., 2014). This synthetic approach gave a direct demonstration of Freeman's (1999) conclusions regarding gas selectivity and polymer chain rigidity. A spirobifluorene-based PIM within which the 1,1′-spirobisindane units are replaced with more rigid 9,9′-spirobifluorene units, showed an enhanced permeability data for PIM-SBF relative to that of PIM-1 (Bezzu et al., 2012).

The rigid V-shaped bridged bicyclic unit 2,8-dimethyl-6H,12H-5,11-methanodibenzo[b,f][1,5]diazocine, known as Tröger's base (TB), demonstrated attractive features for the synthesis of novel polymers. Figure 7.4 reports some representative PIM structures that include the TB unit with a progressively more bulky section that was obtained passing from EA, to the triptycene (Trip), to the benzotriptycene (B-Trip). The design of these materials satisfies the requirement of microporous polymers of extreme rigidity in order to combine high permeability with selectivity for gas separation. It has been recognized that highly rigid bridged bicyclic structural units such as Trip and TB enhance the gas selectivity of the polymer (Guiver and Lee, 2013).

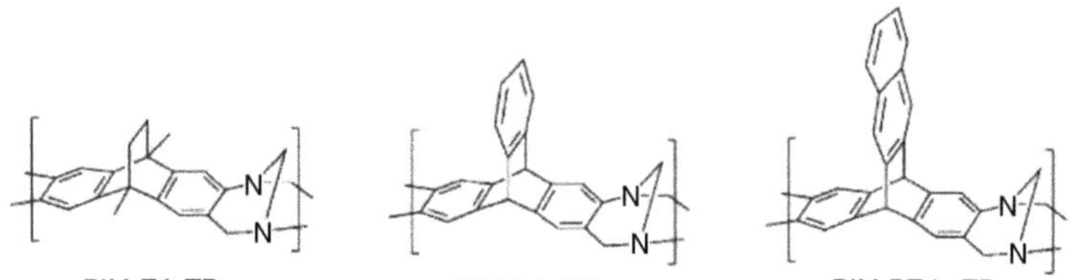

Figure 7.4 Repeat unit of PIM-EA-TB (Carta et al., 2013), PIM-Trip-TB (Carta et al., 2014d), and PIM-B-Trip-TB (Rose et al., 2015).

PIM-EA-TB is one of the most interesting novel structures (Carta et al., 2013). It has a shape-persistent monomer consisting of benzene rings fused together by inflexible bridged bicyclic units, EA and TB. The polymer's contorted shape ensures a microporosity characterized by very high values of the internal surface area, greater than 1000 m^2/g. This value is larger than that quoted for PIM-1 (700–800 m^2/g) (Budd and McKeown, 2010). Solution-cast robust films demonstrate exceptional performance as molecular sieves. Indeed, they have high gas permeability and good selectivity for smaller gas molecules, such as hydrogen and oxygen, over larger molecules, such as nitrogen and methane. Because of an enhanced size-sieving behavior, different from PIM-1, in PIM-EA-TB, the most permeable molecule is the small H_2. As already observed in the case of the amine-modified PIM-1 (Mason et al., 2014), the high solubility selectivity for the CO_2/N_2 gas pair did not give rise to a high ideal selectivity in PIM-EA-TB. The nitrogen atoms of the tertiary amine in the TB unit could strongly interact with CO_2 owing to the Lewis acid–base interaction and/or affinity between polymer and CO_2 molecules with a quadrupole moment, penalizing the gas diffusion.

Figure 7.5 indicates how the torsional flexibility decreases passing from the spirocenter to other units, particularly when bridged byciclics are involved. The PIM-1 polymer contains SBI and dioxane linkages that are quite flexible. Instead, the PIM-EA-TB has a more rigid architecture that resists rotation, maintaining a more uniform pore structure (Carta et al., 2013). This arrangement imparts high gas permeability in tandem with size discrimination between gases having slight differences in molecular dimensions, such as O_2 and N_2.

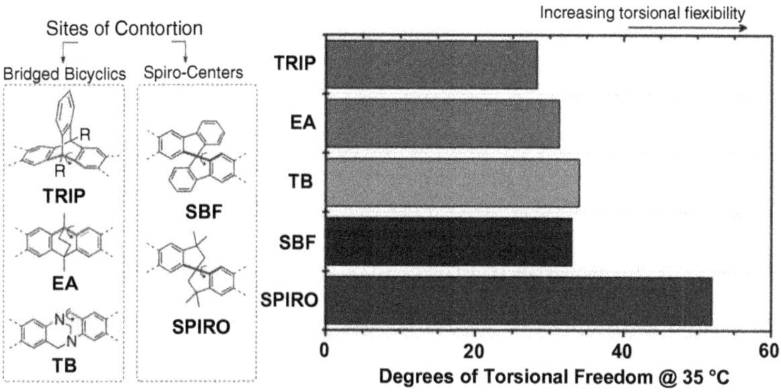

Figure 7.5 Molecular simulation of dihedral angle distributions for bridged bicyclic and spiro-based sites of contortion. The range of accessible angles is reported for each, assuming an available thermal energy of 3RT. Adapted with permission from Swaidan et al. (2014). Copyright (2014) American Chemical Society.

As expected because of the reduced flexibility (Swaidan et al., 2014), the use of Trip in place of the EA unit of the PIM-EA-TB provided an interesting evolution in PIMs (Fig. 7.4). The lack of methyl groups in PIM-Trip-TB (Carta et al., 2014d) may enhance performance as compared to PIM-EA-TB, within which methyl rotary motion may reduce diffusivity selectivity. PIM-Trip-TB provides highly impressive data for a number of commercially relevant gas pairs, including O_2/N_2, H_2/N_2, and CO_2/CH_4 in terms of permeability, selectivity, and position relative to the Robeson upper bound (Carta et al., 2014d). This performance of gas transport relies on exceptional diffusivity selectivity. Indeed, as PIM-Trip-TB is a ladder polymer composed of only benzene rings fused together by rigid bridged bicyclic units and contains no groups that can act as rotors, it results in optimal performance as a molecular sieve.

On attempts to increase intrinsic microporosity, the rigid structure of the Trip unit that is used to introduce nonlinearity into the PIM structure was extended to B-Trip (Rose et al., 2015). The polymer is highly permeable (e.g., P_H_2 = 9980 Barrer; P_O_2 = 3290 Barrer) with moderate selectivity (e.g., H_2/N_2 = 11.0; O_2/N_2 = 3.6) so that its data lie over the 2008 Robeson upper bounds for the H_2/N_2, O_2/N_2, and H_2/CH_4 gas pairs and on the upper bound for CO_2/CH_4. On aging, the polymer demonstrates a drop in permeability, which is typical for ultrapermeable polymers, but with a significant increase in gas selectivities (e.g., PO_2 = 1170 Barrer; PO_2/PN_2 = 5.4). PIM-BenzoTrip-TB (Fig. 7.4) exceeds the overall gas permeability performance of previously reported PIM-TBs apparently due to the additional free volume generated by the B-Trip component (Rose et al., 2015).

7.2.1.3 Free volume in PIMs

As shown in the previous section, the amount and size distribution of the free volume elements have an important role in determining the achievable performance of a nanostructured material. Detailed studies are available on the free volume of some PIMs investigated by combining experimental techniques and molecular modeling. Force field-based molecular mechanics and molecular dynamic methods were also applied to model the polymer structure and the spatial arrangement of the free volume in PIMs (Heuchel et al., 2008; Tocci et al., 2014).

A positron annihilation lifetime spectroscopy (PALS) analysis carried out on PIM-1 samples provided evidence of relatively large microcavities (R = 5 Å) (Budd and McKeown, 2010). The sample soaked with methanol contained very little of "residual solvent," as confirmed by the larger elements. The samples that included larger quantity of "residual solvents" (water and chloroform) are characterized by smaller radii of free volume elements. The size distribution function of free volume elements derived from the models of PIM-1 (Heuchel et al., 2008) indicates a range of pore width of 5–15 Å, in agreement with results obtained from experimental sorption data and analyzed with the Horvath–Kawazoe method. The Gusev–Suter transition state theory was used to calculate gas solubility and diffusion coefficients for He, H_2, Ar, O_2, N_2, CH_4, and Xe. Good agreement with experimental data (factors 1–4) was found. The calculated nitrogen adsorption isotherm at 77 K shows properties typical for sorption on microporous materials.

The obtained microporous volume is in good agreement with the respective experimental value.

Different techniques such as PALS, ^{129}Xe NMR, and molecular modeling were combined to investigate the size distribution of the free volume elements in PIM-EA-TB (Tocci et al., 2014). The extensive investigation of the structural, mechanical, and transport properties of this polymer gave deep insight into the behavior of this polymer, clarifying its remarkable performance as a membrane material (Tocci et al., 2014). An alcohol treatment for PIM-EA-TB produced a much stronger permeability increase than the archetypal PIM-1, with performance above the Robeson upper bound for several gas pairs. This is in part due to an extremely high gas solubility in PIM-EA-TB, even higher than in PIM-1. Modeling confirms that the high gas permeability depends on the large fractional free volume of the polymer. The simulated free volume size distribution in PIM-EA-TB is in agreement with the average experimental free volume elements size determined by PALS and 129Xe NMR. The modeled spatial arrangement of the free volume revealed a slightly lower interconnectivity of the FV elements in PIM-EA-TB compared to PIM-1. Along with its higher chain rigidity, determined by analysis of the torsion angles in the polymer model, this was identified as the main reason for its stronger size-sieving behavior and relatively high permselectivity (Tocci et al., 2014).

7.2.1.4 PIM-PIs

Polyimides (PIs) are an important class of polymers for membrane-based gas separations (Liaw et al., 2012). PIs, obtained by combining a dianhydride and a diamine, offer good thermal and chemical stability, as well as excellent mechanical properties. When applied to the separation of gaseous mixtures, PIs show high gas selectivity, but a low gas permeability (typically less than 10 Barrer for CO_2) (Cecopieri-Gomez et al., 2007).

The concept of intrinsic microporosity was applied also to this class of polymeric materials, producing semi-ladder polymers referred to as PIM-PIs, which demonstrated a high gas permeability (Ghanem et al., 2008b; Ghanem et al., 2009; Rogan et al., 2013). Different PIM–PIs with bulk and kinked structure were synthesized, containing dianhydrides derived from spirobisindane, Trip, EA, or diamines from spirobifluorene, spirobisindane, fluorene, binaphthalene, and Trip.

These PIM-PIs exhibit both high apparent surface areas and high gas permeability coefficients, greater than those of most polyimides studied so far and only slightly smaller than the permeability coefficients of the archetypal PIM, PIM-1. PIM-PIs are distinguished by extremely high gas solubility coefficients, a property characteristic to all PIMs. AA PIM–PI (PIM-EA-PI) derived from a dianhydride based on the bridged bicyclic EA (Fig. 7.6) demonstrated enhanced gas separation performance (Rogan et al., 2014). Similar performance was shown using a novel spirobifluorene-based dianhydride (SBFDA) (Ma et al., 2013).

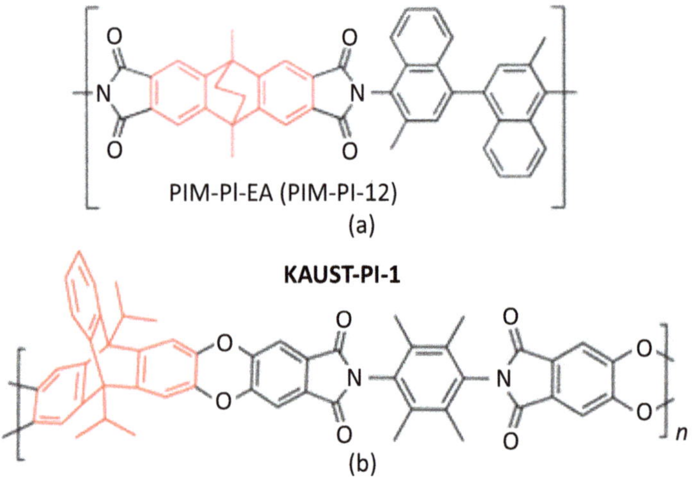

Figure 7.6 Repeat unit of semi-ladder PIM-EA-PI (a) and TPDA-TMPD or KAUST-PI-1 (b).

Novel TB-derived PIMs (PIM–TB–PI) were recently reported (Lee et al., 2016; Wang et al., 2014). A TB diamine monomer (4MTBDA) with four methyl groups was considered in order to restrict rotation about the C–N imide bonds in the resulting polymers. Gas permeability measurements demonstrated enhanced performance over previously reported polyimide based on TB, confirming the benefit of the additional methyl groups within the TB diamine monomer. In particular, the PI derived from 4MTBDA and pyromellitic anhydride (PMDA) demonstrates gas permeability data above the 2008 upper bounds for important gas pairs such as O_2/N_2, H_2/N_2, and H_2/CH_4. The solubility effect of the TB unit on CO_2 behaves differently for PIs with different free volume (Wang et al., 2014). Indeed, CO_2 solubility originated from TB units has a positive effect on the polymer with large free volume and thus enhances permeability, while has a negative effect on the polymer with small free volume, thus decreasing permeability.

The interesting features of trypticene units observed in the ladder PIMs were confirmed by the exceptional performance of polyimides derived from a 9,10-di-iso-propyltriptycene-based dianhydride (e.g., KAUST-PI-1, Fig. 7.6b), with data that lie well above the 2008 Robeson upper bounds for several important gas pairs (Ghanem et al., 2014). Interesting results were also reported for the challenging separation of propylene/propane (Swaidan et al., 2015a).

The thermal conversion approach (Park et al., 2007; Han et al., 2012) was also investigated for modifying the properties of PIM-PIs. Two spirocenter containing dianhydrides were utilized in the preparation of hydroxyl-functionalized polyimides with significant internal surface area and with interconnected free volume (Shamsipour et al., 2014). The PIM-PI-OH materials treated at 450°C yield TR polymers with enhanced performance for gas separation membranes and increased solvent resistance. The transport properties of the TR polymers can be tailored through the choice of the diamine used to prepare the precursor. For all polymers, the TR reaction and the ethanol treatment of the membranes enhance drastically the diffusion coefficient, while the solubility increases moderately, proving the key role of free volume and microporosity on gas diffusion (Shamsipour et al., 2014).

7.2.1.5 Shift in the state of the art by ladder and semi-ladder polymers

Novel upper bounds were recently defined (Swaidan et al., 2015b) considering representative ladder and semi-ladder PIMs. They were developed for air separation (O_2/N_2) and hydrogen recovery (H_2/N_2 and H_2/CH_4). The case of membrane-based air separation is reported in Fig. 7.7, showing the gain achievable in performance by using these nanostructured membranes, which is even larger than that reached passing from the upper bound defined in 1991 (Robeson, 1991) to the updated version of 2008 (Robeson, 2008).

A substantial shift in the permeability is evidenced with respect to the polymers still used at the industrial level as polysulfone (PSf). Therefore, these materials display an excellent potential for making membranes suitable for large-scale gas separations of commercial relevance.

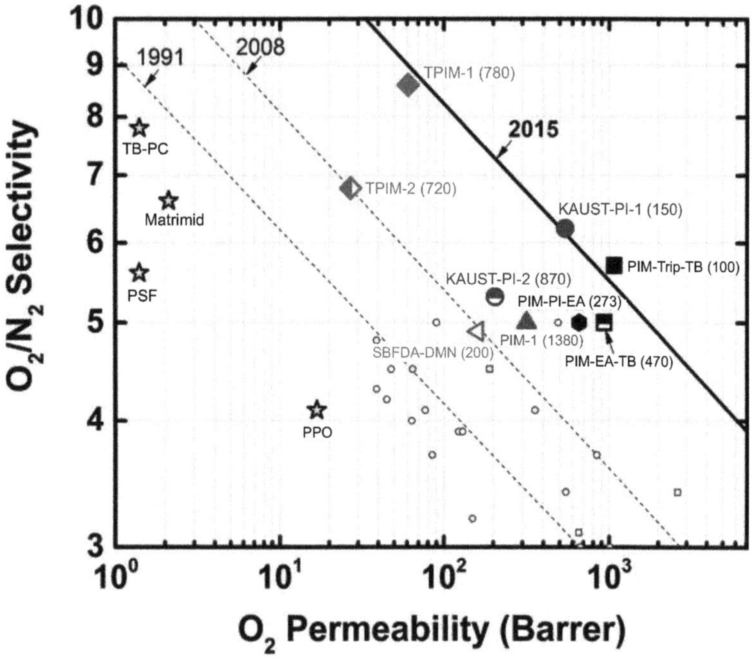

Figure 7.7 The 2015 "upper bound" to polymer membrane performance in O_2/N_2 separation, defined by outstanding PIMs. Previous upper bounds (1991 and 2008) are included as dashed lines. The "age" of the polymer in days after methanol treatment is reported in parentheses. The stars indicate commercially used membrane materials. The open circles and squares represent other semi-ladder PIM polyimides and ladder PIMs. Reprinted with permission from Swaidan et al. (2015b). Copyright (2015) American Chemical Society.

7.2.1.6 Mixed matrix membranes

Mixed matrix membranes represent alternate and cost-effective membrane systems with separation properties well above the upper bound limit between permeability and selectivity (Chung et al., 2007). They were initially developed in order to exploit gas transport mechanisms that differ from the classical solution-diffusion for dense polymer films. Simple models of the gas transport in these hybrid matrices (e.g., the Maxwell model) predict that the addition of impermeable nanofillers reduces permeability compared to the neat polymer. However, in different cases, the addition of nonporous fillers to glassy polymers was reported to increase permeability, as found by Pinnau and He (2001) for some high free volume polymers. The presence of a nanofiller surface restricts the conformational freedom of polymer chains in its vicinity, which may further hinder the chain packing, thus increasing free volume and enhancing permeability. In addition, void formation at the interface polymer/filler or within filler aggregates can occur, resulting in additional paths for the transport of gases. This behavior was confirmed for nonporous inorganic fillers, such as MgO or TiO_2 in various polymers (Matteucci et al., 2008; Moghadam et al., 2011) and for fumed silica dispersed in PIM-1 (Ahn et al., 2010).

Different studies have focused on the combination of PIM-1 polymer and fillers of diverse type. MMMs based on PIM-1 and the zeolitic imidazolate framework ZIF-8 (up to 43 vol.%) were studied by permeation and sorption (Bushell et al., 2013a). For "as-cast" films, an increase in ZIF-8 loading results in enhanced permeability and diffusion coefficients as well as in enhanced separation factors for the H_2/N_2, H_2/CH_4, He/N_2, O_2/N_2, and CO_2/CH_4 gas pairs. Permeability is further improved by a treatment with ethanol. The resulting data points are located above the 2008 upper bound on several Robeson diagrams. PALS and radiothermoluminescence (RTL) methods demonstrated that the introduction of ZIF-8 nanoparticles into the PIM-1 matrix results in an increase in free volume. Therefore, both the MOF cavities and the more loosely packed polymer chains at the ZIF-8/PIM-1 interface contribute to the transport.

The potential of purely organic dispersed phases, comprising only C, H, N, and O atoms, which is largely unexplored, was demonstrated in a study on MMMs based on PIM-1 filled with porous organic cages (Bushell et al., 2013a). These porous fillers should show better compatibility with a continuous polymeric matrix. A novel route to MMMs was demonstrated in which the dispersed phase is generated by in situ crystallization of porous organic cage molecules from a single homogeneous, molecular solution. The crystallizable precursor is CC3, which has approximately triangular windows of effective diameter 0.6 nm, large enough to admit gases and small organic molecules (Bushell et al., 2013b). The incorporation of the porous cages significantly enhances permeability of PIM-1, whereas chemically reduced nonporous cage molecules have an opposite effect.

Another possibility to exploit the properties of 2D materials is to use them as additives for polymeric matrices. The effect of graphene on the gas permeability of PIM-1 was recently reported (Althumayri et al., 2016). The data were compared with those for fused silica nanofillers (Ahn et al., 2010) and multiwalled carbon nanotubes functionalized with poly(ethylene glycol) (f-MWCNTs) (Khan et al., 2013b). This study proves that the curvature of the surface at the nanoscale of a nanofiller is an additional property that affects MMM properties, besides its available surface area (Althumayri et al., 2016). While graphene sheets are planar, multiwalled carbon nanotubes are cylindrical and fumed silica nanoparticles are almost spherical. For graphene, a significant enhancement in permeability is observed at very low content (0.05 vol.%), which may be attributed to the effect of the nanofiller on the packing of the polymer chains. At higher graphene content, permeability decreases, as expected for the addition of an impermeable filler. Other nanofillers were reported to enhance the permeability, but at substantially higher loadings: the highest measured permeability was found at 1 vol.% for f-MWCNTs and 24 vol.% for fused silica. These results are consistent with the assumption that packing ability of the polymer chains could be influenced by the curvature of the nanofiller surface at the nanoscale, with an increasingly pronounced effect on moving from a more-or-less spherical nanoparticle morphology (fused silica), to a cylindrical morphology (f-MWCNT), to a planar morphology (graphene). In addition, despite the permeability decrease over time through physical aging for a high-free-volume polymer such as PIM-1, the PIM-1/graphene MMMs showed that a significant permeability enhancement was retained after eight months storage.

7.2.1.7 Physical aging

Physical aging is typically encountered in highly permeable polymers that are in a glassy state and thus in a non-equilibrium condition. According to the free volume theory, the free volume can diffuse in polymeric membranes via the thermal motion of the polymer segments (Rowe et al., 2009). The relaxation of the polymer chains in the glassy state results in a decrease in permeability over time (Harms et al., 2012). However, the performance of an aged polymer typically follows the usual selectivity/permeability trade-off, with the reduction in permeability associated to a higher selectivity.

The long-term aging of PIM-1 membranes was tracked studying the variations in the permeation and sorption properties over time (Bernardo et al., 2012). The PIM-1 films were tested in different states: "as cast" (i.e., with residual solvent), after soaking in ethanol, and after soaking in ethanol plus a thermal treatment at 75°C or 125°C. The observed reduction in permeability during time depends mainly on the decrease in the diffusion coefficient. However, a thermal treatment, by accelerating the aging and slowing down the further aging rate, can be adopted to effectively control the extent of the aging-induced changes in the transport properties.

The addition of different fillers to PIM-1 could be an effective strategy in controlling this issue, as in the case of the porous organic cages (Bushell et al., 2013b) and of graphene (Althumayri et al., 2016). Appropriate amounts of the filler can increase the permeability over that of the pristine polymer, leading also to an improved stability over time.

Recently, it was reported that addition of a polymeric filler could reduce aging while improving selectivity manifold as reported by Lau et al. (2014, 2015a, 2015b). This filler was a microporous organic polymer, referred to as PAF-1, which was incorporated at a loading of 10 wt.%. However, its synthesis requires relatively expensive building blocks, air- and moisture-sensitive organometallic coupling agents (Mitra et al., 2016). A water- and acid-tolerant hyper-crosslinked polymer, referred to as HCP-24, which is synthesized from cheap starting materials using relatively inexpensive and air-tolerant catalysts, was dispersed in PIM-1 (Mitra et al., 2016). The resulting MMMs were found to age faster than those containing the PAF-1, but the hypercrosslinked nanoparticle fillers are cost-effective and scalable (Mitra et al., 2016).

7.3 Conclusion

In the last few years, novel microporous materials demonstrated outstanding properties as membrane materials for gas separation.

Among them, PIMs represent an implementation of the synthetic strategies to be pursued in order to improve the performance of polymeric materials for gas separation. Indeed, they combine increasing interchain distances that would result in polymers with higher internal free volume, with an enhanced backbone rigidity that could improve selectivity via molecular sieving. A significant evolution was achieved in the field of PIMs, resulting in a platform with excellent potential for making membranes suitable for large-scale gas separations of commercial and environmental relevance. A rational design enabled tailoring the gas transport properties to combine high permeability with high selectivity for gas separations. Especially, the replacement of the spirobisindane contortion centers present in the original PIMs by using bridged bicyclics (e.g., EA, TB, and Trip) enhanced the polymer intra-chain rigidity. Accordingly, a substantial improvement in gas transport performance was demonstrated, thus contributing to redefine the state of the art in membrane-based air and hydrogen separations. The same synthetic concept applied to the structural design of polyimides produced semi-ladder PIM-PIs with significantly enhanced permeability.

A number of peculiarities in the behavior of PIMs were outlined by different studies, highlighting the effect of the thermomechanical history, residual solvent, aging, or conditioning, on the free volume amount and size distribution and, thus, on the resulting transport properties.

The addition of suited nanofillers to these high free volume polymers demonstrated to increase in some cases the permeability of the membranes, but, more interestingly, can lead to more stable membranes toward physical aging.

The synergy of experimental and modeling studies in developing structure–property relationships has a key role in exploiting completely the potential of these materials, as well as the engineering of membranes as thin-film composites for practical applications.

Abbreviations

BTRIP	benzotrypticene
EA	ethanoanthracene
FFV	fractional free volume
HCP	hypercrosslinked polymer

MOF	metal organic framework
PAF	porous aromatic framework
PALS	positron annihilation lifetime spectroscopy
PI	polyimide
PIM	polymer of intrinsic microporosity
PSf	polysulfone
PTMSP	poly[1-(trimethylsilyl)-1-propyne]
SBI	spirobisindane
TB	Tröger's base
TRIP	triptycene

References

Alentiev, A. Y., Yampolskii, Y. P., Shantarovich, V. P., Nemser, S. M., and Plate, N. A. (1997). High transport parameters and free volume of perfluorodioxole copolymers, *J. Membr. Sci.*, **126**, pp. 123–132.

Aitken, C. L., Koros, W. J., and Paul, D. R. (1992). Effect of structural symmetry on gas transport properties of polysulfones, *Macromolecules*, **25**(13), pp. 3424–3434.

Ahn, J., Chung, W.-J., Pinnau, I., Song, J., Du, N., Robertson, G. P., and Guiver, M. D. (2010). Gas transport behavior of mixed-matrix membranes composed of silica nanoparticles in a polymer of intrinsic microporosity (PIM-1), *J. Membr. Sci.*, **346**(2), pp. 280–287.

Althumayri, K., Harrison, W. J., Shin, Y., Gardiner, J. M., Casiraghi, C., Budd, P. M., Bernardo, P., Clarizia, G., and Jansen, J. C. (2016). The influence of graphene and other nanofillers on the gas permeability of the high-free-volume polymer PIM-1, *Phil. Trans. R. Soc. A*, **374**(2060), art. no. 20150031.

Bernardo, P., Bazzarelli, F., Jansen, J. C., Clarizia, G., Tasselli, F., Mason, C. R., Maynard-Atem, L., Budd, P. M., Friess, K., Lanč, M., Yampolskii, Y. P., and Shantarovich, V. (2012). Physical ageing study of post-treated PIM-1 membranes: Effect on gas transport properties, *Procedia Eng.*, **44**, pp. 874–876.

Bernardo, P., Drioli, E., and Golemme, G. (2009). Membrane gas separation: A review/state of the art, *Ind. Eng. Chem. Res.*, **48**, pp. 4638–4663.

Berry, V. (2013). Impermeability of graphene and its applications, *Carbon*, **62**, pp. 1–10.

Bezzu, C. G., Carta, M., Tonkins, A., Jansen, J. C., Bernardo, P., Bazzarelli, F., and McKeown, N. B. (2012). A spirobifluorene-based polymer of intrinsic microporosity with improved performance for gas separation, *Adv. Mater.*, **24**, pp. 5930–5933.

Budd, P. M., Ghanem, B. S., Makhseed, S., McKeown, N. B., Msayib, K. J., and Tattershall, C. E. (2004a). Polymers of intrinsic microporosity (PIMs): Robust, solution-processable, organic nanoporous materials, *Chem. Commun.*, pp. 230–231.

Budd, P. M., Makhseed, S. M., Ghanem, B. S., Msayib, K. J., Tattershall, C. E., and McKeown, N. B. (2004b). Microporous polymeric materials, *Mater. Today*, **7**(4), pp. 40–46.

Budd, P. M., Msayib, K. J., Tattershall, C. E., Ghanem, B. S., Reynolds, K. J., McKeown, N. B., and Fritsch, D. (2005). Gas separation membranes from polymers of intrinsic microporosity, *J. Membr. Sci.*, **251**, pp. 263–269.

Budd, P. M., McKeown, N. B., Ghanem, B. S., Msayib, K. J., Fritsch, D., Starannikova, L., Belov, N., Sanfirova, O., Yampolskii, Y., and Shantarovich, V. (2008). Gas permeation parameters and other physicochemical properties of a polymer of intrinsic microporosity: Polybenzodioxane PIM-1, *J. Membr. Sci.*, **325**, pp. 851–860.

Budd, P. M. and McKeown, N. B. (2010). Highly permeable polymers for gas separation membranes, *Polym. Chem.*, **1**, pp. 63–68.

Bunch, J. S., Verbridge, S. S., Alden, J. S., van der Zande, A. M., Parpia, J. M., Craighead, H. G., and McEuen, P. L. (2008). Impermeable atomic membranes from graphene sheets, *Nano Lett.*, **8**(8), pp. 2458–2462.

Bushell, A. F., Attfield, M. P., Mason, C. R., Budd, P. M., Yampolskii, Y., Starannikova, L., Rebrov, A., Bazzarelli, F., Bernardo, P., Jansen, J. C., Lanč, M., Friess, K., Shantarovich, V., Gustov, V., and Isaeva, V. (2013a). Gas permeation parameters of mixed matrix membranes based on the polymer of intrinsic microporosity PIM-1 and the zeolitic imidazolate framework ZIF-8, *J. Membr. Sci.*, **427**, pp. 48–62.

Bushell, A. F., Budd, P. M., Attfield, M. P., Jones, J. T. A., Hasell, T., Cooper, A. I., Bernardo, P., Bazzarelli, F., Clarizia, G., and Jansen, J. C. (2013b). Nanoporous organic polymer/cage composite membranes, *Angew. Chem., Int. Ed.*, **52**, pp. 1253–1256.

Car, A., Stropnik, C., Yave, W., and Peinemann, K.-V. (2008). PEG modified poly(amide-b-ethylene oxide) membranes for CO_2 separation, *J. Membr. Sci.*, **307**(1), pp. 88–95.

Carta, M., Msayib, K. J., Budd, P. M., and McKeown, N. B. (2008). Novel spirobisindanes for use as precursors to polymers of intrinsic microporosity, *Org. Lett.*, **10**, pp. 2641–2643.

Carta, M., Msayib, K. J., and McKeown, N. B. (2009). Novel polymers of intrinsic microporosity (PIMs) derived from 1,1-spiro-bis(1,2,3,4-tetrahydronaphthalene)-based monomers, *Tetrahedron Lett.,* **50**, pp. 5954–5957.

Carta, M., Malpass-Evans, R., Croad, M., Jansen, J. C., Bernardo, P., Bazzarelli, F., and McKeown, N. B. (2013). An efficient polymer molecular sieve for membrane gas separations, *Science,* **339**, pp. 303–307.

Carta, M., Malpass-Evans, R., Croad, M., Rogan, Y., Lee, M., Rose, I., and McKeown, N. B. (2014a). The synthesis of microporous polymers using Tröger's base formation, *Polym. Chem.,* **5**, pp. 5267–5272.

Carta, M., Croad, M., Jansen, J. C., Bernardo, P., Clarizia, G., and McKeown, N. B. (2014b). Synthesis of cardo-polymes using Tröger's base formation, *Polym. Chem.,* **5**, pp. 5255–5261.

Carta, M., Bernardo, P., Clarizia, G., Jansen, J. C., and McKeown, N. B. (2014c). Gas permeability of hexaphenylbenzene based polymers of intrinsic microporosity, *Macromolecules,* **47**(23), pp. 8320–8327.

Carta, M., Croad, M., Malpass-Evans, R., Jansen, J. C., Bernardo, P., Clarizia, G., Friess, K., Lanč, M., and McKeown, N. B. (2014d). Triptycene induced enhancement of membrane gas selectivity for microporous Tröger's base polymers, *Adv. Mater.,* **26**(21), pp. 3526–3531.

Cecopieri-Gomez, M. L., Palacios-Alquisira, J., and Dominguez, J. M. (2007). On the limits of gas separation in CO_2/CH_4, N_2/CH_4 and CO_2/N_2 binary mixtures using polyimide membranes, *J. Membr. Sci.,* **293** (1-2), pp. 53–65.

Chung, T.-S., Jiang, L. Y., Li, Y., and Kulprathipanja, S. (2007). Mixed matrix membranes (MMMs) comprising organic polymers with dispersed inorganic fillers for gas separation, *Prog. Polym. Sci.,* **32**(4), pp. 483–507.

Du, N. Y., Robertson, G. P., Song, J. S., Pinnau, I., Thomas, S., and Guiver, M. D. (2008). Polymers of intrinsic microporosity containing trifluoromethyl and phenylsulfone groups as materials for membrane gas separation, *Macromolecules,* **41**, pp. 9656–9662.

Du, N., Robertson, G. P., Song, J., Pinnau, I., and Guiver, M. D. (2009). High-performance carboxylated polymers of intrinsic microporosity (PIMs) with tunable gas transport properties, *Macromolecules,* **42**, pp. 6038–6043.

Du, N. Y., Robertson, G. P., Pinnau, I., and Guiver, M. D. (2010). Polymers of intrinsic microporosity with dinaphthyl and thianthrene segments, *Macromolecules,* **43**, pp. 8580–8587.

Du, N., Park, H. B., Robertson, G. P., Dal-Cin, M. M., Visser, T., Scoles, L., and Guiver, M. D. (2011). Polymer nanosieve membranes for CO_2-capture applications, *Nat. Mater.,* **10**, pp. 372–375.

Emmler, T., Heinrich, K., Fritsch, D., Budd, P. M., Chaukura, N., Ehlers, D., Ratzke, K., and Faupel, F. (2010). Free volume investigation of polymers of intrinsic microporosity (PIMs): PIM-1 and PIM1 copolymers incorporating ethanoanthracene units, *Macromolecules,* **43**, pp. 6075–6084.

Friess, K., Jansen, J. C., Bazzarelli, F., Izák, P., Jarmarova, V., Kacırkova, M., Schauer, J., Clarizia, G., and Bernardo, P. (2012). High ionic liquid content polymeric gel membranes: Correlation of membrane structure with gas and vapour transport properties, *J. Membr. Sci.,* **415–416**, pp. 801–809.

Fritsch, D., Bengtson, G., Carta, M., and McKeown, N. B. (2011). Synthesis and gas permeation properties of spirobischromane-based polymers of intrinsic microporosity, *Macromol. Chem. Phys.,* **212**, pp. 1137–1146.

Freeman, B. D. (1999). Basis of permeability/selectivity tradeoff relations in polymeric gas separation membranes, *Macromolecules,* **32**(2), pp. 375–380.

Ghanem, B. S., McKeown, N. B., Budd, P. M., and Fritsch, D. (2008a). Polymers of intrinsic microporosity (PIMs) derived from bis(phenazyl) monomers, *Macromolecules,* **41**, pp. 1640–1646.

Ghanem, B. S., McKeown, N. B., Budd, P. M., and Fritsch, D. (2008b). High performance membranes from polyimides with intrinsic microporosity, *Adv. Mater.,* **20**, pp. 2766–2771.

Ghanem, B. S., McKeown, N. B., Budd, P. M., Al-Harbi, N. M., Fritsch, D., Heinrich, K., Starannikova, L., Tokarev, A., and Yampolskii, Y. (2009). Synthesis, characterization, and gas permeation properties of a novel group of polymers: PIM-polyimides, *Macromolecules,* **42**, pp. 7881–7888.

Ghanem, B. S., Swaidan, R., Litwiller, E., and Pinnau, I. (2014). Ultra-microporous triptycene-based polyimide membranes for high-performance gas separation, *Adv. Mater.,* **26**, pp. 3688–3692.

Guiver, M. D. and Lee, Y. M. (2013). Polymer rigidity improves microporous membranes, *Science,* **339**(6117), pp. 284–285.

Han, S. H., Kwon, H. J., Kim, K. Y., Seong, J. G., Park, C. H., Kim, S., Doherty, C. M., Thornton, A. W., Hill, A. J., Lozano, A. E., Berchtold, K. A., and Lee, Y M. (2012). Tuning microcavities in thermally rearranged polymer membranes for CO_2 capture, *Phys. Chem. Chem. Phys.,* **14**(13), pp. 4365–4373.

Harms, S., Raetzke, K., Faupel, F., Chaukura, N., Budd, P. M., Egger, W., and Ravelli, L. (2012). Aging and free volume in a polymer of intrinsic microporosity (PIM-1), *J. Adhes.*, **88**, pp. 608–619.

Heuchel, M., Fritsch, D., Budd, P. M., McKeown, N. B., and Hofmann, D. (2008). Atomistic packing model and free volume distribution of a polymer with intrinsic microporosity (PIM-1), *J. Membr. Sci.*, **318**(1), pp. 84–99.

Khan, M. M., Bengtson, G., Shishatskiy, S., Gacal, B. N., Rahman, Md. M., Neumann, S., Filiz, V., and Abetz, V. (2013a). Cross-linking of polymer of intrinsic microporosity (PIM-1) via nitrene reaction and its effect on gas transport property, *Eur. Polym. J.*, **49**(12), pp. 4157–4166.

Khan, M. M., Filiz, V., Bengtson, G., Shishatskiy, S., Rahman, M. M., Lillepaerg, J., and Abetz, V. (2013b). Enhanced gas permeability by fabricating mixed matrix membranes of functionalized multiwalled carbon nanotubes and polymers of intrinsic microporosity (PIM), *J. Membr. Sci.*, **436**, pp. 109–120.

Kim, H. W., Yoon, H. W., Yoon, S.-M., Yoo, B. M., Ahn, B. K., Cho, Y. H., Shin, H. J., Yang, H., Paik, U., Kwon, S., Choi, J.-Y., and Park, H. B. (2013). Selective gas transport through few-layered graphene and graphene oxide membranes, *Science*, **342**(6154), pp. 91–95.

Koschine, T., Raetzke, K., Faupel, F., Khan, M. M., Emmler, T., Filiz, V., Abetz, V., Ravelli, L., and Egger, W. (2015). Correlation of gas permeation and free volume in new and used high free volume thin film composite membranes, *J. Polym. Sci. B Polym. Phys.*, **53**, pp. 213–217.

Lau, C. H., Li, P., Li, F., Chung, T.-S., and Paul, D. R. (2013). Reverse-selective polymeric membranes for gas separations, *Prog. Polym. Sci.*, **38**, pp. 740–766.

Lau, C. H., Nguyen, P. T., Hill, M. R., Thornton, A. W., Konstas, K., Doherty, C. M., Mulder, R. J., Bourgeois, L., Liu, A. C. Y., Sprouster, D. J., Sullivan, J. P., Bastow, T. J., Hill, A. J., Gin, D. L., and Noble, R. D. (2014). Forever young: Ending aging in super glassy polymer membranes, *Angew. Chem., Int. Ed.*, **53**, pp. 5322–5326.

Lau, C. H., Konstas, K., Doherty, C. M., Kanehashi, S., Ozcelik, B., Kentish, S. E., Hill, A. J., and Hill, M. R. (2015a). Tailoring physical aging in super glassy polymers with functionalized porous aromatic frameworks for CO_2 capture, *Chem. Mater.*, **27**, pp. 4756–4762.

Lau, C. H., Konstas, K., Thornton, A., Liu, W. A. C. Y., Mudie, S., Kennedy, D. F., Howard, S. C., Hill, A. J., and Hill, M. R. (2015b). Gas-separation membranes loaded with porous aromatic frameworks that improve with age, *Angew. Chem., Int. Ed.*, **54**, pp. 2669–2673.

Lee, M., Bezzu, G. C., Carta, M., Bernardo, P., Clarizia, G., Jansen, J. C., and McKeown, N. B. (2016). Enhancing gas permeability of polyimides of intrinsic microporosity containing Tröger's base, *Macromolecules,* **49**(11), pp. 4147–4154.

Liaw, D.-J., Wang, K.-L., Huang, Y.-C., Lee, K.-R., Lai, J.-Y., and Ha, C.-S. (2012). Advanced polyimide materials: Syntheses, physical properties and applications, *Prog. Polym. Sci.,* **37**(7), pp. 907–974.

Lin, H. and Freeman, B. D. (2004). Gas solubility, diffusivity and permeability in poly(ethylene oxide), *J. Membr. Sci.,* **239**(1), pp. 105–117.

Lin, H. and Freeman, B. D. (2005). Materials selection guidelines for membranes that remove CO_2 from gas mixtures, *J. Mol. Struct.,* **739**, pp. 57–74.

Liu, G., Jin, W., and Xu, N. (2016). Two-dimensional-material membranes: A new family of high-performance separation membranes, *Angew. Chem. Int. Ed.,* **55**, pp. 2–16.

Ma, X., Salinas, O., Litwiller, E., and Pinnau, I. (2013). Novel spirobifluorene- and dibromospirobifluorene-based polyimides of intrinsic microporosity for gas separation applications, *Macromolecules,* **46**, pp. 9618–9624.

Mason, C. R., Maynard-Atem, L., Al-Harbi, N. M., Budd, P. M., Bernardo, P., Bazzarelli, F., Clarizia, G., and Jansen, J. C. (2011). Polymer of intrinsic microporosity incorporating thioamide functionality: Preparation and gas transport properties, *Macromolecules,* **44**, pp. 6471–6479.

Mason, C. R., Maynard-Atem, L., Heard, K. W. J., Satilmis, B., Budd, P. M., Friess, K., Lanc, M., Bernardo, P., Clarizia, G., and Jansen, J. C. (2014). Enhancement of CO_2 affinity in a polymer of intrinsic microporosity by amine modification, *Macromolecules,* **47**, pp. 1021–1029.

Matteucci, S., Kusuma, V. A., Kelman, S. D., and Freeman, B. D. (2008). Gas transport properties of MgO filled poly(1-trimethylsilyl-1-propyne) nanocomposites, *Polymer,* **49**, pp. 1659–1675.

McKeown, N. B., Budd, P. M., Msayib, K. J., Ghanem, B. S., Kingston, H. J., Tattershall, C. E., Makhseed, S., Reynolds, K. J., and Fritsch, D. (2005). Polymers of intrinsic microporosity (PIMs): Bridging the void between microporous and polymeric materials, *Chem. Eur. J.,* **11**, pp. 2610–2620.

McKeown, N. B. (2012). Polymers of intrinsic microporosity, *ISRN Mater. Sci.,* 2012, Article ID 513986.

Merkel, T. C., Bondar, V. I., Nagai, K., Freeman, B. D., and Pinnau, I. (2000). Gas sorption, diffusion, and permeation in poly(dimethylsiloxane), *J. Polym. Sci. B: Polym. Phys.,* **38**, pp. 415–434.

Mitra, T., Bhavsar, R. S., Adams, D. J., Budd, P. M., and Cooper, A. I. (2016). PIM-1 mixed matrix membranes for gas separations using cost-effective hypercrosslinked nanoparticle fillers, *Chem. Commun.*, **52**, pp. 5581–5584.

Moe, M., Koros, W. J., Hoehn, H. H., and Husk, G. R. (1988). Effects of film history on gas transport in a fluorinated aromatic polyimide, *J. Appl. Polym. Sci.*, **36**, pp. 1833–1841.

Moghadam, F., Omidkhah, M. R., Vasheghani-Farahani, E., Pedram, M. Z., and Dorosti, F. (2011). The effect of TiO_2 nanoparticles on gas transport properties of Matrimid5218-based mixed matrix membranes, *Sep. Purif. Technol.*, **77**, pp. 128–136.

Novoselov, K. S., Geim, A. K., Morozov, S. V., Jiang, D., Zhang, Y., Dubonos, S. V., Grigorieva, I. V., and Firsov, A. A. (2004). Electric field effect in atomically thin carbon films, *Science,* **306**(5296), pp. 666–669.

Park, H. B., Jung, C. H., Lee, Y. M., Hill, A. J., Pas, S. J., Mudie, S. T., Wagner, E. V., Freeman, B. D., and Cookson, D. J. (2007). Polymers with cavities tuned for fast selective transport of small molecules and ions, *Science,* **318**(5848), pp. 254–258.

Patel, H. A. and Yavuz, C. T. (2012). Noninvasive functionalization of polymers of intrinsic microporosity for enhanced CO_2 capture, *Chem. Comm.,* **48**, pp. 9989–9991.

Pinnau, I. and He, Z. (2001). Filled superglassy membrane. US patent 6,316,684.

Puleo, A. C., Paul, D. R., and Kelley, S. S. (1989). The effect of degree of acetylation on gas sorption and transport behavior in cellulose acetate, *J. Membr. Sci.*, **47**(3), pp. 301–332.

Ritter, N., Antonietti, M., Thomas, A., Senkovska, I., Kaskel, S., and Weber, J. (2009). Binaphthalene-based, soluble polyimides: The limits of intrinsic microporosity, *Macromolecules,* **42**, pp. 8017–8020.

Robeson, L. M. (1991). Correlation of separation factor versus permeability for polymeric membranes, *J. Membr. Sci.,* **62**, pp. 165–185.

Robeson, L. M. (2008). The upper bound revisited, *J. Membr. Sci.,* **320**, pp. 390–400.

Rogan, Y., Starannikova, L., Ryzhikh, V., Yampolskii, Y., Bernardo, P., Bazzarelli, F., Jansen, J. C., and McKeown, N. B. (2013). Synthesis and gas permeation properties of novel spirobisindane-based polyimides of intrinsic microporosity, *Polym. Chem.*, **4**(13), pp. 3813–3820.

Rogan, Y., Malpass-Evans, R., Carta, M., Lee, M., Jansen, J. C., Bernardo, P., Clarizia, G., Tocci, E., Friess, K., Lanc, M., and McKeown, N. B. (2014). A highly permeable polyimide with enhanced selectivity for membrane gas separations, *J. Mater. Chem. A,* **2**, pp. 4874–4877.

Rose, I., Carta, M., Malpass-Evans, R., Bernardo, P., Clarizia, G., Jansen, J. C., and McKeown, N. B. (2015). A highly permeable benzotryptcene-based polymer of intrinsic microporosity, *ACS Macro Lett.,* **4**, pp. 912–915.

Rowe, B. W., Freeman, B. D., and Paul, D. R. (2009). Physical aging of ultrathin glassy polymer films tracked by gas permeability, *Polymer,* **50**(23), pp. 5565–5575.

Shamsipour, H., Dawood, B. A., Budd, P. M., Bernardo, P., Clarizia, G., and Jansen, J. C. (2014). Thermally rearrangeble PIM-polyimides for gas separation membranes, *Macromolecules,* **47**(16), pp. 5595–5606.

Song, J., Du, N., Dai, Y., Robertson, G. P., Guiver, M. D., Thomas, S., and Pinnau, I. (2008). Linear high molecular weight ladder polymers by optimized polycondensation of tetrahydroxytetramethylspirobisindane and 1,4-dicyanotetrafluorobenzene, *Macromolecules,* **41**, pp. 7411–7417.

Srinivasan, R., Auvil, S. R., and Burban, P. M. (1994). Elucidating the mechanism(s) of gas transport in poly(1-(trimethylsilyl)-1-propyne) (PTMSP) membranes, *J. Membr. Sci.,* **86**, pp. 67–86.

Swaidan, R., Ghanem, B. S., Litwillwer, E., and Pinnau, I. (2011). Pure- and mixed-gas CO_2/CH_4 separation properties of PIM-1 and an amidoxime-functionalized PIM-1, *J. Membr. Sci.,* **457**, pp. 95–102.

Swaidan, R., Al-Saeedi, M., Ghanem, B., Litwiller, E., and Pinnau, I. (2014). Rational design of intrinsically ultramicroporous polyimides containing bridgehead-substituted triptycene for highly selective and permeable gas separation membranes, *Macromolecules,* **17**, pp. 5104–5114.

Swaidan, R. J., Ghanem, B., Swaidan, R., Litwiller, E., and Pinnau, I. (2015a). Pure- and mixed-gas propylene/propane permeation properties of spiro- and triptycene-based microporous polyimides, *J. Membr. Sci.,* **492**, pp. 116–122.

Swaidan, R., Ghanem, B., and Pinnau, I. (2015b). Fine-tuned intrinsically ultramicroporous polymers redefine the permeability/selectivity upper bounds of membrane-based air and hydrogen separations, *ACS Macro Lett.,* **4**(9), pp. 947–951.

Thomas, S., Pinnau, I., Du, N., and Guiver, M. D. (2009). Hydrocarbon/hydrogen mixed-gas permeation properties of PIM-1, an amorphous microporous spirobisindane polymer, *J. Membr. Sci.,* **338**, pp. 1–4.

Tocci, E., De Lorenzo, L., Bernardo, P., Clarizia, G., Bazzarelli, F., McKeown, N. B., Carta, M., Malpass-Evans, R., Friess, K., Pilnácek, K., Lanč, M.,

Yampolskii, Y. P., Strarannikova, L., Shantarovich, V., Mauri, M., and Jansen, J. C. (2014). Ethanoanthracene-based polymer of intrinsic microporosity from Tröger's base formation: Detailed permeation properties and molecular modelling, *Macromolecules*, **47**(22), pp. 7900–7916.

Yave, W., Car, A., and Peinemann, K.-V. (2010). Nanostructured membrane material designed for carbon dioxide separation, *J. Membr. Sci.*, **350**(1–2), pp. 124–129.

Yoon, H. W., Cho, Y. H., and Park, H. B. (2016). Graphene-based membranes: Status and prospects, *Phil. Trans. R. Soc. A*, **374**, 20150024.

Wang, Z., Wang, D., Zhang, F., and Jin, J. (2014). Tröger's base-based microporous polyimide membranes for high-performance gas separation, *ACS Macro Lett.*, **3**(7), pp. 597–601.

Weber, J., Du, N., and Guiver, M. D. (2011). Influence of intermolecular interactions on the observable porosity in intrinsically microporous polymers, *Macromolecules*, **44**, pp. 1763–1767.

Zhang, Y., Musselman, I. H., Ferraris, J. P., and Balkus, K. J. (2008). Gas permeability properties of mixed-matrix matrimid membranes containing a carbon aerogel: A material with both micropores and mesopores, *Ind. Eng. Chem. Res.*, **47**, pp. 2794–2802.

Zhuang, G.-L., Wey, M.-Y., and Tseng, H.-H. (2015). The density and crystallinity properties of PPO-silica mixed-matrix membranes produced via the in situ sol-gel method for H_2/CO_2 separation. II: Effect of thermal annealing treatment, *Chem. Eng. Res. Des.*, **104**, pp. 319–332.

Chapter 8

Nano-Based Catalytic Membrane Reactors for Hydrogen Production

Adele Brunetti and Giuseppe Barbieri
Institute on Membrane Technology of the National Research Council of Italy (ITM-CNR), Via P. Bucci, Cubo 17C, 87036 Rende (CS), Italy
a.brunetti@itm.cnr.it; adele.brunetti@cnr.it

Today, the use of membrane reactors for hydrogen production is becoming more and more a reality with various pilot installations all over the world, and increasing literature studies focused on the development of mechanically and chemically stable membranes with high permselectivity further improved in terms of stability and hydrogen flux. The possibility to combine reaction and separation in the same unit, reducing the whole volume of the plant and increasing its efficiency, is the main asset promoting the development of this technology. The scope of this chapter is to highlight the main findings about membranes and membrane reactors used in hydrogen production from reforming processes. A deep discussion is presented about the general classification of the membranes, with special attention being paid toward palladium-based membranes, as well as on the different types of membrane reactors available (from

Functional Nanostructured Membranes
Edited by Enrico Drioli, Lidietta Giorno, and Annarosa Gugliuzza

ISBN 978-981-4774-79-6 (Hardcover), 978-1-351-13511-5 (eBook)
www.panstanford.com

packed bed to fluidized bed to micro-reactors). In addition, the application of membrane reactors for hydrogen production is also discussed with a short overview on some representative results on the upgrading of syngas via water gas shift reaction.

8.1 Introduction

As indicated in the EU-White Paper[1] on renewable and alternative energy and according to leading energy scenarios, the use of hydrogen as an energetic vector is expected to become more and more important. In July 2014, British Petroleum and General Electrics announced plans to jointly develop up to 15 new hydrogen power plants to generate electricity over the coming decade.[2] The hydrogen required is planned to be derived from fossil fuels, including coal and natural gas, and the plants will include carbon capture technologies, reducing the CO_2 emissions by 90%.[3] If one considers that, at present, 96% of hydrogen is directly produced from fossil fuels and that 4% ca. is produced indirectly by using electricity generated through them (Kothari et al., 2008), the need to find new efficient and competitive technologies able to maximize hydrogen production from fossil fuels while lowering emissions of CO_2 and other pollutants appears evident. Up to now, most H_2 is obtained industrially from natural gas via steam reforming, although other alternative sources (i.e., bio-alcohol steam reforming) have been attracting great attention owing to the use of biomass as a renewable source (Abanades et al., 2013; Vizcaıno et al., 2012; Sanz et al., 2014; Iulianelli et al., 2014). The outlet stream of a reformer or a coal gasification plant contains around 50% hydrogen (on a dry basis), which must be recovered, and between 40 and 45% CO, which is usually reduced in an upgrading stage with the aim to produce more hydrogen at the same time. Traditionally, a multistage CO-shift process based on a series of catalytic reactors (Fig. 8.1) is used for the upgrading of reformate streams: the first, operating at high temperatures (about 350–400°C) and taking advantage of the high reaction rate, converts a large portion of the carbon monoxide

[1]http://ec.europa.eu/research/energy/pdf/efchp_hydrogen13.pdf

[2]www.bp.com/content/dam/bp/pdf/investors/BP_Annual_Report_and_Form_20F_2014.pdf

[3]http://www.worldwatch.org/node/4516

to give hydrogen and CO_2; the other, operating at a low temperature (around 220–300°C), refines the carbon monoxide conversion, thus allowing a lower final concentration of CO (less than 1% molar).

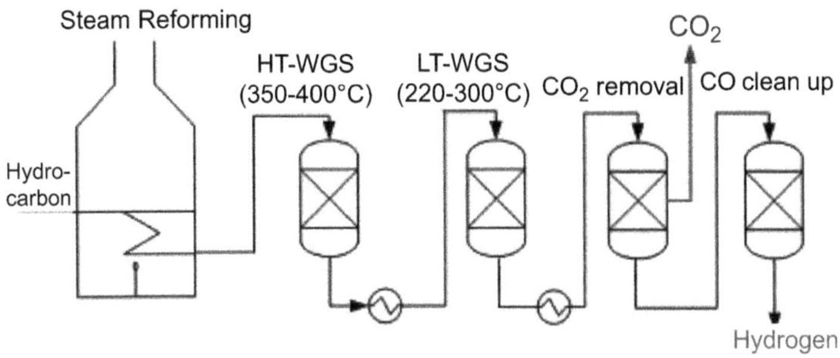

Figure 8.1 Scheme of the traditional process for hydrogen production from light hydrocarbons. Reproduced from Brunetti et al. (2011b) with permission from The Royal Society of Chemistry.

This H_2-rich stream coming out from the last reactor is fed to a pressure swing adsorption (PSA) unit for H_2 separation from other gases. The use of H_2 as feed in fuel cells for mobile power sources requires the anode inlet gas to have a CO concentration lower than 10–20 ppm to avoid catalyst poisoning with consequent drops in the fuel cell efficiency. Hence, the purification step of the H_2 produced from hydrocarbon must be very efficient to fulfil the fuel cell requirements. Because of this in some cases, another reaction unit is added for oxidizing CO in CO_2. One of the main challenges in the next few years will be the identification of new technologies able to provide a better exploitation of fossil fuels, e.g., hydrocarbons, in order to improve yield, energy saving, and so on. The reduction in the reaction/separation/purification stages, which means lower footprint area occupied by the whole plant, less auxiliary devices required, reduction in the energetic load, and so forth, are fundamental aspects to be taken into account in the redesigning of hydrogen production processes. A promising approach for concretizing these technological aspects in the field of hydrogen production is the use of membrane reactors (MRs), combining the reaction and H_2 separation by means of a selective membrane. Many studies are now focused on the analysis of MR performance where light hydrocarbon reforming or water gas shift (WGS) reaction is carried out. In these cases, for both these reactions the presence of the membrane allows the recovery of a hydrogen-rich stream, which does not require further separation. Moreover, the removal of the H_2, reaction product, from the reaction volume shifts the reaction toward further conversion. This means the possibility of having an intensified process with a reduced plant size and higher yield. The traditional process can thus be redesigned as more compact and efficient pursuing the logic of the process-intensification strategy, which is an innovative methodology for process and plant design, proposing a new design philosophy to achieve significant reductions (by factors of 10 to 100 or more) in plant volume at the same production capacity, improving in the meantime the overall efficiency.

As Fig. 8.2 shows, the integrated membrane system can be constituted of less reaction/separation units than the conventional one (Fig. 8.1). A first MR can be used for carrying out the reforming of the light hydrocarbons and another one for the WGS reaction.

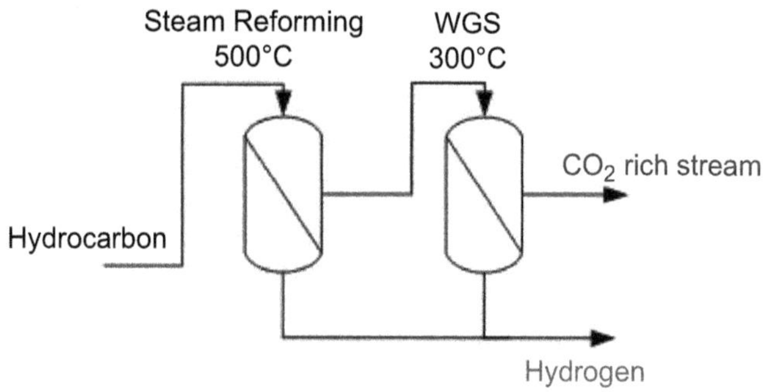

Figure 8.2 Scheme of the integrated membrane process for hydrogen production from light hydrocarbons. Reproduced from Brunetti et al. (2011b) with permission from The Royal Society of Chemistry.

The presence of the membrane in both the reactors allows the separation of a hydrogen-rich stream from the two reaction volumes as well as improvement in the conversion of both stages. Obviously, the H_2 purity level strictly depends on the membrane type used in both MRs. Actually, membranes can be distinguished by their selectivity, which can be infinite or finite. The first ones, traditionally Pd-based, allow a pure hydrogen stream to be obtained, whereas the others provide a hydrogen-rich stream with a variable purity. If the recovered H_2 stream does not have the purity required, the latter can be increased by adding another purification unit on the basis of the final use of the H_2 stream. Selective CO oxidation is known as an interesting and economical approach for CO removal from H_2-rich gas streams. Also in this field, new studies proposed in the literature have demonstrated how the use of an MR can improve the process by increasing CO conversion as well as the purity of the hydrogen stream. In this context, membrane engineering plays a fundamental role in the integration of these units in a single plant and, at the same time, in the definition of the knowledge necessary to drive the process by maximizing the gains both in terms of efficiency and plant size reduction. The synergic effects offered by MRs through combining reaction and separation in the same unit, their simplicity, and the possibility of advanced levels of automation and control offer an attractive opportunity to redesign industrial processes.

8.2 Membrane Reactors

8.2.1 Configuration and Utilization

According to the IUPAC definition, an MR is a device capable of carrying out a chemical reaction and separation in the same unit. The functions of a membrane in an MR, all operating at the nanoscale level, can be basically distinguished into:

- **Extractor or separator**: for the selective removal of one or more products from the reaction mixture
- **Distributor**: for distributing and controlling the addition of reactants in the reactor volume
- **Contactor**: for optimizing the contact between reactants and catalyst or between two immiscible phases.

Various combinations of these functions are also possible. Depending on these combinations and, thus, on its role in an MR, the membrane can be selective or nonselective, with or without catalytic activity. A membrane acting as a catalyst is called "catalytic membrane"; otherwise, it is called "inert membrane." A membrane can either be catalytic by itself or act as a support for catalytically active-phase deposited on its surface or inside its pores. A further distinction has to be made when the membrane acts as the sole catalyst, and when a conventional catalyst is present in addition to the catalytic membrane. A detailed description on various concepts of MRs on the basis of the membrane role was given by Dittmeyer et al. (2001) and Shu et al. (1991).

The membrane can have a cylindrical or flat sheet geometry. The cylindrical membranes are subdivided according to their dimensions: "tubular membranes" with a diameter higher than around 10 mm, "hollow fibers" with a diameter of a few hundreds microns, and "capillary membranes" with intermediate sizes (>1 mm). When the membranes are packed closely together in a module in tubes-and-shell geometry, a fiber diameter between 5 and 0.05 mm corresponds to a surface membrane area per volume of membrane module between 360 and 36000 m^2/m^3, respectively. Tubular membranes are placed inside a pressure-resistant tube. The capillary and hollow-fiber membranes are assembled in a tubes-and-shell module with the free ends of the fibers potted with, e.g., epoxy resins or silicone rubber. The flat sheet membranes are usually assembled in a spiral wound or plate-and-frame configuration, where sets of two membranes are placed in a sandwich-like configuration with their feed sides facing each other and separated by spacers. When such a plate-and-frame module is wrapped around a central collector pipe, a spiral wound module results. Capillary, spiral wound and plate, and frame configurations are most often used in low-temperature MRs such as membrane bioreactors.

In most of the high-temperature applications, when the membrane acts as separator with or without catalytic function, the membrane module has a tube-in-tube configuration with, in the former case, the catalytic bed placed in the annulus or in the core of the inner tube. The same configuration can be assumed if the membrane is catalytic.

All these systems can be operated in continuous mode, with co-current or counter-current configuration. In addition to the catalyst/membrane arrangement, the reactor module configuration depends also on the reaction phases and on the function of the membrane as well as on economic considerations, with the correct engineering parameters being utilized for achieving the target (Vankelecom et al., 2002).

The packed bed MR is the most studied typology of MR; it can be operated in both tubular and plane configurations. In particular, in the packed bed MRs, the catalyst is packed in a fixed volume and generally, it is in contact with the membrane surface (Fig. 8.3). In tubular configuration, the catalyst can be packed in the membrane lumen but also in the annulus, depending on which type of inorganic membrane is allocated inside the MR.

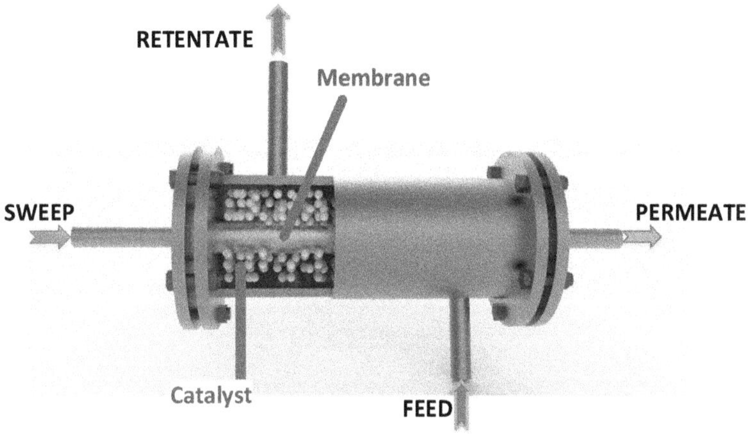

Figure 8.3 Packed bed MR. Reproduced from Brunetti et al. (2015) with the permission from Elsevier.

As reported in the recent literature, the fluidized-bed MR is an emerging technology and a typical configuration consists of a membrane immersed in a catalytic bed operated in bubbling or turbulent regime (Adris et al., 1997) (Fig. 8.4). In particular, a fluidized-bed MR makes possible a higher catalyst effectiveness and an enhanced mass and heat transfer rates with respect to a packed bed MR, favoring a more uniform temperature profile. On the contrary, its main drawback is the minimum fluidization velocity, which has to be guaranteed for keeping the catalyst bed in suspension (Deshmukh et al., 2007).

Membrane micro-reactors or micro-membrane reactors (MMRs) are a recent evolution of MRs. They can be defined as micro-reactors reinforced by membrane separation/purification or MRs miniaturized into characteristic size of 1–1000 μm, which combine the advantages of both MRs and micro-reactors, leading to a greatly intensified operation unit (Fig. 8.5) (Alfadhel et al., 2005; Gallucci et al., 2013; Jensen, 1999; Kim et al., 2009; Kolb, 2013; Kothare, 2006; Tan and Li, 2013b).

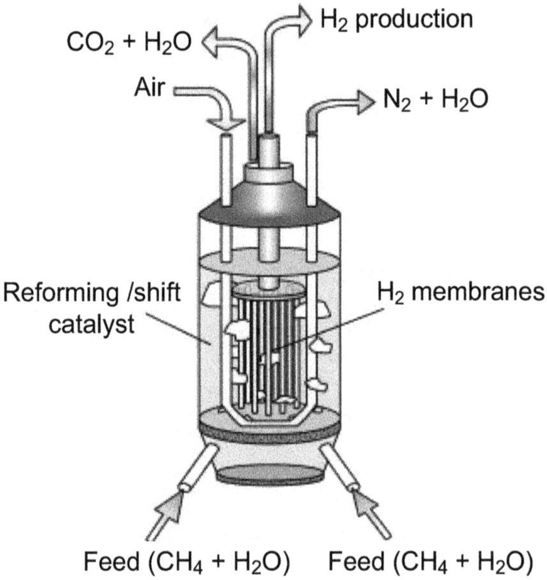

Figure 8.4 Fluidized-bed MR scheme. Reprinted from Patil et al. (2007) with permission of Elsevier.

The MMRs with hydrogen separation function are under prevailing investigation and have found a number of applications such as hydrogen production from WGS reaction, hydrogen production from methanol steam-reforming reaction, onboard fuel processing for portable fuel cells, production of moisture-free formaldehyde by the dehydrogenation of methanol, and dehydrogenation of cyclohexane to benzene (Karnik et al., 2003). Although MMRs are very promising devices for hydrogen production, the related research is still very limited. The pioneering investigations on MMRs were focused on microchannels in a planar configuration based on micro-electromechanical systems (MEMS).

When the diameter of a membrane tube is reduced to a certain level, i.e., ID < 1000 μm, it becomes a hollow fiber and the fiber lumen may take effects of a microchannel. Catalysts can be coated on the inner surface of the hollow fibers or impregnated inside the porous wall, while the separation can be achieved by the porous hollow fibers themselves or by the membrane formed on the outer surface of the hollow fibers. This kind of hollow-fiber MR can also be attributed to MMRs, then called hollow-fiber MMRs (Kothare, 2006). Hollow-fiber MMRs have been investigated in detail by Li's group for hydrogen production (García-García et al., 2010, 2011; García-García and Li, 2013; Kothare 2006; Rahman et al., 2011, 2012).

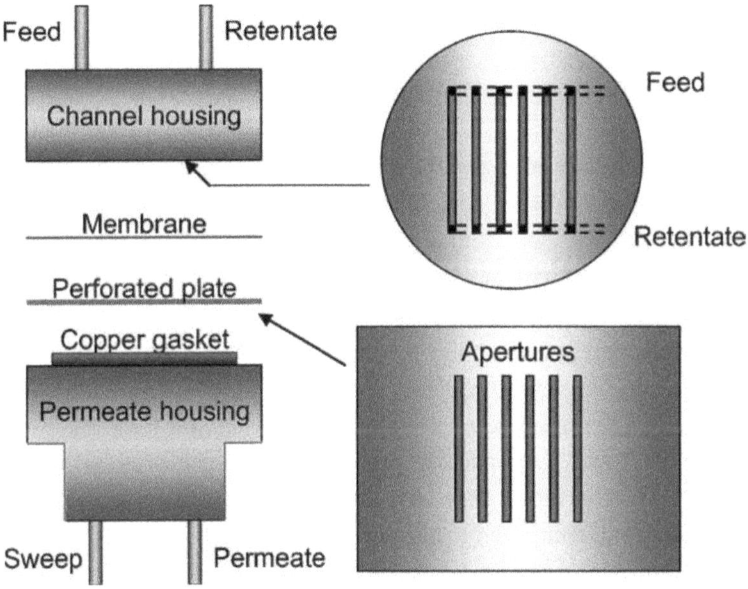

Figure 8.5 Sketch of the microchannel MMR configuration. Reprinted from Mejdell et al. (2009) with permission of Elsevier.

Honeycomb or straight-channel monoliths (Fig. 8.6) can provide an inexpensive and rapid means of constructing scalable two-dimensional arrays of identical square microchannels with diameters 500–5000 μm and wall thickness 200–2000 μm (Kim, 2009). This type of structures can be formed from a variety of porous ceramic materials such as cordierite, mullite, and alumina, which can realize large networks of MMRs. The monolithic MMRs can provide much better mechanical stability than the hollow-fiber MMRs and much higher intensification than the planar microchannel MMRs.

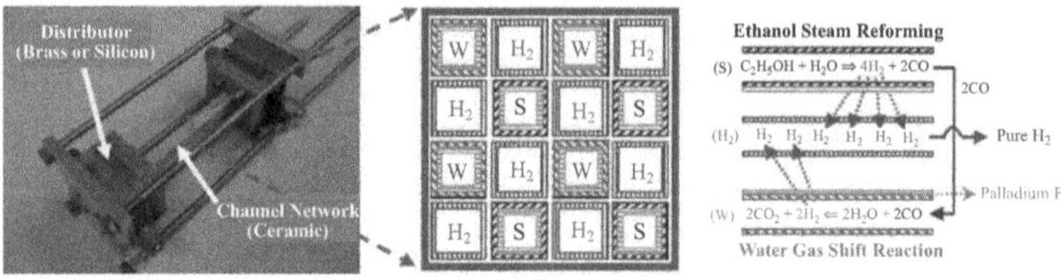

Figure 8.6 Prototype of honeycomb-monolith MMRs, with inset showing schematic of integration schemes. Reprinted from Kim (2009) with permission of Elsevier.

8.3 Membranes for MR

Inorganic membranes are the most used in MRs because of their high resistance toward solvents, well-defined stable pore structure (in the case of porous inorganic membranes), high mechanical stability, and, in particular, elevated resistance at high operating temperatures. They can be constituted by different materials such as metals (palladium, silver, niobium, titanium, etc., and their alloys), zeolite, ceramic, carbon, silica, and oxides (alumina, titania, zirconia). Commonly, they are stable in the temperature range of 300–800°C and in particular cases (ceramic membranes), up to 1000°C. Furthermore, they also show high resistance to chemical degradation if compared to organic membranes.

In hydrogen production, dense or microporous membranes can be used depending on the role of the membrane, whether it be for H_2 separation or purification. Most studies in the open literature show that the membranes can be distinguished in dense inorganic Pd-based membranes and ceramic membranes (silica, zeolite, etc.). The former show a permselective transport governed by a solution–diffusion mechanism. Microporous ceramic membranes can have both the permselective and non-permselective transport, depending on the size of the permeating molecules with respect to the membrane pore size as well as on the chemical nature of the permeating molecules and the membrane material (Dittmeyer et al., 2001).

8.3.1 Pd Alloy

Commonly, dense metal membranes based on Pd, Ti, Ta, V, Nb, etc. are applied in the field of hydrogen separation from gas mixtures and/or in MR applications. Owing to the characteristics of hydrogen permselectivity with respect to all other gases, palladium and its alloys play the role of the dominant material in this field. The removal of a product such as hydrogen from the reaction volume implies a series of advantages:

- Conversion enhancement of equilibrium-limited reactions
- Depletion of undesired secondary reactions
- Recovery of a concentrated rich stream
- Coupling of two or more reactions, e.g., dehydrogenation (endothermic) with a hydrogenation (exothermic) on the two membrane sides
- Better operating conditions (e.g., temperature)
- Higher neat reaction rate
- Higher residence time

The removal of a product from the reaction volume allows the thermodynamic equilibrium limit of a traditional reactor (TR) to be exceeded, obtaining higher conversion in analogous operating conditions. In other words, for endothermic reactions, this allows the MR to achieve the same conversion of a TR, but at significantly lower temperatures. Another interesting aspect of the MR use is the positive effect that the reaction pressure can have on the process, also for reactions taking place without mole number variation (e.g., WGS) or with an increase in mole number (e.g., methane steam reforming), because of the favored removal of a product from the reaction volume.

The main advantage of the dense Pd-based membranes is their infinite selectivity versus H_2. In specific case, the important advantages of the Pd-alloy MR are the following:

- Production of pure H_2 permeate stream (when a pressure difference is used instead of a sweep gas to create the permeation driving force)
- Retentate streams concentrate in the other species (e.g., CO_2)
- Positive effect of reaction pressure on the equilibrium conversion also for reactions characterized by an increase in the number of moles

This also affects the performance of the catalyst in terms of activity and lifetime, also avoiding possible catalyst deactivation with significant advantages in terms of equipment costs. However, the high cost is actually one of the main problems that limits their diffusion at the industrial level. In order to overcome this drawback, several studies have presented new composite membranes consisting of a thin Pd-based layer deposited, with different techniques, on porous supports that can be ceramic or stainless steel. In this way, the Pd content is reduced as well as the related cost. Another typical problem of these dense membranes is the inhibiting or poisonous effects that some species such as carbon monoxide (Barbieri et al., 2008; Caravella et al., 2009) or sulfur can exercise on the membrane, consequently reducing the permeance. On the contrary, the meso- and microporous membranes (alumina, silica, titania, zirconia, zeolites, etc.) are not affected by poisoning and are cheaper than the metallic membranes. However, they show finite selectivity versus all the chemical species; therefore, it is not possible to obtain pure streams as permeate, as when using dense membranes.

Different companies are working on the commercialization of dense metal membranes for hydrogen separation/production. Some examples are shown in Figs. 8.7 and 8.8.

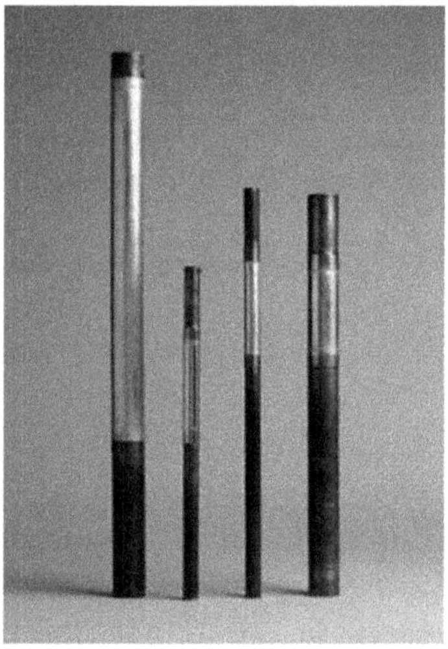

Figure 8.7 High-temperature Pd-based composite commercial hydrogen separation membranes manufactured by CRI/Criterion (http://www.cricatalyst.com/).

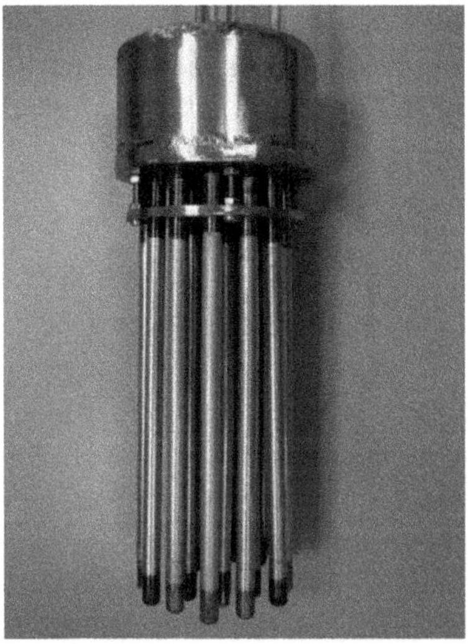

Figure 8.8 Pd-based composite membrane module manufactured by Green Hydrotech (www.grnhydrotec.com).

CRI/Criterion (a company owned by Shell) has produced Pd and Pd-alloy membranes (Fig. 8.7) on sintered porous metal supports as published in a report (Shell Impact, 2010). H_2 purity of more than 99% was demonstrated for a time exceeding 4000 h in high-temperature gas separations.

Green Hydrotech has developed Pd and Pd–Cu membranes on porous stainless steel tubes (Fig. 8.8) with an extremely high H_2/N_2 permselectivity (4100), and these membranes provide in situ purification of hydrogen in a steam reformer for high-purity hydrogen (99.996%). Other examples can be found in Gallucci et al. (2013).

8.3.2 Ceramic

Ceramic membranes commonly constituted by alumina, titanium, or silica oxides are chemically inert and stable at high temperatures (Smart et al., 2013). They can be used for microfiltration and ultrafiltration processes (Kim and van der Bruggen, 2010; Fan et al., 2014), gas separation (Hashim et al., 2010; Adewole et al., 2013; Hu et al., 2011) besides applications in MRs (Wu et al., 2010; Volkov et al., 2012; Jiang et al., 2013). Most of the drawbacks of ceramic membranes consist of the difficulties in their proper sealing modules when operated at high temperature as well as the high sensitivity to temperature gradient as the main cause of membrane cracking.

8.3.3 Zeolite

As a new class of inorganic membranes and unlike the most representative microporous metal oxides such as silica, alumina, and titania, zeolites can be described as microporous crystalline alumina silicate with uniform pore size (Caro et al., 2000). They are relatively stable at high temperatures, acidic or basic in nature, and can show hydrophilic or organophilic characteristics. The molecular sieving abilities of zeolite membranes combined to selective sorption properties and catalytic activity, in addition to their thermal and chemical stability, make them attractive for gas separation (Wang et al., 2014), pervaporation (Bowen et al., 2004), and several MR applications (Caravella et al., 2015a, 2015b, 2016; Fong et al., 2008; Kosinov et al., 2016; McLeary et al., 2006; Zhang et al., 2012; Zito et al., 2015, 2016).

One of the main drawbacks related to these membranes is represented by their relatively low gas fluxes compared to other inorganic membranes (ceramic). Furthermore, the zeolites suffer the thermal effect; indeed, at high temperature, the zeolite layer can show negative thermal expansion (shrinkage), while the support continuously expands, involving in thermal stress at the attachment within the zeolite layer and the support as well as for the connection of the individual microcrystals within the zeolite layer.

8.3.4 Carbon

Carbon membranes are currently categorized on the basis of the pore size in carbon molecular sieve (CMS) and selective surface flow membranes. Furthermore, carbon membranes can be subdivided into two categories: supported and unsupported.

CMS membranes favor the transport of small molecules through the pores, blocking the passage of larger molecules. This characteristic (small pore size) makes CMS membranes highly permselective for separation of gas mixtures containing small gas species (Deng and

Hägg, 2015; He and Hägg, 2013). On the contrary, SSF membranes possess pores larger than the size of the molecules passing through them, and, as a consequence, they are based on the preferential adsorption of some components in the gas mixture, which is followed by surface diffusion in the carbon matrix. In particular, CMS membranes were identified as very interesting candidates for gas separation, both in terms of separation properties and stability. CMS membranes can be obtained by pyrolysis of many thermosetting polymers such as polyvinylidene chloride, polyfurfuryl alcohol, cellulose triacetate, polyacrylonitrile, and phenol formaldehyde. Today, carbon membranes are also studied in MR applications for hydrogen production from methanol steam reforming (Briceño et al., 2012a; Sá et al., 2011) or, more generally, in dehydrogenation reaction process (Hirota et al., 2013; Sznejet and Sheintuch, 2004).

8.3.5 Perovskites and MIEC

During the last decade, powerful ceramic materials selectively permeable to oxygen at high temperatures, i.e., dense mixed ion–electron conducting (MIEC) perovskites, have been developed, making the realization of MRs increasingly possible.

Research in the field of MIEC membranes has increased substantially in the last decade, which is proved by the filing of numerous patents in this area. The best membranes are made from alkaline earth doped, oxygen-deficient perovskite-type or perovskite-related oxides, exhibiting both ionic and electronic conductivity. The transport of oxygen anions, which requires elevated temperatures typically above 800°C, is charge-balanced by a simultaneous transport of electronic charge carriers. It was already found in the 1980s that some of the acceptor-doped, oxygen-deficient perovskite materials show considerable proton conductivity (Tan and Li, 2013a). Typical examples are doped $BaCeO_3$, $SrCeO_3$, and related doped $BaZrO_3$ and $SrZrO_3$ showing also improved stability, e.g., in CO_2. The protons are embedded in these materials in the electron cloud of an oxygen ion, forming hydroxide defects. They migrate by hopping between oxide ions (Grotthuss mechanism). As for the oxygen conductors, temperatures sufficiently high are required for high proton conductivity. However, the electron conductivity of these materials is very low, making the use of an external electrical circuit a must for their application. This is a serious drawback. In order to enhance the electronic conductivity, multi-element doping was explored, however, with only minor effect. Recently a new class of crystalline oxide-based ceramics was found with sufficient mixed electron–proton conductivity at intermediate temperatures: the doped lanthanide tungstates $LnWO_{12}$ (Haugsrud, 2007). These materials also show considerable chemical stability, making them interesting candidates for nano-based MRs.

8.3.6 Composite

The necessity to reduce the amount of Pd has led to the development of composite Pd-based membranes. Indeed, as addressed in an extensive literature, to pursue the scope of reducing the cost of Pd-based membranes, recently there has been a strong interest toward the development of membranes constituted of a thin Pd-layer coating (ranging from submicron to a few microns) on porous supports (ceramic, Vycor, or porous stainless steel) (Chen et al., 2010; David and Kopac, 2011; Itoh et al., 2014; Moparthi et al., 2010; Pinacci and Drago, 2012; Wei et al., 2012). On this route, the benefits of composite Pd-based membrane utilization take into account a reduced material cost, an enhanced resistance to mechanical strength, and higher permeating flux, even if the hydrogen permselectivity of these membranes shows a finite value.

Commonly, composite membranes are constituted of a thin layer of palladium or its alloy deposited on porous supports such as porous Vycor glass (silica gel), SiO_2, Al_2O_3 and B_2O_3, or PSS. High mechanical resistance and easy gas sealing are allowed by supports having a thermal expansion coefficient close to palladium like stainless steel. However, porous stainless steel support shows, as a main issue, the intermetallic diffusion of palladium at relatively high temperatures, inducing a depletion of hydrogen permselectivity.

8.4 Membrane Reactors: Applications

In the past, MR performances were studied by carrying out several gaseous phase reactions and using different membrane types, in particular for high-temperature operations. Table 8.1 summarizes the main gaseous phase reactions for H_2 production carried out in MRs. It reports the membrane type used as well as the temperature range investigated in the experiments performed at the laboratory scale.

Table 8.1 Main gaseous phase reactions for H_2 production in MRs

Process	**Membrane type**	**Temperature range (°C)**
	Hydrogen production	
Steam reforming of light hydrocarbons	Pd–Ag	400–600
Water gas shift	Pd–Ag; Pd on silicalite; Pd on ceramics	200–500
CO clean-up	Silica; ceramic; Pd–Cu; catalytic zeolite	180–250
	Syngas production	
Partial oxidation of light hydrocarbons	Composite Pd–ceramic; silica-supported alumina; perovskite ceramics	800
	Pre-combustion capture	
Oxy-fuel combustion		800
Chemical looping	Pd–Ag; composite Pd-alloy	400–600
	Crossing cut	
Oxidative dehydrogenation of light hydrocarbons	Composite Pd-Alloy; silica oxide ceramic; metal-modified alumina; Pd-alloy; zeolite	350–600
Oxidative coupling		

Most of the studies carried out so far on MRs were focused on equilibrium-limited reactions, where the permeation of a product enhances the conversion with respect to a TR. Other new applications propose the use of the membrane as a contactor for catalyst and reactants. However, even MR studies also on the pilot plant gave promising results, indicating wide-ranging potential for this technology; there are currently no large-scale applications for their application. The next sections will focus on providing an overview of the MR technology application, considering only the main reaction stages constituting the traditional plant for hydrogen production from the reforming of light hydrocarbons (Fig. 8.1).

8.4.1 Steam Reforming

Steam methane reforming (SMR) is the most common and cost-effective method for hydrogen production:

$$2H_2O + CH_4 = CO_2 + 4H_2; \Delta H^0 = 191 \text{ kJ/mol}$$

SMR on a very large scale is carried out commercially in externally heated fixed-bed reactors. It is a reversible process usually operated at 700–900°C using supported nickel catalysts (Li et al., 2008). Owing to mass and heat transfer issues, the packed bed SMR suffers from significant disadvantages, such as low catalyst effectiveness factors and large temperature gradients. In addition, in most cases, to obtain hydrogen of the purity required, the reactor off-gas must go through a series of steps, such as high and low WGS reactions and PSA.

For the last 20 years, membrane technology has been proposed as an alternative to improving the performance of the conventional process. Hydrogen can be produced in membrane steam reformer units. Owing to the selective and continuous removal of hydrogen in situ from the reaction zone, SMR reaction shifts toward the product side. Therefore, also the feed pressure, which has a negative effect from a thermodynamic point of view, acts positively on methane conversion, so that compared with TRs, MRs can achieve higher conversions at the same temperature or the same conversion at lower temperatures (Shu et al., 1994). Moreover, the use of dense Pd-based membranes allows a pure hydrogen stream to be obtained so that there is no need for additional purification as in the conventional process. Up to now, many researchers have proposed applying Pd-based MRs for methane steam-reforming reaction operating at milder conditions than the TRs (Amanipour et al., 2016;

Barbieri et al., 1997; Bernardo et al., 2010; Borgognoni et al., 2011; Chiappetta et al., 2010; Castro-Dominguez et al., 2016; El Hawa et al., 2016; Hara et al., 2008, 2010; Iulianelli et al., 2010; Marigliano et al., 2003; Patrascu and Sheintuch, 2015; Roses et al., 2013; Sato et al., 2010). The operating temperature of 500–550°C used in most experiments is a compromise among several factors. Both membrane permeance and the SMR thermodynamics and kinetics are favored by higher temperatures, but the membrane is more stable at lower temperatures. The promising results obtained at the laboratory scale encouraged application on a larger scale. MRT Inc.[4] has developed a proven technology based on a patented fluidized-bed MR for high-purity hydrogen. The process combines hydrocarbon reforming, shift conversion, and hydrogen purification in a single step. A unit supplied to Tokyo Gas gave improved performance in comparison with fixed-bed membrane reformers. Another unit was also provided to BOC in a project sponsored by the S-DoE. Actually, although the units are initially geared to capacities in the 15–50 N m^3/h range, the technology can be readily scaled to higher or lower capacity as applications require.

The Shell Oil Company has recently patented (Cui et al., 2009) a process and apparatus for the production of pure hydrogen by steam reforming. This process integrates steam reforming and shift reaction to produce pure hydrogen with minimal production of CO and virtually no CO in the hydrogen stream, provides for CO_2 capture by sequestration, uses a steam-reforming MR, and is powered by heat from a heater convection section.

Hydrogen production by steam reforming of methanol, ethanol, and other light hydrocarbons has become an attractive alternative to traditional operations. Especially attractive is its use in the decentralized production of clean electrical energy from fuel cells. The main differences as against the reforming of light hydrocarbons are the catalyst types used and the product distribution in the two reaction systems. Recent studies concern MR use in these reaction types. The main advantage is in performing both reaction and pure hydrogen recovery in the same device, thus replacing the conventional system (reformer and gas cleaning unit) with one MR (Han et al., 2000). In both conventional and membrane systems, the main reaction products are hydrogen, carbon monoxide, and

[4]http://www.membranereactor.com

carbon dioxide. Depending on the fuel used, e.g., ethanol, bioethanol, methane, acetaldehyde, and ethylene could be also present.

8.4.1.1 Autothermal reforming

Autothermal reforming (ATR) can be achieved as oxygen (or air) is supplied to the reactor, providing the necessary heat required by the reforming of methane. However, with a fluidized-bed reactor, hotspot formation can be avoided even for very exothermic reactions (Abashar, 2004; Abashar et al., 2008; Arratibel et al., 2015; Chen et al., 2007). ATR with integrated CO_2 capture can be obtained in fluidized bed when heat is supplied either by burning a small part of the recovered hydrogen (Gallucci et al., 2008a; 2008b) or by burning part of the feed with oxygen fed to the reactor via oxygen-selective membranes (Gallucci et al., 2010a; 2010b).

In a conventional reactor system, the methane ATR is performed at high temperatures (>850–900°C). However, this reaction can be carried out in MRs below 600°C. A way to obtain ATR is by using an external heat carrier. In this case, molten salts at temperatures up to 550°C can be used to supply the required heat for methane reforming (Giaconia et al., 2008). Consequently, a tubes-in-shell configuration reactor can be used with molten salts flowing in the reactor shell to supply the necessary heat to drive reforming reaction.

Other possibilities for supplying the necessary heat are using indirect coupling reactors. Exothermic and endothermic reactions take place in different chambers, which are separated by heat-conductive walls. Heat transfer happens via dividing walls from the chamber where the exothermic reaction (methane oxidation) is taking place to another chamber where the endothermic reaction (methane reforming) is carried out (Patel and Pant, 2007). Two important parameters need to be finely tuned. One is the ratio between the oxidation of methane and steam-reforming reaction, which defines the temperature profile inside the membrane reformer. However, because the two reactions generally occur with different kinetic rates, the flow rates and residence times in the two chambers need to be controlled to avoid hot/cold spots along the reactor.

8.4.2 Water Gas Shift

Successful examples of the use of Pd-based MRs are the high-temperature reactions for hydrogen production. The possibility to selectively remove hydrogen from reaction volume leads to significant advantages such as the production of a pure H_2 stream, the enhancement of the conversion, the deletion of secondary reactions, the increase in residence time of reactants, the reduction in reaction volume, etc. (Barbieri, 2015a). The upgrade of syngas by means of WGS reaction was widely investigated both experimentally and by simulation (Barbieri, 2015b; Barbieri et al., 2005, 2008, 2011; Brunetti et al., 2007a, 2007b, 2007c, 2008, 2009a, 2009b, 2011, 2012a, 2012b).

WGS reaction is industrially carried out in two fixed-bed adiabatic reactors, connected in series by a cooler (heat exchanger) between them. The first reactor operates at a high temperature (HT-WGS) ranging from 300 to 500°C using Fe–Cr-based catalysts. The second reactor (LT-WGS, low-temperature WGS) uses CuO–ZnO-based catalysts and operates at lower temperatures (180–300°C) in order to displace the equilibrium, since WGS reaction is exothermic. The whole cycle has the big disadvantage of being accompanied by large emissions of CO_2. The use of this three–four stages of reaction-purification can be replaced by a single stage, the MR, in which reaction and separation occur in the same vessel and conversion significantly higher than the traditional system can be reached (Fig. 8.9).

Figure 8.10 compares CO conversion as a function of the temperature obtained for the MR and the traditional process operating at the same inlet conditions, i.e., same gas hourly space velocity (20,000 h^{-1}) and temperature for the MR and the traditional process (first stage). The CO conversion achieved by MR is around 10% higher than the overall conversion of the traditional process; it also exceeds significantly (ca. 25–30%) the traditional reactor equilibrium conversion (TREC). Hydrogen removal from the reaction side owed to the permeation shifts the reaction toward further conversion. This effect is well operated in this MR since a reaction pressure of 15 bar well promotes the hydrogen permeation.

This gain is clearer considering that the MR conversion is ca. 33% higher than that achieved by the first stage of the traditional process (HT-WGS).

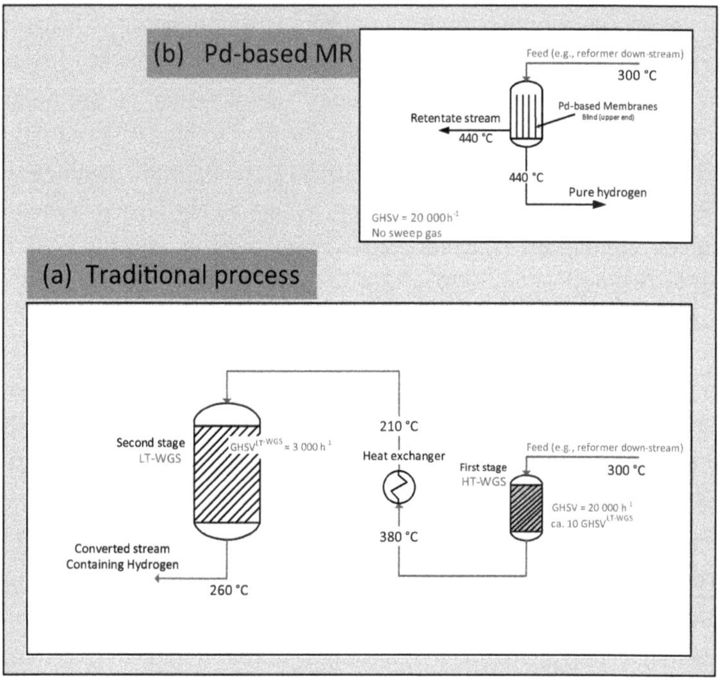

Figure 8.9 Schemes of "Pd-based MR" and "traditional process" for WGS reaction. The temperature values reported are indicative of a typical operation. Reproduced from Barbieri et al. (2011) with the permission from The Royal Society of Chemistry.

The upgrade of a syngas stream was recently investigated by Brunetti et al. (2015) in a WGS-MR operated in a high temperature range with an ultra-thin supported membrane (3.6 micron thick).

Figure 8.11 shows the CO conversion and hydrogen recovery as a function of gas hourly space velocity (GHSV) for different feed pressures. The conversion decreases with space velocity because of the lower residence time, whereas it is higher at the higher feed pressure owing to the positive effect of the pressure on the hydrogen removal through the membrane. In this figure, the conversion limit for MR in equilibrium conditions (MREC) is also shown along with the TREC for indicating how far/close the MR is operating both from its maximum achievable performance and from the limit of TR. The CO conversion in MR exceeds the TREC, the maximum conversion achievable by TR, operating up to 10,700 h^{-1} for both values of feed pressure considered. In addition, the CO conversion obtained at 5000 h^{-1} and 5 bar is quite close to the MREC, indicating a very good performance of the MR.

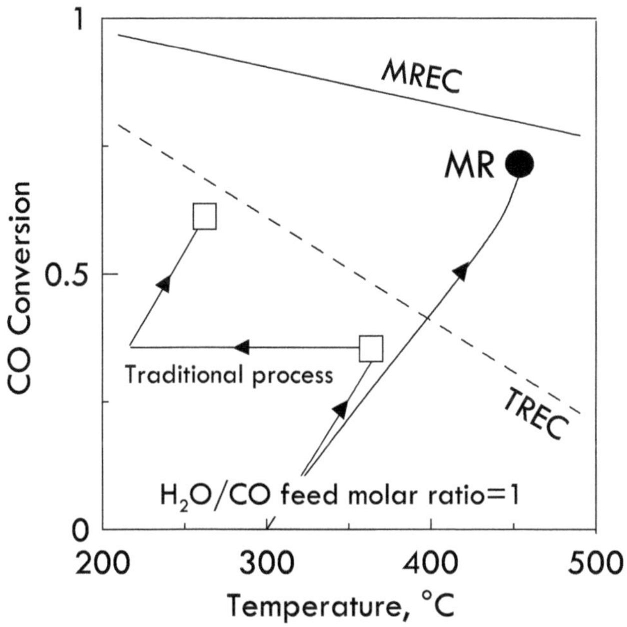

Figure 8.10 CO conversion as a function of temperature for MR and traditional process. Reproduced from Barbieri et al. (2011) with the permission from The Royal Society of Chemistry.

Hydrogen recovery also follows the same trend as the conversion, occurring for the same reasons as those described above. At 5000 h^{-1} and 5 bar, more than 90% of hydrogen is recovered as a pure stream in the permeate. A higher GHSV means a lower CO conversion and, thus, less hydrogen produced and that can be recovered. However, it has to be pointed out that even in the worse conditions (i.e., 15,000 h^{-1} and 4 bar), around 50% of hydrogen is recovered.

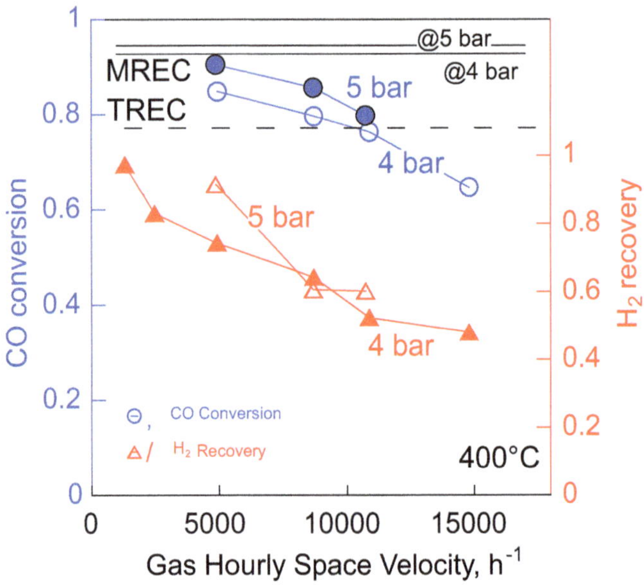

Figure 8.11 CO conversion (blue symbols) and H_2 recovery (red symbols) as functions of gas hourly space velocity (GHSV) at different feed pressures. Reproduced from Brunetti et al. 2015 with the permission from Elsevier.

The use of sweep gas was found to further promote the MR performance, improving the CO conversion of around 16% with respect to the case without sweep gas and gaining a hydrogen recovery of an additional 28%, passing from 52% to 74% (Fig. 8.12). The positive effect of the temperature on permeation and kinetics, contrasting with its negative effect on thermodynamics, was clearly evident in terms of recovery, which increased from 40% at 360°C up to 78% at 400°C at 4500 h^{-1}. A similar increase occurred at 10,070 h^{-1}, at which the recovery was double at 400°C with respect to that at 360°C.

The advantage offered by MR can be quantified also in terms of volume index. In fact, at 5 bar, the MR required a reaction volume of 85% less than TR operated in the same conditions (Fig. 8.13). However, already at 3 bar, the volume index was only 30%, which means, in other words, that a flow rate three times higher than that considered for TR could be treated in an MR to achieve the same conversion of a TR using the same amount of catalyst.

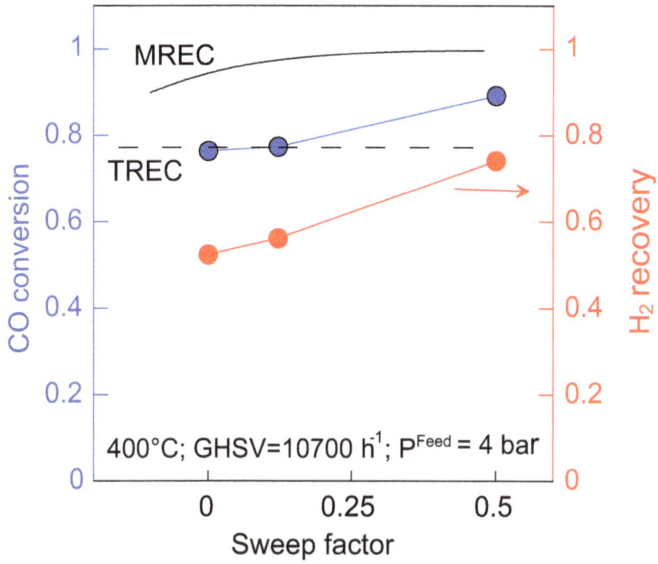

Figure 8.12 CO conversion (blue symbols) and H_2 recovery (red symbols) as function of sweep factor at 4 bar. Reproduced from Brunetti et al. 2015 with the permission from Elsevier.

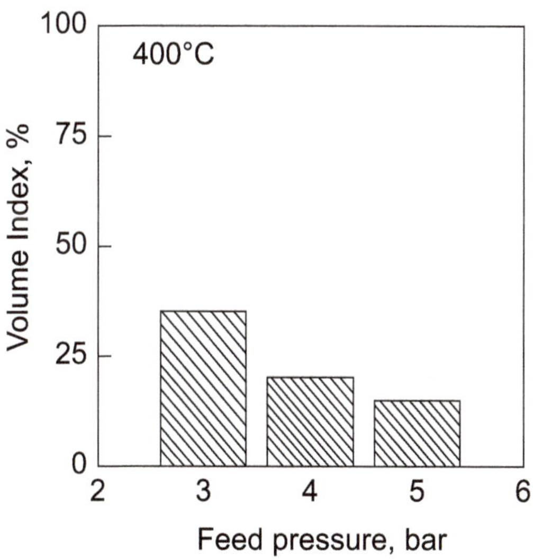

Figure 8.13 Volume index ratio as functions of feed pressure. Reproduced from Brunetti et al. 2015 with the permission from Elsevier.

As in the case of SMR, the interesting results achieved at the laboratory level arose in various, also industrial, patents. Among them, the United Technologies Corporation (Gummalla et al., 2010) patented the use of a WGS-MR, comprising a WGS reaction region and a permeate volume, separated by an H_2-separation membrane, which allows H_2 formed over a catalyst in the reaction region to be passed selectively to the permeate region for delivery to a point of use (such as the fuel cell of a fuel cell power plant). ExxonMobil (Deckman et al., 2005) developed and patented a heat-exchanged MR for electric power generation comprising an MR system that utilizes catalytic or thermal SMR and a WGS on one side of the membrane and hydrogen combustion on the other side for producing electricity. Also, the General Electric Company (Wei, 2009) patented a polygeneration system including (i) a syngas generator for producing a syngas, (ii) a syngas enrichment unit for removing undesired species from the syngas for producing an enriched syngas, and (iii) a syngas utilization system that utilizes the enriched syngas to produce useful products. In some embodiments, the polygeneration system includes MR, catalytic burner, and a power-generation unit.

8.5 CO Clean-Up

Selective or preferential CO oxidation (CO-selox) is usually considered a suitable process for CO removal from H_2-rich gas streams produced by reforming processes or available in petrochemical plants (e.g., in ethylene process). Traditionally, the reaction is carried out in conventional reactors; however, the presence of H_2O and CO_2 in the feed causes a significant decrease in the activity of the catalyst (Avgouropoulos, 2005). In an integrated plant for hydrogen production, depending on the types of membranes (e.g., of silica) used in the SMR and WGS-MRs, the H_2 stream recovered as permeate could require further purification to reduce the CO level (Fig. 8.14).

The CO-selox thus becomes a fundamental stage also in integrated membrane systems for H_2 production. The use of a catalytic membrane for this reaction stage can improve the depletion of CO content. In this case, the membrane is constituted of a ceramic tube (most often zeolite) in which the catalyst is opportunely deposited in the pores and does not have the function of separating a the stream components. It is instead used for improving the reactant/catalytic phase contact, reducing by-passing and misdistribution generally shown in a packed bed and providing a hotspot control to limit reverse WGS reaction (Quyang et al., 2005). Bernardo et al. (2008) successfully carried out the CO-selox in the presence of large quantities of hydrogen in continuous flow-through MRs using different Pt-loaded catalytic zeolite membranes.

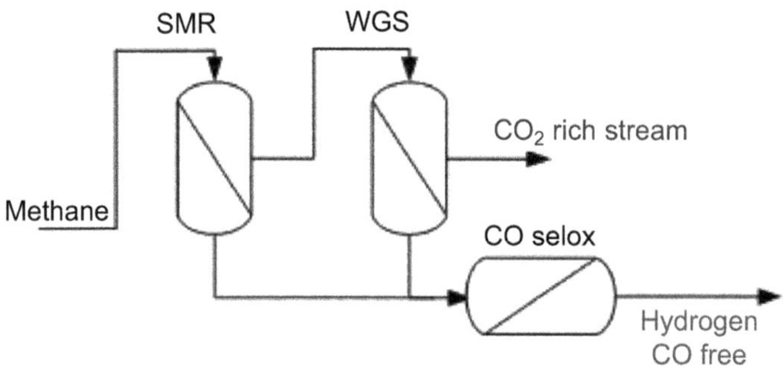

Figure 8.14 Scheme of the integrated membrane process for hydrogen production from light hydrocarbons with a CO-selox membrane unit for permeate purification. Reproduced from Brunetti et al. (2011b) with permission from The Royal Society of Chemistry.

The catalytic MRs succeeded in reducing the outlet CO concentration from 10,000 ppm (1% molar) down to 10–50 ppm. These results confirmed the good potentiality of catalytic membranes for deep purification of H_2-rich streams, allowing the hydrogen final use, e.g., also in fuel cell applications.

Symbols

GHSV	gas hourly space velocity, h^{-1}
P	pressure, Pa
T	temperature, °C

Abbreviations

ATR	autothermal reforming
CMS	carbon molecular sieve
HT-WGS	high temperature water gas shift

LT-WGS	low temperature water gas shift
MIEC	mixed ion-electron conducting
MMR	micro membrane reactor
MR	membrane reactor
MREC	MR equilibrium conversion
PSA	pressure swing adsorption
SMR	steam methane reforming
TR	traditional reactor
TREC	TR equilibrium conversion
WGS	water gas shift

References

Abanades, A., Rubbia, C., and Salmieri, D. (2013). Thermal cracking of methane into hydrogen for a CO_2-free utilization of natural gas. *Int. J. Hydrogen Energy*, **38**(20), 8491–8496.

Abashar, M. E. E. (2004). Coupling of steam and dry reforming of methane in catalytic fluidized bed membrane reactors. *Int. J. Hydrogen Energy*, **29**, 799e808.

Abashar, M. E. E., Alhabdan, F. M., and Elnashaie, S. S. E. H. (2008). Discrete injection of oxygen enhances hydrogen production in circulating fast fluidized bed membrane reactors. *Int. J. Hydrogen Energy*, **33**, 2477e2488.

Abu El Hawa, H. W., Lundin, S.-B., Patki, N. S., and Douglas Way, J. (2016). Steam methane reforming in a Pd–Au membrane reactor: Long-term assessment. *Int. J. Hydrogen Energy*, **41**(24), 10193–10201. doi:10.1016/j.ijhydene.2016.04.244.

Adewole, J. K., Ahmad, A. L., Ismail, S., Leo, C. P. (2013). Current challenges in membrane separation of CO_2 from natural gas: A review. *Int. J. Greenhouse Gas Contr.*, **17**, 46–65.

Adris, A. M. and Grace, J. R. (1997). Characteristics of fluidized-bed membrane reactors: Scale-up and practical issues. *Ind. Eng. Chem. Res.*, **36**, 4549–4556.

Alfadhel, K. and Kothare, M. V. (2005). Modeling of multicomponent concentration profiles in membrane microreactors. *Ind. Eng. Chem. Res.*, **44**, 9794–9804.

Amanipour, M., Towfighi, J., Zamaniyan, A., Babakhani, E. G., and Heidari, M. (2016). Performance of a nickel–alumina catalytic layer for simultaneous production and purification of hydrogen in a tubular membrane reactor. *RSC Adv.*, **6**(79), 75686–75692. doi:10.1039/c6ra12870j.

Arratibel, A., Tanaka, D. A. P., van Sint Annaland, M., and Gallucci, F. (2015). Membrane reactors for autothermal reforming of methane, methanol, and ethanol, in *Membrane Reactors for Energy Applications and Basic Chemical Production*, A. Basile, L. Di Paola, F. Hai, and V. Piemonte (eds.), pp. 61–98, Woodhead Publishing Series in Energy.

Avgouropoulos, G., Ioannides, T., and Matralis, H. (2005). Influence of the preparation method on the performance of CuO–CeO_2 catalysts for the selective oxidation of CO. *Appl. Catal. B*, **56**, 87–93.

Barbieri, G. (2015a). Hydrogen production by membrane reactors, in *Encyclopedia of Membranes*, E. Drioli and L. Giorno (eds.). Springer-Verlag GmbH, Berlin Heidelberg.

Barbieri, G. (2015b). Water gas shift (WGS), in *Encyclopedia of Membranes*, E. Drioli and L. Giorno (eds.). Springer-Verlag GmbH, Berlin Heidelberg.

Barbieri, G., Brunetti, A., Caravella, A., and Drioli, E. (2011). Pd-based membrane reactors for one-stage process of water gas shift. *RSC Adv.*, **1**(4), 651–661.

Barbieri, G., Brunetti, A., Granato, T., Bernardo, P., and Drioli, E. (2005). Engineering evaluations of a catalytic membrane reactor for water gas shift reaction. *Ind. Eng. Chem. Res.*, **44**, 7676–7683.

Barbieri, G., Brunetti, A., Tricoli, G., and Drioli, E. (2008a). An innovative configuration of a Pd-based membrane reactor for the production of pure hydrogen. Experimental analysis of water gas shift. *J. Power Sources*, **182**(1), 160–167.

Barbieri, G., Scura, F., Lentini, F., De Luca, G., and Drioli, E. (2008b). A novel model equation for the permeation of hydrogen in mixture with carbon monoxide through Pd–Ag membranes. *Sep. Purif. Technol.*, **61**, 217–224.

Barbieri, G., Violante, V., Di Maio, F. P., Criscuoli, A., and Drioli, E. (1997). Methane steam reforming analysis in a palladium-based catalytic membrane reactor. *Ind. Eng. Chem. Res.*, **36**(8), 3369–3374.

Bernardo, P., Algieri, C., Barbieri, G., and Drioli, E. (2008). Hydrogen purification from carbon monoxide by means of selective oxidation using zeolite catalytic membranes. *Sep. Purif. Technol.*, **62**, 629–635.

Bernardo, P., Barbieri, G., and Drioli, E. (2010). Evaluation of membrane reactor with hydrogen-selective membrane in methane steam reforming. *Chem. Eng. Sci.*, **65**(3), 1159–1166.

Borgognoni, F., Tosti, S., Vadrucci, M., and Santucci, A. (2011). Pure hydrogen production in a Pd–Ag multi-membranes module by methane steam reforming. *Int. J. Hydrogen Energy*, **36**(13), 7550–7558.

Bowen, T. C., Noble, R. D., and Falconer, J. L. (2004). Fundamentals and applications of pervaporation through zeolite membranes. *J. Membr. Sci.*, **245**, 1–33.

Briceño, K., Iulianelli, A., Montané, D., Garcia-Valls, R., and Basile, A. (2012a). Carbon molecular sieve membranes supported on non-modified ceramic tubes for hydrogen separation in membrane reactors. *Int. J. Hydrogen Energy*, **37**, 13536–13544.

Brunetti, A., Barbieri, G., and Drioli, E. (2008). A PEM-FC and H2 membrane purification integrated plant. *Chem. Eng. Process.*, **47**(7), 1081–1089.

Brunetti, A., Barbieri, G., and Drioli, E. (2009a). Upgrading of a syngas mixture for pure hydrogen production in a Pd–Ag membrane reactor. *Chem. Eng. Sci.*, **64**, 3448–3454.

Brunetti, A., Barbieri, G., and Drioli, E. (2009b). Pd-based membrane reactor for syngas upgrading. *Energy Fuel*, **23**, 5073–5076.

Brunetti, A., Barbieri, G., and Drioli, E. (2011a). Integrated membrane system for pure hydrogen production: A Pd–Ag membrane reactor and a PEMFC. *Fuel Process. Technol.*, **92**, 166–174.

Brunetti, A., Barbieri, G., and Drioli, E. (2011b). Membrane reactors in hydrogen production, in *Membrane Engineering for the Treatment of Gases: Gas-Separation Problems Combined with Membrane Reactors*, E. Drioli and G. Barbieri (eds.), 2, pp. 87–109.

Brunetti, A., Barbieri, G., Drioli, E., Lee, K.-H., Sea, B., and Lee, D. W. (2007a). WGS reaction in a membrane reactor using a porous stainless steel supported silica membrane. *Chem. Eng. Process.*, **46**, 119–126.

Brunetti, A., Barbieri, G., Drioli, E., Granato, T., and Lee, K.-H. (2007b). A porous stainless steel supported silica membrane for WGS reaction in a catalytic membrane reactor. *Chem. Eng. Sci.*, **62**, 5621–5626.

Brunetti, A., Caravella, C., Barbieri, G., and Drioli, E. (2007c). Simulation study of water gas shift in a membrane reactor. *J. Membr. Sci.*, **306**(1–2), 329–340.

Brunetti, A., Drioli, E., and Barbieri, G. (2012a). Medium/high temperature water gas shift reaction in a Pd–Ag membrane reactor: An experimental investigation. *RSC Adv.*, **2**(1), 226–233.

Brunetti, A., Caravella, A., Drioli, E., and Barbieri, G. (2012b). Process intensification by membrane reactors: High temperature water gas shift reaction as single stage for syngas upgrading. *Chem. Eng. Technol.*, **35**, 1238–1248.

Brunetti, A., Caravella, A., Fernandez, E., Pacheco Tanaka, D. A., Gallucci, F., Drioli, E., Curcio, E., Viviente, J. L., and Barbieri, G. (2015). Syngas upgrading in a membrane reactor with thin Pd-alloy supported membrane. *Int. J. Hydrogen Energy*, **40**(34), 10883–10893.

Caravella, A., Barbieri, G., and Drioli, E. (2009). Concentration polarization analysis in self-supported Pd-based membranes. *Sep. Purif. Technol.*, **66**, 613–624.

Caravella, A., Zito, P. F., Brunetti, A., Drioli, E., and Barbieri, G. (2015a). Evaluation of pure-component adsorption properties of DD3R zeolite. *J. Chem. Eng. Data*, **60**(8), 2343–2355.

Caravella, A., Zito, P. F., Brunetti, A., Drioli, E., and Barbieri, G. (2015b) Evaluation of pure-component adsorption properties of silicalite based on the Langmuir and Sips models. *AIChE J.*, **61**(11), 3911–3922.

Caravella, A., Zito, P., Brunetti, A., Drioli, E., and Barbieri, G. (2016). A novel modelling approach to surface and Knudsen multicomponent diffusion through NaY zeolite membranes. *Microporous Mesoporous Mater.*, **235**, 87–99.

Caro, J., Noack, M., Kölsch, P., and Schäfer, R. (2000). Zeolite membranes: State of their development and perspective. *Microporous Mesoporous Mater.*, **38**, 3–24.

Castro-Dominguez, B., Mardilovich, I. P., Ma, L., Ma, R., Dixon, A. G., Kazantzis, N. K., and Ma, Y. H. (2016). Integration of methane steam reforming and water gas shift reaction in a Pd/Au/Pd-based catalytic membrane reactor for process intensification. *Membranes*, **6**(3) doi:10.3390/membranes6030044.

Chen, W., Hu, X., Wang, R., and Huang, Y. (2010). On the assembling of Pd/ceramic composite membranes for hydrogen separation. *Sep. Purif. Technol.*, **72**, 92–97.

Chen, Z., Grace, J. R., Jim Lim, C., and Li, A. (2007). Experimental studies of pure hydrogen production in a commercialized fluidized-bed membrane reactor with SMR and ATR catalysts. *Int. J. Hydrogen Energy*, **32**, 2359e2366.

Chiappetta, G., Barbieri, G., and Drioli, E. (2010). Pd/Ag-based membrane reactors on small scale: Assessment of the feed pressure and design parameters effect on the performance. *Chem. Eng. Process.*, **49**(7), 722–731.

Cui, M., Joshi, L., and Wellington, S. L. (2009). System and process for making hydrogen from a hydrocarbon stream, U. S. Patent 2009/0180949 Al.

David, E. and Kopac, J. (2011). Development of palladium/ceramic membranes for hydrogen separation. *Int. J. Hydrogen Energy*, **36**, 4498–4506.

Deckman, H. W., Fulton, J. W., Grenda, J. M., and Hershkowitz, F. (2005). European Patent 1294637.

Deng, L. and Hägg, M. (2015). Fabrication and evaluation of a blend facilitated transport membrane for CO_2/CH_4 separation. *Ind. Eng. Chem. Res.*, **54**(44), 11139–11150.

Deshmukh, S. A. R. K., Heinrich, S., Mörl, L., van Sint Annaland, M., and Kuipers, J. A. M. (2007). Membrane-assisted fluidized bed reactors: Potentials and hurdles. *Chem. Eng. Sci.*, **62**, 416–436.

Dittmeyer, R., Höllein, V., and Daubb, K. (2001). Membrane reactors for hydrogenation and dehydrogenation processes based on supported palladium. *J. Mol. Catal. A: Chem.*, **173**, 135–184.

Fan, X., Tao, Y., Wang, L., Zhang, X., Lei, Y., Wang, Z., and Noguchi, H. (2014). Performance of an integrated process combining ozonation with ceramic membrane ultra-filtration for advanced treatment of drinking water. *Desalination*, **335**, 47–54.

Fong, Y. Y., Abdullah, A. Z., Ahmad, A. L., and Bhatia, S. (2008). Development of functionalized zeolite membrane and its potential role as reactor combined separator for para-xylene production from xylene isomers. *Chem. Eng. J.*, **139**, 172–193.

Gallucci, F., Fernandez, E., Corengia, P., and van Sint Annaland, M. (2013). Recent advances on membranes and membrane reactors for hydrogen production. *Chem. Eng. Sci.*, **92**, 40–66.

Gallucci, F., van Sint Annaland, M., and Kuipers, J. A. M. (2008a). Autothermal reforming of methane with integrated CO_2 capture in a novel fluidized bed membrane reactor. Part 1: Experimental demonstration. *Topics Catal.*, **51**, 133e145.

Gallucci, F., van Sint Annaland, M., and Kuipers, J. A. M. (2008b). Autothermal reforming of methane with integrated CO_2 capture in a novel fluidized bed membrane reactor. Part 2: Comparison of reactor configurations. *Topics Catal.*, **51**, 146e157.

Gallucci, F., van Sint Annaland, M., and Kuipers, J. A. M. (2010a). Pure hydrogen production via autothermal reforming of ethanol in a fluidized bed membrane reactor: A simulation study. *Int. J. Hydrogen Energy*, **35**, 1659e1668.

Gallucci, F., van Sintannaland, M., and Kuipers, J. A. M. (2010b). Theoretical comparison of packed bed and fluidized bed membrane reactors for methane reforming. *Int. J. Hydrogen Energy*, **35**, 7142e7150.

García-García, F. R. and Li, K. (2013). New catalytic reactors prepared from symmetric and asymmetric ceramic hollow fibers. *Appl. Catal. A Gen.*, **456**, 1–10.

García-García, F. R., Rahman, M. A., González-Jiménez, I. D., and Li, K. (2011). Catalytic hollow fiber membrane microreactor: High purity H_2 production by WGS reaction. *Catal. Today*, **171**, 281–289.

García-García, F. R., Rahman, M. A., Kingsbury, B. F. K., and Li, K. (2010). A novel catalytic membrane microreactor for Cox free H_2 production. *Catal. Commun.*, **12**, 161–164.

Giaconia, A., De Falco, M., Caputo, G., Grena, R., Tarquini, P., and Marrelli, L. (2008). Solar steam reforming of natural gas for hydrogen production using molten salt heat carriers. *Am. Inst. Chem. Eng. J.*, **54**, 1932e1944.

Gummalla, M., Vanderspurt, T. H., She, Y., Dardas, Z., and Olsommer, B. (2010). U. S. Patent 20100104903.

Han, J., Kim, I.-S., and Choi, K.-S. (2000). Purifier-integrated methanol reformer for fuel cell vehicles. *J. Power Sources*, **86**, 223–227.

Hara, S., Haraya, K., Barbieri, G., and Drioli, E. (2008). Reaction rate profiles in long palladium membrane reactors for methane steam reforming. *Desalination*, **233**(1–3), 359–366.

Hara, S., Haraya, K., Barbieri, G., and Drioli, E. (2010). Estimating limit conversion for methane steam reforming in a palladium membrane reactor using countercurrent sweep gas. *Asia-Pacific J. Chem. Eng.*, **5**(1), 48–59.

Hashim, S. M., Mohamed, A. R., and Bhatia, S. (2010). Current status of ceramic-based membranes for oxygen separation from air. *Adv. Colloid Interf. Sci.*, **160**, 88–100.

Haugsrud, R. (2007). Defects and transport properties in Ln_6WO_{12} (Ln = La, Nd, Gd, Er). *Solid State Ionics*, **178**, 555–560.

He, X. and Hägg, M. (2013). Hollow fiber carbon membranes: From material to application. *Chem. Eng. J.*, **215–216**, 440–448.

Hirota, Y., Ishikado, A., Uchida, Y., Egashira, Y., and Nishiyama, N. (2013). Pore size control of microporous carbon membranes by post-synthesis activation and their use in a membrane reactor for dehydrogenation of methylcyclohexane. *J. Membr. Sci.*, **440**, 134–139.

Hu, X., Yu, J., Song, J., Wang, X., and Huang, Y. (2011). Toward low-cost Pd/ceramic composite membranes for hydrogen separation: A case study on reuse of the recycled porous Al_2O_3 substrates in membrane fabrication. *Int. J. Hydrogen Energy*, **36**, 15794–15802.

Itoh, N., Suga, E., and Sato, T. (2014). Composite palladium membrane prepared by introducing metallic glue and its high durability below the critical temperature. *Sep. Purif. Technol.*, **121**, 46–53.

Iulianelli, A., Manzolini, G., De Falco, M., Campanari, S., Longo, T., Liguori, S., and Basile, A. (2010). H_2 production by low pressure methane steam reforming in a Pd–Ag membrane reactor over a Ni-based catalyst: Experimental and modeling. *Int. J. Hydrogen Energy*, **35**(20), 11514–11524.

Iulianelli, A., Pirola, C., Comazzi, A., Galli, F., Manenti, F., and Basile, A. (2014). Water gas shift membrane reactors, in *Membrane Reactors for Energy Applications and Basic Chemical Production*, A. Basile, F. Hai, L. Di Paola, and V. Piemonte (eds.), Woodhead.

Jensen, K. F. (1999). Microchemical systems: Status, challenges and opportunities. *AIChE J.*, **45**(10), 2051–2054.

Jiang, H., Meng, L., Chen, R., Jin, W., Xing, W., and Xu, N. (2013). Progress on porous ceramic membrane reactors for heterogeneous catalysis over ultrafine and nano-sized catalysts. *Chinese J. Chem. Eng.*, **21**, 205–215.

Karnik, S. V., Hatalis, M. K., and Kothare, M. V. (2003). Towards a palladium micro-membrane for the water gas shift reaction: Microfabrication approach and hydrogen purification results. *J Microelectromech Syst.*, **12**(1), 93–100.

Kim, D., Kellogg, A., Livaich, E., and Wilhite, B. A. (2009). Towards an integrated ceramic micro-membrane network: Electroless plated palladium membranes in cordierite supports. *J. Membr. Sci.*, **340**, 109–116.

Kim, J. and van der Bruggen, B. (2010). The use of nanoparticles in polymeric and ceramic membrane structures: Review of manufacturing procedures and performance improvement for water treatment. *Environ. Pollut.*, **158**, 2335–2349.

Kolb, G. (2013). Review: Microstructured reactors for distributed and renewable production of fuels and electrical energy. *Chem. Eng. Process.*, **65**, 1–44.

Kosinov, N., Gascon, J., Kapteijn, F., and Hensen, E. J. M. (2016). Recent developments in zeolite membranes for gas separation. *J. Membr. Sci.*, **499**, 65–79.

Kothare, M. V. (2006). Dynamics and control of integrated microchemical systems with application to micro-scale fuel processing. *Comput. Chem. Eng.*, **30**, 1725–1734.

Kothari, R., Buddhi, D., and Sawhney, R. L. (2008). Comparison of environmental and economic aspects of various hydrogen production methods. *Ren. Sust. En. Rev.*, **12**, 553–563.

Li, A., Lim, C. J., and Grace, J. R. (2008). Staged-separation membrane reactor for steam methane reforming. *Chem. Eng. J.*, **138**, 452–459.

Marigliano, G., Barbieri, G., and Drioli, E. (2003). Equilibrium conversion for a Pd-based membrane reactor. Dependence on the temperature and pressure. *Chem. Eng. Process.*, **42**(3), 231–236.

McLeary, E. E., Jansen, J. C., and Kapteijn, F. (2006). Zeolite based films, membranes and membrane reactors: Progress and prospects. *Microporous Mesoporous Mater.*, **90**, 198–220.

Mejdell, A. L., Jøndahl, M., Peters, T. A., Bredesen, R., and Venvik, H. J. (2009). Experimental investigation of microchannel membrane configuration with a 1.4 μm Pd/Ag 23wt.% membrane: Effects of flow and pressure. *J. Membr. Sci.*, **327**, 6–10.

Moparthi, A., Uppaluri, R., and Gill, B. S. (2010). Economic feasibility of silica and palladium composite membranes for industrial dehydrogenation reactions. *Chem. Eng. Res. Des.*, **88**, 1088–1101.

Ouyang, X., Bednarova, L., Besser, R. S., and Ho, P. (2005). Preferential oxidation (PrOx) in a thin-film catalytic microreactor: Advantages and limitations. *AIChE J.*, **51**, 1758–1772.

Patel, S. and Pant, K. K. (2007). Selective production of hydrogen via oxidative steam reforming of methanol using Cu–Zn–Ce–Al oxide catalysts. *Chem. Eng. Sci.*, **62**, 5436e5443.

Patil, C. S., van Sint Annaland, M., and Kuipers, H. A. M. (2007). Fluidised bed membrane reactor for ultrapure hydrogen production via methane steam reforming: Experimental demonstration and model validation. *Chem. Eng. Sci.*, **62**(11), 2989–3007.

Patrascu, M. and Sheintuch, M. (2015). Design concepts of a scaled-down autothermal membrane reformer for onboard hydrogen production. *Chem. Eng. J.*, **282**, 123–136.

Pinacci, P. and Drago, F. (2012). Influence of the support on permeation of palladium composite membranes in presence of sweep gas. *Catal. Today*, **193**, 186–193.

Rahman, M. A., García-García, F. R., and Li, K. (2012). Development of a catalytic hollow fiber membrane microreactor as a microreformer for automotive application. *J. Membr. Sci.*, **390–391**, 68–75.

Rahman, M. A., García-García, F. R., Irfan Hatim, M. D., Kingsbury, B. F. K., and Li, K. (2011). Development of a catalytic hollow fiber membrane microreactor for high purity H_2 production. *J. Membr. Sci.*, **368**, 116–123.

Roses, L., Gallucci, F., Manzolini, G., and van Sint Annaland, M. (2013). Experimental study of steam methane reforming in a Pd-based fluidized bed membrane reactor. *Chem. Eng. J.*, **222**, 307–320.

Sá, S., Silva, H., Sousa, J. M., and Mendes, A. (2011). Steam reforming of methanol over a $CuO/ZnO/Al_2O_3$ catalyst. Part II: A carbon membrane reactor. *Chem. Eng. Sci.*, **66**, 5523–5530.

Sanz, R., Calles, J. A., Alique, D., and Furones, L. (2014). H_2 production via water gas shift in a composite Pd membrane reactor prepared by the pore-plating method. *Int. J. Hydrogen Energy*, **39**, 4739–4748.

Sato, T., Suzuki, T., Aketa, M., Ishiyama, Y., Mimura, K., and Itoh, N. (2010). Steam reforming of biogas mixtures with a palladium membrane reactor system. *Chem. Eng. Sci.*, **65**(1), 451–457.

Shu, J., Grandjean, B. P. A., and Kaliaguine, S. (1994). Methane steam reforming in asymmetric Pd- and Pd–Ag/porous SS membrane reactor. *Appl. Catal. A: Gen.*, **119**, 305–325.

Shu, J., Grandjean, B. P. A., van Neste, A., and Kaliaguine, S. (1991). Catalytic palladium-based MRs: A review. *Can. J. Chem. Eng.*, **69**, 1036–1060.

Smart, S., Liu, S., Serra, J. M., Diniz Da Costa, J. C., Iulianelli, A., and Basile, A. (2013). Porous ceramic membranes for membrane reactors, in *Handbook of Membrane Reactors*: *Fundamental Materials Science, Design and Optimisation*, A. Basile (ed.), pp. 298–336, Woodhead Publishing Ltd. Series in Energy, Cornwall.

Sznejer, G. and Sheintuch, M. (2004). Application of a carbon membrane reactor for dehydrogenation reactions. *Chem. Eng. Sci.*, **59**, 2013–2021.

Tan, X. and Li, K. (2013a). Dense ceramic membranes for membrane reactors, in *Handbook of Membrane Reactors: Fundamental Materials Science, Design and Optimisation*, A. Basile (ed.), pp. 271–297, Woodhead Publishing Ltd. Series in Energy.

Tan, X. Y. and Li, K. (2013b). Membrane microreactors for catalytic reactions. *J. Chem. Technol. Biotechnol.*, **88**, 1771–1779.

Vankelecom, I. F. J. (2002). Polymeric membranes in catalytic reactors. *Chem. Rev.*, **102**, 3779–3810.

Vizcaıno, A. J., Lindo, M., Carrero, A., and Calles, J. A. (2012). Hydrogen production by steam reforming of ethanol using Ni catalysts based on ternary mixed oxides prepared by coprecipitation. *Int. J. Hydrogen Energy*, **37**(2), 1985–1992.

Volkov, V. V., Novitskii, E. G., Dibrov, G. A., Samokhin, P. V., Kipnis, M. A., and Yaroslavtsev, A. B. (2012). Catalytic conversion of methanol to dimethyl ether on polymer/ceramic composite membranes. *Catal. Today*, **193**, 31–36.

Wang, H., Dong, X., and Lin, Y. S. (2014). Highly stable bilayer MFI zeolite membranes for high temperature hydrogen separation. *J. Membr. Sci.*, **450**, 425–432.

Wei, L., Yu, J., Hu, X., and Huang, Y. (2012). Fabrication of H_2-permeable palladium membranes based on pencil-coated porous stainless steel substrate. *Int. J. Hydrogen Energy*. **37**, 13007–13012.

Wei, W. (2009). Polygeneration systems, U. S. Patent 20090084035.

Wu, Z., Wang, B., and Li, K. (2010). A novel dual-layer ceramic hollow fibre membrane reactor for methane conversion. *J. Membr. Sci.*, **352**, 63–70.

Zhang, Y., Wu, Z., Hong, Z., Gu, X., and Xu, N. (2012). Hydrogen-selective zeolite membrane reactor for low temperature water gas shift reaction. *Chem. Eng. J.*, **197**, 314–321.

Zito, P. F., Caravella, A., Brunetti, A., Drioli, E., and Barbieri, G. (2015). Estimation of Langmuir and Sips models adsorption parameters for NaX and NaY FAU zeolites. *J. Chem. Eng. Data*, **60**(10), 2858–2868.

Zito, P. F., Caravella, A., Brunetti, A., Drioli, E., and Barbieri, G. (2016). Knudsen and surface diffusion competing for gas permeation inside silicalite membranes. *J. Membr. Sci.*, DOI: 10.1016/j.memsci.2016.10.016.

Chapter 9

Nanostructured Functional Membranes for Self-Cleaning Separations

Abaynesh Yihdego Gebreyohannes, Annarosa Gugliuzza, and Lidietta Giorno
Institute on Membrane Technology of the National Research Council of Italy (ITM-CNR), Via P. Bucci, Cubo 17C, 87036 Rende (CS), Italy
l.giorno@itm.cnr.it

Driven by economic factors, environmental concern, and technological advancement, membrane technology has been playing a major role in various industries. However, membrane productivity and lifetime are severely limited by membrane fouling. Although various methods ranging from process optimization and feed stream pre-treatment to physical and harsh chemical cleaning have been employed to limit its extent, none of these methods was found reliable.

An alternative yet very attractive approach to tackle the challenges faced due to membrane fouling is the development of bio-inspired self-cleaning membranes. These membranes take inspiration from nature such as lotus plant and the wings of butterfly. In this chapter, various strategies that can be used to create membrane

Functional Nanostructured Membranes
Edited by Enrico Drioli, Lidietta Giorno, and Annarosa Gugliuzza

ISBN 978-981-4774-79-6 (Hardcover), 978-1-351-13511-5 (eBook)
www.panstanford.com

with self-cleaning capacity are discussed. These methods are further illustrated with examples where they have shown effectiveness as self-cleaning surfaces. The relation between membrane wettability and self-cleaning property is also highlighted.

9.1 Introduction

Membrane has a versatile application in the separation of particulate matter and biomolecules. However, membrane bio(fouling) severely limits its productivity and lifetime. Depending on the process stream, membrane fouling could be due to salt residue (inorganic fouling), organic fouling due to macromolecules, or biofilm due to microorganisms. Methods to mitigate membrane fouling ranges from feed stream pre-treatment and optimization of operational parameters to application of post-process physical, chemical, or biological membrane cleanings. In particular, for organic fouling and biofilm growth, use of oxidative chemical cleaning reagents such as hypochlorite is the most common approach. Although oxidative reagents proved to be successful in degrading the organic foulant and restoring the membrane performance to a higher degree, they are often nonselective and aggressive to the polymeric constituent of the membrane. Hence depending on the exposure duration and concentration (ppm*h), they tend to degrade the membrane and result in premature membrane module disposal/replacement. Furthermore, since membrane cleaning is a post-damage action, which already allowed direct-membrane foulant interaction, the presence of residual effects even after cleaning is inevitable. In addition, the use of chemical cleaning incurs cost through reduced production during periodic shutdown for cleaning and acquisition and disposal of cleaning reagents.

An alternative yet very attractive approach for tackling the challenges faced due to membrane fouling is the development of bio-inspired self-cleaning surfaces. Self-cleaning membranes have recently drawn an ever-growing interest both in academics and industry since they have broad applications ranging from window glass, solar panel cleaning to cements and textiles (Ganesh et al., 2011). The self-cleaning capacity of lotus, often termed as the lotus effect, is hypothesized to be the result of the complex surface structure of the leaf. Examination of the leaf structure through a scanning electron microscope (SEM) shows two levels of surface structure: micro-scale mound-like structures protruding from the leaf and nanoscale hair-like structures covering the leaf surface (Fig. 9.1). The multi-level surface roughness together with its waxy coatings is thought to be responsible for the lotus leaf's water repellency. The rolling water on the surface of the leaf collects dirt and other particles along its way, hence self-cleaning. Studies also showed that altering the surface structure without affecting the surface composition, in particular modifying the nanoscale hair-like structure, significantly hampered the membrane surface property and wettability. This clearly indicates the important role of the leaf's nanoscale hair-like structure on its self-cleaning ability (Cheng et al., 2006).

Figure 9.1 Image and SEM images of lotus leaf surface. (a) A fresh lotus leaf in nature; (b) the micro-structure of lotus leaf; (c) the nanostructure of lotus leaf; (d) the micro-structure of annealed lotus leaf; (e) the nanostructure of annealed lotus leaf; (f) a droplet placed on an untreated lotus leaf; and (g) a droplet placed on an annealed lotus leaf, then tilted to an angle of 90°. Scale bar: (b and d) 10 μm, (c and e) 3 μm. Reprinted from Zhang et al. (2016) with permission from Elsevier.

Likewise, playing on the membrane property at a micro/nano level is a strategy that should be explored to develop the next generation of bio-inspired self-cleaning membranes. Development of membranes with self-cleaning capacity is essential for the rational design and reproducible manufacturing of membranes with higher productivity and longer working lifetime (Ganesh et al., 2011).

Self-cleaning membranes can be produced through the four main known concepts that are being practiced to create self-cleaning surfaces:

1. TiO_2-based superhydrophilic self-cleaning
2. Lotus-effect self-cleaning (superhydrophobicity with a small sliding angle)
3. Gecko setae-inspired self-cleaning
4. Underwater organisms-inspired antifouling self-cleaning

Advances in nanoscience and technology are providing new opportunities to develop cost-effective and environmentally benign membranes. Nanostructured adsorbents such as mesoporous titania beads, alumina, metal or metal oxides, polyethylenimine/polyvinyl alcohol nanofibers, etc. have recently demonstrated remarkable advancement in this regard due to their higher surface areas and surface-active sites. Hence nanotechnology has evolved to commercial reality, which created ease of availability to most of the nanofunctional materials with versatile functionality. The concept of nanotechnology has already led to the development of innovative nanotechnology membranes that exceed the state-of-the-art perm-selectivity (Pendergast and Hoek, 2011). These nanofunctional materials can further be exploited in creating nature-inspired self-cleaning/antifouling membranes. Nanofunctional membranes include zeolite and catalytic nanoparticle coated membranes, hybrid inorganic–organic nanocomposite membranes, vertically aligned nanotube membranes, aquaporin membranes, isoporous block co-polymer membranes, and multi-level bio-hybrid membranes (see Schematic 9.1).

9.2 Self-Cleaning Related Wetting

Hydrophilic–hydrophobic transitions are regarded as powerful routes for locally changing surface properties and, where any, producing in situ self-cleaning actions. Patterned surfaces can modify their own wetting properties when variations take place in the surrounding environment. Sometimes, it is enough to apply external triggers such as temperature, light, pH, and molecular recognition or, more simply, mechanical, electrical, and magnetic stimulation to induce changes enabling the surface to pull out or prevent foulant adhesion through controlling forces such as surface free tension (γ) and amplified rugosity factor (R). Before examining some explicative induced self-cleaning events, it would be better to examine in depth some basic concepts concerning surface wetting and related chemical and morphological parameters frequently involved in adhesion of foulants on membrane surface. This undesired event could also be controlled and limited through the switch of surface properties, which are strongly dependent on chemical and topographical features.

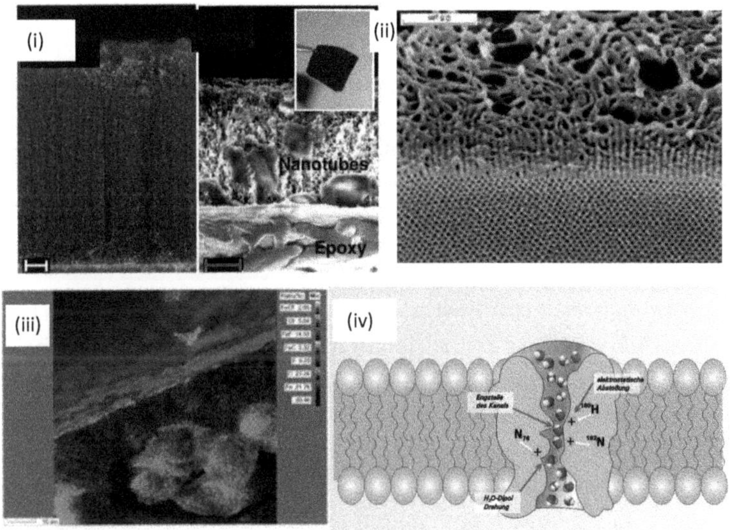

Schematic 9.1 (i) Scanning electron microscope image of vertically aligned nanotubes (the scale bar is 10 μm and the nanotubes are approximately 150 μm tall and nanotube arrays following epoxy impregnation to form a free-standing membrane; (ii) isoporous self-assembled block co-polymer membrane; (iii) nanobiocatalyst-coated membrane; and (iv) water transport through selective channels of aquaporin.

9.2.1 Wetting Concepts

Wetting (S) takes place when liquids, approaching the surface, spread on it under controlling interfacial forces that control adhesion events at the solid/liquid/air interfaces (Gugliuzza, 2014; Gugliuzza, 2016). These forces are well quantified by the surface free energy (γ_S), which is proportional to the cohesive energy required to minimize the free surface area at the interface (Gugliuzza et al., 2006). So wetting and surface free energy are correlated by the following mathematical equation:

$$S = \gamma_{SV} - (\gamma_{SL} + \gamma_{LV}) \tag{9.1}$$

where γ_{SV}, γ_{SL}, and γ_{LV} are the free tensions at solid–vapor, solid–liquid, and liquid–vapor interfaces, respectively.

Depending on the spreading values, it is possible to establish if a surface is hydrophilic or hydrophobic. Positive values are expression of hydrophilic surface, whereas negative values are indicative of surface hydrophobic behavior. Very negative values of spreading give a clear indication about superhydrophobic surface character. Empirically, the hydrophilic/hydrophobic character of the surface is estimated by measuring the apparent contact angle (θ_a). This is a thermodynamic parameter, which is usually expressed by Young's equation when ideal and almost flat surfaces are considered:

$$\cos(\theta_a) = \frac{\gamma_{SV} - \gamma_{SL}}{\gamma_{LV}} \tag{9.2}$$

Although it is not fully truthful, there is a general convention that values of $\theta_a < 90°$ are expression of hydrophilic membrane surfaces, while values of $\theta_a > 90°$ indicate hydrophobic surfaces. Values of $\theta_a > 150°$ are emblematic of ultraphobic surfaces, whereon liquid flow over the surfaces is the roll-off angle or sliding angle.

Unfortunately, completely flat surfaces are not found in nature, because tailed and loopy materials, irregularities, grooves, and pores are well distributed, which cause important changes in liquid–solid adhesion (Thomas, 1999). This implies that the morphological factor, expressed as the surface roughness parameter (R), must be considered together with the chemical one (Perrotta et al., 2017; Speranza et al., 2010). With this regard, Wenzel and Cassie–Baxter's equations are considered to be more suitable than Young's equation for a more accurate description of real surfaces, including membranes.

The Wenzel equation assumes that the liquid fills the grooves through the surfaces. In this case, the contact angle is expressed as:

$$\cos(\theta_w) = R\left[\frac{\gamma_{SV} - \gamma_{SL}}{\gamma_{LV}}\right] \tag{9.3}$$

Combined with Young's equation, the Wenzel contact angle (θ_w) is correlated to the apparent contact angle by the following relation:

$$\cos(\theta_w) + R \cdot \cos(\theta_a) \tag{9.4}$$

where R is the rugosity factor that quantifies the irregularities of the membrane surface, which are responsible for increased contact angle values. The Cassie–Baxter model amplifies this event according to this equation:

$$\cos(\theta_{CB}) = f_s \cdot \cos(\theta_a) + f_v \cdot \cos(\theta_v) \tag{9.5}$$

where θ_a is the apparent contact angle measured on the flat surface, whereas f_s and f_v are the fractions of the solid and vapor on the surface (i.e., $f_s + f_v = 1$), respectively. According to this model, the air fills the surface grooves so that the liquid–surface adhesion is significantly reduced, giving rise to a very high hydrophobic behavior. The role of chemistry on surface wettability would be properly evaluated through correction of the apparent contact angle for the roughness factor according to the zigzag model proposed by Taniguchi and Belfort (2002).

In this context, it is clear that rugosity becomes a powerful tool to make surfaces highly hydrophobic and, hence, unwettable (Chen at al., 2011). Multiscale surface roughness can amplify the wettability causing typical rolling water drops across the surfaces. Natural objects are an extraordinary source of inspiration for many scientists, including membranologists, which try to realize bio-inspired architectures and supramolecular structures, enabling contaminant resistance, antifogging, antifouling, and self-cleaning under different local environmental conditions.

9.2.2 Responsive Materials–Induced Surface Cleaning via Wetting

Self-cleaning actuation is frequently obtained with the switch of the wetting properties of the membrane surface through reversible hydrophilic/hydrophobic transitions. With this intent, membrane surfaces are often functionalized with compounds responsive to temperature, pH, light, electrical/magnetic fields, mechanical and molecular stimulation (Verplanck, 2007). Hereafter, we report just some relevant examples, which underscore the great opportunity to use smart materials to get morphological and chemical changes in the surface with the target to modify wetting properties and, consequently, self-cleaning capabilities.

9.2.2.1 Temperature and pH stimulation

Poly(*N*-isopropylacrylamide) (PNIPAM) hydrogels are thermally responsive materials that enable one to reversibly control wetting adhesion moving from a superoleophobic state, below the PNIPAM lower critical solution temperature (LCST, 32°C), to an oleophobic state, above its LCST, causing changes of about 24° in the contact angle value. This behavior is due to the contraction and expansion on the local surface topography induced by temperature (Chen, 2010).

Wetting triggered by pH is observable when polymers such as polyacrylic acid/polyvinylpyrrolidone (PAA/PVP) and polyacrylic acid/poly(diallyldimethylammonium chloride) (PAA/PDDA) polyacrylic acid/poly(allylamine hydrochloride) (PAA/ PAH) and polyvinylpyrrolidone/carboxyl-terminated polyether dendrimer (PVP/DEN-COOH) multilayer films are used (Connal et al., 2008; Fu et al., 2002; Gugliuzza et al., 2015; Pingitore and Gugliuzza, 2013; Synytska et al., 2007; Zhang et al., 2003). In this case, the ionic charge can be controlled and modulated in order to promote expansion or shrinking of polymer segments, which are responsible for changes in liquid–solid adhesion events. For example, a transition from superoleophilicity to superoleophobicity state in aqueous media has been achieved by grafting pH-responsive poly(2-vinylpyridine) and oleophilic/hydrophobic polydimethylsiloxane (P2VP-b-PDMS) (Zhang et al., 2007).

9.2.2.2 Light irradiation

Light is another powerful trigger to stimulate changes in both the surface polarity and texture of membrane surfaces, thereby providing amplified wetting. Conformational transitions can be induced reversibly by using classes of compounds including azobenzenes, spiropyrans, dipyridylethylenes, stilbenes, and pyrimidines (Nicoletta et al., 2012). The switchable transition from an isomer to another under UV/light irradiation makes them interesting materials to functionalize membrane surfaces with controllable wetting properties. For example, biodegradable polycarilactone (PLC) nanofibers have been modified with azobenzene to get trans-to-cis isomerization under UV irradiation. A change in the static contact angle from 132.2 ± 2.8° to 53.1 ± 3.2° has been obtained, while full reversibility has been observed under visible light (Chen and Besenbacher, 2011). Likewise, an increase in hydrophilicity (112° to 90°) has been achieved, changing micro-structurally rough polymer surfaces containing spiropyran molecules (1′,3′-dihydro-1′,3′,3′-trimethyl-6-nitro-spiro[2H-1-benzopyran-2,2′-(2H)-indole]) (Anastasiadis et al., 2008). In this case, the reversible change in wettability was caused by a transition from non-polar spiropyran molecules to polar merocyanine isomers (Nayak et al., 2006; Wang et al., 2012). The effect of isomerization from spiropyran (non-polar closed form) to merocyanine (polar open form) on protein adsorption has been studied evaluating the permeation rate. A reduction of 26% of protein adsorption has been quantified on the open configuration of the merocyanine surface with resulting higher permeation rate. As discussed hereafter, wetting with resulting effects on cleaning ability can be amplified by using metal oxides when patterned in intricate structures. In this case, semiconductor properties can be combined with complex light-responsive topographies, yielding in situ self-cleaning behaviors as a result of photocatalysis.

9.2.2.3 Mechanical stress and solvent annealing

Mechanical stress can also dramatically change wetting properties. An elastic polyamide ($[-(CH_2)_{11}-CO-NH-]_n$) film exhibiting a triangular net-like structure demonstrated how the topography can be modified under biaxial extension and unloading, yielding reversible variation of the contact angle values from 151° to 0° (Zhang et al., 2005).

Also, stimulating interactions between a vapor phase (solvent) and a solid phase (polymeric surface) can induce local molecular rearrangement in order to make smoother or rougher surfaces.

Poly(methyl acrylate)/polystyrene/poly(methyl acrylate) (PMA-*block*-PS-*block*-PMA) anchored on an Si/SiO_2 surface has shown contact angle of 69° after annealed with dichloromethane, while a local collapse of the polymer structure has been induced causing contact angle values of 90° when cyclohexane has been used in place of dichloromethane. In the first case, a smoother surface has been obtained; in the second case, a high degree of roughness has been achieved (Boyes et al., 2002).

Another class of interesting compounds enabling changes in wetting properties is that of ionic liquids (ILC) (Liu et al., 2015). The liquid spreading on the surface can be obtained reversibly by dipping the samples in solution containing different hydrophilic (Cl^-, Br^-, and BF_4^-) or hydrophobic anions (PF_6^- and Tf_2N^-) (He et al., 2008).

9.3 Inorganic–Organic Nanocomposite Membranes

The concept of mixed matrix membranes (MMMs) where an inorganic filler (in the form of nano- or microparticles) was introduced to polymeric continuous phase was first introduced by Zimmerman et al. in the 1990s, as a strategy to overcome the limitations of polymeric membranes for gas separation. MMMs combine the low cost and ease of fabrication of polymeric membranes and mechanical strength and functional properties of inorganic materials (Dong et al., 2013). The most commonly exploited inorganic fillers are zeolites, alumina, metal–organic frameworks (MOFs), carbon nanotubes, iron–palladuim particles, silica, and titania. The particle size covers broader ranges from 0.5 to 300 nm, while the general composition ranges from 0.01 to 40 wt% (Pendergast and Hoek, 2011). Changes imparted by the inclusion of these inorganic fillers into the polymer matrix are hydrophilicity/hydrophobicity, roughness, permeability, and selectivity depending on the type, size, and weight fraction of the inorganic particle, improved thermal and mechanical stability as well as enhanced antifouling property.

Compared to microparticles, inclusion of nanoparticles increases permeability as nanoparticles tend to increase the characteristic pore size. Overall, the addition of nanoparticles can be tailored to particular membrane applications with the selection of nanoparticle size, shape, and type. For instance, inclusion of silver nanoparticle could inhibit biofouling growth due to the antimicrobial property of nano-silver (Basri et al., 2010).

With the notion of employing nanofunctional membranes to impart self-cleaning behaviors, specific cases where various strategies has been used to create self-cleaning nanofunctional membranes will be further illustrated in the subsequent sections.

9.3.1 TiO_2-Loaded Self-Cleaning Membrane

Various approaches have been studied in order to reduce/minimize membrane fouling either by improving or optimizing operating process conditions or by tailoring membrane surface properties. The affinity between solute and membrane such as sorption and the aggradations of solute on the membrane surface and also in and around the pores are the main factors that affect membrane fouling. Therefore, modifying the membrane surface property to more hydrophilic appears to be a better way to improve the membrane resistance to fouling.

Although membrane surface modification showed better membrane performance in reducing fouling, it requires higher energy and production cost and it can also block surface pores and reduce the intrinsic membrane integrity.

Recently, membrane surface modification such as preparation of organic–inorganic composite membrane by mixing inorganic materials has gained much importance due to simplicity, mild conditions, and stable performance. For instance, many studies have focused on the preparation of TiO_2-entrapped or -coated or -self-assembled membranes, which can provide excellent self-cleaning/antifouling, bactericidal properties. These membranes could also exhibit photocatalytic property based on UV/TiO_2 photocatalytic mechanism. TiO_2 has gained much attention as a nanofiller since it exhibits great stability, superhydrophilicity, lower cost, commercial availability, excellent photocatalytic, antibacterial, and self-cleaning properties by absorbing UV rays (Linsebigler et al., 1995). Therefore, incorporating TiO_2 nanoparticles over polymeric membrane can be used to create nanofunctional self-cleaning membrane. The self-cleaning of TiO_2 membranes mimics the natural self-cleaning strategies through photocatalysis and increased surface hydrophilicity.

For photocatalysis, TiO_2 absorbs light with energy equal to or greater than its band-gap energy (3.2 eV) generating valence band holes (h^+) and conduction band electrons (e^-). Though most of these charges undergo recombination, a few of them migrate to the surface and react with species adsorbed on the membrane surface. Holes react with water (H_2O/OH^-) to generate hydroxyl radicals, while electrons reduce adsorbed molecular oxygen to superoxide radical anion, which, in turn, reacts with protons to form peroxide radicals. The generated hydroxyl radicals oxidize organic foulant into CO_2 and water at room temperature and also disinfect certain bacteria and viruses (Linsebigler et al., 1995; Maness et al., 1999).

For the photoinduced self-cleaning superhydrophilic property of TiO_2-modified membrane surface, photogenerated electrons tend to reduce Ti^{4+} cation to Ti^{3+} cation state and holes oxidize O_2^- anions to evolved O_2 resulting in a group of oxygen vacancies on the surface of the membrane. Subsequently, water molecule on the surface occupies empty sites and forms OH groups, which will increase the membrane surface hydrophilicity (Maness et al., 1999).

For instance, PVDF/TiO_2 composite membranes containing 0–4 wt% TiO_2 prepared via phase inversion method resulted in different hydrophilicity and pore size with varying amount of TiO_2. The membrane showed better bactericidal effects under UV light, increased permeability, higher Rose Bengal degradation (use as a model dye) while also showing a great self-cleaning capacity when the membrane was fouled by a 1% bovine serum albumin (BSA) solution (Damodar et al., 2009).

Free-standing self-cleaning membranes for size-selective macromolecular separation using nanochannel TiO_2 membrane were also prepared by Roy et al. (2010). The nanochannel TiO_2 membrane fabricated in a single-step complete anodic oxidation of thin Ti metal foils exhibited 8–12 nm channel diameter with a narrow size distribution. Apart from size-selective separation of macromolecules, the main advantage of these nanochannel membranes is the inherent photocatalytic property of Ti, which gave the membrane a self-cleaning property. The formed nanochannels revealed excellent permeability for cytochrome C (12 kDa), 1.63 nm; and BSA (69 kDa), 3.62 nm, while it completely rejected galactosidase (116 kDa), 6.86 nm. During long-term filtration of BSA over 120 h, the membrane pores were clogged causing an almost zero flux. Thanks to the photocatalytic property of the Ti-based nanomembranes, authors also demonstrated the possibility to self-clean the membrane and re-establish its initial perm-selectivity with a brief UV exposure (Roy et al., 2010).

9.3.2 Nano-silica Fabricated with Silver Nanoparticle Antifouling Membranes

Among all types of fouling, biofilm formation (membrane biofouling) causes reduction in pore size and surface roughness and blockage of surface functional groups. This leads to higher energy consumption, premature membrane module replacement, and may be generation of antibiotic resistant strains that significantly hamper the treated effluent quality.

The main strategy through which one can prevent biofouling is to prevent bacterial adhesion, which is a critical step of colonization. A successful nanomaterial coated with antimicrobial agent may thus be used to form antifouling membrane surface, which may prevent biofilm formation. In this regard since nanoscale silver materials (AgNPs) exhibit a wide range of antimicrobial activity with low toxicity to mammalian cells, fabrication of nanocomposite membrane containing AgNPs could offer a potential solution to inhibit biofilm formation. Over the last decades, several methods have been employed to fabricate AgNP nanocomposite membranes. The mechanisms that are most often used to fabricate AgNP nanocomposite membranes are impregnation and coprecipitation, while hydrothermal reactions and sol–gel methods have also been used.

In addition to the chemical methods, inspired by natural biomineralization processes, nanocomposites able to function as biocidal/antifoulant have been synthesized using various types of biomolecules such as nucleic acids, proteins, peptide, bacteria, and fungi (Das et al., 2010; Das et al., 2012; Gao and Matsui, 2005). For instance, Das et al. (2013) prepared protein-coated AgNPs on the surface of nano-silica to form a stable functional composite material. The synthesized AgNPs tested as a novel antifouling nanomaterial for efficient removal of dyes and pathogenic microorganisms from waterbodies efficiently killed bacteria and prevented biofilm formation.

Various studies have also reported the development of antimicrobial membranes containing immobilized nano-Ag in a cellulose, polyethersulfone, and PVDF membrane through physical entrapment or by in situ formation, which involves Ag^+ capture followed by reduction to nano-Ag (De Gusseme et al., 2011; He et al., 2016; Liu et al., 2017; Nisola et al., 2012).

AgNP-loaded membranes were prepared on a graphene oxide and sulfonated polyanion co-doped dual-layered polymeric membrane. The dual-layered membrane was first prepared by coating a top layer of graphene oxide and a sulfonated polyanion co-doped hydrogel thin film on a bottom PES membrane substrate. Ag^+ ions were introduced into the membrane through electrostatic interaction when the dual-layered membrane was immersed in an $AgNO_3$ solution for 5 h. Subsequently, Ag nanoparticles were formed through an in situ growth by immersing the membranes in vitamin C solution for 1 h. The AgNP-loaded composite membranes showed effective antibacterial properties for *E. coli* and *S. aureus* (He et al., 2016). Graphene oxide nano-sheets embedded in a membrane coating have demonstrated to exhibit self-cleaning properties by both hydrophilic properties (that prevent bacteria attachment on the membrane surface) and mechanical damage of the bacteria cell membrane (thus avoiding bacteria growth).

9.3.3 Carbon Nanotubes–Coated Self-Cleaning Membranes

Aquaporin are protein channels that control water flux across biological membranes. Water movement in aquaporin is mediated by selective, rapid diffusion caused by osmotic gradients (Agre, 2006; Meinild et al., 1998). Carbon nanotubes (CNTs) are known to have a comparable fast mass transfer to aquaporin water channels (Holt et al., 2006). Molecular dynamic simulation studies ascribed this excellent performance of CNTs to atomic smoothness and molecular ordering. In particular, water molecules permeate through the CNTs in a one-dimensional single file resulting in very small driving force use. Moreover, CNT-based membrane represents excellent mechanical properties that impart longer lifetime than conventional membrane materials. CNTs have also been proven to have good electrical conductivity (De Volder et al., 2013). This property was recently exploited to create membranes with self-cleaning capacity using the concept of electrolytic cleaning in an electrolysis process, in which a direct voltage is applied between an anode and cathode dipped in an electrolyte containing free ions for conducting electric current (Hakimhashemi et al., 2012).

To prove this concept, Hashaikeh et al. (2014) coated a 0.2 μm flat sheet membrane (9.34 cm^2 active area) with 0.05 wt% CNTs, which were dispersed in 1 wt% sodium dodecyl sulfate in water using vacuum filtration at 30 kPa pressure. Subsequently, silver electrodes were printed on the CNT-coated membrane to improve the electric charge distribution of the membranes. Based on SEM images, the total thickness of the CNT layer was 3–4 μm. This coating layer over the surface of the polymeric membrane served as an electrode during the periodic electrolysis cleaning of the membrane. Hence subsequent to filtration of $CaCO_3$ or yeast suspension used as model inorganic and organic foulant, respectively, it was possible to restore about 80% of the original flux through the designed electrolysis cleaning method. During electrolysis of NaCl solution, hydrogen gas is produced at the cathode (the conductive membrane) and chlorine gas is produced at the anode (a stainless steel counter electrode). The produced hydrogen microbubbles were able to disperse the foulant away from the membrane surface into the bulk stream. The presence of a thin layer of CNTs over the surface of microporous membrane thus helped with in situ membrane self-cleaning, thus providing a non-destructive, continuous, and renewable approach to mitigate the different types of membrane fouling (Hashaikeh et al., 2014).

9.4 Biomimetic and Bio-inspired Self-Cleaning Membranes

Biomimetic nanostructured membranes are formed by either directly embedding biomolecules into synthetic materials (bio-hybrid) or using functional molecules to modify synthetic materials to impart specific biological property. These hybrid biomimetic membranes combine the accurate structure of a biological pore with the durability, robustness, and the possibility to control the pore size and shape of solid-state nanopore membrane (Shen et al., 2014). Bio-inspired nanostructured membranes more generally refer to structures and/or functions inspired by biological world. Usually they are made of synthetic systems.

9.4.1 Enzyme-Loaded Self-Cleaning Membranes

Of the various types of membrane fouling, the occurrence of organic fouling is prominent in most municipal wastewater treatment and food-processing industries. Typical food industries that use single or hybrid membrane operations include concentration of milk and whey, and fractionation of milk in the dairy industry; fruit juice, wine, and vinegar clarification in the beverage industry; concentration of egg and egg white; recovery of starches; decolorization/demineralization/clarification of sugar and sweeteners; production of soy protein concentrates and isolates; and applications in vegetable oil processes. Although with varying degree, all these streams are known to cause severe membrane fouling (Li and Chen, 2010). The majority of these fouling are macromolecular fouling due to protein, polysaccharides, lipids, and starches in which most often more than one component is available. In particular, polysaccharide-rich streams are notorious as they form compact gel layer due to the crosslinking between their surface functional groups and divalent cations while those rich in protein cause complex fouling phenomena due to the labile and dynamic nature of the protein. Moreover, biofouling on membrane surfaces is common due to cell growth and produced by-products such as soluble microbial products (SMPs) or extracellular polymeric substances (EPSs). SMPs and EPSs are large molecular weight compounds excreted by bacteria and composed mostly of proteins, lipids, and polysaccharides (Besha et al., 2017).

An interesting aspect of the various macromolecules mentioned above is the fact that they are easily hydrolyzable with enzymes. Hence, development of membranes functionalized with enzymes that can hydrolyze the specific foulant on those streams is a very suitable strategy that can be used to develop bio-hybrid and biomimetic self-cleaning nanofunctional membranes that mimic the biological membrane compartmentalization (Mazzei et al., 2017).

In our recent work, a biomimetic membrane with covalently immobilized pectinase was used for in situ degradation of pectin during the microfiltration of wastewater obtained from edible oil production industry (Gebreyohannes et al., 2013). Coupling of microfiltration with in situ degradation of pectins with immobilized pectinase at the membrane–solution interface and simultaneous separation of the hydrolysis product opened the pathway for continuous treatment of real industrial wastewater, which is a complex mixture containing oil, salts, polysaccharides, and polyphenols. This wastewater is already known to cause up to 80% flux loss short time after the onset of filtration. The membrane with immobilized enzyme gave 50% higher permeate flux compared to the inert membrane. In addition to the enzymatic degradation, this approach is robust as it avoids direct interaction between the membrane and the foulant; hence, it prevents irreversible or irrecoverable damage to the membrane. This self-cleaning action can be implemented during the fluid treatment when the obtained products are nontoxic or non-desired to be collected in the permeate. For example, in the treatment of wastewaters coming from agro-food industry, where fouling is caused by pectins and proteins, mainly galacturonic acid and amino acids, respectively, are produced. These components can sometimes even increase the value of the collected fractions of biocative molecules that can be valorized.

In another study, a self-cleaning polyethersulfone membrane was created with covalently immobilized pancreatin, which is a mixture of digestive enzymes such as protease, lipase, and amylase. Fouling tests with solutions of protein, oil, and their mixtures showed that the immobilized enzyme actively degraded a fouling layer on its surface and regained up to 90% initial permeability after "switching on" the respective enzyme by adjusting pH and temperature. To activate the self-cleaning property of the membrane, the fouled membrane was soaked overnight in PBS buffer at pH 8 and 37°C to activate the enzymes (Schulze et al., 2017).

9.4.2 Bio-hybrid Magnetic-Responsive Membranes

Of the various interesting phenomena, imparting stimuli-responsive properties to membrane in order to create biomimetic self-cleaning membrane is appealing. One of the main strategies that is used to impart stimuli-responsive property to a membrane is incorporation of superparamagnetic nanoparticles (NP^{SP}) to form organic–inorganic (O/I) hybrid membranes (Daraei et al., 2013; Sanchez et al., 2011). NP^{SP} are most often superparamagnetic iron oxide with zero memory of their magnetic property in the absence of an external magnetic field (Yeon et al., 2009). Therefore, the presence of NP^{SP} leads to stimuli-responsive "smart" polymeric membrane that can be modulated in a reversible manner. For instance, this concept was used to induce magnetic-responsive micro-mixing nanofiltration membrane. The membrane was prepared by grafting magnetic-responsive nanolayers consisting hydrophilic poly(2-hydroxyethyl methacrylate) (polyHEMA) flexible polymer chains with NP^{SP} attached to the chain ends. The chain oscillates in an oscillating magnetic field, induced mixing at the membrane–fluid interface that maximized the disruption of concentration polarization layer (Himstedt et al., 2011).

NP^{SP} have also been used as inorganic nanofillers in a PVDF membrane to impart controlled porosity, texture, and chemical composition that can provide real prospect to efficient membrane filtration (Huang et al., 2012). The presence of NP^{SP} also makes the membrane superparamagnetic, which can be turned on/off by an external magnetic field (Daraei et al., 2013; Thevenot et al., 2013). Moreover, inclusion of NP^{SP} can enhance the membranes' hydrophilicity, mechanical strength, compaction resistance, improved perm-selectivity, and antifouling property (Daraei et al., 2013; Huang et al., 2012). Therefore, these membranes have broad functions ranging from environmental remediation and bio-commodity production to smart product manufacturing. Particularly, the synergistic antifouling property and improved perm-selectivity give the membrane a big prospect to be utilized in water purification. NP^{SP}-coated membranes, for example, have effectively removed natural organic matter (Yao et al., 2009), arsenic (Sabbatini et al., 2010), and copper (Daraei et al., 2012) during water treatment.

Coating of NP^{SP} with a polymer, which has amine or carboxylic surface functional groups, facilitates the immobilization of biomolecules on the surface of these particles (Brullot et al., 2012; Miguel-Sancho et al., 2012; Xiao-Ming and Wainer, 1993). The resulting nanosized bioactive particles represent an important material with a versatile application in biotechnology, fine chemicals, drug delivery, cell transplantation, or cell immobilization. Due to the high surface-to-volume ratio they generally exhibit, it is possible to achieve higher enzyme loading and enhanced mass transfer efficiency. Hence, these materials with excellent mechanical, optical, electrical, ionic, and catalytic properties are a good candidate to reversibly immobilize enzyme on the surface of a polymeric membrane to create a bio-inspired self-cleaning biomimetic membrane.

Immobilization of enzymes on membranes to form biocatalytically active biomimetic membranes has already shown an excellent performance when tested for self-cleaning applications. However, direct integration of enzymes on the membrane has two critical challenges: over-accumulation of foulant due to the unbalanced ratio of mass transfer to reaction rate and denaturation due to the sensitivity of enzymes to any change in their microenvironment. Both conditions demand chemical cleaning of the membrane and enzyme replenishment, respectively. Since covalently bonded to the membrane, they are highly sensitive to chemical cleanings. In addition, when membrane-immobilized enzymes get denatured or oversaturated, it is very difficult (if not impossible) to replenish with fresh enzymes. Consequently, the ultimate fate of biocatalytic membranes is only disposal. Hence, a novel immobilization technique that fulfils the conditions of BMRs but still facilitates membrane chemical cleaning and enzyme renewal is urgently required.

An alternative to direct integration of the enzyme within the polymer matrix, stimulus-responsive programmable layers on a membrane can be formed through the attraction of NP^{SP} through reversible magnetic force. These biomimetic nanolayers will enable development of adaptive biocatalytic membranes. Hence, when an enzyme is immobilized on NP^{SP}, they can be easily dispersed in a reaction medium in the absence of an external magnetic field. Subsequently, by applying an external magnetic field, it is possible to attract these particles toward the surface of a membrane using an external magnetic field, thus physical immobilization of enzyme on the membrane using a reversible magnetic field. The method is promising to resolve pertinent issues related to the direct integration of enzyme within the polymeric matrix. It also facilitates enzyme recovery for reuse and enzyme removal for membrane cleaning. Moreover, the formed dynamic layer helps to avoid direct-membrane foulant interaction.

This approach has recently been demonstrated to form a biomimetic self-cleaning membrane (Gebreyohannes et al., 2015). The method is novel as it uses NP^{SP} as a carrier to immobilize the various enzymes and also as a nanofiller to form a magnetic-responsive hybrid membrane. When an external magnetic field is applied, the magnetic-responsive polymeric membrane gets magnetized. Subsequently, NP^{SP} with the immobilized enzyme, which are dispersed in the upper stream, are attracted toward the surface of the membrane. The magnetic-responsive membrane acts as a magnetic field actuator, which eventually helps with uniform dispersion of NP^{SP} on the surface of the membrane.

The biomimetic membrane formed by magnetically depositing 1–2 g/m^2 of pectinase-activated NP^{SP} reduced fouling caused by pectin by 40–100%. Owing to the facile removal of the immobilized NP^{SP}, it was possible to use both membrane and immobilized enzyme over several cycles (Gebreyohannes et al., 2015; Gebreyohannes et al., 2017; Gebreyohannes et al., 2016).

Fouling is mostly caused by a mixture of macromolecules. Apparently, instead of a single enzyme, a mixture of enzymes is known to work better in degrading macromolecules. Hence, the immobilization technique used for creating biomimetic self-cleaning membrane should provide the degree of freedom to co-immobilize multiple biomolecules in the same compartment in order to address various foulants simultaneously (Allie et al., 2003).

The possibility to mix different enzymes either coated as an enzyme mixture per batch of NP^{SP} or combined as a mixture of single enzyme immobilized NP^{SP} can be a strategy to cope with multiple foulants without the need for using multistage reactors.

To prove this concept, a mixture of xylanase-and-pectinase-activated NP^{SP} was magnetically attracted to the membrane surface to create a self-cleaning membrane. The performance of this system was tested using two streams rich in pectin and arabinoxylan, mimicking wastes that are often encountered in fruit juice clarification, breweries, and bakeries.

Figure 9.2 represents the pressure increase over time while feeding a mixture of pectin and arabinoxylan at 40°C. The pressure required to keep the flux constant at about 20 L/m^2 h while using the self-cleaning membrane was significantly lower than the neutral membrane due to the excellent membrane cleansing potential of the enzyme blend, resulting in up to 40% saving in the energy required to run the filtration.

An additional flexibility that could be imparted by using this approach is that whenever the biocatalytic layer gets oversaturated, one can opt either to re-disperse the particle back to the bulk medium (which comes with an additional benefit of dispersing accumulated foulant) or to add a fresh batch of enzymatically activated NP^{SP} over the foulant layer. In the latter case, the biocatalyst facing down would work to degrade the accumulated foulant, while the side facing upward can continue to degrade newly arriving foulant without the need to interrupt the filtration. Figure 9.3 depicts that the dynamics of TMP required to keep the flux constant at 20 L/m^{2}h for normal membrane shows a sharp rise. On the contrary, owing to the upward and downward degradation of the foulant by the enzymatically active NP^{SP} layer, the rate of fouling was very low.

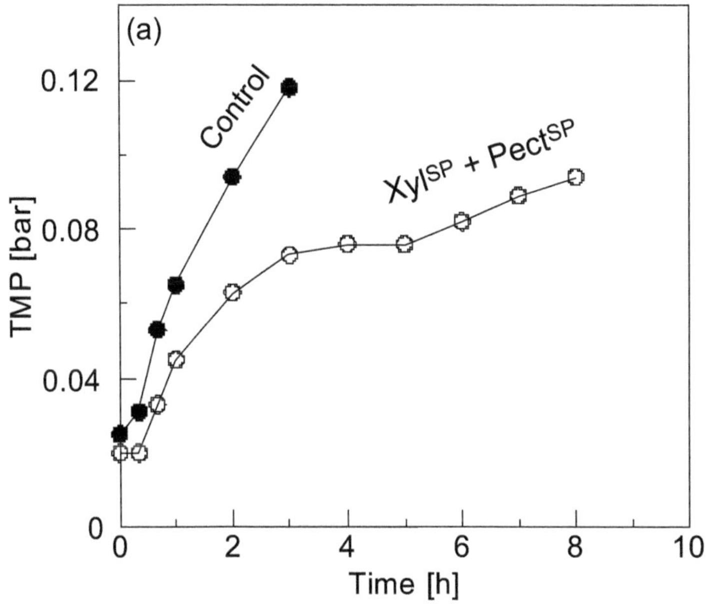

Figure 9.2 Effectiveness of using enzyme blend on degrading mixed pectin and arabinoxylan fouling layers (feed containing 0.3 mg/mL wheat arabinoxylan and 0.3 mg/mL pectin mixed in equal ratio, 17 L/m^{2}h flux at 40°C). Reprinted from Gebreyohannes et al. (2015) with permission from Elsevier.

9.4.3 Polyoxometalates-Loaded Self-Cleaning Membranes

Heteropoly acids are known active catalysts in both homogeneous and heterogeneous acid-catalyzed reactions. Owing to their ease of handling and removal, reusability, limited side reactions, strong Brønsted acidity approaching the superacid region, high proton mobility, stability, and catalytic activity, they are considered environmentally benign and economically feasible (Kozhevnikov, 1998). Thanks to their large versatility in terms of tunable structural features and properties, polyoxometalates (POMs) are generally used for the preparation of hybrid materials. POMs, especially Keggin heteropolyanion, which have high redox potential, have shown a very high antimicrobial property when used as antimicrobial films. Owing to this redox potential, bamboo charcoal/POM composite showed significant efficacy against bacteria, especially against *S. aureus*, MRSA, and *E. coli* JM109, while an increased efficacy was observed with increasing POM content (Fig. 9.4) (Yang et al., 2009). Their biological activity arises from the electrostatic interactions with biological molecules and from their redox behavior. This property has already been exploited to develop nanohybrid membranes based on the Keggin-type polyoxometalate (POM) $H_5PV_2Mo_{10}O_{40}$ and a polyvinyl alcohol/polyethyleneimine (PVA/PEI) blends, which can oxidize organics (Wu et al., 2009).

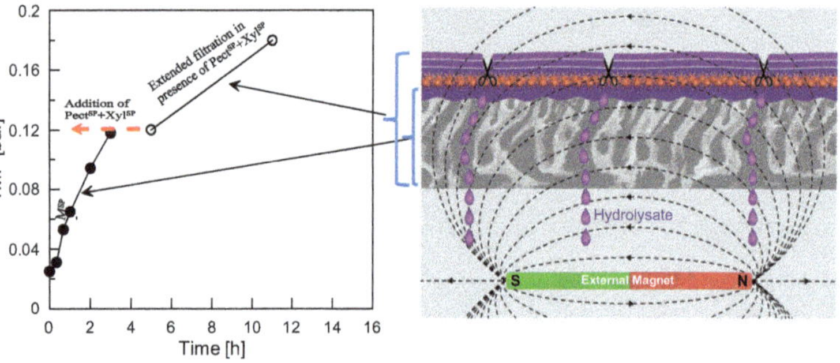

Figure 9.3 (a) Filtration on a dynamic layer formed from a xylanase and pectinase immobilized NP^{SP} mixture; (b) two-step filtration, involving the first step on bare M^{SP} (during first 3 h), followed by dispersion of a mixture of xylanase and pectinase immobilized NP^{SP} mixture recovered from part (a) on the pre-fouled M^{SP} and subsequent filtration (3–11 h). Reprinted from Gebreyohannes et al. (2015) with permission from Elsevier.

Hence, depending on the type of foulant, one can easily vary the type of POM that would be used to impart a desired membrane surface functionality. Therefore, membranes coated with "smart" POMs can exhibit a stimuli-responsive dynamics/function that can be used to create self-cleaning hybrid membranes.

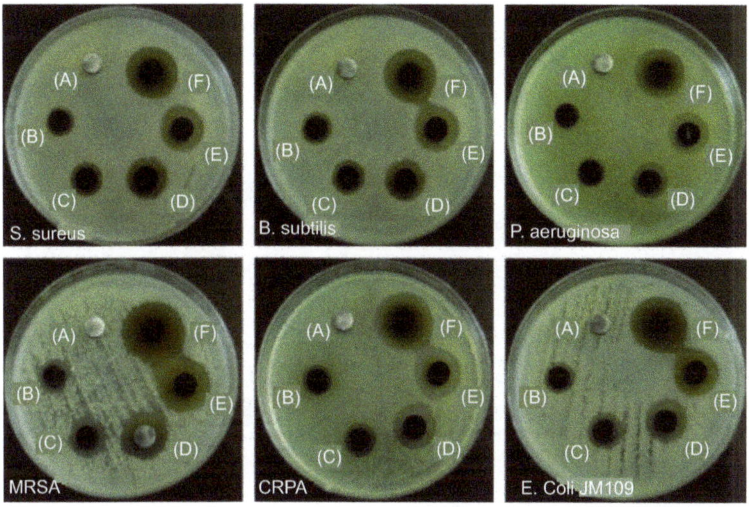

Figure 9.4 Photographs of the antibacterial test results on these microbes in the systems of (A) BC, (B) BC/POM-0.25, (C) BC/POM-0.5, (D) BC/POM-1, (E) BC/POM-2, and (F) BC/POM-3. Reprinted from Yang et al. (2009) with permission from Elsevier.

In a recent work, a hybrid membrane with POMs embedded in a polyether ether ketone membrane was exploited to create a biomimetic self-cleaning membrane through mechanical sweeping or decontamination of the local areas (Squarcina et al., 2015). The used POM, which is tetra-ruthenium-substituted polyoxometalate $[Ru_4(H_2O)_4(\mu\text{-}O)_4(\mu\text{-}OH)_2(\gamma\text{-}SiW_{10}O_{36})_2]^{10-}(Ru_4(SiW_{10})_2$, has catalase-like behavior. The presence of this POM in the polymer matrix enabled to produce a stimuli-responsive membrane that can produce nascent oxygen at the membrane pore-mouth level in the presence of H_2O_2 as a chemical trigger. Hence, it was possible to produce a bio-inspired membrane fouling mitigation mechanism involving natural catalase enzymes for H_2O_2 dismutation and endogenous oxygen production (Fig. 9.5). The evolved oxygen induces a localized micro-mixing that helps with the displacement of foulant located both inside the pore and on the membrane surface. An additional benefit of these membranes is their antioxidant behavior, mimicking the catalase enzymes, which offers a protection barrier against hydroxyl radicals and reactive oxygen species, which enhance the material stability under operating conditions.

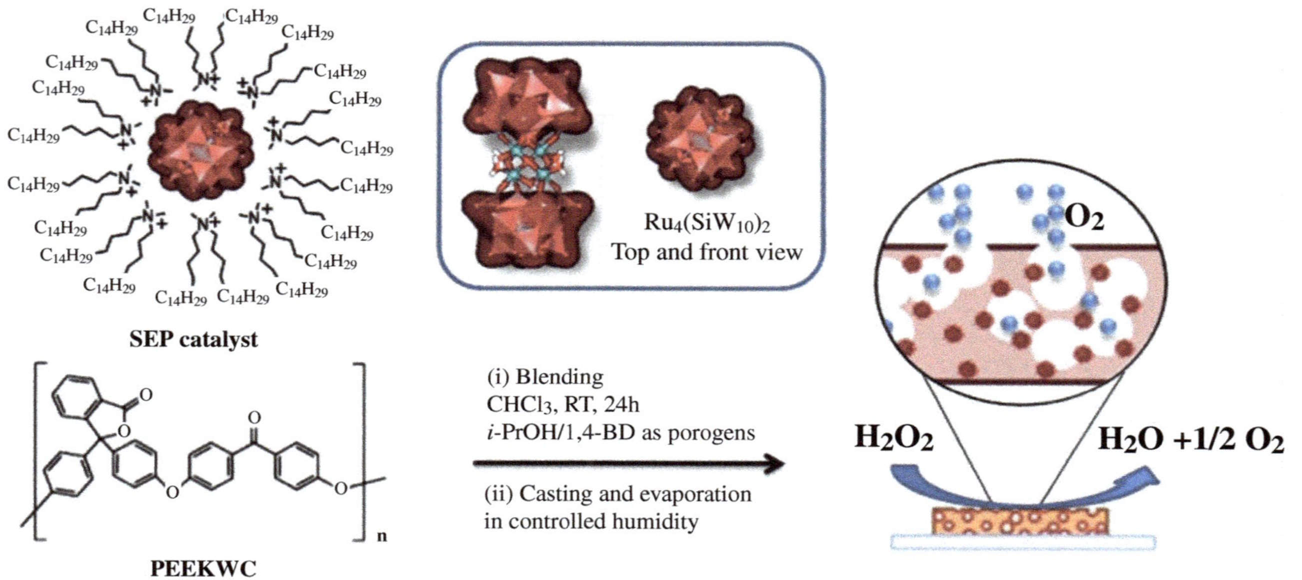

Figure 9.5 Molecular components and membrane fabrication scheme: blending of a $CHCl_3$ solution containing the surfactant-encapsulated catalyst (SEP catalyst) DODA)$_{10}$/$Ru_4(SiW_{10})_2$, the PEEK-WC polymer, and porogens (*i*-propanol and 1,4-butanediol) was followed by casting and the solvent evaporation under controlled humidity (U = 75%). A porous texture is obtained with embedded oxygenic catalysts. The catalyst structure, $Ru_4(SiW_{10})_2$, is highlighted in the box. Reprinted from Squarcina et al. (2015) with permission from John Wiley and Sons.

Abbreviations

DEN-COOH	carboxyl-terminated polyether dendrimer
f_s	fraction of the solid on the surface
f_v	fraction of the vapor on the surface
ILs	ionic liquids
LCST	lower critical solution temperature
PAA	polyacrylic acid
PDDA	poly(allylamine hydrochloride)
PDMS	polydimethylsiloxane
PLC	polycarilactone
PMA	poly(methyl acrylate)
PS	polystyrene
P2VP	poly(2-vinylpyridine)
PVP	polyvinylpyrrolidone
PNIPAM	poly(*N*-isopropylacrylamide)
R	rugosity factor
S	wetting

Symbols

γ	surface free energy
γ_{SV}	free tensions at solid–vapor interface
γ_{SL}	free tensions at solid–liquid interface
γ_{LV}	free tensions at liquid–vapor interface
θ_a	contact angle
θ_W	contact angle according to Wenzel's model
θ_{CB}	contact angle according to Cassie–Baxter's model

References

Agre, P. 2006. The aquaporin water channels. *Proceedings of the American Thoracic Society*, **3**(1), 5–13.

Allie, Z., Jacobs, E. P., Maartens, A., and Swart, P. 2003. Enzymatic cleaning of ultrafiltration membranes fouled by abattoir effluent. *Journal of Membrane Science*, **218**(1), 107–116.

Anastasiadis, S. H., Lygeraki, M. I., Athanassiou, A., Farsari, M., and Pisignano, D. 2008. Reversibly photo-responsive polymer surfaces for controlled wettability. *Journal of Adhesion Science and Technology*, **22**, 1853–1868.

Basri, H., Ismail, A. F., Aziz, M., Nagai, K., Matsuura, T., Abdullah, M. S., and Ng, B. C. 2010. Silver-filled polyethersulfone membranes for antibacterial applications: Effect of PVP and TAP addition on silver dispersion. *Desalination*, **261**(3), 264–271.

Besha, A. T., Gebreyohannes, A. Y., Tufa, R. A., Bekele, D. N., Curcio, E., and Giorno, L. 2017. Removal of emerging micropollutants by activated sludge process and membrane bioreactors and the effects of micropollutants on membrane fouling: A review. *Journal of Environmental Chemical Engineering*, **5**(3), 2395–2414.

Brullot, W., Reddy, N. K., Wouters, J., Valev, V. K., Goderis, B., Vermant, J., and Verbiest, T. 2012. Versatile ferrofluids based on polyethylene glycol coated iron oxide nanoparticles. *Journal of Magnetism and Magnetic Materials*, **324**(11), 1919–1925.

Boyes, S. G., Brittain, W. J., Wang, X., and Cheng, S. Z. D. 2002. Synthesis, characterization, and properties of ABA type triblock copolymer brushes of styrene and methyl acrylate prepared by atom transfer radical polymerization. *Macromolecules*, **35**, 4960–4967.

Chen, L., Liu, M., Lin, L., Zhang, T., Ma, J., Song, Y., and Jiang, L. 2010. Thermal-responsive hydrogel surface: Tunable wettability and adhesion to oil at the water/solid interface. *Soft Matter*, **6**, 2708–2712.

Cheng, Q., Li, M., Zheng, Y., Su, B., Wang, S., and Jiang, L. 2011. Janus interface materials: Superhydrophobic air/solid interface and superoleophobic water/solid interface inspired by a lotus leaf. *Soft Matter*, **7**, 5948–5051.

Chen, M. and Besenbacher, F. 2011. Light-driven wettability changes on a photoresponsive electrospun mat. *ACS Nano*, **5**, 1549–1555.

Connal, L. A., Li, Q., Quinn, J. F., Tjipto, E., Caruso, F., and Qiao, G. 2008. pH-responsive poly(acrylic acid) core cross-linked star polymers: Morphology transitions in solution and multilayer thin films. *Macromolecules*, **41**, 2620–2626.

Corbatón-Báguena, M.-J., Gugliuzza, A., Cassano, A., Mazzei, R., and Giorno, L. 2015. Evaluation of the effectiveness of salt solutions to clean a hydrophilic polyethersulfone ultrafiltration membrane fouled with enzymatic solutions. *Journal of Membrane Science*, **486**, 207–214.

Cheng, Y. T., Rodak, D. E., Wong, C. A., and Hayden, C. A. 2006. Effects of micro- and nano-structures on the self-cleaning behaviour of lotus leaves. *Nanotechnology*, **17**(5), 1359.

Damodar, R. A., You, S.-J., and Chou, H.-H. 2009. Study the self-cleaning, antibacterial and photocatalytic properties of TiO_2 entrapped PVDF membranes. *Journal of Hazardous Materials*, **172**(2), 1321–1328.

Daraei, P., Madaeni, S. S., Ghaemi, N., Khadivi, M. A., Astinchap, B., and Moradian, R. 2013. Fouling resistant mixed matrix polyethersulfone membranes blended with magnetic nanoparticles: Study of magnetic field induced casting. *Separation and Purification Technology*, **109**(0), 111–121.

Daraei, P., Madaeni, S. S., Ghaemi, N., Salehi, E., Khadivi, M. A., Moradian, R., and Astinchap, B. 2012. Novel polyethersulfone nanocomposite membrane prepared by PANI/Fe_3O_4 nanoparticles with enhanced performance for Cu(II) removal from water. *Journal of Membrane Science*, **415–416**(0), 250–259.

Das, S. K., Das, A. R., and Guha, A. K. 2010. Microbial synthesis of multishaped gold nanostructures. *Small*, **6**(9), 1012–1021.

Das, S. K., Dickinson, C., Lafir, F., Brougham, D. F., and Marsili, E. 2012. Synthesis, characterization and catalytic activity of gold nanoparticles biosynthesized with *Rhizopus oryzae* protein extract. *Green Chemistry*, **14**(5), 1322–1334.

Das, S. K., Khan, M. M. R., Parandhaman, T., Laffir, F., Guha, A. K., Sekaran, G., and Mandal, A. B. 2013. Nano-silica fabricated with silver nanoparticles: Antifouling adsorbent for efficient dye removal, effective water disinfection and biofouling control. *Nanoscale*, **5**(12), 5549–5560.

De Gusseme, B., Hennebel, T., Christiaens, E., Saveyn, H., Verbeken, K., Fitts, J. P., Boon, N., and Verstraete, W. 2011. Virus disinfection in water by biogenic silver immobilized in polyvinylidene fluoride membranes. *Water Research*, **45**(4), 1856–1864.

De Volder, M. F. L., Tawfick, S. H., Baughman, R. H., and Hart, A. J. 2013. Carbon nanotubes: Present and future commercial applications. *Science*, **339**(6119), 535.

Dong, G., Li, H., and Chen, V. 2013. Challenges and opportunities for mixed-matrix membranes for gas separation. *Journal of Materials Chemistry A*, **1**(15), 4610–4630.

Fu, Y., Bai, S. L., Cui, S., Qiu, D., Wang, Z., and Zhang, X. 2002. Hydrogen-bonding-directed layer-by-layer multilayer assembly: Reconformation yielding microporous films. *Macromolecules*, **35**, 9451–9458.

Ganesh, V. A., Raut, H. K., Nair, A. S., and Ramakrishna, S. 2011. A review on self-cleaning coatings. *Journal of Materials Chemistry*, **21**(41), 16304–16322.

Gao, X. and Matsui, H. 2005. Peptide-based nanotubes and their applications in bionanotechnology. *Advanced Materials*, **17**(17), 2037–2050.

Gebreyohannes, A. Y., Bilad, M. R., Verbiest, T., Courtin, C. M., Dornez, E., Giorno, L., Curcio, E., and Vankelecom, I. F. J. 2015. Nanoscale tuning of enzyme localization for enhanced reactor performance in a novel magnetic-responsive biocatalytic membrane reactor. *Journal of Membrane Science*, **487**(0), 209–220.

Gebreyohannes, A. Y., Giorno, L., Vankelecom, I. F. J., Verbiest, T., and Aimar, P. 2017. Effect of operational parameters on the performance of a magnetic responsive biocatalytic membrane reactor. *Chemical Engineering Journal*, **308**, 853–862.

Gebreyohannes, A. Y., Mazzei, R., Curcio, E., Poerio, T., Drioli, E., and Giorno, L. 2013. Study on the in situ enzymatic self-cleansing of microfiltration membrane for valorization of olive mill wastewater. *Industrial and Engineering Chemistry Research*, **52**(31), 10396–10405.

Gebreyohannes, A. Y., Mazzei, R., Poerio, T., Aimar, P., Vankelecom, I. F. J., and Giorno, L. 2016. Pectinases immobilization on magnetic nanoparticles and their anti-fouling performance in a biocatalytic membrane reactor. *RSC Advances*, **6**(101), 98737–98747.

Gugliuzza, A., Fabiani, R., Garavaglia, M. G., Spisso, A., and Drioli, E. 2006. Study of the surface character as responsible for controlling interfacial forces at membrane-feed interface. *Journal of Colloid and Interface Science*, **303**, 388–403.

Gugliuzza, A. 2016. Membrane wettability, in: *Encyclopedia of Membranes* (Drioli, E., Giorno, L., Eds.), Springer References.

Gugliuzza, A. 2014. Smart membrane surface, in: *Smart Membranes and Sensors: Synthesis, Characterization and Applications* (Gugliuzza, A., Ed), John Wiley & Sons-Scrivener Publishing, Salem, Massachusetts, USA, pp. 1–448.

Gugliuzza, A., Pingitore, V., Miriello, D., and Drioli, E. 2015. Functional carbon nanotubes for high-quality charge transfer and moisture regulation through membranes: Structural and functional insights. *Physical Chemistry Chemical Physics*, **17**, 12919–12926.

Hakimhashemi, M., Gebreyohannes, A. Y., Saveyn, H., Van der Meeren, P., and Verliefde, A. 2012. Combined effects of operational parameters on electro-ultrafiltration process characteristics. *Journal of Membrane Science*, **403–404**(0), 227–235.

Hashaikeh, R., Lalia, B. S., Kochkodan, V., and Hilal, N. 2014. A novel in situ membrane cleaning method using periodic electrolysis. *Journal of Membrane Science*, **471**, 149–154.

He, C., Shi, Z.-Q., Cheng, C., Lu, H.-Q., Zhou, M., Sun, S.-D., and Zhao, C.-S. 2016. Graphene oxide and sulfonated polyanion co-doped hydrogel films for dual-layered membranes with superior hemocompatibility and antibacterial activity. *Biomaterials Science*, **4**(10), 1431–1440.

He, X., Yang, W., and Pei, X. 2008. Preparation, characterization, and tunable wettability of poly(ionic liquid) brushes via surface-initiated atom transfer radical polymerization. *Macromolecules*, **41**, pp. 4615–4621.

Himstedt, H. H., Yang, Q., Dasi, L. P., Qian, X., Wickramasinghe, S. R., and Ulbricht, M. 2011. Magnetically activated micromixers for separation membranes. *Langmuir*, **27**(9), 5574–5581.

Holt, J. K., Park, H. G., Wang, Y., Stadermann, M., Artyukhin, A. B., Grigoropoulos, C. P., Noy, A., and Bakajin, O. 2006. Fast mass transport through sub-2-nanometer carbon nanotubes. *Science*, **312**(5776), 1034–1037.

Huang, Z.-Q., Zheng, F., Zhang, Z., Xu, H.-T., and Zhou, K.-M. 2012. The performance of the PVDF-Fe_3O_4 ultrafiltration membrane and the effect of a parallel magnetic field used during the membrane formation. *Desalination*, **292**(0), 64–72.

Kozhevnikov, I. V. 1998. Catalysis by heteropoly acids and multicomponent polyoxometalates in liquid-phase reactions. *Chemical Reviews*, **98**(1), 171–198.

Liu, D., Liu, L., Onck, P. R., and Broer, D. J. 2015. Reverse switching of surface roughness in a self-organized polydomain liquid crystal coating. *PNAS*, **112**, 133880–133885.

Li, H. and Chen, V. 2010. *Membrane Fouling and Cleaning in Food and Bioprocessing*. Oxford, UK: Butterworth-Heinemann.

Linsebigler, A. L., Lu, G., and Yates, J. T. 1995. Photocatalysis on TiO_2 surfaces: Principles, mechanisms, and selected results. *Chemical Reviews*, **95**(3), 735–758.

Liu, S., Chu, M., Zhu, Y., Li, L., Wang, L., Gao, H., and Ren, L. 2017. A novel antibacterial cellulose based biomaterial for hernia mesh applications. *RSC Advances*, **7**(19), 11601–11607.

Maness, P.-C., Smolinski, S., Blake, D. M., Huang, Z., Wolfrum, E. J., and Jacoby, W. A. 1999. Bactericidal activity of photocatalytic TiO(2) reaction: Toward an understanding of its killing mechanism. *Applied and Environmental Microbiology*, **65**(9), 4094–4098.

Mazzei, R., Emma, P., Yihdego, G. A., and Lidietta, G. 2017. Membrane bioreactors in food, pharmaceutical and biofuel applications: State of the art, progresses and perspectives. *Current Organic Chemistry*, **21**, 1–31.

Meinild, A. K., Klaerke, D. A., and Zeuthen, T. 1998. Bidirectional water fluxes and specificity for small hydrophilic molecules in aquaporins 0-5. *Journal of Biological Chemistry*, **273**(49), 32446–32451.

Miguel-Sancho, N., Bomati-Miguel, O., Roca, A. G., Martinez, G., Arruebo, M., and Santamaria, J. 2012. Synthesis of magnetic nanocrystals by thermal decomposition in glycol media: Effect of process variables and mechanistic study. *Industrial and Engineering Chemistry Research*, **51**(25), 8348–8357.

Nisola, G. M., Park, J. S., Beltran, A. B., and Chung, W.-J. 2012. Silver nanoparticles in a polyether-block-polyamide copolymer towards antimicrobial and antifouling membranes. *RSC Advances*, **2**(6), 2439–2448.

Nayak, A., Liu, H., and Belfort, G. 2006. An optically reversible switching membrane surface. *Angewandte Chemie International Edition*, **45**, 4094–4409.

Nicoletta, F. P., Cupelli, D., Formoso, P., Filpo, G. D., Colella, V., and Gugliuzza, A. 2012. Light responsive polymer membranes: A review. *Membranes*, **2**, 134–197.

Pendergast, M. M. and Hoek, E. M. V. 2011. A review of water treatment membrane nanotechnologies. *Energy and Environmental Science*, **4**(6), 1946–1971.

Perrotta, M. L., Saielli, G., Casella, G., Macedonio, F., Giorno, L., Drioli, E., and Gugliuzza, A. 2017. An ultrathin suspended hydrophobic porous membrane for high-efficiency water desalination. *Applied Materials Today*, **9**, 1–9.

Pingitore, V. and Gugliuzza, A. 2013. Fabrication of porous semiconductor interfaces by pH-driven assembly of carbon nanotubes on honeycomb structured membranes. *Journal of Physical Chemistry C*, **117**, 26562–26572.

Roy, P., Dey, T., Lee, K., Kim, D., Fabry, B., and Schmuki, P. 2010. Size-selective separation of macromolecules by nanochannel titania membrane with self-cleaning (declogging) ability. *Journal of the American Chemical Society*, **132**(23), 7893–7895.

Sabbatini, P., Yrazu, F., Rossi, F., Thern, G., Marajofsky, A., and Fidalgo de Cortalezzi, M. M. 2010. Fabrication and characterization of iron oxide ceramic membranes for arsenic removal. *Water Research*, **44**(19), 5702–5712.

Sanchez, C., Belleville, P., Popall, M., and Nicole, L. 2011. Applications of advanced hybrid organic–inorganic nanomaterials: From laboratory to market. *Chemical Society Reviews*, **40**(2), 696–753.

Schulze, A., Breite, D., Kim, Y., Schmidt, M., Thomas, I., Went, M., Fischer, K., and Prager, A. 2017. Bio-inspired polymer membrane surface cleaning. *Polymers*, **9**(3), 97.

Shen, Y.-X., Saboe, P. O., Sines, I. T., Erbakan, M., and Kumar, M. 2014. Biomimetic membranes: A review. *Journal of Membrane Science*, **454**, 359–381.

Speranza, V., Trotta, F., Drioli, E., and Gugliuzza, A. 2010. High-definition polymeric membranes: Construction of 3D lithographed channel arrays through controlling natural building blocks dynamics. *ACS Applied Materials and Interfaces*, **2**, 459–466.

Squarcina, A., Fortunati, I., Saoncella, O., Galiano, F., Ferrante, C., Figoli, A., Carraro, M., and Bonchio, M. 2015. Oxygen evolution: Dynamic antifouling of catalytic pores armed with oxygenic polyoxometalates. *Advanced Materials Interfaces*, **2**(7).

Synytska, A., Diez, M. S., and Lonov, L. 2007. Simple and fast method for the fabrication of switchable bicomponent micropatterned polymer surface. *Langmuir*, **23**, 5205–5209.

Taniguchi, M. and Belfort, G. 2002. Correcting for surface roughness: Advancing and receding contact angles. *Langmuir*, **18**, 6465–6467.

Thevenot, J., Oliveira, H., Sandre, O., and Lecommandoux, S. 2013. Magnetic responsive polymer composite materials. *Chemical Society Reviews*, **42**(17), 7099–7116.

Thomas, T. R. 1999. *Rough Surfaces*, 2nd ed. London: Imperial College Press.

Verplanck, N., Coffinier, Y., Thom, V., and Boukherroub, R. 2007. Wettability switching techniques on superhydrophobic surfaces. *Nanoscale Research Letters*, **2**, 577–596.

Wang, D., Jiao, P., Wang, J., Zhang, Q., Feng, L., and Yang, Z. 2012. Fast photo-switched wettability and color of surfaces coated with polymer brushes containing spiropyran. *Journal of Applied Polymer Science*, **125**, 870–875.

Wu, K. H., Yu, P. Y., Yang, C. C., Wang, G. P., and Chao, C. M. 2009. Preparation and characterization of polyoxometalate-modified poly(vinyl alcohol)/polyethyleneimine hybrids as a chemical and biological self-detoxifying material. *Polymer Degradation and Stability*, **94**(9), 1411–1418.

Xiao-Ming, Z. and Wainer, I. W. 1993. On-line determination of lipase activity and enantioselectivity using an immobilized enzyme reactor coupled to a chiral stationary phase. *Tetrahedron Letters*, **34**(30), 4731–4734.

Yang, F.-C., Wu, K.-H., Lin, W.-P., and Hu, M.-K. 2009. Preparation and antibacterial efficacy of bamboo charcoal/polyoxometalate biological protective material. *Microporous and Mesoporous Materials*, **118**(1), 467–472.

Yao, P., Choo, K. H., and Kim, M. H. 2009. A hybridized photocatalysis-microfiltration system with iron oxide-coated membranes for the removal of natural organic matter in water treatment: Effects of iron oxide layers and colloids. *Water Research*, **43**(17), 4238–4248.

Yeon, K. M., Lee, C. H., and Kim, J. 2009. Magnetic enzyme carrier for effective biofouling control in the membrane bioreactor based on enzymatic quorum quenching. *Environmental Science and Technology*, **43**(19), 7403–7409.

Zhang, M., Feng, S., Wang, L., and Zheng, Y. 2016. Lotus effect in wetting and self-cleaning. *Biotribology*, **5**, 31–43.

Zhang, H., Fu, Y., Wang, D., Wang, L., Wang, Z., and Zhang, X. 2003. Hydrogen-bonding-directed layer-by-layer assembly of dendrimer and poly(4-vinylpyridine) and micropore formation by post-base treatment. *Langmuir*, **19**, 8497–8502.

Zhang, J., Lu, X., Huang, W., and Han, Y. 2005. Reversible superhydrophobicity to superhydrophilicity transition by extending and unloading an elastic polyamide film. *Macromolecular Rapid Communication*, **26**, 477–480.

Zhang, M., Liu, L., Wu, C., Fu, G., Zhao, H., and He, B. 2007. Synthesis, characterization and application of well-defined environmentally responsive polymer brushes on the surface of colloid particles. *Polymer*, **48**, 1989–1997.

Chapter 10

Innovative Coating Membranes for Water Treatment

Alberto Figoli, Claudia Ursino, and Francesco Galiano
Institute on Membrane Technology of the National Research Council of Italy (ITM-CNR), Via P. Bucci, Cubo 17C, 87036 Rende (CS), Italy
a.figoli@itm.cnr.it

The functionalization of membranes via coating technique represents an interesting and advanced approach to prepare tailored membranes. The most common methods for the preparation of coated membranes, such as solution coating and polymerization reactions, have been reported and discussed in the chapter. The coatings are mainly applied for enhancing either hydrophilicity or hydrophobicity of membranes depending on the field of application. In particular, two case studies of hydrophobic membrane coatings and one case study of hydrophilic coating, developed at ITM-CNR, are reported and discussed for specific application in water treatment.

Functional Nanostructured Membranes
Edited by Enrico Drioli, Lidietta Giorno, and Annarosa Gugliuzza

ISBN 978-981-4774-79-6 (Hardcover), 978-1-351-13511-5 (eBook)
www.panstanford.com

10.1 Introduction

During the last decades, membrane processes have been widely used for water treatment in different applications, such as desalination, micro- and ultrafiltration, reverse osmosis, and pervaporation. In order to achieve high-performance membranes and overcome problems determined, for example, by irreversible membrane fouling, several methods have been developed and studied for modifying and tailoring membrane surfaces. In fact, membrane functionalization could be the way to minimize undesired interactions, which reduce the membrane performance, or for introducing additional functional groups for improving selectivity, creating an entirely novel separation functional layer (Ulbricht, 2006), improving chemical and thermal stability. In particular, coated membranes are prepared combining two or more different materials with different properties and they are also named (thin film) (nano-) composite membranes. Different techniques can be used to functionalize the membranes with tailored properties, and the chapter illustrates the innovative coating materials employed recently in membrane fabrication. The general focus is on the studies of coating methods used to prepare both hydrophobic and hydrophilic membranes. The first part gives a general introduction to typical coating methods, while the second part focuses on case studies based on the research performed at the Institute on Membrane Technology (ITM-CNR). The first case study reports innovative materials for producing hydrophobic coating on porous membranes of different configuration (flat and hollow fiber) by using both perfluoropolyether compounds (Figoli et al., 2017) and ethylene chlorotrifluoroethylene (ECTFE) copolymer (Santoro et al., 2016). In the second case study, a novel hydrophilic coating, made of a polymerizable bicontinuous microemulsion (Galiano et al., 2015) on an ultrafiltration (UF) membrane, was developed to reduce fouling and enhance the antimicrobial resistance of commercial membranes (Scheme 10.1).

In general, the preparation of these coatings can be summarized in two basic steps using (i) non-woven or porous membrane polymer support and (ii) deposition/coating of one or more functionalized materials (Susanto et al., 2009). An important advantage is that each layer can be optimized independently, modifying the thickness, the type of polymer employed both as the porous support and the selective barrier layer.

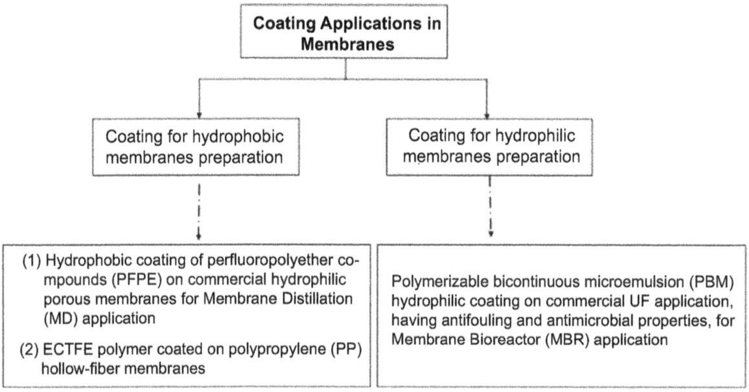

Scheme 10.1 Structure of the chapter including the case studies on hydrophobic and hydrophilic coatings developed at the Institute on Membrane Technology (ITM-CNR).

Several techniques can be applied for preparing these tailor-coated membranes and two main coating categories can be identified (Table 10.1): (i) solution coating and (ii) polymerization reactions.

Table 10.1 Principal coating methods

Coating methods	
Solution coating	***Polymerization reactions***
Dip coating	Interfacial polymerization
Spin coating	In situ polymerization
Spray coating	Plasma polymerization
Casting coating	Grafting

10.2 Solution Coating

10.2.1 Dip Coating

The dip coating technique is widely used in making composite membranes thanks to its simplicity. In general, the membrane is immersed in a precursor solution. After a fixed time, the membrane is recovered and the excess of solution is removed and dried. After immersion, the membrane can be exposed to UV or IR radiation if the precursor solution contained a reactive pre-polymer (or monomer) (Susanto et al., 2009). Membranes obtained by this method present a very thin layer but often dense. The technique is shown schematically in Fig. 10.1a.

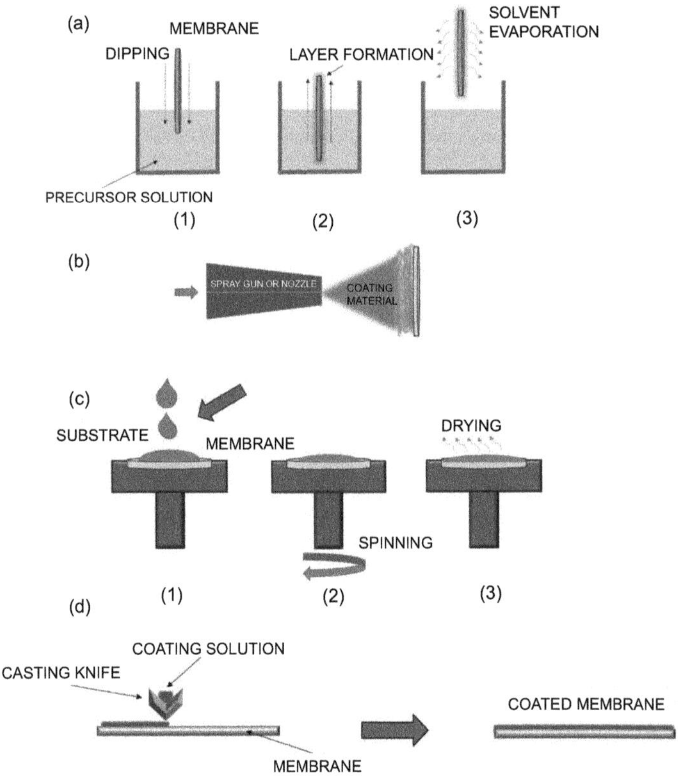

Figure 10.1 Simplified scheme of solution coating by (a) dip coating, (b) spray coating, (c) spin coating, and (d) casting coating.

The final thickness of the coated layer can be controlled by the withdrawal speed (immersion time) and the polymer concentration (or pre-polymer or monomer) (Santoro et al., 2016). An equation, derived by the Navier–Stokes equation, can be utilized for calculating the final thickness of the coating layer:

$$T_{\text{equ.}} = \frac{2}{3}\sqrt{\frac{\eta v}{\rho g}} \tag{10.1}$$

where $T_{\text{equ.}}$ is the equilibrium thickness, η is dynamic viscosity, v is the withdrawal speed, ρ is the solution density, and g represents gravity.

10.2.2 Spin and Spray Coating

Spin coating is usually applied for functionalizing flat membrane/support type. The procedure can be summarized in three simple steps: (i) deposition of coating solution on membrane/support surface; (ii) rotation speed and acceleration, which defines the final thickness; and (iii) solvent evaporation. This process relies on centrifugal forces to spread the liquid phase evenly across a spinning substrate (Faure et al., 2013). The final thickness depends on the rotation speed and on the viscosity of the coating solution. The main obstacle is caused by the centrifugal force during acceleration (spinning), and the final layer could be non-uniform with striation. Finally, spray coating exhibits some advantages, when the uniformity of the thickness is not required. In fact, spray coating allows the high-speed deposition also on different shapes of substrate, thanks to its high flexibility (Aegerter et al., 2004). However, the main limitation of the spray coating technique is that the solution viscosity must be low to facilitate the nebulization process (Schwartz et al., 2008).

10.2.3 Casting Coating

In the casting coating method, the coating solution is cast directly on the membrane surface, generally using a casting knife. This technique is usually applied on flat sheet membranes. Both surfaces can be coated, and several layers can be subsequently applied (Siemann, 2005).

A summary of the advantages and limitations of various solution coating methods is provided in Table 10.2.

Table 10.2 Summary of advantages and limitation of solution coating (adapted from Schwartz et al., 2008)

Coating technique	Thickness	Advantages	Limitation
Dip	20 nm to 50 μm	Complex and irregular shapes can be coated, can be a continuous process, very simple	Thickness variations, voids
Spin	10 nm to 5 μm	Uniformity, reproducibility, excellent thickness control, low cost	Requires flat substrate, high material loss
Spray	2 nm to 20 μm	Fast and adaptable to complex shapes and sizes, conformal step coverage, high efficiency	Expensive, low viscosity coating solution
Casting	10 nm to 50 μm	Uniformity, very simple, thickness control, low cost	Requires flat substrate

10.3 Polymerization Reactions

10.3.1 Interfacial Polymerization

The first interfacial polymerization on membrane surface was developed by Cadotte et al. (1980). The reaction occurs between two reactive monomers at the interface of two liquids, which are insoluble to each other. The initial support layer is, in general, a microporous membrane, such as ultra- or microfiltration membrane. This membrane is immersed in an aqueous solution consisting of reactive monomer, prior to immersion in second organic solution containing a water-immiscible solvent, in which the other reactive monomer was previously dissolved. The membrane is thus usually crosslinked by heat treatment (70–110°C) (Kochkodan et al., 2015; Mulder, 1996). The schematic procedure of interfacial polymerization formation is given in Fig. 10.2. The sequence can be reversed while using hydrophobic substrates (first immersion in the organic solution). The advantages of interfacial polymerization include rapid and self-inhibiting reaction and low requirement of reactants' purity (Kao et al., 2010).

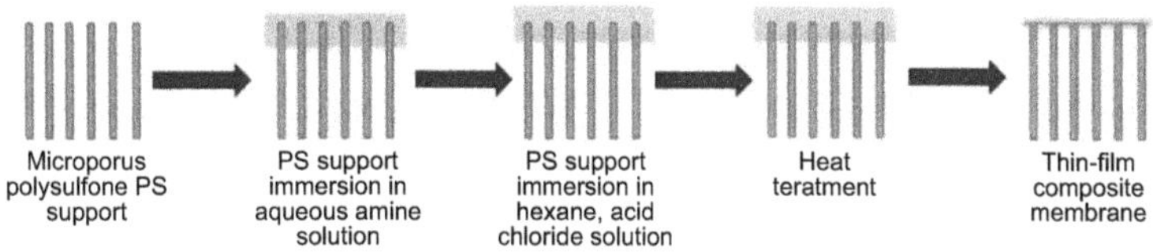

Figure 10.2 Interfacial polymerization procedure. Reprinted from Cadotte et al. (1980) with permission from Elsevier.

10.3.2 In situ Polymerization

In situ polymerization is an easy technique where the reactive monomer (or oligomer, or reactive particles) is placed in direct contact with the membrane by casting or immersion to initiate polymerization directly on the membrane surface (Fig. 10.3). Polymerization can occur by irradiation, heating, organic initiator, or by a catalyst present in the initial solution (Mulder, 1996).

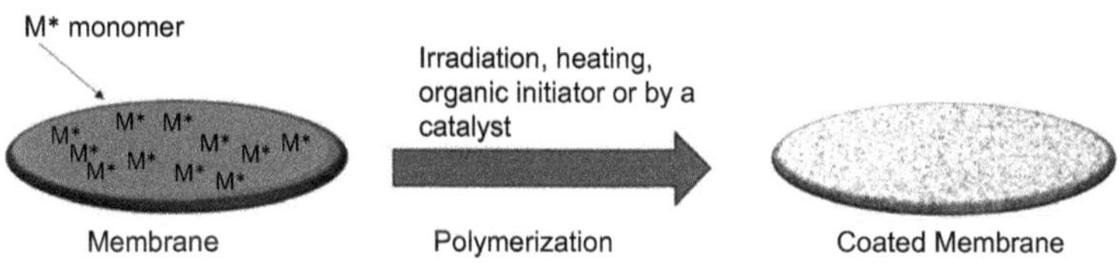

Figure 10.3 In situ polymerization procedure.

10.3.3 Plasma Polymerization

Plasma treatment involves physical and chemical interactions of the active species with the polymer surface. The reaction is heterogeneous, and it does not involve polymerization but depolymerization and modification of functional groups (Strathmann, 1990). Plasma coating leads to the formation of a uniform layer owing to both outstanding chemical resistance and mechanical strength (Fig. 10.4). Plasma is obtained from the ionization of a gas present in the reactor, by means of an electrical discharge at high frequencies. Different reactant gases are usually employed, including inert gases (He, Ar), reductive gases (CF_3Cl, NH_3), and oxidative gases (CO_2, H_2O) (Khulbe et al., 2010; Steen et al., 2001; Tyszler et al., 2006). This process can be used to produce a polymeric film, using organic compounds, that does not polymerize under normal chemical polymerization conditions. Different parameters play a crucial role in plasma polymerization, such as monomer reactivity, system pressure and discharge power, frequency of the excitation signal, and the substrate temperature (Lin and Yasuda, 1996). The thickness of the coated layer depends on the polymerization time, vacuum pressure, gas flow and pressure, and obviously on the monomer concentration in the reactor (Mulder, 1996).

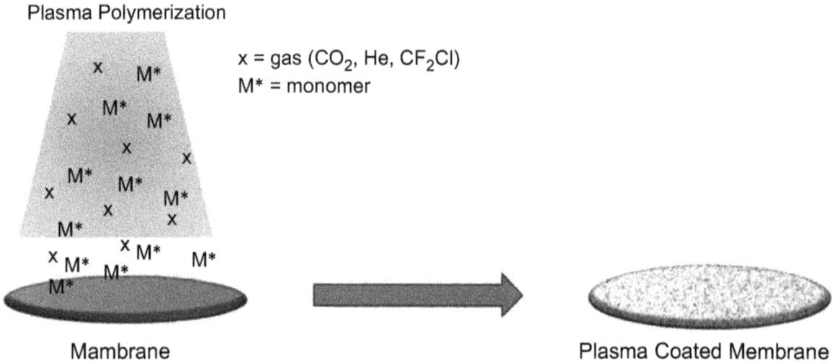

Figure 10.4 Simplified scheme of plasma polymerization.

10.3.4 Grafting

Grafting polymerization is a chemical modification method. This method allows grafting of different chemical groups on membranes via a covalent bond, maintaining the chemical stability of the modified surface (Bhattacharya and Misra, 2004). Two alternative approaches can be used for grafting: "grafting-to" and "grafting-from" processes. As reported in Fig. 10.5, the grafting-to method is characterized by the attachment of functional groups on the membrane surface in one step (covalently binding) (Susanto et al., 2009; Xu et al., 2009). The advantage is that the structure and property of the monomer or polymer used can be well controlled and characterized in detail (Ulbricht, 2006). The grafting-from method involves active sites existing on the membrane surface in order to promote the polymerization of monomers from the surface. Usually, the membrane surface is pre-activated by treating a substrate with plasma or UV treatment (Xu et al., 2009). Molecules of a monomer easily penetrate through the already activated membrane surface and significant amounts of polymers can be grafted.

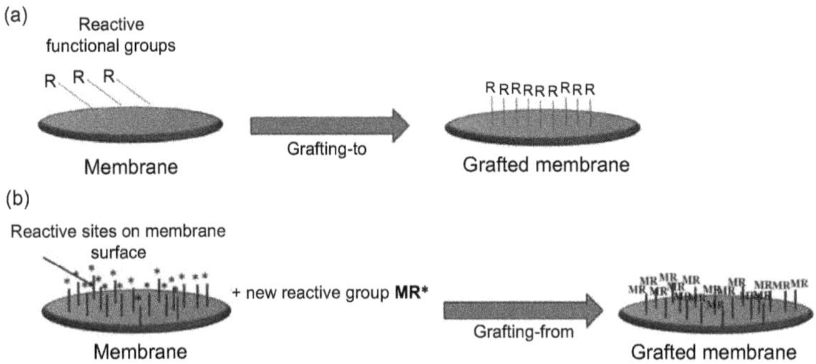

Figure 10.5 Functionalized polymer membranes by (a) "grafting-to" and (b) "grafting-from."

10.4 Hydrophobic Coating Application

The material of a membrane determines its interactions with water or other chemical species, thus affecting its wettability. While hydrophilic membranes are characterized by the presence of active groups on their surface that have the ability to form "hydrogen bonds" with water, hydrophobic membranes present the opposite response to water interaction (water repellent). Both hydrophobic and hydrophilic dense membranes can be applied in pervaporation (PV) for the dehydration of organic solvents and the separation of organic mixtures (Figoli et al., 2015). However, hydrophobic membranes with high performances in terms of flux and selectivity and high resistance to aggressive chemicals, are required and represent the preferred choice in the case of removal of organic solvents from water, the recovery of volatile organic compounds (VOCs) from aqueous mixtures or for organic/organic separation (Chovau et al., 2010; Simone et al., 2012). Hydrophobic membranes are also preferable in membrane distillation (MD) applications (Garcìa-Fernàndez. et al., 2015), membranes contactors (Drioli et al., 2006), oil/water separations (Carnevale et al., 2016; Daiminger et al., 1995), gas absorption (Ahmad et al., 2015) or stripping (Li et al., 1999), and enzyme immobilization (Xu et al., 2009). The hydrophobic polymer/monomer can be used as a coating material to improve membrane properties such as membrane hydrophobicity, mechanical strength, and liquid entry pressure (LEP), and in the case of fluoropolymers, to improve the chemical resistance as well (Ursino et al., 2016).

Many of the membranes, despite satisfactory properties such as selectivity and permeability, are tailored only for some of the processes they are employed. Considering, e.g., the MD processes, different authors reported that the typical hydrophobic porous membranes used are not specifically designed for this application (Criscuoli et al., 2008; Figoli et al., 2017; Qtaishat et al., 2009a). In fact, in the case of MD, the membranes employed are usually of micro- and ultrafiltration type and only recently most of the researches have been devoted to the development of membranes specifically designed for MD applications to increase both the stability and performance of membranes (Figoli et al., 2014; Simone et al., 2010).

In this direction, several studies have been reported in the literature, focusing on developing novel hydrophilic/hydrophobic composite membranes for MD application (Fig. 10.6).

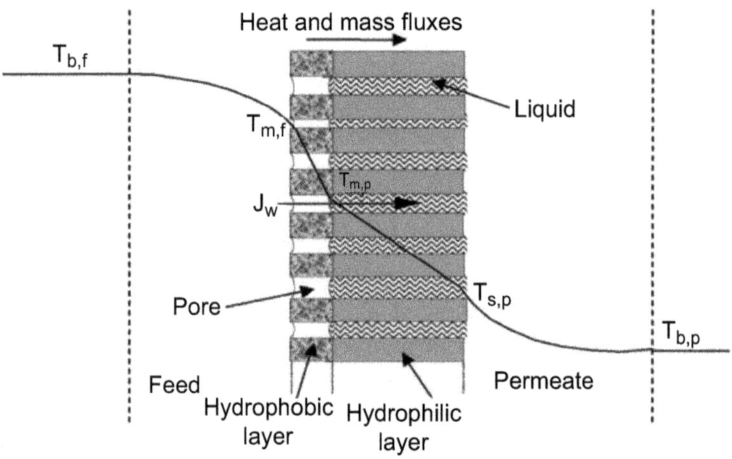

Figure 10.6 Scheme of the direct contact membrane distillation (DCMD) mechanism of transport through a porous composite hydrophobic/hydrophilic membrane reported by Qtaishat et al. (2009a). Reprinted with permission from Elsevier.

Several authors have employed hydrophobic surface-modifying macromolecules (SMMs) in order to change the hydrophilic surfaces of polyetherimide (PEI) membranes in a single casting step by the phase inversion method (Khayet et al. 2003; Khayet et al., 2005; Qtaishat et al. 2009b). Prince et al. (2013) prepared a novel triple layer hydrophilic–hydrophobic composite membrane for desalination, while Bonyadi and Chung (2007) made dual layer hydrophilic/hydrophobic hollow fibers by co-extrusion. However, only a few studies were conducted in order to enhance the hydrophobicity by means of coated treatment. Wei et al. (2012) investigated tetrafluoromethane (CF_4) plasma modification of a hydrophilic polyethersulfone (PES) membrane, for MD applications. Hydrophilic PES flat sheet and hollow-fiber membranes were exposed to plasma treatment. The surface treatment effects were characterized by contact angle (CA) measurements, LEP, and X-ray photoelectron spectroscopy.

Membrane performance was evaluated in direct contact membrane distillation (DCMD) salty solutions. Plasma modification allowed conversion of the hydrophilic PES membranes into hydrophobic PES, and depending on treatment time (from 5 min to 40 min), CA increased from about 110° to 125° for top side and from 118° to 124° for bottom side. DCMD tests were carried out using 4% NaCl as a feed and varying the feed temperature from 45°C to 73°C. For the hollow fibers, the flux increased from 20.4 to 66.7 kg/m^{2}h, when the temperature increased from 45°C to 75°C, about 50% higher with respect to flat sheet membranes. The salt rejection was as high as 99.97%, indicating a nearly complete rejection of NaCl. Following the increase in the NaCl concentration from 4% to 16%, the water flux declined from 45.4 to 30.1 kg/m^{2}h (experiment performed at 63.3°C).

Dumée et al. (2011) reported the coating of carbonanotube composite Bucky-papers (CNT BPs) on poly(tetra-fluoro-ethylene) by sputtering. The composite membrane was composed of four different layers of (i) PTFE, (ii) CNT BPs, (iii) the other PTFE layer, and (iv) the polyether (PE) support. The CA increased from 118° to 155° with respect to the untreated Bucky-paper membranes, and it was up 145° for the commercial PTFE membrane. It was reported that BP-PTFE-coated membranes, tested in DCMD, exhibited a permeability (3.05 kg/m sPa (*10^{-12})), lifespan (>40 h), and salt rejection (99.9%) higher than the BP uncoated (1.92 kg/m sPa ($\times 10^{-12}$), 10 h, salt rejection 97.5). In another work, Shang et al. (2012) described an in situ polymerization approach for oil–water separation using nanofiber membranes. Cellulose acetate (CAc) nanofibers were coated by in situ polymerization using fluorinated polybenzoxazine (FPBZ) incorporating SiO_2 nanoparticles (FPBZ/SiO_2 NPs), as shown Fig. 10.7.

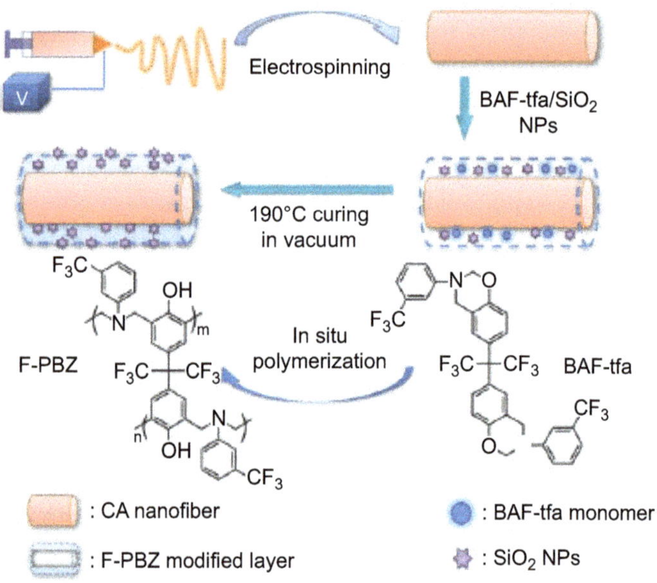

Figure 10.7 Illustration of the strategy using the in situ polymerization approach to the synthesis of FPBZ/SiO_2 NPs modified CA nanofibrous membranes (Shang et al., 2012). Reprinted from Shang et al. (2012) with permission from Royal Society of Chemistry.

The CA nanofiber membranes have shown superhydrophilicity with a CA of less than 5°, while depending on BAF-tfa 2,2-bis(3-m-trifluouromethylphenyl-l-3,4-dihydro-2H-1,3-benzoxazinyl) propane concentrations as monomer, 0.01, 0.02, 0.05, 0.1 wt.% of BAF-tfa, coated membranes presented a linear increase in CA of 68°, 90°, 121° up to 136°, respectively. The oil–water separation experiment procedure demonstrated that oil quickly permeated through the membranes due to the superoleophilicity, while water was retained above the membranes because of the superhydrophobic and low water adhesion properties of the membranes.

Hydrophilic cellulose acetate and cellulose nitrate membranes were modified in hydrophobic-coated membranes via grafting and plasma polymerization by Wu et al. (1992). Two different experiments were carried out: (i) radiation polystyrene-grafting modified membranes of cellulose acetate, using grafting liquid composed of styrene (St)-pyridine (Pyd)-carbon tetrachloride(CC14) and (ii) plasma polymerization modified membranes of cellulose nitrate, using two monomer systems, vinyltrimethylsilicon (VTMS)/ carbon tetrafluoride (CF4) and octafluoro-cyclobutane (OFCB). The performance of membranes was evaluated in the MD process. In case the amount of pyridine in the grafting liquid was fixed at 10%, the maximum CA obtained was 96° (St 85 wt.% : CCl4 5 wt.%), with a flux of 1.64 kg/m^{2}h and a rejection of about 89.6% (Warm side: 0.3 M NaC1 solution, 50°C; cold side: distilled water, 20°C). However, the rejection decreased over time, probably because the membrane was wetted during prolonged MD test. Plasma polymerization of the VTMS/CF4 system was similar to that of the OFCB system, and depending on the discharge power (*W*), the CA was up to 115° (60 W) for the coated membranes using the VTMS/CF4 system with a flux of 37.2 kg/m^{2}h and rejection 96.4%, and up to 120° (200 W) for the OFCB system, with a flux around 30 kg/m^{2}h and salty rejection of 92% (0.5 M NaCl solution).

10.4.1 Case Studies on Hydrophobic Coating Membrane Formation at ITM-CNR

The first case study reports a recent research in which an innovative hydrophobic coating has been used to modify commercial hydrophilic membranes (Figoli et al., 2017). The research has been made in the logic of making membranes specifically designed for MD application. Only a few studies report the preparation of hydrophobic membranes starting from hydrophilic polymer as already mentioned in the introduction. Often the techniques used, such as plasma polymerization, involve relatively high initial equipment costs and a lack of adequate cost estimation for the entire process (Yasuda and Matsuzawa, 2005). Therefore, Figoli et al. (2017) proposed the coating of UV-curable perfluoropolyether compounds (PFPEs) (Fluorolink® AD1700, Solvay Specialty Polymers, Italy) as an easier alternative way to produce hydrophobic composite coating membranes. The possibility of modifying the surface of hydrophilic commercial membranes (such as polyamide (PA) or PES) in comparison to more expensive hydrophobic membranes, made of polypropylene (PP), PTFE, and ECTFE, usually applied in MD, has been proved.

PFPEs are a class of partially fluorinated oligomers characterized by perfluorinated carbon units, as $-CF_2-$ and $-CF_2-CF_2-$, separated by oxygen atoms, with urethane (meth)acrylates blocks that can be either mono- or bifunctional, which make them UV or thermally crosslinkable (Fig. 10.8).

$$R_H\text{-}CF_2O\text{-}(CF_2CF_2O)_m\text{-}(CF_2O)_n\text{-}CF_2\text{-}R_H$$

Figure 10.8 Chemical structure of Fluorolink® AD1700, where R_H are urethane (meth)acrylates blocks that can be either mono- or bifunctional.

In this way, hydrophilic membranes with specific morphology can be selected "a priori" and it is possible to coat whatever microfiltration membranes (flat sheet or hollow fiber). In detail, hydrophilic polyamide (PA) membranes (0.45–0.22–0.1 μm), purchased from Sterlitech (USA), were used as support. Different oligomer concentrations (5–10–20 wt.%), and Darocure 1173 used as photoinitiators, were employed to prepare coating solutions. The coating technique applied is a two-step procedure consisting of dip coating and in situ polymerization (Fig. 10.9).

Commercial membranes were initially dipped into the oligomer solution for 10 min and dried overnight. The dip-coated membranes were exposed for 5 min at UV lamp in order to promote the oligomer polymerization on membrane surface. In order to remove the un-polymerized oligomer, the coated membranes were washed and finally dried. Using this procedure, composite hydrophilic/hydrophobic-coated membranes were obtained.

The membranes were characterized in terms of surface morphology by using scanning electron microscopy (SEM) and atomic force microscopy (AFM), CA, porosity, pore size, and LEP. Coating stability and performance were evaluated by DCMD configuration experiments, carried out with both deionized water and a salty solution (0.6M NaCl) as feed. Initially PA membranes, with and without coating, were analyzed using SEM and AFM. The images showed that on increasing the oligomer concentration (from 5 to 20 wt.%), the coating becomes more visible even if Fluorolink®

coating did not significantly modify the original commercial membrane morphology (see Fig. 10.10).

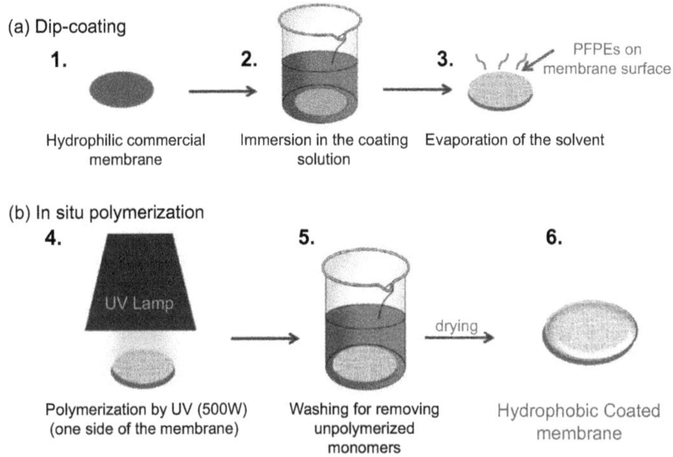

Figure 10.9 Coating procedure using Fluorolink® AD1700. Adapted from Figoli et al. (2017) with permission from Elsevier.

The study of pore size and porosity of coated and uncoated membranes indicates that while porosity decreases in relation to oligomer concentration (e.g., from 66 to 54%, for 5 and 20 wt% of oligomer concentration respectively, at the same membrane pore size), only the bubble point (up to 2.05 bar) registered an increase due to the reduction in the biggest membrane pores. Porosity results were explained due to the partial penetration of the coating within the membrane sublayer. The LEP_W values obtained, in the range of 2–3.5 bar, demonstrated that the Fluorolink®-coated membranes were comparable with most of the commercial hydrophobic membranes commonly used in MD.

CA measurements confirmed the success of the coating process: starting from hydrophilic membranes with a CA of about 40°, highly hydrophobic membranes were obtained. In fact, the CA increased from 120° up to 155° at higher oligomer concentration (from 5 to 20 wt.%), and CA experiments conducted in time also proved the coating stability. The water drop was quickly absorbed with the pristine hydrophilic membranes; CA decreased from 46.27° to 26.3° in only 5 s, while the water drop was not absorbed at all using the novel PFPE-coated membranes. A representative picture of the CA results is show in Fig. 10.11.

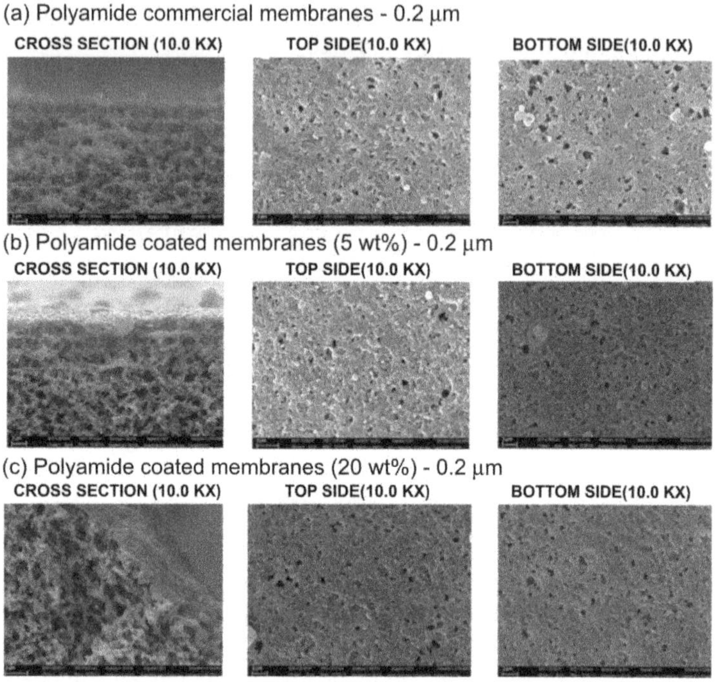

Figure 10.10 SEM morphology of coated polyamide (PA) membranes (pore size, 0.22 μm) with different Fluorolink® AD1700 concentrations (0–5–20 wt.%). Adapted from Figoli et al. (2017) with permission from Elsevier.

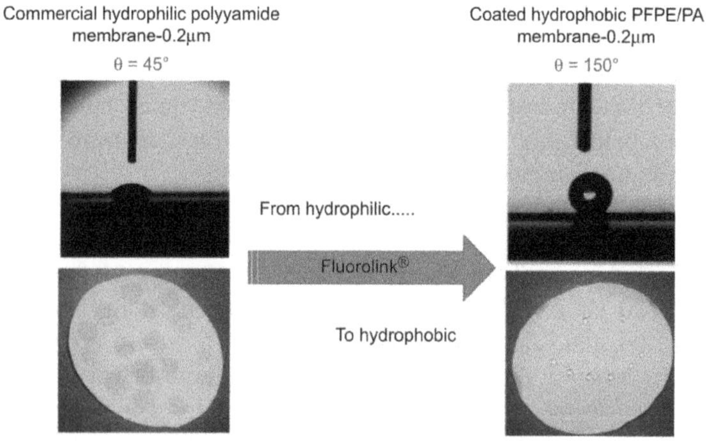

Figure 10.11 Contact angle measurements of the commercial polyamide (PA) membranes without and with PFPE coating.

The performance of the produced membranes has been tested in DCMD using deionized water and salty solution as feed. Feed temperature was varied from 40°C to 60°C, and in all cases, the flux increased with the feed temperature. Depending on temperature and pore size used, the flux obtained varied from 4.6 to 22.8 L/m^{2}h. The flux, during the salty solution test, reported a lower constant value of 11 kg/m^{2}h with a salt rejection of 99.6% (5 wt.% oligomer coating on the 0.45 μm PA membrane; T_{feed} = 50°C; Q_{feed} = 80 L/h; T_p = 14°C; Q_p = 60 L/h), as shown in Fig. 10.12. The DCMD results, similar to those reported in the literature, confirmed that Fluorolink®-coated membranes could be a valid alternative in producing hydrophobic membranes.

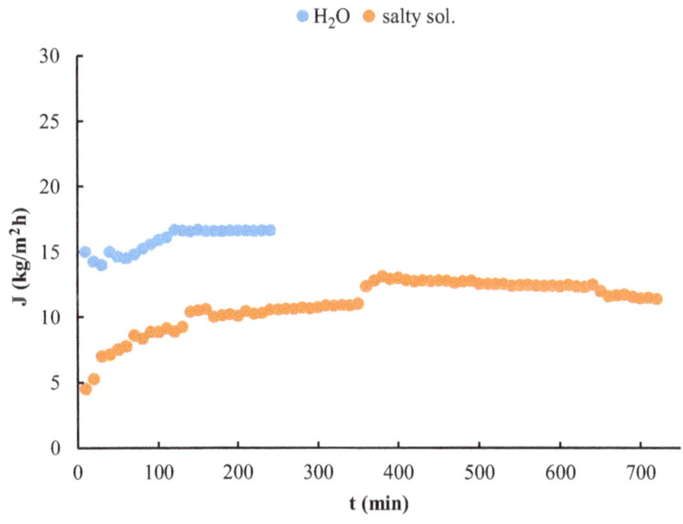

Figure 10.12 Comparison among permeate fluxes as a function of different feed (deionized water and salty solution 0.6 M); T_{feed} = 50°C; membrane tested: 5 wt.% PFPE coating on the 0.45 μm PA membrane. Reprinted from Figoli et al. (2017) with permission from Elsevier.

The second case study focuses on the preparation of composite PP hollow fibers through dip coating using ethylene–chlorotrifluoroethylene copolymer (ECTFE). ECTFE is a semi-crystalline, hydrophobic fluoropolymer; the structure is shown in Fig. 10.13. Thanks to its properties, in terms of hydrophobicity, mechanical and thermal resistance, and its excellent resistance to corrosive chemicals and organic solvents, ECTFE membranes have been studied and applied in several processes (Drioli et al., 2014; Simone et al., 2012; Ursino et al., 2016). In this study, the ECTFE polymer (HALAR® grade 901) was supplied by Solvay Specialty Polymer, Bollate, Italy.

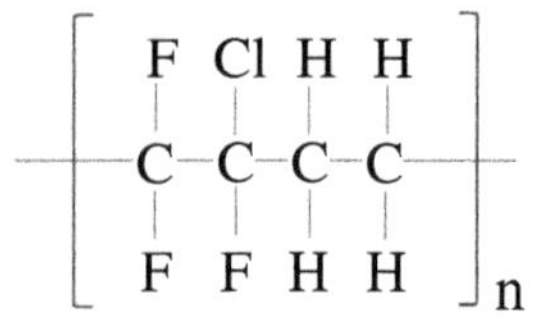

Figure 10.13 ECTFE structure.

In particular, Santoro et al. (2016) developed a novel composite hollow-fiber membrane made of a porous PP support coated with a thin resistant active layer of ECTFE in order to improve the chemical resistance of the pristine porous membranes. The study evaluated the effects of ECTFE concentration and the operating conditions of dip coating on the properties and performance, in terms of water transport, of the novel coated PP/ECTFE hollow fibers. Coating solution was prepared using different ECTFE concentrations (1, 3, and 5 wt.%) and dip coating exposure time varied from 1 s to 5 s. The dip coating procedure was performed at 135°C (coating solution temperature) due to the low ECTFE solubility at room temperature. The procedure is drawn in Fig. 10.14.

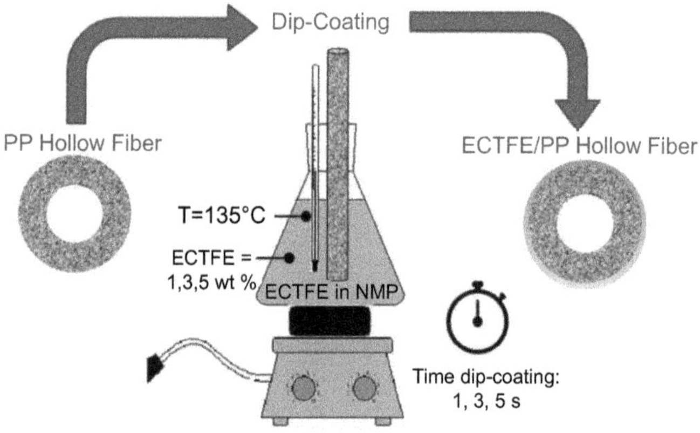

Figure 10.14 Scheme of dip coating procedure. Reprinted from Santoro et al. (2016) under Creative Commons Attribution License.

Composite hollow fibers were characterized in terms of morphology, pore size, hydrophobicity, and water permeability. The PP hollow fibers were immersed in a polymeric solution of ECTFE (from 1 wt.% to 5 wt.%) in *N*-methyl pyrrolidone (NMP) at 135°C. This temperature did not compromise the shape and morphology of the PP support. The thickness of the ECTFE-coated layer increased with the dip coating time in the polymeric solution and with the concentration of ECTFE, which led to a decrease in the mean pore size (from 0.4 μm to 0.04 μm) and porosity (Fig. 10.15).

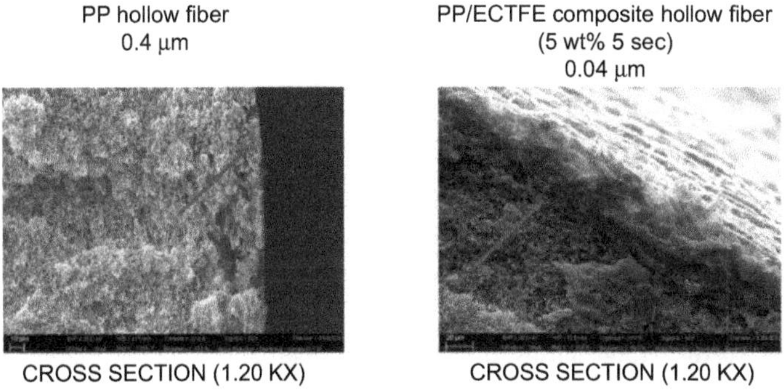

Figure 10.15 SEM pictures of PP hollow fiber, with and without ECTFE coating.

The CA measurements showed that CA decreased (from 126° to 93°) with increasing ECTFE concentration (from 0 wt.% to 5 wt.%) mainly due to a decrease in the roughness of the external surface. Water flux also decreased due to the extra-coating layer of ECTFE on the hollow fiber (Fig. 10.16). However, in all the cases, the water flux was constant, demonstrating an optimal adhesion of ECTFE on PP hollow fibers as well as good stability of ECTFE coating. Therefore, the ECTFE coating has the function of both improving the chemical stability of the starting PP membrane for specific applications in harsh environments and tailoring the pore size (depending on the process parameters and ECTFE concentration) of the newly coated composite hollow fiber.

10.5 Hydrophilic Coating Applications

Surface hydrophilization is beneficial, and necessary in some cases, for all those polymeric materials that are applied in aqueous systems.

Thus, for instance, membranes applied for water desalination and wastewater treatment must be as wettable as possible in order to be able to guarantee high performance in terms of permeability. Hydrophilic surfaces, in fact, exhibit chemical water affinity. This means they can establish hydrogen bonds with the surrounding water molecules, facilitating and promoting water permeation through the membrane.

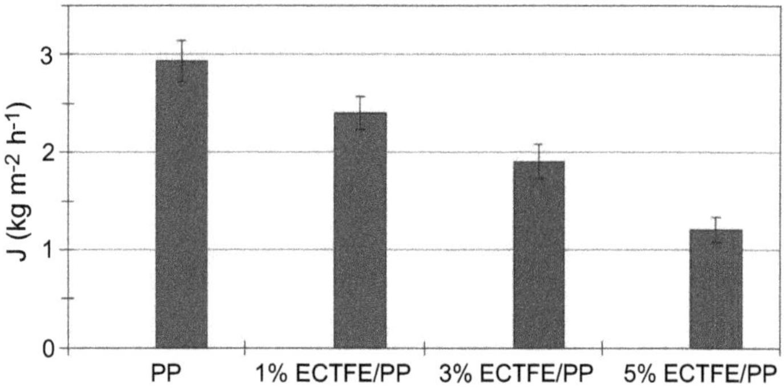

Figure 10.16 Water fluxes of PP hollow fiber and PP/ECTFE composite hollow fibers. Reprinted from Santoro et al. (2016) under Creative Commons Attribution License.

Moreover, hydrophilic surfaces are less prone to the fouling phenomenon (Galiano et al., 2015). Organic compounds, in fact, are mostly responsible for fouling as they tend to interact with membrane surfaces through hydrophobic interactions, leading to their accumulation and formation of a cake layer.

Hydrophilic surfaces are also widely appreciated in the biomedical field due to their higher compatibility with blood and tissue fluids, limiting the onset of undesired body reactions.

The hydrophilization of membranes is generally based on chemical and/or physical processes. The majority of methods adopted are based on the formation of hydrophilic groups such as hydroxyl, carboxyl, and keto groups on the surface of polymeric membranes (Anders et al., 1999). The physical approaches generally used for the hydrophilization of hydrophobic surfaces consist in the deposition of hydrophilic materials on the membrane surface by adsorption (Schroen et al., 1993). The chemical methods include plasma treatment (able to generate hydrophilic groups from the polymer itself) (Yu et al., 2005), polymerization (Steuck,

1986), chemical oxidation (Dimov and Islam, 1990), and surface grafting polymerization (Carrol et al., 2002). The introduction of hydrophilic modifiers into a polymeric membrane can be carried out at three different stages: (i) membrane polymer modification (pre-modification); (ii) addition of modifying agents (additives) to membrane polymer; and (iii) surface modification after membrane preparation (post-modification) (Susanto and Ulbricht, 2009).

Surface coating is one of the easiest hydrophilization techniques. However, it has some drawbacks that can hinder its application in some cases. First, a coating layer on the original membrane represents an extra mass resistance negatively affecting the overall membrane permeability. Second, the coating layer, under particular conditions, can be released in time, resulting in the loss of membrane hydrophilicity. Therefore, several parameters have to be taken into account, such as the substrate on which the coating has to adhere and the chemical, environmental, and mechanical stress that the coating has to experience. Some substrates are ideal for the adhesion of hydrophilic coatings bearing, for instance, –OH or $–NH_2$ polar groups; other substrates, however, have to be previously functionalized by the introduction of polar groups by plasma or coronal treatments to make the coating bind to the surface.

This section aims to present the approaches used so far for the hydrophilization of membrane surfaces by using different coating techniques. In particular, nanostructured hydrophilic coatings for water treatment are illustrated and described in details.

Susanto et al. (2012) studied the hydrophilization effect of a PES membrane by using three different approaches: (i) coating by photo-graft copolymerization with the water-soluble monomer polyethylene glycol methacrylate (PEGDMA); (ii) blending the PES polymer with hydrophilic additives during the phase separation process; (iii) blending the PES polymer with hydrophilic additives by reactive phase separation. The resulting membranes were characterized by FTIR, CA, environmental SEM (ESEM) measurements, and antifouling potential activity comparing the results with the pristine unmodified PES membrane. With the first approach, it was found that the fouling propensity of the membrane could be significantly reduced by the graft copolymerization process. The technique adopted was stable in time, even if the process was accompanied by a reduction in permeate flux. In the second case, where the polymer was blended during phase separation, a membrane with a higher fouling tendency, above all for long-term applications, was obtained. This was mainly due to the leaching of the hydrophilic additives from the polymer matrix. Finally, the last case, where the polymer was blended during reactive phase separation, combined the benefits of the two previous approaches (Fig. 10.17), resulting in the most suitable method.

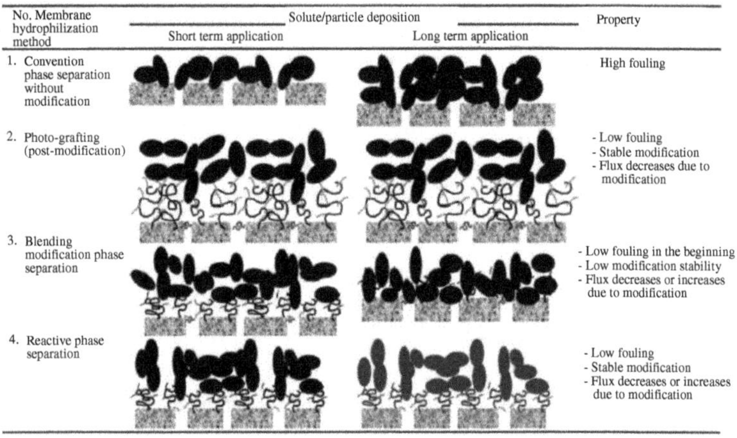

Figure 10.17 Illustration of hydrophilic structure for different hydrophilization methods and their corresponding characteristic and fouling description. Reprinted from Susanto et al. (2012) with permission from Elsevier.

Surface hydrophilization through coating techniques has also been explored for producing membranes that can limit the biofouling phenomenon caused by the adsorption of proteins or adhesion and colonization of bacteria at the membrane surface. Yang et al. (2010), for instance, grafted a zwitterionic polymer such as polysulfobetaine methacrylate (polySBMA) on the surface of microporous PP membranes. In the first step, a photoinitiator was entrapped on the surface of a PP membrane by UV irradiation. After that, the membrane was dipped into an SBMA aqueous solution and again exposed to UV light. The formation of a polySBMA layer was confirmed by FTIR analyses. The water CA of the grafted membranes decreased (from more than 140° to about 20°) with the increase in grafting density. Also the water permeability was positively influenced by the higher hydrophilicity degree with an increase of about four times with respect to the original PP membrane. The grafted membranes were, moreover, just slightly influenced by fouling formation when tested with the two model proteins: bovine serum albumin (BSA) and lysozyme. Their anti-biofouling potential was also assessed with different bacterial strains, proving the low adhesion tendency of the modified membranes (up to 97%) in comparison to pristine PP membranes.

Guo et al. (2016) developed a hydrophilic coating for NF membranes based on the use of polydopamine (PDA) for the removal of endocrine-disrupting compounds (EDCs) from aqueous streams. EDCs are a class of micropollutants usually found in wastewater and cause negative health effects on the endocrine system of humans and wildlife. PDA was widely investigated for membrane antifouling modification for its ability to easily form stable and performing coatings (Lee et al., 2007; McCloskey et al., 2012). The coating was realized through the self-polymerization of dopamine on the active layer of the pristine NF membrane. Almost no modifications of membrane morphology were observed by SEM after the deposition of the PDA coating. The water CA values decreased from 69° (pristine membrane) up to 60° (coated membrane). The use of PDA hydrophilic coating proved that the rejection of EDCs was much more enhanced in comparison to uncoated membrane, avoiding the undesired hydrophobic interactions between the compounds and the membrane surface and responsible of fouling formation.

Fouling prevention in nanostructured membranes, used for water treatment or water desalination, is at the basis of the several hydrophilization processes developed for hydrophobic surfaces (see Table 10.3). In this context, chitosan (CHT) is a biopolymer well known and appreciated for its high chemical and thermal stability, antibacterial properties, and elevated degree of hydrophilicity. Properties of CHT were exploited, for instance, by Akbari et al. (2015) for the modification of a PA membrane surface. The coating was deposed by interfacial polymerization of the amine and hydroxyl groups of CHT and the free acylchloride groups of PA membrane. CHT-coated membranes presented more than 52% water flux in comparison to unmodified pristine PA membranes with enhanced properties also in terms of antifouling properties. These results were attributed to the higher level of hydrophilicity exhibited by CHT-coated membranes and to their lower degree of surface roughness.

In the context of fouling remediation, Ba et al. (2010) modified the surface of a nanofiltration P84 copolyimide-polyethylenimine (P84-PEI) membrane by applying three different coatings: polyvinyl alcohol (PVA), polyacrilic acid (PAA), and polyvinyl sulfate (PVS). Unmodified P84-PEI membranes exhibit on their surface different functional groups (such as carbonyl, amine, ammonium, and phenyl groups), giving an overall positive charge and thus promoting interactions with the negatively charged foulant particles. Each coating was deposited by filtration and adsorption of a 50 mg/L polyelectrolyte (PVA, PAA, and PVS) aqueous solution on the membrane surface.

The effect of the coating on membrane surface was investigated in terms of hydrophilicity, roughness, surface charge, and pore size. The CA decreased in all cases (from about 86° for the uncoated membrane to a minimum of about 75° for the PVS-coated membrane) when coating was applied as a layer on the surface of P84-PEI membranes. Also the surface roughness decreased (for PVA- and PAA-coated membranes) or just slightly increased (for PVS-coated membrane) in comparison to P84-PEI uncoated membrane. In terms of surface charge, PAA polyelectrolyte, being neutral, did not modify the charge of the original positive P84-PEI membrane; while PAA and PVS polyelectrolytes, being negatively charged, counterbalanced the positive charge of the P84-PEI membrane leading to the formation of a close to neutral surface. Hydrophilicity and roughness both contributed to the improvement in the antifouling properties of coated membranes when tested with BSA, humic acid, and sodium alginate used as model foulants. However, the major role in fouling prevention was played by the surface charge. The PVA-coated membrane was most affected by the deposition of foulant particles due to the persisting interactions of the negatively charged particles with the positively charged coated surface. PAA and PVS, on the contrary, were less susceptible to fouling thanks to the neutralization of the charge interactions. In all cases, coated membranes were more resistant toward fouling in comparison to uncoated P84-PEI membranes.

For all the coatings evaluated, the pore size decreased as a result of the coating applied with a consequent decrease in permeate flux and increase in rejection of salts.

The hydrophilization process through coating technique is often carried out through the application of hydrophilic materials, which are generally represented by water-soluble compounds. In order to stabilize them avoiding their excessive swelling or release in aqueous environments, a crosslinking procedure is required. Covalent crosslinking makes the coating material more stable and resistant to swelling, chemical cleanings, and hydraulic pressure. Zhu et al. (2016), for instance, coated polysulfone (PS) membranes with disulfonated poly(arylene ether sulfone) containing pendant amine groups (SPES-NH_2) by covalent crosslinking with glutaric dialdehyde (GA) or polyethylene glycol diglycidyl ether (PEGDGE) (Fig. 10.18). Nanofiltration (NF) membranes have been, thus, obtained by using 1 wt.% SPES-NH_2 polymer concentration and 90°C as curing temperature (optimal conditions). The hydrophilization of the surface was proved by CA measurements, testifying a decrease in CA value from 89° (for the PS membrane) to 66°. The more hydrophilic surface resulted in the production of an NF membrane less prone to fouling when humic acid was applied to simulate fouling formation.

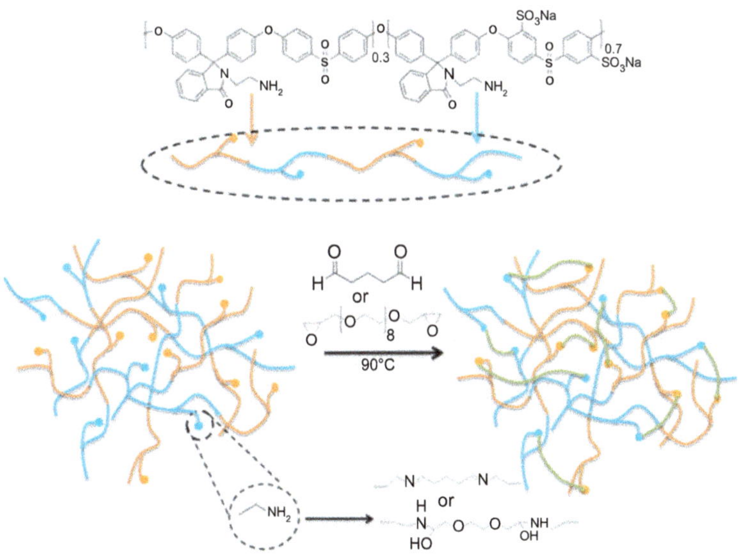

Figure 10.18 In situ crosslinking process of SPES-NH_2-70. Reprinted from Zhu et al. (2016) with permission from Elsevier.

Table 10.3 Summary of membrane surface hydrophilization by applying nanostructured coatings

Starting material	Coating technique applied	Coating material	Scope of hydrophylization	Reference
PS	Dip coating	PVA + sodium alginate	Nanofiltration application	Bano et al. (2014)
PA	Dip coating	PEGDE	Fouling remediation	Weinman and Husson (2016)
PA	Interfacial polymerization	CHT	Fouling remediation	Akbari et al. (2015)
P84 copolyimide-polyethylenimine	Dip coating	PVA PAA PVS	Fouling remediation	Ba et al. (2010)
PS	Dip coating	SPES-NH_2	Fouling remediation	Zhu et al. (2016)
PA	Dip coating	PVA	Removal of pesticides, bentazone, and tebuconazole	Dražević et al. (2013)
PA	Dip coating	PEBAX 1657 Polyethylene glycol (PEG)	Fouling remediation	Louie et al. (2006)

PEBAX: copolymer of Nailon-6

10.5.1 Case Study on Hydrophilic Coating Membrane Formation at ITM-CNR

As reported in Section 10.5, most of the times, the hydrophilization of membrane surfaces is change oriented, which makes membranes more resistant to fouling and biofouling phenomena. The hydrophilization process is generally carried out by the dip coating technique through the deposition of a hydrophilic polymer (such as PVA) on the surface of a hydrophobic membrane. In this paragraph, an exhaustive case study of hydrophilic coating applications, by using a novel technique, is reported and described in details.

Recently, Galiano et al. (2015) investigated a novel coating material based on the polymerization of a polymerizable bicontinuous microemulsion (PBM), for the production of membranes more resistant and less susceptible to fouling and biofouling to be applied in membrane bioreactor (MBR) technology for wastewater treatment. Microemulsions are isotropic, transparent systems made up of an oil phase and a water phase stabilized by the presence of a surfactant. Often a co-surfactant is added to the system in order to further improve the dispersion of the two immiscible phases. Despite emulsions, microemulsions are transparent, thermodynamically stable, and they form spontaneously. Among the different types of microemulsions, bicontinuous systems are formed at specific oil/water/surfactant ratios, and they are characterized by an interconnected network of oil and water channels. A ternary phase diagram reporting oil, water, and surfactant helps to identify the different microemulsion structures that can be encountered. As shown in Fig. 10.19, microemulsion systems can be constituted by oil swollen micelles dispersed in an aqueous continuous phase (O/W microemulsions), water swollen droplets dispersed in an oil continuous phase (W/O microemulsions), or can be formed by the alternation of oil and water domains forming the bicontinuous structure.

Oil channels are generally represented by a monomer (constituting the oil phase of the system), and they can be polymerized to form the matrix of the membrane. The water channels are constituted by the aqueous phase, and, not being polymerizable, they represent the porous part of the final membrane. In the specific case, methylmethacrylate (MMA) was used as a monomer constituting the oil phase, water as aqueous phase, and both stabilized by the polymerizable lab-made surfactant acryloyloxyundecyltriethylammonium bromide (AUTEAB). The short-chain polymerizable alcohol 2-hydroxyethyl methacrylate (HEMA) was used as a co-surfactant and ethylene glycol dimethacrylate (EGDMA) as a crosslinking agent. The microemulsion structure and the bicontinuous range, considering different water/monomer/surfactant ratios, were identified by ternary phase diagram and conductivity measurements. Consequently to the addition of redox initiators, in order to start the polymerization process, the PBM system was cast on a commercial ultrafiltration PES membrane, where it was left to polymerize in situ in inert atmosphere.

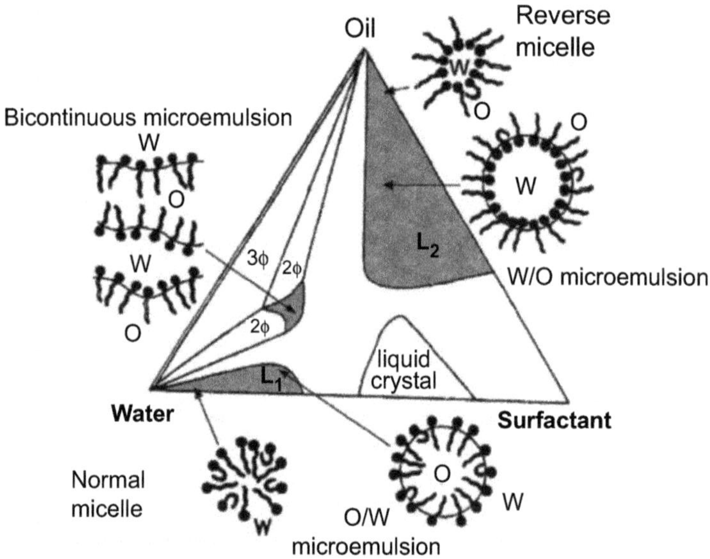

Figure 10.19 Schematic ternary phase diagram of a mixture of water, oil, and surfactant. L_1 : oil-in-water (O/W) microemulsion; L_2 : water-in-oil microemulsion (W/O), and bicontinuous microemulsion regions are reported. 2ϕ and 3ϕ are two-phase and three-phase regions. Reprinted from Acharya and Hartley (2012) with permission from Elsevier.

Characterization tests were then carried out in order to study and investigate the morphology and the performances of the novel PBM membranes so produced. From SEM, the presence of the PBM coating on the PES membrane was directly and clearly observed. As shown in Fig. 10.20, in fact, the surface of the PBM appeared to be made up of white islets (representing the oil-polymerized phase) and dark islets (representing the unpolymerized water phase). The two phases appeared interconnected and randomly distributed as the typical PBM structure.

A nanostructured coating with a particular nanosized channel-like morphology was, thus, obtained. The surface properties of the PBM coating produced were determined by AFM and CA measurements. AFM results showed approximately 17-fold decrease in surface roughness in comparison to uncoated PES membrane.

From the study of the wettability of coated and uncoated membranes, the PBM ones presented an enhanced hydrophilic moiety of about 30% (CA decreased from 68° to 47°) in comparison to PES membranes. The higher hydrophilicity of the PBM membranes was, then, attributed to (i) their smoother surface as proved by AFM measurements; (ii) the presence, in the polymerized coating, of the hydrophilic –OH functional groups provided by the HEMA alcohol used as a co-surfactant and by the presence of the quaternary ammonium group (positively charged) borne by the AUTEAB surfactant.

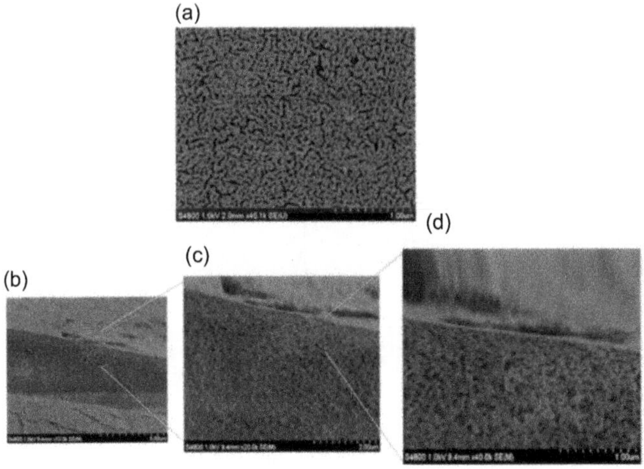

Figure 10.20 SEM surface image (a) and cross sections (b–d) of PBM membrane on PES support with magnification (from ×10,000 to ×40,000). Reprinted from Galiano et al. (2015) with permission from Elsevier.

Antifouling properties of the produced PBM membranes were measured by using humic acid (HA) as a model foulant. The remarkable results obtained, in comparison to uncoated PES membranes, in terms of fouling formation, permeability reduction, and permeability recovery after membrane washing, were attributed to the smoother and more hydrophilic surface of the coating as well as to its particular channel-like nanostructure. PBM membranes were subjected to a decrease in permeability, after HA filtration, of about 45% against the 81% reduction in the PES uncoated membranes. Moreover, the water permeability recovery after washing with water was more pronounced for PBM membranes (54%) in comparison to PES ones (9%) due to the different types of fouling (reversible) occurring on the surface of the previous ones.

In another work, Deowan et al. (2016) tested the PBM nanostructured membranes in MBR for model textile wastewater treatment. PBM membranes were assembled into a module with an active area of 0.33 m^2 and tested in a submerged MBR, and the results were compared with the PES uncoated membranes tested under identical conditions. As shown in Fig. 10.21, reporting the water permeability for both membranes, during 105 days of operation, PES membranes showed several problems related to fouling and biofouling formation on their surface leading to different cleaning protocols (in order to restore as much as possible the permeability) and to the final complete replacement of the membranes as a consequence of the severe fouling degree occurred.

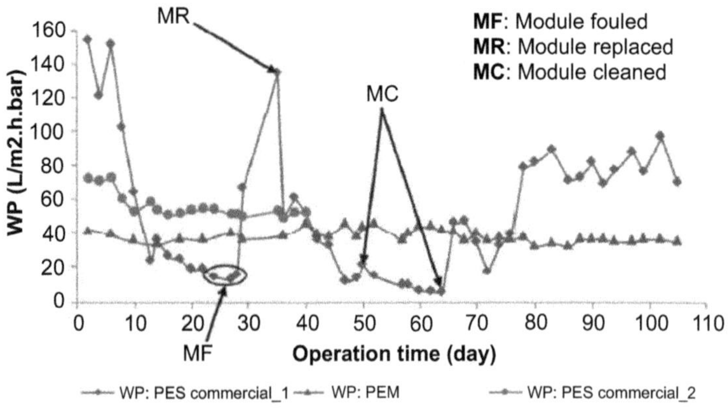

Figure 10.21 Water permeability for PES uncoated commercial and PBM-coated membranes. Reprinted from Deowan et al. (2016) with permission from Elsevier.

During the same interval of time, PBM membranes, on the contrary, did not show particular problems related to fouling formation and no cleaning procedures were required. From the calculations, it was proven, in fact, that PBM membranes presented an antifouling activity from 2.5 to 5 times higher than uncoated PES ones. These results were also clearly visible by observing both membranes before and after the pilot PBM trials (Fig. 10.22). PBM membranes appeared, in fact, less fouled and much cleaner than the PES commercial ones.

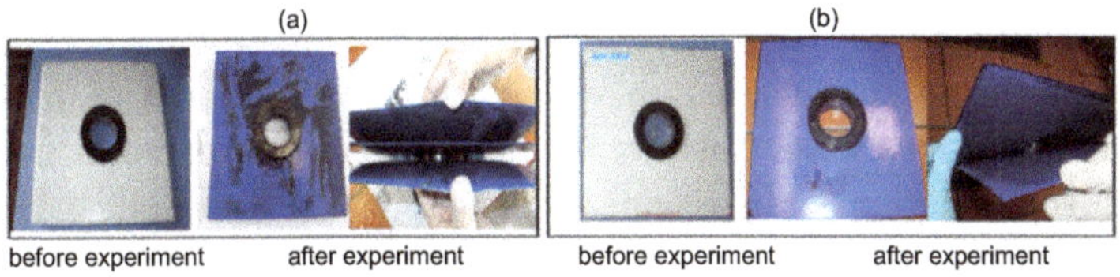

Figure 10.22 (a) Commercial PES and (b) PBM MBR modules before and after pilot trials. Reprinted from Deowan et al. (2016) with permission from Elsevier.

Moreover, the chemical oxygen demand (COD) removal efficiency for PBM membranes was very stable and about 95%, whereas the average COD for PES commercial membranes was around 90%. The rejection of the colorants applied to simulate textile wastewater was in the range of 40–50% for both membranes for the red dye, while it was 55% for the PBM membranes for the blue dye (higher than 10% with respect to PES membranes). Even in this case, the good performances of the PBM membranes in terms of antifouling and anti-biofouling activity were related to the important properties exhibited by the PBM coating, such as the relative high degree of hydrophilicity, the smoother surface, the channel-like structure typical of bicontinuous microemulsions, and the intrinsic antimicrobial activity.

10.6 Conclusion

The chapter aimed to illustrate coating techniques as a viable and valid approach for modifying polymeric membrane surfaces. Different types of coating, including solution coating and polymerization coating, have been treated and discussed. In particular, special attention has been paid on the hydrophobization/hydrophilization of membranes, resulting in the production of surfaces owing to particular and specific properties that can be exploited in different applications. The hydrophobization of membrane surfaces is particularly relevant, for instance, for the preparation of composite hydrophilic/hydrophobic membranes applied in DCMD. The hydrophilic native membrane limits the temperature polarization effect and acts as a mechanical support for the overall membrane; the hydrophobic coating deposited on its surface, on the contrary, can prevent the wetting of the membrane, hampering the entrance of the liquid into the pores and making the process feasible. Hydrophobic coatings are, however, highly appreciated in other membrane processes such as PV for the recovery of VOCs from aqueous solutions or in organic/organic separations.

The application of coating techniques in the hydrophilization of membrane surfaces is more common. The literature documents very well the possibility to limit and mitigate the fouling phenomenon by making the membrane surface more hydrophilic. A higher degree of hydrophilicty, in fact, results in the fabrication of membranes less prone to be affected by fouling and, at the same time, more permeable to water. Therefore, this type of coating is highly appreciated for the modification of MF/UF membranes used in water treatment processes (such as MBR).

So far, coating represents an important and indispensable tool in the hands of membrane producers in order to tailor the surface chemistry of membranes for meeting the requirements of the market and final end users.

Appendix

Notation

g	Gravity (m/s^2)
M	Molar (mol/L)
Q	Flux (L/h)
T	Temperature (°C)
T_{equ}	Equilibrium thickness
wt.%	Weight percentage

Greek symbols

η	Dynamic viscosity (PI)
ρ	Solution density (kg/m^3)
v	Withdrawal speed (m/s)

Subcripts

feed	Feed or upstream side of the membrane
perm	Permeate or downstream side of the membrane

Abbreviations

AFM	atomic force microscopy
AUTEAB	acryloyloxyundecyltriethylammonium bromide
BP	bucky-papers
BSA	bovine serum albumin
CA	contact angle
CAc	cellulose acetate
CHT	chitosan
CNT BPs	carbonanotube composite bucky-papers
COD	chemical oxygen demand
DCMD	direct contact membrane distillation
ECTFE	ethylene–chlorotrifluoroethylene copolymer
EDCs	endocrine disruptive compounds
EGDMA	ethylene glycol dimethacrylate
ESEM	environmental scanning electron microscopy
FPBZ	fluorinated polybenzoxazine
GA	glutaric dihaldehyde
HA	humic acid
HEMA	2-hydroxyethyl methacrylate
IP	interfacial polymerization
LEP	liquid entry pressure
LEP_w	water liquid entry pressure
MBR	membrane bioreactor
MD	membrane distillation
MMA	methyl methacrylate
NF	nanofiltration

NMP	*N*-methyl pyrrolidone
P84-PEI	P84 copolyimide-polyethylenimine
PA	polyamide
PAA	polyacrilic acid
PBM	polymerizable bicontinuous microemulsion
PDA	polydopamine
PE	polyether
PEGDGE	polyethylene glycol diglycidyl ether
PEI	polyetherimide
PES	polyethersulfone
PFPEs	perfluoropolyether compounds
PolySBMA	polysulfobetaine methacrylate
PP	polypropylene
PS	polysylfone
PTFE	polytetrafluorethylene
PV	pervaporation
PVA	polyvinyl alcohol
PVS	polyvinyl sulfate
SEM	scanning electron microscopy
SMMs	surface modifying macromolecules
UF	ultrafiltration
VTMS	vinyltrimethylsilicon
WP	water permeability

References

Acharya, D. P. and Hartley, P. G. (2012). Progress in microemulsion characterization, *Curr. Opin. Colloid Interface,* **17**, pp. 274–280.

Aegerter, M. A., Puetz, J., Gasparro, G., and Al-Dahoudi, N. (2004). Versatile wet deposition techniques for functional oxide coatings, *Opt. Mater.*, **26**, pp. 155–162.

Ahmad, N. A., Leo, C. P., Ahmad, A. L., and Ramli, W. K. W. (2015). Membranes with great hydrophobicity: A review on preparation and characterization, *Sep. Purif. Rev.*, **44**, pp. 109–134.

Akbari, A., Derikvandi, Z., and Rostami, S. M. M. (2015). Influence of chitosan coating on the separation performance, morphology and anti-fouling properties of the polyamide nanofiltration membranes, *J. Ind. Eng. Chem.*, **28**, pp. 268–276.

Anders, C., Hoecker, H., Klee, D., and Lorenz, G. (1999). Hydrophilic coating of surfaces of polymeric substrates, Patent n. 5,871,823.

Ba, C., Ladner, D. A., and Economy, J. (2010). Using polyelectrolyte coatings to improve fouling resistance of a positively charged nanofiltration membrane, *J. Membr. Sci.*, **347**, pp. 250–259.

Bano, S., Mahmood, A., Kim, S. J., and Lee, K.-H. (2014). Chlorine resistant binary complexed NaAlg/PVA composite membrane for nanofiltration, *Sep. Purif. Technol.*, **137**, pp. 21–27.

Bhattacharya, A. and Misra, B. N. (2004). Grafting: A versatile means to modify polymers—Techniques, factors and applications, *Prog. Polym. Sci.*, **29**, pp. 767–814.

Bonyadi, S. and Chung, T. S. (2007). Flux enhancement in membrane distillation by fabrication of dual layer hydrophilic–hydrophobic hollow fiber membranes, *J. Membr. Sci.*, **306**, pp. 134–146.

Cadotte, J. E., Petersen, R. J., Larson, R. E., and Erickson, E. E. (1980). New thin-film composite seawater reverse-osmosis membrane, *Desalination*, **32**, pp. 25–31.

Carnevale, M. C., Gnisci, E., Hilal, J., and Criscuoli, A. (2016). Direct contact and vacuum membrane distillation application for the olive mill wastewater treatment, *Sep. Purif. Technol.*, **169**, pp. 121–127.

Carroll, T., Booker, N. A., and Meier-Haack, J. (2002). Polyelectrolyte-grafted microfiltration membranes to control fouling by natural organic matter in drinking water, *J. Membr. Sci.*, **203**, pp. 3–13.

Chovau, S., Dobrak, A., Figoli, A., Galiano, F., Simone, S., Drioli, E., Sikdar, S. K., and Bruggen, B. V. (2010). Pervaporation performance of unfilled and filled PDMS membranes and novel SBS membranes for the removal of toluene from diluted aqueous solutions, *Chem. Eng. J.*, **159**, pp. 37–46.

Criscuoli, A., Carnevale, M. C., and Drioli, E. (2008). Evaluation of energy requirements in membrane distillation, *Chem. Eng. Process.*, **47**(7), pp. 1098–1105.

Daiminger, U., Nitsch, W., Plucinski, P., and Hoffmann, S. (1995). Novel techniques for oil/water separation, *J. Membr. Sci.*, **99**, pp. 197.

Deowan, S. A., Galiano, F., Hoinkis, J., Johnson, D., Altinkaya, S. A., Gabriele, B., Hilal, N., Drioli, E., and Figoli, A. (2016). Novel low-fouling membrane bioreactor (MBR) for industrial wastewater treatment, *J. Membr. Sci.*, **510**, pp. 524–532.

Dimov, A. and Islam, M. A. (1990). Hydrophilization of polyethylene membranes, *J. Membr. Sci.*, **50**, pp. 97–100.

Dražević, E., Košutić, K., Dananić, V., and Mutavdžić Pavlovic, D. (2013). Coating layer effect on performance of thin film nanofiltration membrane in removal of organic solutes, *Sep. Purif. Technol.*, **118**, pp. 530–539.

Drioli, E., Criscuoli, A., and Curcio, E. (2006). *Membrane Contactors: Fundamentals, Applications and Potentialities*. Membrane Science and Technology Series, 11, Elsevier, Amsterdam, ISBN: 0-444-52203-4.

Drioli, E., Santoro, S., Simone, S., Barbieri, G., Brunetti, A., Macedonio, F., and Figoli, A. (2014). ECTFE membrane preparation for recovery of humidified gas streams using membrane condenser, *React. Funct. Polym.*, **79**, pp. 1–7.

Dumée, L., Campbell, J. L., Sears, K., Schütz, J., Finn, N., Duke, M., and Gray, S. (2011). The impact of hydrophobic coating on the performance of carbon nanotube bucky-paper membranes in membrane distillation, *Desalination*, **283**, pp. 64–67.

Faure, B., Salazar-Alvarez, G., Ahniyaz, A., Villaluenga, I., Berriozabal, G., De Miguel, Y. R., and Bergström, L. (2013). Dispersion and surface functionalization of oxide nanoparticles for transparent photocatalytic and UV-protecting coatings and sunscreens, *Sci. Technol. Adv. Mater.*, **14**, pp. 023001.

Figoli, A., Simone, S., Criscuoli, A., AL-Jlil, S. A., Al Shabouna, F. S., Al-Romaih, H. S., Al- Harbi, O. A., Di Nicolò, E., and Drioli, E. (2014). Hollow fibers for seawater desalination from blends of PVDF with different molecular weights: Morphology, properties and VMD performance, *Polymer*, **55**, pp. 1296–1306.

Figoli, A., Ursino, C., Galiano, F., Di Nicolò, E., Campanelli, P., Carnevale, M. C., and Criscuoli, A. (2017). Innovative hydrophobic coating of perfluoropolyether (PFPE) on commercial hydrophilic membranes for DCMD application, *J. Membr. Sci.*, **522**, pp. 192–201.

Figoli, A., Santoro, S., Galiano, F., and Basile, A. (2015). Pervaporation membranes: preparation, characterization and applications. In: *Pervaporation, Vapour Permeation and Membrane Distillation*, Basile, A., Figoli, A., and Khayet, M. (eds.) (Woodhead Publishing, United Kingdom), pp.107–140.

Galiano, F., Figoli, A., Deowan, S. A., Johnson, D., Altinkaya, S. A., Veltri, L., De Luca, G., Mancuso, R., Hilal, N., Gabriele, B., and Hoinkis J. (2015). A step forward to a more efficient wastewater treatment by membrane surface modification via polymerizable bicontinuous microemulsion, *J. Membr. Sci.*, **482**, pp. 103–114.

Garcìa-Fernàndez, L., Khayet, M., and Garcìa-Payo, M. C. (2015). Membrane used in membrane distillation: preparation and characterization. In: *Pervaporation, Vapour Permeation and Membrane Distillation*, Basile, A., Figoli, A., and Khayet, M. (eds.) (Woodhead Publishing, United Kingdom), pp. 317–357.

Guo, H., Deng, Y., Tao, Z., Yao, Z., Wang, J., Lin, C., Zhang, T., Zhu, B., and Tang, C. Y. (2016). Does hydrophilic polydopamine coating enhance membrane rejection of hydrophobic endocrine-disrupting compounds? *Environ. Sci. Technol. Lett.*, **3**, pp. 332–338.

Kao, S. T., Huang, S. H., Liaw, D. J., Chao, W. C., Hu, C. C., Li, C. L., Wang, D. M., Lee, K. R., and Lai, J. Y. (2010). Interfacially polymerized thin-film composite polyamide membrane: Positron annihilation spectroscopic study, characterization and pervaporation performance, *Polym. J.*, **42**, pp. 242–248.

Khayet, M., Mengual, J. I., and Matsuura, T. (2005). Porous hydrophobic/hydrophilic composite membranes: Application in desalination using direct contact membrane distillation, *J. Membr. Sci.*, **252**, pp. 101–113.

Khayet, M., Suk, D. E., Narbaitz, R. M., Santerre, J. P., and Matsuura, T. (2003). Study on surface modification by surface-modifying macromolecules and its applications in membrane-separation processes, *J. Appl. Polym. Sci.*, **89**, pp. 2902–2916.

Khulbe, K. C., Feng, C., and Matsuura, T. (2010). The art of surface modification of synthetic polymeric membranes, *J. Appl. Polym. Sci.*, **115**, pp. 855–895.

Kochkodan, V. and Hilal, N. (2015). A comprehensive review on surface modified polymer membranes for biofouling mitigation, *Desalination*, **356**, pp. 187–207.

Lee, H., Dellatore, S. M., Miller, W. M., and Messersmith, P. B. (2007). Mussel-inspired surface chemistry for multifunctional coatings, *Science*, **318**, pp. 426–430.

Li, K., Kong, J. F., and Teo, W. K. (1999). Tailor made asymmetric hollow fibers for soluble gas removal, *AIChE J.*, **45**, pp. 1211.

Lin, Y. and Yasuda, H. (1996). Effect of plasma polymer deposition methods on copper corrosion protection, *J. Appl. Polym. Sci.*, **60**, pp. 543.

Louie, J. S., Pinnau, I., Ciobanu, I., Ishida, K. P., Ng, A., and Reinhard, M. (2006). Effects of polyether–polyamide block copolymer coating on performance and fouling of reverse osmosis membranes, *J. Membr. Sci.*, **280**, pp. 762–770.

McCloskey, B. D., Park, H. B., Ju, H., Rowe, B. W., Miller, D. J., and Freeman, B. D. (2012). A bioinspired fouling-resistant surface modification for water purification membranes, *J. Membr. Sci.*, 413–414, pp. 82–90.

Mulder, M. (1996). *Basic Principles of Membrane Technology*, 2nd Ed. (Kluwer Academic Publishers, The Netherlands).

Prince, J. A., Anbharasi, V., Shanmugasundaram, T. S., and Singh, G. (2013). Preparation and characterization of novel triple layer hydrophilic–hydrophobic composite membrane for desalination using air gap membrane distillation, *Sep. Purif. Technol.*, **118**, pp. 598–603.

Qtaishat, M., Khayet, M., and Matsuura, T. (2009a). Guidelines for preparation of higher flux hydrophobic/hydrophilic composite membranes for membrane distillation, *J. Membr. Sci.*, **329**, pp. 193–200.

Qtaishat, M., Rana, D., Khayet, M., and Matsuura, T. (2009b). Preparation and characterization of novel hydrophobic/hydrophilic polyetherimide composite membranes for desalination by direct contact membrane distillation, *J. Membr. Sci.*, **327**, pp. 264–273.

Santoro, S., Drioli, E., and Figoli, A. (2016). Development of novel ECTFE coated PP composite hollow-fiber membranes, *Coatings*, **6**, pp. 40.

Schroen, C. G. P. H., Wijers, M. C., Cohen-Stuart, M. A., Vander Padt, A., and Van't Riet, K. (1993). Membrane modification to avoid wettability changes due to protein adsorption in an emulsion/membrane bioreactor, *J. Membr. Sci.*, **80**, pp. 265–274.

Schwartz, W. and Narayanan, M. (2008). Chemical solution deposition: Basic principles. In: *Solution Processing of Inorganic Materials*, Mitzi, D. B. (ed.) (John Wiley & Sons, Inc., USA), pp. 33–76.

Shang, Y., Si, Y., Raza, A., Yang, L., Mao, X., Ding, B., and Yu, J. (2012). An in situ polymerization approach for the synthesis of superhydrophobic and superoleophilic nanofibrous membranes for oil–water separation, *Nanoscale*, **4**, pp. 7847.

Simone, S., Figoli, A., Santoro, S., Galiano, F., Alfadul, S. M., Al-Harbi, O. A., and Drioli, E. (2012). Preparation and characterization of ECTFE solvent resistant membranes and their application in pervaporation of toluene/water mixtures, *Sep. Purif. Technol.*, **90**, pp. 147–161.

Steen, M. L., Hymas, L., Havey, E. D., Capps, N. E., Castner, D. G., and Fisher, E. R. (2001). Low temperature plasma treatment of asymmetric polysulfone membranes for permanent hydrophilic surface modification, *J. Membr. Sci.*, **188**, pp. 97–114.

Steuck, M. (1986). Porous membrane having hydrophilic surface and process, Patent n. US 4618533 A.

Strathmann, H. (1990). Synthetic membranes and their preparation. In: *Handbook of Industrial Membrane Technology,* Porter, M. C. (ed.) (Noyes Publication Westwood, New Jersey, USA), pp. 1–56.

Susanto, H. and Ulbricht, M. (2009). Polymeric membranes for molecular separations. In: *Membrane Operations Innovative Separations and Transformations,* Drioli, E. and Giorno, L. (eds.) (Wiley-VCH Verlag GmbH & Co. KGaA, Weinheim), pp. 19–44.

Susanto, H., Roihatin, A., Aryanti, N., Anggoro, D. D., and Ulbricht, M. (2012). Effect of membrane hydrophilization on ultrafiltration performance for biomolecules separation, *Mater. Sci. Eng. C*, **32**, pp. 1759–1766.

Tyszler, D., Zytner, R. G., Batsch, A., Brügger, A., Geissler, S., Zhou, H., Klee, D., and Melin, T. (2006). Reduced fouling tendencies of ultrafiltration membranes in wastewater treatment by plasma modification, *Desalination*, **189**, pp. 119–129.

Ulbricht, M. (2006). Advanced functional polymer membranes, *Polymer*, **47**, pp. 2217–2262.

Ursino, C., Simone, S., Donato, L., Santoro, S., De Santo, M. P., Drioli, E., Di Nicolò, E., and Figoli, A. (2016). ECTFE membranes produced by non-toxic diluents for organic solvent filtration separation, *RSC Adv.*, **6**, pp. 81001.

Wei, X., Zhao, B., Li, X. M., Wanga, Z., He, B. Q., He, T., and Jiang, B. (2012). CF4 plasma surface modification of asymmetric hydrophilic polyethersulfone membranes for direct contact membrane distillation, *J. Membr. Sci.*, **407**, pp. 164–175.

Weinman, S. T. and Husson, S. M. (2016). Influence of chemical coating combined with nanopatterning on alginate fouling during nanofiltration, *J. Membr. Sci.*, **513**, pp. 146–154.

Wu, Y., Kong, Y., Lin, X., Liu, W., and Xu, J. (1992). Surface-modified hydrophilic membranes in membrane distillation, *J. Membr. Sci.*, **72**, pp. 189–196.

Xu, Z., Huang, J. X., and Wan, L. (2009). *Surface Engineering of Polymer Membranes* (Zhejiang University Press, Hangzhou and Springer-Verlag GmbH Berlin Heidelberg).

Yang, Y.-F., Li, Y., Li, Q. L., Wan, L.-S., and Xu, Z.-K. (2010). Surface hydrophilization of microporous polypropylene membrane by grafting zwitterionic polymer for anti-biofouling, *J. Membr. Sci.*, **362**, pp. 255–264.

Yasuda, H. and Matsuzawa, Y. (2005). Economical advantages of low-pressure plasma polymerization coating, *Plasma Process. Polym.*, **2**, pp. 507–512.

Yu, H. Y., Xie, Y. J., Hu, M. X., Wang, J. L., Wang, S. Y., and Xu, Z. K. (2005). Surface modification of polypropylene microporous membrane to improve its antifouling property in MBR: CO_2 plasma treatment, *J. Membr. Sci.*, **254**, pp. 219–227.

Zhu, J., Zhang, Q., Li, S., and Zhang, S. (2016). Fabrication of thin film composite nanofiltration membranes by coating water soluble disulfonated poly(arylene ether sulfone) and in situ crosslinking, *Desalination*, **387**, pp. 25–34.

Chapter 11

Nano-assembled Membranes for Smart Textiles Solutions

Annarosa Gugliuzza
Institute on Membrane Technology of the National Research Council of Italy (ITM-CNR), Via P. Bucci, Cubo 17C, 87036 Rende (CS), Italy
a.gugliuzza@itm.cnr.it

Nanostructured membranes have great potential in the development of new-generation textiles for indoor and outdoor solutions. These membranes provide concrete opportunity for allocating different and complementary functions in the same device and provide new smart behaviors through cooperative mechanisms. Indeed, they can work as a permselective interface through which mass, heat, charge, and other signals can be transferred under specific working conditions. This chapter provides basic concepts related to membranes and their potential roles in textiles, with a special focus on voltage-activated membranes for microclimate regulation. Future directions in sustainable marketplaces have also been discussed.

Functional Nanostructured Membranes
Edited by Enrico Drioli, Lidietta Giorno, and Annarosa Gugliuzza

ISBN 978-981-4774-79-6 (Hardcover), 978-1-351-13511-5 (eBook)
www.panstanford.com

11.1 Introduction

Together with climatic disorders and worldwide population growth, the recent increase in the costs of raw materials due to the global crisis and drastic contraction of the production of natural fibers in favor of bio-energy crops has caused significant stress in the textile industry. The volatility of cotton prices in the world has further caused trade disruption throughout the entire textile supply chain. Cotton consumption decreased by 1.2 million tons (–4.9%) to 23.8 million tons in 2011, while non-cotton fiber consumption increased by 2.6 million tons (5.5%) to 50.1 million tons in 2011. In 2010 and 2011, the price of cotton relative to other fibers shot up to 1.41, i.e., cotton became 41% more expensive than polyester. In 2015, the price of cotton and wool continued to fall by 15% and 7%, respectively.[1] In December 2016, prices were between 70 and 72 cents/lb.

Within this frame, the demand for alternative materials has recorded a huge hike, with particular emphasis on using advanced engineered materials in clothing, architecture, cultural heritage, agriculture, automobiles, and aerospace (Gugliuzza and Drioli, 2013). Nevertheless, there are some common concepts regarding protection, comfort, sensing, and aesthetic (Paul, 2014), which would be integrated for making any artifact appealing and competitive (Kwon et al., 1998; Levin and Maibach, 2005). Since 1960, technical textiles have increased five times faster than traditional textiles, covering one-third of the global production. Asia covers 45% of the global consumption alone. Medical textiles and protective apparel from the Chinese technical textile market are indeed experiencing high demand and accelerating at a fast pace. Contrary to traditional textiles that fulfill climatic and aesthetic sense, technical textiles provide practical performance and functional properties rather than aesthetic or decorative features. In this context, nanotechnologies are expected to bring revolution and innovative solutions to the textile market (Yetisen, 2016). The integration of nanomaterials in textiles can be regarded as a suitable strategy to meet the increasing demand for durable and functional but also sustainable apparels. This means a multidisciplinary approach, including chemistry, physics, and engineering, is necessary to develop new devices (De Rossi et al.,

[1]http://www.eulerhermes.com

2003), enabling one to face key challenges such as implementation at the industrial scale, T-integration, and competitive production processes. However, some current bottlenecks have to be taken into account, such as shortage of resources, recycling, biodegradability, and costs of raw materials.[2]

Nevertheless, featured solutions such as actuating, generating/storing power together with communicating and long-distance monitoring represent other ambitious challenges that can transform textile markets and everyday life (Kristy et al., 2014; Zysset et al., 2013). In this context, membrane science may provide several interesting solutions, because membranes can work as permselective and responsive interfaces under different working conditions (Gugliuzza and Drioli, 2014).

11.2 Role of Membranes in Textiles Applications

Before discussing the potential of membranes in textiles, it is appropriate to provide a definition of membrane. A "membrane" is a selective or non-selective barrier that separates and/or contacts two adjacent phases and allows or promotes the exchange of matter, energy, charge, and information between phases in a specific or non-specific manner. More specifically, a membrane enables some species to pass through it while others are stopped according to various mechanisms (Fig. 11.1), which depend on intrinsic structural and chemical features of the membrane (Bitter, 1991). Generally speaking, the choice of membrane-type depends on the final application (Baker, 2012).

This means that the membranes must be designed keeping in mind the surrounding environment coming in contact with as well as the working conditions to be applied for desired applications. Both the porous hydrophobic or dense hydrophilic membranes can be designed for textiles (Fig. 11.1). The first ones combine the great ability to promote large water vapor transport (breathability), conferring a degree of waterproofness to the artifact as pore size and hydrophobicity are well balanced (Gugliuzza and Drioli, 2007;

[2]http://www.ecowatch.com

Speranza et al., 2010). The second ones provide, instead, lower vapor exchange but preserve the textiles from undesired wetting and breeze events (Gugliuzza et al., 2007). For both kinds of membranes, the basic idea is to provide protection against external agents such as rain, snow, and wind, keeping optimal comfort. Today, the role of membranes in textiles is beyond the traditional vision of the breathable protective layer (Fig. 11.2).

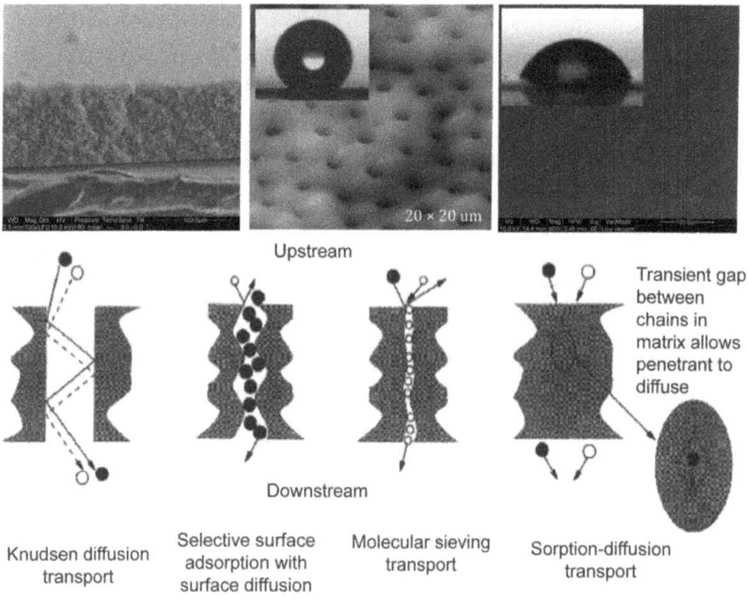

Figure 11.1 SEM and AFM micrographs of porous and dense membranes (a); schematic mechanisms of transport through membranes. Adapted from Baker (2012), with permission from John Wiley and Sons.

Due to its ability to accommodate different and complementary functions, membrane is a key element for making the textiles industry more competitive and innovative for practical solutions[3] (Gugliuzza, 2014). Traditional protective action can be, in fact, associated with sensing and adaptive functions, which can also accelerate the transition from traditional to ultra-smart textiles to create devices with programmable and predictable utilities for detection, long-distance monitoring, and actuation, and capabilities

[3]http://www.freedoniagroup.com

to be exploited in environment, energy, automotive, biotechnology, and medicine (Hu et al., 2012; Karpagam et al., 2016; Massaroni et al., 2015; Narayanan et al., 2015; Schubert and Werner, 2006).

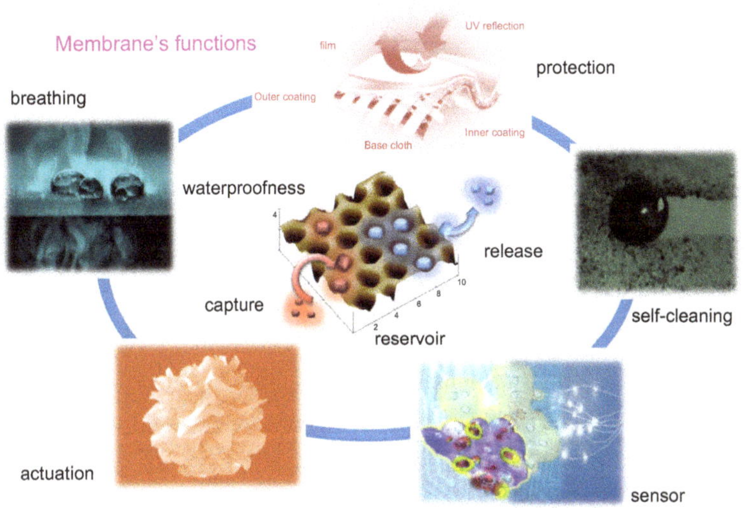

Figure 11.2 Scheme of multiple role of membranes in textiles.

11.3 Concepts of Breathing, Heat Exchange, and Waterproofness

"Breathability" is the ability of a membrane to let water vapor pass through it while preventing entry of liquids (Qui et al., 2004; Wu, 1994). In general, a membrane is said to be breathable if the water permeability exceeds the water vapor transmission rate (WVTR) of 400 g/m^{2}day (Brewer, 2011); otherwise, it works as a barrier. There are many approaches for measuring this intrinsic membrane property (Gibson et al., 2000; McCullough et al., 2003), which can be termed as flux, energy, or resistance depending on the testing method used (Table 11.1).

However, the major concern is the difficulty to pass from a rate value to another by using a univocal statistical correlation. This implies that the same membrane can exhibit different WVTR depending on the method used, thus making a comparison between the various artifacts difficult (Fig. 11.3).

Table 11.1 Traditional testing methods for measuring breathability of textiles

Test method	Description	Regulation	WVTR units
Upright cup	A cup of water is sealed with a fabric and an air gap is left between liquid and fabric. The cup is weighed before and after to determine how much vapor has been lost.	ASTM E96	g/m^2day
Inverted cup	A cup of water is sealed with a fabric and inverted, keeping the water in direct contact with fabric. This causes absorption of large amount of moisture, increasing up to five times the breathability.	ASTM E96	g/m^2day
The desiccant inverted cup	A liquid desiccant is used to move water through the fabric. The cup is weighed before and after to quantify the permeability that can be 20 times higher than that produced by the upright cup.	JIS 1099	g/m^2day
Dynamic moisture permeation cell	A dry and water-saturated nitrogen stream passing over each side of the fabric is used. The control of flow rates, temperature, and humidity of the two flows, and the measurement of the humidity afterward, give indication of the water vapor transmission rate of the fabrics' resistance. The breathability rate can be four to six times those of the upright cup.	ASTM F 2298	g/m^2day

Test method	Description	Regulation	WVTR units
Evaporative resistance	A "sweating hot plate," electrically powered and controlled by a thermostat, is used to maintain a steady temperature. The amount of power necessary to maintain that temperature defines how much energy has been lost through the fabric covering it by evaporation.	ISO 11092	m^2Pa/W

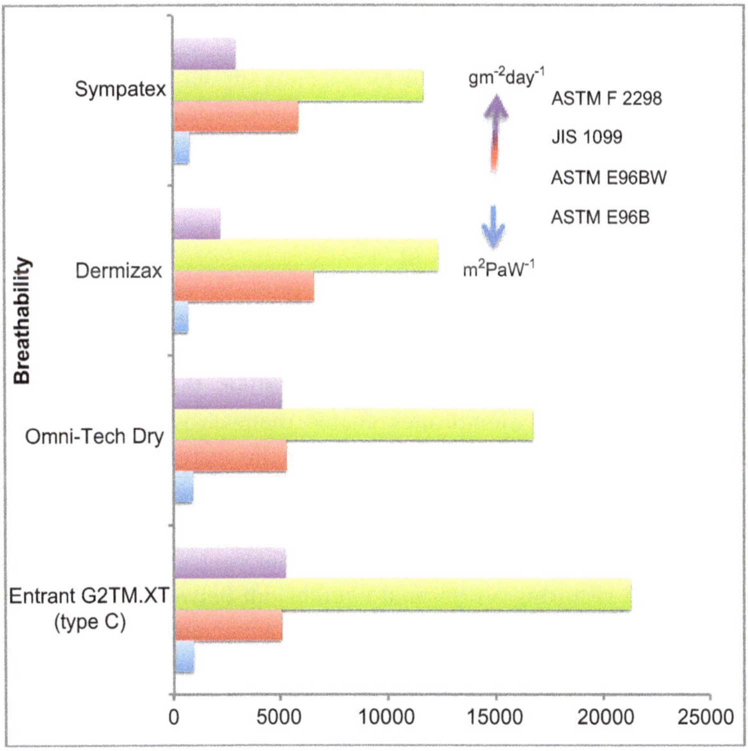

Figure 11.3 A comparison of breathability values measured through porous and dense membranes according to different ASTM testing methods. Data adapted from Gugliuzza and Drioli, 2013.

Figure 11.3 highlights how difficult it is to get comparable values of breathability for the same artifact when different ASTM methods are used. Generally, breathability is expressed as:

$$P = 22.414\frac{(W_\mathrm{f} - W_0)}{18At}\frac{l_\mathrm{m}}{\Delta P} \tag{11.1}$$

where P is the permeability, W_f and W_o are the final and initial weights of a cup hermetically sealed by the membrane, A is the effective membrane area, t is the time of a measurement, and ΔP is the transmembrane water vapor pressure.

When the transport obeys Fick's first law, permeability can be related to flux (J) according to the following equation:

$$J = P\frac{(p_\mathrm{h} - p_\mathrm{l})}{l_\mathrm{m}} \tag{11.2}$$

where l_m is the membrane thickness, p_h is the upstream pressure (high pressure), and p_l is the downstream pressure (low pressure).

However, under the dependence of diffusion coefficient, the flux becomes:

$$J = \frac{D_0}{\beta l_\mathrm{m}}\left(e^{\beta C_0} - 1\right) \tag{11.3}$$

where β is a constant and D_0 is the relationship between the diffusion coefficient as the concentration is close to zero.

Within the frame of application in textiles, the basic concept is, however, to get membranes enable themselves to change the degree of permeability as a function of the water concentration gradient. This implies an intrinsic ability of the materials forming the membrane to sense changes taking place in the surrounding environment and then adapt themselves for contrasting undesired events and restoring an optimal equilibrium between the outside and inside local environment. This accomplishment is well known as "microclimate regulation."

Nevertheless, the establishment of comfort conditions depends also on the membrane capacity to promote exchange of heat (Fig. 11.4).

From the viewpoint of mechanisms, heat transfer through porous membranes is the result of different contributions (Macedonio et al., 2013), including the following:

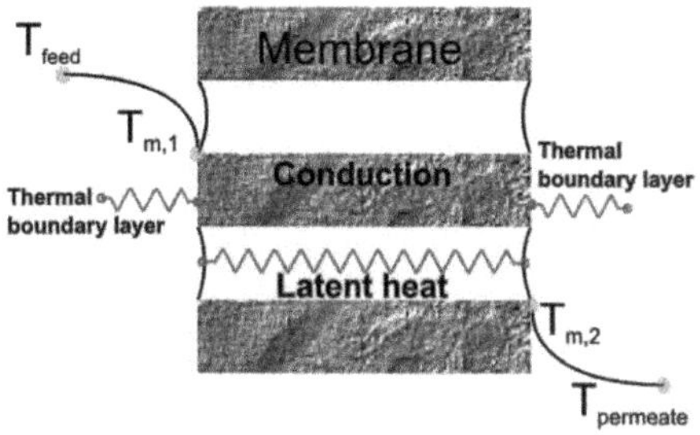

Figure 11.4 Scheme describing thermal conduction through porous membranes.

- The convective heat transfer (Q) from the feed (heated upstream side) across the boundary layer to the membrane surface

$$Q = h_f(T_f - T_1) \tag{11.4}$$

where h_f is the heat transfer coefficient in the feed boundary layer, T_f is the temperature of the bulk feed stream, and T_1 is the temperature at the feed membrane surface.

- The heat transfer across the membrane by conduction and heat assembling vapor flow through the membrane pores

$$Q = N\Delta H_v + \frac{k_m}{l_m}(T_1 - T_2) \tag{11.5}$$

where N is the permeation flux, ΔH_v is the heat of vaporization, k_m is the thermal conductivity of the membrane, and T_2 is the temperature at the permeate membrane surface.

- The convective heat transfer from the membrane surface of the permeate side across the boundary layer to the bulk of the permeate (downstream side)

$$Q = h_p(T_2 - T_p) \tag{11.6}$$

where h_p is the heat transfer coefficient in the permeate boundary layer and T_p is the temperature of the bulk permeate stream.

The heat transfer coefficients in the feed and permeate boundary layers can be expressed, respectively, as

$$h_f = \frac{Nu_f k_f}{d_h}, \quad h_p = \frac{Nu_p k_p}{d_h} \tag{11.7}$$

where the values of T_1 and T_2 can be calculated under steady state according to the following equations (11.8), respectively:

$$T_1 = \frac{h_\mathrm{m}(T_\mathrm{p} + (h_\mathrm{f}/h_\mathrm{p})T_\mathrm{f}) + h_\mathrm{f}T_\mathrm{f} - N\Delta H_\mathrm{v}}{h_\mathrm{m} + h_\mathrm{f}(1 + h_\mathrm{m}/h_\mathrm{p})}$$

$$T_2 = \frac{h_\mathrm{m}(T_\mathrm{f} + (h_\mathrm{p}/h_\mathrm{f})T_\mathrm{p}) + h_\mathrm{p}T_\mathrm{p} - N\Delta H_\mathrm{v}}{h_\mathrm{m} + h_\mathrm{p}(1 + h_\mathrm{m}/h_\mathrm{f})} \quad (11.8)$$

A frequent shortcoming is the temperature polarization (τ) due to heat absorption from aqueous solution and/or sweat. This takes place when the temperature near the membrane is lower than the main flow temperature and can be expressed as

$$\tau = \frac{T_1 - T_2}{T_\mathrm{f} - T_\mathrm{p}} \quad (11.9)$$

The membrane works well when the temperature polarization approaches unity, whereas it takes its suitability away when the value is close to zero.

It is clear that both the moisture and heat have to be transferred through the membranes in order to re-establish microclimate conditions under adverse external or internal climatic conditions. So a future challenge is to get membranes with internal thermoregulation function.

Another important concept is the membrane's ability to work as a barrier against undesired physical, chemical, and biological agents while working as a microclimate regulator.

Although resistance to chemical and biological agents represents a key function for the textile world, a description of the ability to contrast liquid penetration into the layers is given here. The expression "waterproofness" means an attempt to prevent spreading and penetration of liquids such as water. When using porous membranes, it is necessary to choose highly hydrophobic materials for water repulsion. However, this prerogative is not enough to avoid wetting due to the lack of resistance to liquid entry pressure (LEP) (Houtman and Orpana, 2000). The latter indicates the capacity of a membrane to resist intrusion of liquid inside the pore under a determined pressure. The mathematical expression, as well known as Laplace (Cantor) equation, that well quantifies this feature is the following:

$$p_{\mathrm{b}} = \frac{-2\gamma\cos\theta}{r} \tag{11.10}$$

where the breakthrough pressure (p_b) is defined as the pressure exceeding the pressure of the liquid phase and is expressed as a function of the surface free tension (γ) and the contact angle (θ) between the membrane and the liquid. This pressure is inversely proportional to the maximum pore radius (r). It is clear that chemical and structural features of the membrane play a key role in establishing the related waterproof character of the film. Smaller pores increase the resistance to liquid penetration as well as low surface free tensions prevent liquid spreading.

Indeed, the degree of wetting is also strongly dependent on the surface free tensions at the different interfaces according to this relation:

$$W = \gamma_{\mathrm{SV}} - (\gamma_{\mathrm{SL}} + \gamma_{\mathrm{LV}}) \tag{11.11}$$

where γ_{SV} is the surface free tension between the solid and vapor phases, γ_{SL} is the surface free tension between the solid and liquid phases, and γ_{LV} is the surface free tension between the liquid and vapor phases. Positive values of wetting ($S>0$) mean high hydrophilicity, while negative values ($S<0$) are typical of hydrophobic surface. While discussing the wetting phenomena, it is appropriate to dwell upon the role of the surface roughness factor in the wetting resistance (Gugliuzza, 2015).

In this regard, Wenzel and Cassie–Baxter's models (Quéré, 2002; Quéré and Aussillous, 2002) well describe how grooves and irregularities of the surface can contrast the spread of liquid and make the film highly waterproof. Specifically, Wenzel incorporated the factor R in the Young equation, considering the surface irregularities filled with liquid:

$$\cos(\theta_{\mathrm{w}}) = R\left[\frac{\gamma_{\mathrm{SV}} - \gamma_{\mathrm{SL}}}{\gamma_{\mathrm{LV}}}\right] \tag{11.12}$$

On the other hand, Cassie and Baxter considered pillar surfaces where the grooves are completely filled with air so that the contact angle measured on the flat surface (θ_{a}) and the fractions of the solid and vapor on the surface (f_{s} and f_{v}) are included in the equation:

$$\cos(\theta_{\mathrm{CB}}) = f_{\mathrm{s}}\cos(\theta_{\mathrm{a}}) + f_{\mathrm{v}}\cos(\theta_{\mathrm{v}}) \tag{11.13}$$

The increase in surface roughness can bring about roll-to-roll effects, which are regarded as the self-cleaning ability of the membrane surface. This ability can be further amplified if light-sensing functions are inserted in the membrane surface (Nicoletta et al., 2012).

11.4 Membrane Actuators for Smart Textiles

Smart membranes designed from adaptive materials can open new horizons toward the development of new-generation interactive textiles, wherein "*the actuation follows the sensing*." In membranes, the actuation is envisaged as a capability to oppose an action to an induction through changes in their own structure and, hence, of specific properties (Akle et al., 2006). This takes place generally under external solicitations, including reversible polymer segment chain relaxations, variation in temperature, pH, light, or specific molecular interaction (Fig. 11.5).

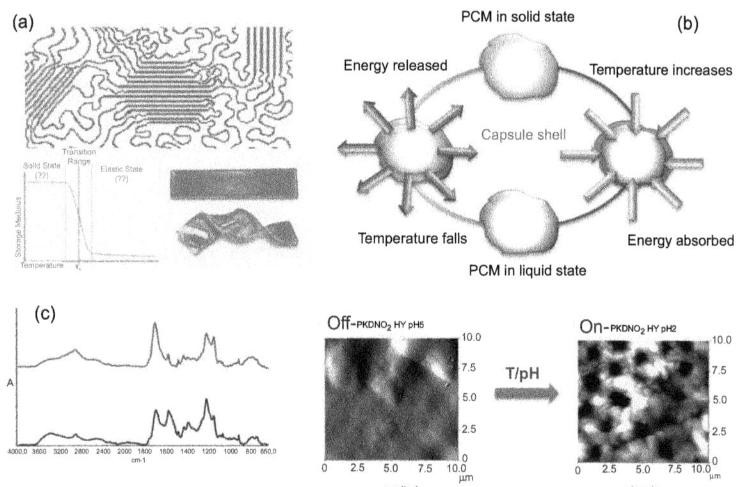

Figure 11.5 Examples of adaptive materials for smart textiles solutions: (a) segregated structure of a memory-shape polyurethane; (b) capsules containing phase change materials working as thermo-regulators; (c) pH and T-responsive polyelectrolyte for on/off pores in membranes of modified NO_2-polyetherketone ($PKDNO_2HY$).

Among the various categories of smart materials, polyurethanes are particularly interesting for adaptive membranes to be used in microclimate regulation (Hager et al., 2015; Han et al., 2012). Indeed, a temporary deformation of the soft micro-phase (Fig. 11.5a), which controls the elastic behavior of the polymer, can open additional free pathways through which water vapor molecules can diffuse, whereas a return to the memorized permanent shape takes place via the hard phase (Biju and Nair, 2013; Gök et al., 2015; Meng and Hu, 2009). This transition can cause an increase in breathability from 700 to 1600 g/m^2day within the range of 30–50°C (Xie and Rousseau, 2009). A current concern is the present low availability of memory-shape polymers, which exhibit a transition at temperatures lower than 50°C, whereas textiles applications require sensing within the range of at least 10–40°C.

Dynamic phase change materials (PCMs) form another interesting class of adaptive systems due to their ability to store and release heat through a change in phase (Socaciu, 2012). Moving from solid to liquid phase and back, these materials absorb or release thermal energy at constant temperature (Arjun and Hayavadana, 2014). During melting, heat is pulled out of the surrounding environment, while the surplus of energy is stored. As a result, a cooling effect is sensed. On the contrary, heat is obtained when the stored energy is released (Fig. 11.5b).

There are different classes of materials regarded as smart. They have ability to response to physical and chemical triggers, including temperature, light, electrical and magnetical fields, mechanical stress, pH, solvents, and molecular interactions (Hu et al., 2012). The mechanisms for adjusting membrane properties normally include conformational changes in the molecular structure, local rearrangement of polymer segment chains, textures, polarity, or breaking and formation of chemical bonds (Fig. 11.5c). Generally, membranes embedding these materials can bring about significant changes in mass or charge transfer together with variations of surface features through amplification events.

Concerning surface wetting, light-responsive inorganic and organic materials are frequently proposed to amplify wettability through reversible hydrophilic–hydrophobic transitions (Khranovskyy et al., 2012; Nicoletta et al., 2012).

Also, thermal-responsive hydrogels such as poly(N-isopropylacrylamide) (PNIPAM) allow switching from superoleophobic and low adhesive state, below the critical solution temperature (LCST, 32°C), to oleophobic and high adhesive state, above its LCST (Chen et al., 2010; Fu et al., 2004). Superoleophilicity to superoleophobicity and oleophilic to hydrophobic switching is, instead, proposed by using pH-responsive polyelectrolytes such as poly(2-vinylpyridine) and polydimethylsiloxane (P2VP-*b*-PDMS), respectively (Zhang et al., 2012). Another interesting solution to change local surface properties consists of directed block copolymer self-assembly (Albert and Epps, 2010; Shen et al., 2015; Wu et al., 2016), whereas liquid droplets can be manipulated when coming in contact with charged surfaces through electro-wetting processes (Cooney et al., 2006; Cui et al., 2016; You and Steckl, 2014).

Independent of the class of materials used, the challenge is to cause abrupt changes in membranes to manipulate intrinsic properties on different length scales.

11.5 Voltage-Activated Membrane Sensors: A Case Study

Comfortable and livable indoor and outdoor solutions are in great demand in many fields, including health care (Xu and Xia, 2012), cultural heritage conservation (Xia et al., 2012), greenhouse climate control (Weng and Yang, 2006), military (Park et al., 199), aerospace (Laxminarayana and Jalili, 2005), and so on. Long-distance interaction and real-time inspection of the physiological environment are two other challenging aspects that require designing of devices with adjustable functions (Gugliuzza, 2014).

This section gives interesting insights into the development of a new membrane sensor working as a humidity detector and a moisture regulator (Gugliuzza et al., 2016; Pingitore et al., 2015). The approach used is based on self-assembled concepts, which enable one to accommodate different and complementary functions in ordered and well-controlled textures.

Specifically, the hybrid composite membrane is formed by a honeycomb platform on which a better quality electrically conductive pathway is deposited (Fig. 11.6). The manufacturing approach requires two essential steps: the build-up of a well-defined polymeric architecture by means of imprinting water droplets; and the construction of interconnected carbon nanotube (CNT) networks by using layer-by-layer (LBL) assembly (Pingitore and Gugliuzza, 2013).

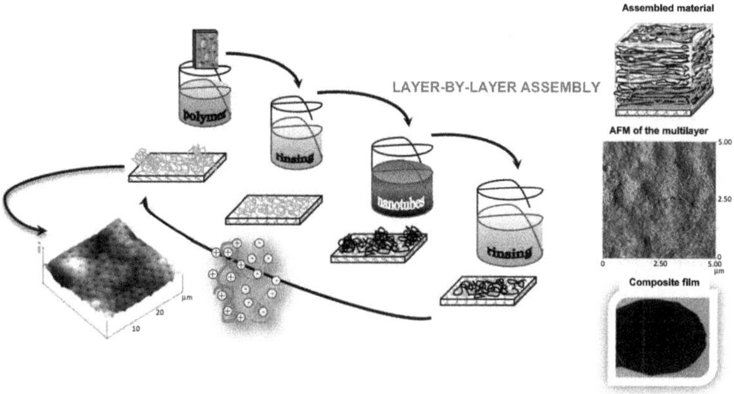

Figure 11.6 Scheme of manufacturing LBL self-assembly approach used to build membrane microclimate regulators.

Different functional single-walled CNTs (SWCNTs) have been selected and integrated into the LBL multilayer by using two polyelectrolytes having opposite ionic charges, i.e., polyacrylic acid (PAA) and poly(diallyldimethylammonium chloride) (PDDA). Due to diverse degree of interconnection between the conductive filaments, a different passage of current can be measured through the electrical pathways (Fig. 11.7).

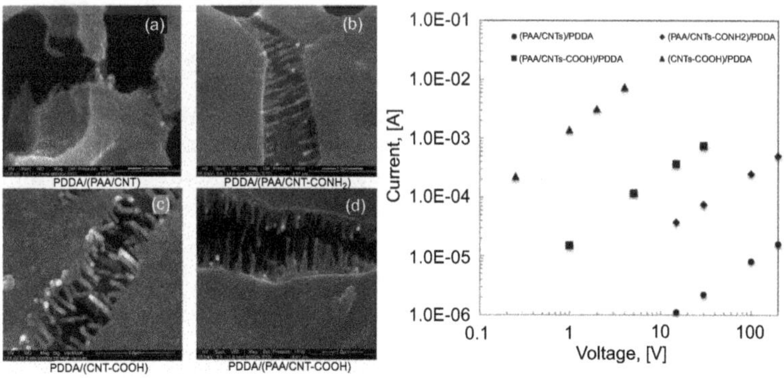

Figure 11.7 SEM micrographs of CNT pathways (a–d) and electrical current versus voltage measured through conductive membranes (e). Reproduced from Gugliuzza et al. (2015) with permission from the Royal Society of Chemistry.

It is relevant to observe how the electrical conduction increases with the increasing ability of the filaments to form hydrogen bonds, while the insulating power of PAA is decreased. Thus, the number of interconnections can be regarded as the result of an intermolecular competition during the build-up of the multilayer (Gugliuzza et al., 2015).

The electrical response of the CNT pathway bringing the PDDA/CNTs–COOH pair, tested under changing local humidity in the surrounding atmosphere (Fig. 11.8a), gives rise to a step-wise behavior with a variation of 40% of the relative conductivity as the relative humidity comes from 38% to 95% at 22°C, thus suggesting water doping mechanisms as also confirmed by Raman spectroscopy (Fig. 11.9). A continuous hydrogen bond is plausibly formed at full saturation. At the same time, an increase in breathability is also appreciated under no favorable environment conditions (T = 40°C, RH = 80%) (Fig. 11.8b), giving a clear indication of an assisted moisture diffusion through membranes. Hopping mechanisms are suggested by swelling events, which can be appreciated by infrared analyses (Fig. 11.9) (Gugliuzza et al., 2015).

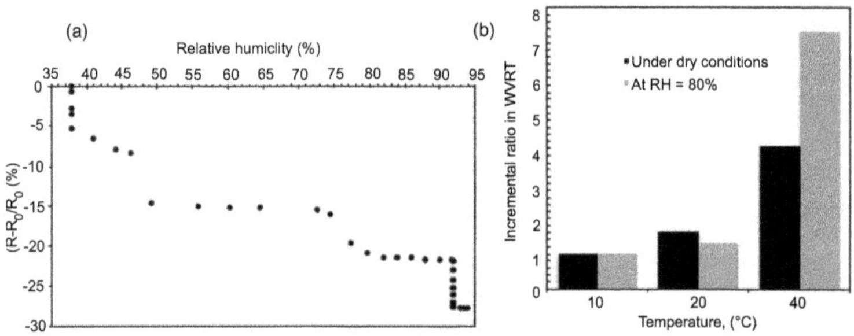

Figure 11.8 Variation of the electrical conductivity with rising relative humidity (a) and WVTR (b) through membranes containing carboxylate–carbon nanotube (CNTs–COOH) networks. Reproduced from Pingitore et al. (2015) with permission from the Royal Society of Chemistry.

To summarize, carboxylate CNTs coupled with PDDA in an LBL film respond to hole and electron doping during water adsorption from the surrounding atmosphere, causing changes in the electrical response. Hence, the action of the sensor is fully carried out. Simultaneously, hydrophilic interactions are well established with water, thus causing free paths through which fast moisture exchange takes place according to hopping mechanisms. In this case, the membrane works as a good moisture exchanger, indicating the ability of this hybrid composite membrane to work as a smart microarray where complementary functions are accommodated and synchronized to yield actuation.

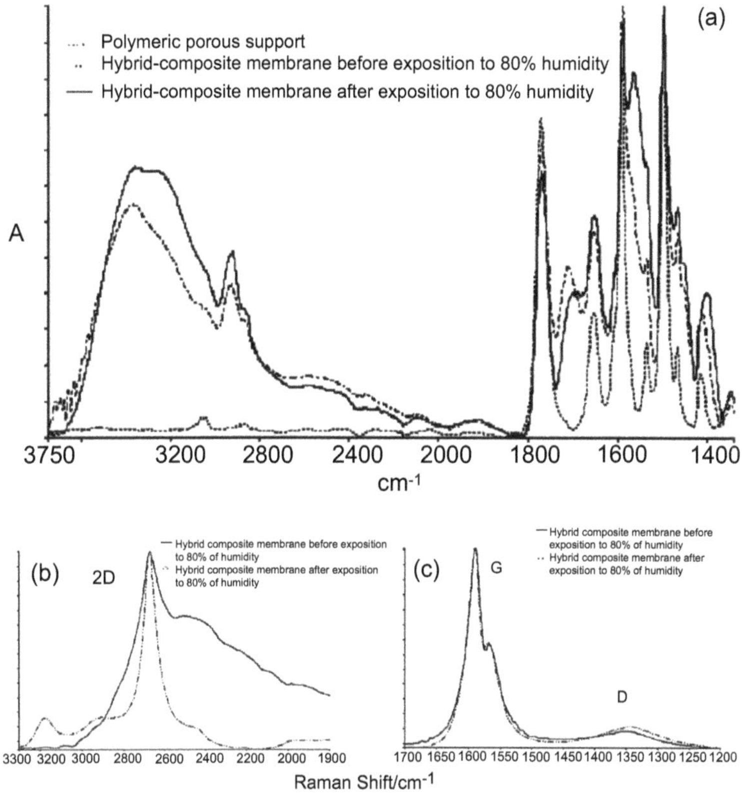

Figure 11.9 Infrared analysis (a) and Raman spectra (b) describing the behavior of membranes containing CNTs–COOH before and after exposition to RH = 80%. Reproduced from Pingitore et al. (2015) with permission from the Royal Society of Chemistry.

11.6 Future Prospects and Concluding Remarks

Nanostructured membranes are expected to give great propulsion to textiles markets, considering the attractive opportunity to create multifunction devices where different functions can be integrated. Breathability, barrier, self-cleaning as well as information/signal transmission can be associated in unique systems, providing for marketplaces, which include healthcare, telemedicine, automobiles, protective clothing for marine, military, and space, the possibility to get devices where a continuum between molecular, biomimetic, and digital world can be strongly established. In this regard, adaptability and auto-adjustment are two imperatives for achieving criteria such as self-maintenance and long-distance interaction with the body and/or the surrounding environment. However, there are still some concerns such as long-term stability, durability, and recyclability. All these targets could be achieved through the build-up of an "integrated industry-driven research approach," enabling one to generate new knowledge and competitiveness and definitely bridging the gap between research and production.

Appendix

A	effective membrane area
D_0	diffusion coefficient close to zero
f_s	fraction of the solid on the surface
f_v	fraction of the vapor on the surface
h_f	heat transfer coefficient in the feed boundary layer
h_p	heat transfer coefficient in the permeate boundary layer
l_m	membrane thickness
P	permeability
p_h	upstream (high pressure)
p_l	downstream pressure (low pressure)
p_b	breakthrough pressure
Q	convective heat transfer
R	rugosity factor
S	wetting
T_f	temperature of the bulk feed stream
T_1	temperature at the feed membrane surface
T_2	temperature at the permeate membrane surface
T_p	temperature of the bulk permeate stream
t	length of time
W_f	final weight of a cup for breathability measurement

W_0	initial weight of a cup for breathability measurement
WVTR	water vapor transmission rate
J	flux
k_m	thermal conductivity of the membrane

Greek symbols

β	a constant
γ	surface free tension
ΔHv	heat of vaporization
θ	the contact angle
τ	temperature polarization

Abbreviations

AFM	atomic force microscopy
CNTs-COOH	carboxylate carbon nanotubes
LCST	critical solution temperature
LBL	layer-by-layer
PAA	polyacrylic acid
PCMs	phase change materials
P2VP-*b*-PDMS	poly(2-vinylpyridine) and polydimethylsiloxane
PDDA	poly(diallyldimethylammonium chloride)
$PKDNO_2HY$	modified NO_2-polyetherketone
PNIPAM	poly(*N*-isopropylacrylamide)
SEM	scanning electronic microscopy
SWCNTs	single walled carbon nanotubes

References

Akle, B. J., Bennett, M. D., and Leo, D. J. (2006). High-strain ionomeric-ionic liquid electro-active actuators, *Sens. Actuators A*, **126**, pp. 173–181.

Albert, J. N. L. and Epps III, T. H. (2010). Self-assembly of block copolymer thin films, *Mater. Today*, **13**, pp. 24–33.

Arjun, D. and Hayavadana, J. (2014). Thermal energy storage materials (PCMs) for textile applications, *JTATM*, **8**, pp. 1–11.

Baker, R. W. (2012). *Membrane Technology and Applications*, 3rd edn., John Wiley & Sons, Chichester.

Biju, R. and Nair, C. P. R. (2013). High transition temperature shape memory polymer composites based on bismaleimide resin, *High Perform. Polym.*, **25**, pp. 464–474.

Bitter, J. G. A. (1991). *Transport Mechanisms in Membrane Separation Processes*, The Plenum Chemical Engineering Series, Springer US, Plenum Press, NY.

Bottino, A., Capannelli, G., Petit-bon, P., Cao, N., Pegoraro, M., and Zoia, G. (1991). Pore size and pore-size distribution in microfiltration membranes, *Sep. Sci. Technol.*, **26**, pp. 10–11.

Brewer, S. A. (2011). Recent advances in breathable barrier membranes for individual protective equipment, *Recent Pat. Mater. Sci.*, **4**, pp. 1–14.

Chen, L., Liu, M., Lin, L., Zhang, T., Ma, J., Song, Y., and Jiang, L. (2010). Thermal-responsive hydrogel surface: Tunable wettability and adhesion to oil at the water/solid interface, *Soft Matter*, **6**, pp. 2708–2712.

Cooney, C. G., Chen, C.-Y., Emerling, M. R., Nadim, A., and Sterling, J. D. (2006). Electrowetting droplet microfluidics on a single planar surface, *Microfluid Nanofluid*, **2**, pp. 435–446.

Cui, W., Ma, H., Tian, B., Ji, Y., and Su, F. (2016). Electrowetting on a multi-walled carbon nanotube membrane with different droplet sizes in an electric field, *J. Mater. Sci.*, **51**, pp. 4031–4036.

De Rossi, D., Carpi, F., Lorussi, F., Mazzoldi, A., Paradiso, R., Scilingo, E. P., and Tognetti, A. (2003). Electroactive fabrics and wearable biomonitoring devices, *AUTEX Res. J.*, **3**, pp. 180–185.

Fu, Q., Rao, G. V. R., Basame, S. B., Keller, D. J., Artyushkova, K., Fulghum, J. E., and Lopez, G. P. (2004). Reversible control of free energy and topography of nanostructured surfaces, *J. Am. Chem. Soc.*, **126**, pp. 8904–8905.

Gibson, P.W., Rivin, D., Kendrick, C. (2000). Convection/diffusion test method for porous textiles, *Int. J. Clothing Sci. Technol.*, 12, pp. 90–113.

Gök, M. O., Bilir, M. Z., and Gürcüm, B. H. (2015). Shape-memory applications in textile design, *Procedia Social Behavioral Sci.*, **195**, pp. 2160–2169.

Gugliuzza, A. (2014). *Smart Membranes and Sensors: Synthesis, Characterization and Applications*, Scrivener Publishing and John Wiley & Sons, Inc., Salem, Massachusetts, USA.

Gugliuzza, A. and Drioli, E. (2014). Membranes for ultra-smart textiles, in: *Smart Membranes and Sensors: Synthesis, Characterization and Applications*, Gugliuzza A. (Ed.), Scrivener Publishing and John Wiley & Sons, Inc., Salem, Massachusetts, USA.

Gugliuzza, A. (2015). Membrane wettability, in: *Encyclopedia of Membranes*, Drioli, E. and Giorno, L. (Eds.), Springer-Verlarg, Berlin.

Gugliuzza, A. and Drioli, E. (2013). A review on membrane engineering for innovation in wearable fabrics and protective textiles, *J. Membr. Sci.*, **446**, pp. 350–375.

Gugliuzza, A., De Luca, G., Tocci, E., De Lorenzo, L., and Drioli, E. (2007). Intermolecular interactions as controlling factor for water sorption into polymer membranes, *J. Phys. Chem. B*, **111**, pp. 8868–8878.

Gugliuzza, A., Pingitore, V., and Drioli, E. (2016). Relationships between structure and electrical sensitivity of breathable membranes, *Mater. Today Proc.*, **3**, pp. 308–312.

Gugliuzza, A., Pingitore, V., Miriello, D., and Drioli, E. (2015). Functional carbon nanotubes for high-quality charge transfer and moisture regulation through membranes: Structural and functional insights, *Phys. Chem. Chem. Phys.*, **17**, pp. 12919–12926.

Hager, M. D., Bode, S., Weber, C., and Schubert, U. S. (2015). Shape memory polymers: Past, present and future developments, *Progr. Polym. Sci.*, **49–50**, pp. 3–33.

Han, H. R., Chung, S. E., and Park, C. H. (2012). Shape memory and breathable waterproof properties of polyurethane nanowebs, *Text. Res. J.*, **83**, pp. 76–82.

Houtman, R. and Orpana, M. (2000). Materials for membrane structures, *Bauen Mit Textilien Heft*, **4**, pp. 1–7.

Hu, J., Meng, H., Li, G., and Ibekwe, S. I. (2012). A review of stimuli-responsive polymers for smart textile applications, *Smart Mater. Struct.*, **21**, pp. 1–23.

Hu, J., Meng, H., Li, G., and Ibekwe, S. I. (2012). A review of stimuli-responsive polymers for smart textile applications, *Smart Mater. Struct.*, **21**, pp. 1–23.

Karpagam, K. R., Saranya, K. S., Gopinathan, J., and Bhattacharyya, A. (2016). Development of smart clothing for military applications using thermochromic colorants, *J. Textile Institute*, **108**, pp. 1122–1127.

Khranovskyy, V., Ekblad, T., Yakimova, R., and Hultman, L. (2012). Surface morphology effects on the light-controlled wettability of ZnO nanostructures, *Appl. Surf. Sci.*, **258**, pp. 8146–8152.

Kristy, J., Genevieve, D., and Yury, G. (2014). Textile energy storage in perspective, *J. Mater. Chem. A*, **2**, pp. 10776–10787.

Kwon, A., Kato, M., Kawamura, H., Yanai, Y., and Tokura, H. (1998). Physiological significance of hydrophilic and hydrophobic textile materials during intermittent exercise in humans under the influence of warm ambient temperature with and without wind, *Eur. J. Appl. Physiol. Occup. Physiol.*, **78**, pp. 487–493.

Laxminarayana, K. and Jalili, N. (2005). Functional nanotube-based textiles: Pathway to next generation fabrics with enhanced sensing capabilities, *Text. Res. J.*, **75**, pp. 670–680.

Levin, J. and Maibach, H. (2005). The correlation between transepidermal water loss and percutaneous absorption: An overview, *J. Controlled Release*, **103**, pp. 291–299.

Macedonio, F., Brunetti, A., Barbieri, G., and Drioli, E. (2013). Membrane condenser as a new technology for water recovery from humidified "waste" gaseous streams, *Ind. Eng. Chem. Res.*, **52**, pp. 1160–1167.

Massaroni, C., Saccomandi, P., and Schena, E. (2015). Medical smart textiles based on fiber optic technology: An overview, *J. Funct. Biomater.*, **6**, pp. 204–221.

McCullough, E. A., Kwon, M., and Shim, H. (2003). A comparison of standard methods for measuring water vapor permeability of fabrics, *Meas. Sci. Technol.*, **14**, pp. 1402–1408.

Meng, Q. and Hu, J. (2009). A review of shape memory polymer composites and blends, *Composites Part A*, **40**, pp. 1661–1672.

Narayanan, S. C., Karpagam, K. R., and Bhattacharyya, A. (2015). Nanocomposite coatings on cotton and silk fibers for enhanced electrical conductivity, *Fibers Polymers*, **16**, pp. 1269–1275.

Nicoletta, F. P., Cupelli, D., Formoso, P., De Filpo, G., Colella, V., and Gugliuzza, A. (2012). Light responsive polymer membranes: A review, *Membranes*, **2**, pp. 134–197.

Park, S., Gopalsamy, C., Rajamanickam, R., and Jayaraman, S. (1999). The wearable motherboard: An information infrastructure or sensate liner for medical applications, *Stud. Health Technol. Inform.*, **62**, pp. 252–258.

Paul, R. (2014). *Functional Finishes for Textiles*, 1st Edn, Woodhead Publishing, Cambridge CN22 3HJ, UK.

Pingitore, V. and Gugliuzza, A. (2013). Fabrication of porous semiconductor interfaces by pH-driven assembly of carbon nanotubes on honeycomb structured membranes, *J. Phys. Chem. C*, **117**, pp. 26562–26572.

Pingitore, V., Miriello, D., Drioli, E., and Gugliuzza, A. (2015). Integrated carboxylic carbon nanotube pathways with membranes for voltage-activated humidity detection and microclimate regulation, *Soft Matter*, **11**, pp. 4461–4468.

Quéré, D. (2002). Rough ideas on wetting, *Physica A*, **313**, pp. 32–46.

Quéré, D. and Aussillous, P. (2002). Non-stick droplets, *Chem. Eng. Technol.*, **25**, pp. 925–928.

Qui, Z. M., Clark, J. C., and Fan, W. W. (2004). Water and oil repellency imparting urethane oligomers comprising perfluoroalkyl moieties, US 6803109.

Schubert, M. B. and Werner, J. H. (2006). Flexible solar cells for clothing, *Mater. Today*, **9**, pp. 42–50.

Shen, L., Zhu, J., and Liang, H. (2015). Heterogeneous patterns on block copolymer thin film via solvent annealing: Effect on protein adsorption, *J. Chem. Phys.*, **142**, pp. 101908-1–101908-7.

Socaciu, L. G. (2012). Thermal energy storage with phase change material, *Leonardo El. J. Pract. Technol.*, **20**, pp. 75–98.

Speranza, V., Trotta, F., Drioli, E., and Gugliuzza, A. (2010). High-definition polymeric membranes: Construction of 3D lithographed channel arrays through controlling natural building blocks dynamics, *ACS Appl. Mater. Interf.*, **2**, pp. 459–466.

Tang, K. P., Fan, J. T., Zhang, J. F., Sarkar, M. K., and Kan, C. W. (2013). Effect of softeners and crosslinking conditions on the performance of easy-care cotton fabrics with different weave constructions, *Fibers Polym.*, **14**, pp. 822–831.

Weng, Q. and Yang, S. (2006). Urban air pollution patterns, land use, and thermal landscape: An examination of the linkage using GIS, *Environ. Monit. Assess.*, **117**, pp. 463–489.

Wu, H. S. (1994). Gas permeable coated porous membranes, US 5286279.

Wu, M.-L., Wang, D., and Wan, L.-J. (2016). Directed block copolymer self-assembly implemented via surface-embedded electrets, *Nat. Commun.*, **7**, 10752.

Xia, M., Dong, Y., Xu, W., Lu, D., Xue, P., and Liu, G. (2012). Long-term microclimate monitoring in wildland cultural heritage sites with wireless sensor networks, *Int. J. High Perform. Comput.*, **7**, pp. 111–122.

Xie, T. and Rousseau, I. A. (2009). Facile tailoring of thermal transition temperatures of epoxy shape memory polymers, *Polymer*, **50**, pp. 1852–1856.

Xu, Y. L. and Xia, Y. (2012). *Structural Health Monitoring of Long-Span Suspension Bridges*, CRC Press, Taylor and Francis Group, Boca Raton, FL, USA.

Yetisen, A. K., Qu, H., Manbachi, A., Butt, H., Dokmeci, M. R., Hinestroza, J. P., Skorobogatiy, M., Khademhosseini, A., and Yun, S. H. (2016). Nanotechnology in textiles, *ACS Nano*, **10**, pp. 3042–3068.

You, H. and Steckl, A. J. (2014). Electrowetting on non-fluorinated hydrophobic surfaces, *J. SID*, **21**(10), pp. 411–416.

Zhang, L., Zhang, Z., and Wang, P. (2012). Smart surfaces with switchable superoleophilicity and superoleophobicity in aqueous media: Toward controllable oil/water separation, *NPG Asia Mater.*, **4**, pp. 1–8.

Zysset, C., Nasseri, N., Büthe, L., Münzenrieder, N., Kinkeldei, T., Petti, L., Kleiser, S., Salvatore, G. A., Wolf, M., and Tröster, G. (2013). Textile integrated sensors and actuators for near-infrared spectroscopy, *Opt. Express*, **21**, pp. 3213–3224.

Chapter 12

Nanostructured Porous Membranes for Membrane Distillation and Membrane Crystallization

Francesca Macedonio,[a,b] **Aamer Ali,**[b] **and Enrico Drioli**[a,b,c,d]
[a]*Institute on Membrane Technology of the National Research Council of Italy (ITM-CNR), Via P. Bucci, Cubo 17C, 87036 Rende (CS), Italy*
[b]*Department of Environmental and Chemical Engineering, University of Calabria, Via P. Bucci, Cubo 17C, 87036 Rende (CS), Italy*
[c]*WCU Energy Engineering Department, Hanyang University, 222 Wangsimni-ro, Sageun-dong, Seoul, South Korea*
[d]*Center of Excellence in Desalination Technology, King Abdulaziz University, Jeddah 21589, Saudi Arabia*
f.macedonio@itm.cnr.it, francesca.macedonio@unical.it

Membrane distillation (MD) and its extension membrane crystallization (MCr) are relatively new thermal membrane processes that have garnered significant interest as a potential low-cost, high-efficiency, and energy-saving alternative to conventional separation processes. The driving force of both these processes is the partial pressure difference across the membrane. Their main advantages are the theoretically complete rejection of nonvolatile solutes, the relatively low operating pressure/temperature and membrane-

Functional Nanostructured Membranes
Edited by Enrico Drioli, Lidietta Giorno, and Annarosa Gugliuzza

ISBN 978-981-4774-79-6 (Hardcover), 978-1-351-13511-5 (eBook)
www.panstanford.com

fouling problem. In the case of MCr, other benefits include the well-controlled nucleation and growth kinetics, the fast crystallization rates and reduced induction time, membrane surface promoting heterogeneous nucleation, and control of supersaturation level and rate. Despite the great potential of MD and MCr, they are still far from fulfilling all the expectations. To overcome the existing barriers, new and better membranes are necessary for further industrial exploitation of these technologies.

Starting with the description of the basic principles of MD and MCr technologies, this chapter summarizes the membranes to be used in these operations together with the important and interesting recent developments in membrane and module fabrication. Future research directions of interest are also pointed out.

12.1 Introduction

Membranes and membrane technologies play a key role in all separation processes due to their capability to work as physical and selective support for the extraction of chemical species from one phase to the other. In the past few decades, numerous research papers have been reported on new membrane materials, aiming at improved separation characteristics (high permeability, high selectivity, high physical/chemical/thermal stability, etc.). Moreover, well-controlled structures combined with process engineering can optimize efficiency and productivity of the separation. The scientific efforts made membrane operations successful substitutes for conventional separation processes. Now membrane processes are extending their application in a wide range of industrial processes and emerging membrane processes, such as membrane distillation (MD) and membrane crystallization (MCr), which are more and more used, e.g., in semiconductors, air and water purification, pharmaceuticals, energy generation, biotechnology (Membrana GmbH, 2006; Li and Sirkar, 2005; Charcosset, 2006; Brunel et al., 2006; Drioli et al., 2006). These techniques belong to the class of membrane contactors in which a non-wetting membrane does not act as a conventional barrier or filter, but promotes mass and energy exchange between two opposite interfaces according to principles of phase equilibrium.

This chapter gives a literature review of promising hydrophobic membranes for MD and MCr applications.

12.2 Membrane Contactors

One of the most crucial aspects of membrane technology is to have disposable membranes with well-controlled properties. Moreover, the final performance of a process is the direct consequence of the structural and physicochemical parameters of the utilized membranes. This aspect gains relevance when the proposed membrane technology is based on advanced systems such as contactor devices, where the use of well-structured and functionalized membranes becomes an imperative.

A lot of membrane operations are based on membrane contactor devices, whose performance is intrinsically affected by the structure of the film in terms of thickness, porosity, mean pore size, pore distribution, and geometry. Thus, the successful outcome of the process reasonably depends on the ability of the membrane to interface two media without dispersing one phase into another and to combine high volumetric mass transfer with high resistance to liquid intrusion in the pores.

Membrane extraction, pertraction, perstraction, gas adsorption, membrane-based solvent extraction, liquid–liquid extraction, membrane-based gas absorption and stripping, MD, and MCr are generally referred to as membrane contactors (MCs). Starting from the description of the basic principles of MD and MCr technologies, the membranes to be used in these operations, together with the important and interesting recent developments in membrane and module fabrication, will be summarized. Future research directions of interest will also be pointed out.

12.2.1 Membrane Distillation Technology

Membrane distillation is an emerging non-isothermal separation technique that uses microporous hydrophobic membrane in contact with a heated aqueous solution on the one hand (feed or retentate) and a condensing phase (permeate or distillate) on the other. The driving force of the process is supplied by a partial pressure difference between both sides of the membrane, caused by temperature gradient imposed between the liquid–vapor interfaces. MD has potential applications in many areas of scientific and industrial interests, yielding highly purified permeate and separating contaminants from liquid solutions. It has been tested in the treatment of thermally sensitive industrial products such as concentrating aqueous solution in fruit juices (Quist-Jensen et al., 2016) and also applied in pharmaceutical industry (El-Bourawi et al., 2006), wastewater treatment (Calabrò et al., 1991), and water desalination (Alklaibi and Lior, 2005; Chouikh et al., 2005). In particular, the MD process can be used as a substitute for conventional desalination processes such as multi-stage flash (MSF), reverse osmosis (RO), and multiple effect distillation (MED). The advantages of MD compared to these processes are as follows:

- Lower operating temperatures and vapor space required than MSF and MED (with consequent reduction in capital and energy costs and allowing efficient recycle of low-grade waste heat streams as well as the use of solar/wind/geothermal energy)
- Lower operating pressure than RO (thus allowing the use of plastic equipment, lowering equipment costs, and increasing process safety)
- 100% (theoretical) rejection of nonvolatile solute
- Performance not limited by high osmotic pressure or concentration polarization (thus preferentially employed whenever elevated permeate recovery factors or high retentate concentrations are requested)

To date, the slow progress of MD has been related with the unavailability of appropriate membranes for this application, high energy consumption, and low flux with respect to RO, membrane wetting, and limited investigations carried out on module designing. For more than 20 years, the available hardware for MD has been microporous hydrophobic membranes (usually made in polypropylene (PP), polyvinylidenefluoride (PVDF), or polytetrafluoroethylene (PTFE)) normally designed for pressure-driven filtration processes rather than for concentration/temperature-driven mass transfer. However, thanks to the recent availability of better membranes and to the growing extensive research activities carried out in various areas of MD, a lot of interest in commercialization has been observed.

12.2.2 Membrane Crystallization Technology

Membrane crystallization is an innovative concept that combines membrane technology and crystallization. It can promote nucleation of crystals and growth in a well-controlled pathway, starting from undersaturated solutions.

As previously anticipated, MCr can be considered an extension of MD. Therefore, as in MD, the membrane does not act as a sieving barrier in MCr for the selective transport of specific components, but as a physical support to promote evaporative mass transfer of volatile solvents through the porous membrane. Solvent evaporation concentrates feed solutions above their saturation limit, thus generating a controlled supersaturated environment in which crystals can nucleate and grow.

Particularly in MCr, a solution containing a nonvolatile solute, which is likely to be crystallized, is in direct contact with the membrane surface (feed or retentate side). On the opposite side, the membrane is in contact with the distillate side. The gradient of vapor pressure between the two subsystems induces evaporation of the volatile components from the feed solution, migration through the porous membrane, and, finally, re-condensation at the distillate side. The continuous removal of solvent from the feed solutions increases solute concentration, thus generating supersaturation. Therefore, the transmembrane flow rate can be controlled by changing the driving force of the process. Moreover, depending on the fluid-dynamic regime and due to the contact between the crystallizing solution and the membrane surface, a solute–membrane interaction is present. The latter is due to (i) the topographical heterogeneities of the membrane surface where solute molecules are physically entrapped, leading locally to enhanced levels of supersaturation and to (ii) the nonspecific and reversible chemical interaction between the membrane and the solute, which can allow concentration and reorientation of molecules on the surface, without loss of mobility, thus facilitating effective interaction among them, proper for crystallization (Drioli et al., 2012). In this way, the membrane surface might operate as a physical substrate for heterogeneous nucleation, by inducing a reduction in the free energy barrier (Gugliuzza and Drioli, 2009). These characteristics make MCr a powerful technology that can speed up crystallization kinetics and modulate the final properties of the produced crystals in terms of both structure (polymorphism) and morphology (habit, shape, size, and size distribution), for both small organic molecules of interest, for example in pharmaceutics, and large bio-macromolecules, such as proteins.

12.3 Membrane for MD and MCr Technology

When producing microporous membranes for MD and MCr operations, the selection of the material is mainly driven by the necessity to achieve high mechanical/chemical/thermal stability, high hydrophobicity, high porosity, and uniform pore size distribution. Since the hydrophobic character of the membrane is an essential requirement of these processes, membranes have to be prepared from polymers with a low value of surface energy.

As previously anticipated, in most of the MD/MCr experiments, membranes typically fabricated from PTFE, PP, or PVDF have been used, which possess high porosity (70–80%), a membrane thickness of 10–300 μm, and provide microfiltration properties with pore sizes of 0.2–1 μm (Table 12.1).

Microporous polymeric membranes are prepared by various techniques: sintering, stretching, and phase inversion.

Sintering is a simple technique for the production of microporous structures having a porosity in the range of 10–40% and rather irregular pore sizes, ranging from 0.2 to 20 μm.

Stretching allows production of microporous membranes with a relatively uniform porous structure, pore size distribution in the range of 0.2–20 mm, and porosity of about 90%.

MD membranes can be prepared (and often done) by the phase inversion technique from polymers that are soluble at a certain temperature in an appropriate solvent or solvent mixture and can be precipitated as a continuous phase by changing temperature and/or composition of the system. These changes aim to create a miscibility gap in the system at a given temperature and composition; from a thermodynamic point of view, the free energy of mixing of the system becomes positive.

Table 12.1 Common commercial membranes for use in membrane distillation.

Membrane type	Trade name	Manufacturer	Material	Pore size (μm)	Porosity
Flat sheet	TF200	Gelman	PTFE	0.20	80
	TF450		supported	0.45	80
	TF1000		by PP	1.0	80
	3MA	3M	PP	0.29	66
	3MB	Corporation		0.40	76
	3MC			0.51	79
	3MD			0.58	80
	3ME			0.73	85
	FGLP	Millipore	PTFE	0.20	70
	FHLP		supported by PE	0.50	80
	GVHP		PVDF	0.22	75
	HVHP			0.45	75
Capillary	Accurel S6/2	AkzoNobel Microdyn	PP	0.20	70
	Accurel BFMF	Enka AG		0.20	

Source: Khayet (2011); Khayet et al. (2005).

The formation of two different phases, that is, a solid phase forming the polymeric structure (symmetric, with porosity almost uniform across the membrane cross section, or asymmetric, with a selective thin skin on a sub-layer) and a liquid phase generating the pores of the membrane, is determined by a few and conceptually simple actions:

- By changing the temperature of the system (cooling of a homogeneous polymer solution, which separates in two phases): temperature-induced phase separation (TIPS)
- By adding non-solvent or non-solvent mixture to a homogeneous solution: diffusion-induced phase separation (DIPS)

In Section 12.4, the main factors affecting the membrane morphology when they are prepared via the phase inversion technique will be presented.

El-Bourawi et al. (2006) pointed that the commercial membranes commonly used in MD are marketed for microfiltration or ultrafiltration purposes, but have been used in MD because of their hydrophobicity and adequate pore sizes. On the other hand, many researches have been performed on new applications, but only a few have ever tried to design and synthesize membranes for membrane contactor processes. Table 12.2 covers most of the studies in which materials other than the typical hydrophobic polymers mentioned above were used for the membrane contactor processes.

In general, two types of porous membranes have been used: (i) homogeneous hydrophobic and (ii) composite hydrophobic/hydrophilic membranes. In both cases, the membranes must be able to separate the contacting fluids. Because hydrophobic membranes possess an inherent potential to be wetted by feed solute precipitation or adsorption onto the membrane surface, a reliable separation of the two phases cannot be guaranteed at long-term operations. For these applications, composite membranes should be preferred (Albrecht et al., 2005). However, an additional membrane resistance will be caused by the composite layer reducing the membrane permeability. Therefore, the thickness of composite layers should be very small in order to keep this resistance to an acceptable limit.

As shown in Table 12.2, in a considerable number of studies, composite hydrophobic/hydrophilic membranes with different hydrophobic surface layer preparation methods have been considered. The composite hydrophobic/hydrophilic membranes had a very thin hydrophobic layer responsible for the mass transfer and a thick hydrophilic layer. This means that those membranes have very short path length between the liquid/vapor interfaces formed at both sides of the hydrophobic membrane pores and better mass transport characteristics. Among the composite hydrophobic/hydrophilic membranes listed in Table 12.2, a number of membranes have been developed and characterized by synthesizing fluorine-containing surface-modifying macromolecules (SMMs) and blending them into different base polymers for surface modification, based on the concept of "surface segregation" in polymer science (Suk et al., 2010). Since the first generation of SMMs was synthesized by Pham et al. (1999) and Ho et al. (2000), a series of SMM researches have been performed to make the SMMs meet the original objective, that is, to prepare membranes with greater surface hydrophobicity and greater

Table 12.2 Studies to develop new membrane materials to be used for membrane contactor

Author(s) (published year)	Materials used in membrane preparation		Surface layer preparation methods
	Base material(s)	Surface-modifying material(s)	
Jansen et al. (2006a)	/	Polyetheretherketone (PEEKWC)	No surface modification (membrane prepared by wet or dry–wet phase inversion)
Simone et al. (2010)	/	Polyvinylidene fluoride (PVDF)(+) [a]	No surface modification (membrane prepared by dry–wet phase inversion)
Jansen et al. (2006b)	Polyetheretherketone (PEEKWC); a cristalline partially oxidized acrylonitrile-vinylacetate	Hyflon AD60x	Coating
Gigliuzza and Drioli (2009)	Polyvinylidene fluoride (PVDF) (+)	Hyflon AD60x	Coating
Gugliuzza et al. (2008)	/	Polyetheretherketone (PEEKWC)	No surface modification (membrane prepared by self-assembly)
Kong et al. (1992)	Cellulose nitrate (−)	Octafluorocyclobutane (OFCB) (+)	Plasma polymerization
Wu et al. (1992)	Cellulose acetate (−)	Vinyltrimethylsilicon (VTMS)/ carbon tetrafluoride (CF4) (+); octafluorocyclobutane (OFCB) (+)	Radiation grafting polymerization

(*Continued*)

Table 12.2 (*Continued*)

Author(s) (published year)	Materials used in membrane preparation		Surface layer preparation methods
	Base material(s)	Surface-modifying material(s)	
Larbot et al. (2002)	Aluminum and zirconium ceramics (−)	Fluoroalkylsilane (FAS) (+)	Grafting
Feng et al. (2004a,b)	/	Poly(vinylidene fluoride-co-hexafluoro propylene) (+)	No surface modification
Khayet et al. (2005)	Polyetherimide (PEI) (−)	Fluoro hydrocarbone endcapped polyurethane (PU: MDIDPS) (+)	Surface segregation
Peng et al. (2005)	Polyvinylidene fluoride (PVDF) (+)	Polyvinyl alcohol (PVA) blended with polyethylene glycol (PEG) (−)	Crosslinking
Krajewski et al. (2006)	Zirconia layer on the microporous alumina support (−)	1H,1H,2H,2H-perfluoro decyltriethoxysilane (C8 compound) (+)	Grafting
Suk et al. (2006)	Polyethersulfone (PES) (−)	Fluoro hydrocarbone endcapped polyurethane-urea (PU: MDI-PDMS) (+)	Surface segregation
Bonyadi and Chung (2007)	Polyacrylonitrile (PAN) with hydrophilic NA+clay particles (−)	Polyvinylidene fluoride (PVDF) with hydrophobic 15A clay particles(+)	Co-extrusion

Author(s) (published year)	Materials used in membrane preparation		Surface layer preparation methods
	Base material(s)	Surface-modifying material(s)	
Song et al. (2007)	Polypropylene (+)	Silicone fluoropolymer (+)	Plasma polymerization
Jin et al. (2008)	Poly(phthalazinone ether sulfone ketone) (PPESK) (−)	Silicone rubber(+); polytrifluoropropyl-siloxane(+)	Coating
Hendren et al. (2009)	Alumina ceramic (−)	1H,1H,2H,2H-perfluoro decyltriethoxysilane (PFS) (+); trichloromethylsilane (TCS) (+); trimethylchlorosilane (TMS) (+)	Grafting
Mourgues et al. (2010)	Stainless steel ASI 316 L (−)	Polydimethylsiloxane (PDMS) (+)	Impregnation and crosslinking
Qtaishat et al. (2009)	Polyvinylidene fluoride (PVDF) (+)	Chitosan (−)	Crosslinking/dip coating

[a]Surface characteristic (hydrophobicity) of materials: (−) hydrophilic material, (+) hydrophobic material.
Source: Developed from Suk et al. (2010).

chemical resistivity at lower cost than the conventional hydrophobic membranes (Suk et al., 2006; Khayet et al., 2003). As well, studies on those hydrophobic SMM-blended membranes covered various membrane applications such as UF, PV (pervaporation), and MD.

For what concerns the commercially available membranes, microfiltration membranes provide high permeability but unfortunately in practice they are unsuitable for coating with thin dense layers (Iverson et al., 1997). Therefore, in principle, only membranes with ultrafiltration separation characteristics can be considered a coatable alternative. However, according to literature data, the surface porosity of common membranes with ultrafiltration profile prepared from hydrophobic polymers is usually lower than the porosity of membranes with microfiltration profile.

Only a few studies (in general applying membranes with microfiltration separation profile) describe the influence of the membrane on the contactor performance (Alves and Coelhoso, 2004; Al-Obaidani et al., 2008). In membrane-based contactors, the membrane is a stagnant gas- or liquid-filled layer between both contacting media. The mass transport in this layer is controlled by an only diffusive transport governed by the partial pressure difference across the membrane. The overall mass transfer of a contactor can be significantly affected by the membrane properties. Therefore, the selection of a suitable membrane, eventually coated with a thin composite layer in order to provide a high water vapor flux, is a crucial factor for optimal contactor performances.

12.4 Membrane Preparation via Phase Inversion: Factors Affecting Membrane Morphology

Polymers that can be dissolved in common organic solvents have a great advantage from a membrane engineering point of view because they can be relatively conveniently formed into membranes by phase inversion processes. These include non-solvent-induced phase separation (NIPS) and TIPS. Generally, phase inversion processes provide a means for a good degree of control over the membrane morphology.

In utilizing phase inversion processes, the factors that have a major effect upon the membrane structure are as follows:

- Choice of polymer and polymer concentration
- Solvent–non-solvent combination selected
- Composition of the casting solution
- Composition of the coagulation bath
- Gelation and crystallization behavior of the polymer
- Location of the liquid–liquid demixing gap
- Temperatures of the casting solution and the coagulation bath
- Evaporation time (before immersion in the coagulation bath)

These factors can have a profound effect on the structure and morphology of the resulting membranes and can be varied to achieve the desired qualities for the intended application.

Some of these parameters are discussed in the following sections:

12.4.1 Choice of the Polymer

Choice of the polymer is an important factor because it limits the solvents and non-solvents that can be used in the phase inversion process. The choice of membrane material is also important with respect to fouling (adsorption effects, hydrophobic/hydrophilic character) and to thermal and chemical stability.

Additionally, the concentration of the polymer in the casting solution influences the ultimate membrane properties. Using high polymer concentrations in the casting solution produces a relatively dense and thick outer layer that may not have the required porosity for high mass transfer.

12.4.2 Choice of Solvent–Non-solvent

One of the main variables in the phase inversion membrane preparation process is the choice of the solvent/non-solvent system. As known from the literature, two distinctive forms of liquid–liquid demixing can occur during membrane formation: (i) instantaneous liquid–liquid demixing and (ii) delayed onset of liquid–liquid demixing, the difference between the two being characterized by the instant at which the liquid–liquid demixing occurs. The two types of demixing processes lead to two different membrane structures: (i) Instantaneous liquid–liquid demixing produces a membrane with a relatively porous outer layer, suitable for MF or UF; (ii) a delayed liquid–liquid demixing process, on the other hand, leads a relatively dense outer layer, producing a membrane that is generally suitable for gas separation or pervaporation.

In both cases, the thickness of the top layer is dependent on all kinds of membrane formation parameters. Where a high mutual affinity exists between solvent and non-solvent, a porous membrane is obtained. On the contrary, in the case of low mutual affinity, a nonporous membrane (or better an asymmetric membrane with a dense nonporous top layer) is obtained.

Although other parameters exist that have an influence on the type of membrane structure, the choice of solvent/non-solvent is crucial. Fixing this parameter still leaves a number of degrees of freedom in the system (such as polymer concentration, addition of solvent to the non-solvent bath, addition of non-solvent to the polymer solution, temperature of the coagulation bath, temperature of the polymer solution, addition of additives to the casting solution or to the coagulation bath).

12.4.3 Composition of Coagulation Bath

The addition of solvent to the coagulation bath is another parameter that strongly influences the structure of the formed membrane. In general, the addition of solvent to the coagulation bath results in a delayed onset of liquid–liquid demixing and to a decrease in the polymer concentration at the film interface. Therefore, increasing solvent content in the coagulation bath leads to two opposite effects:

1. Delayed demixing, which tends to reduce the porosity of the membranes, giving them a relatively thick and dense outer layer
2. Lower polymer concentration at the interface, which tends to cause a more porous outer layer

Moreover, the addition of solvent to the coagulation bath tends to decrease the tendency toward macrovoid formation in the porous top layer.

12.4.4 Temperature of the Coagulation Bath

For further optimization of the membrane performance, the kinetics and thermodynamics of the phase inversion process can be varied by the external conditions. In particular, the coagulation temperature acts upon the physical state of the cast film and it is expected that a lower temperature might cause a decrease in the rate of solvent/non-solvent exchange, causing a delay in polymer precipitation and thus diminish macrovoid formation.

12.5 Desired Membrane Characteristics

It is generally established and well understood that the type of membranes that can be used for MD/MCr needs to have certain characteristics. Above all, the membranes need to be porous and hydrophobic in order to serve their intended function. The membranes should preferably have low thermal conductivity to minimize conduction losses and achieve better thermal efficiency. Additionally, from the practical point of view, the membranes should have good thermal and chemical stability, as well as sufficient mechanical strength. The desired membrane characteristics are elaborated as follows:

- *High liquid entry pressure.* It is the maximum pressure differential that the membrane can withstand before liquid enters its pores and the membrane becomes "wetted." Liquid entry pressure (LEP) is an important membrane characteristic for MD, since it effectively determines how robust the hydrophobic nature of the membranes will be against any possible pressure surges during their service.

  The Laplace (Cantor) equation allows estimating the LEP. It provides a relationship between various membrane properties and the LEP, as follows:

$$\mathrm{LEP} = \frac{B\gamma_{\mathrm{L}}\cos\theta}{d_{\mathrm{max}}} \tag{12.1}$$

  where B is a geometric factor determined by pore structure, γ_{L} is the liquid surface tension, and θ is the liquid/solid contact angle.

When the pressure difference between the feed and permeate sides of the membrane exceeds the LEP, liquid penetrates the pores and it is able to pass through the membrane, thus compromising the separation process. If a membrane becomes wetted during operation, it needs to be dried and cleaned in order to restore its capacity for resuming service.

As Eq. (12.1) suggests, high LEPs require high hydrophobicity and a small maximum pore size. The latter, of course, needs to be balanced with the requirement of high permeability, which requires that the mean pore size is not too low.

- *High permeability*. Transmembrane flux increases with the mean pore size and the porosity of the membranes, and decreases with membrane thickness and pore tortuosity. These factors are summarized in the formula:

$$N \propto \frac{\langle r^{\alpha} \rangle \bullet \varepsilon}{\tau \bullet \delta} \tag{12.2}$$

where $\boldsymbol{\varepsilon}$ is the membrane porosity, $\boldsymbol{\tau}$ is the membrane tortuosity, $\boldsymbol{\delta}$ is the membrane thickness, $\langle r^a \rangle$ is the average pore size; r raised to an exponent α, which depends on the mechanism of transport (for Knudsen diffusion $\alpha = 1$, and for viscous flux $\alpha = 2$).

Equation (12.2) illustrates the importance (in terms of molar flux) of maximizing the membrane porosity and pore size, while minimizing the transport path length through the membrane, $(\tau \cdot \delta)$.

- *Low thermal conductivity*. Heat conduction through the membrane constitutes a loss that lowers the thermal efficiency of the process. Polymeric membrane materials with higher thermal conductivity offer a lower thermal resistance; therefore, the conduction heat transferred through the membrane increases, which in turn reduces the amount of vaporization heat and, ultimately, both flux and thermal efficiency. In principle, the membrane surface should be made of material with small thermal conductivity. However, for common hydrophobic polymers used in MD/MCr (such as PP, PVDF, and PTFE), it is above 0.2 W/mK. Some materials exhibit thermal conductivities in the range of 0.13–0.18 W/mK (cellulose acetate and polyvinylchloride, polystyrene) and even less (polyurethane) (Cheremisinoff, 1997). In order to improve the performance of MD membranes, a possibility comes from microporous hydrophobic/hydrophilic composite membrane, with a top hydrophobic thin layer responsible for the mass transport, and a hydrophilic sub-layer able to provide mechanical support and to reduce the conductive heat loss through the whole membrane matrix. Interesting studies considered: cellulose acetate or cellulose nitrate membranes modified via radiation graft polymerization of vinyltrimethylsilicon/carbon tetrafluoride and octafluoro-cyclobutane (Wu et al., 1992), or surface modification of hydrophilic membranes by adding fluorinated SMMs (Khayet et al., 2005).

- *Low propensity for fouling.* Fouling is generally among the bigger, if not the biggest, challenges of membrane processes. MD is luckily much less prone to fouling than pressure-driven membrane processes. This probably results from a combination of favorable factors, including the use of hydrophobic membranes, which means that the foulant-containing liquid feed does not (normally) enter the membrane pores. In addition, the absence of an applied transmembrane pressure differential, which often contributes to the fouling problem in other processes through a compaction effect, is another favorable factor contributing toward the high fouling resistance of MD. Nevertheless, fouling is still possible in MD; in particular, the precipitation and deposition of crystals resulting from operation at high feed concentrations can cause progressive plugging of membrane pores and an associated decline in flux. In addition, biological and particulate fouling are also possible. The various possible forms of fouling and pore blocking, if they occur, can also lead to wetting of the affected pores and the permeate quality may deteriorate as a result (Khayet, 2011). Thus, some form of pretreatment is usually necessary.
- *High chemical stability.* Membrane contactors, in general, need to be made of materials with high chemical stability in order to maintain good long-term operation. However, in MD the membrane acts merely as a support for a vapor–liquid interface; therefore, they can be fabricated from almost any

chemically resistant polymers with hydrophobic intrinsic properties.

- *High thermal stability.* Under high temperatures, the membrane material may not be able to resist to degradation or decomposition. Continuous stable operation under elevated temperatures requires good thermal stability of the membrane material. Change in the nature of membrane depends on the glass transition temperature T_g for morphous polymers or the melting point T_m for crystalline polymers. Over these temperatures, the properties of the polymers change dramatically. Table 12.3 reports the T_g for the polymers commonly used in membrane contactors.

Table 12.3 Glass transition temperature T_g of polymers

Polymer	T_g (°C)
Polyethylene (PE)	−120
Polypropylene (PP)	−15
Polytetrafluoroethylene (PTFE)	126
Polysulfone	190
Polyether sulfone	230
Polyimide (Kapton)	300

Source: Drioli et al. (2015).

12.6 Promising Membrane Materials

The first step to fabricate a high performance membrane is to choose a correct membrane material. Among the hydrophobic materials applied in the fabrication of membranes for membrane contactors, fluoropolymers constitute a unique class of materials with a combination of interesting properties that attracted significant attention from material researchers over the past few decades (Cui et al., 2014). Generally, these polymers have high thermal stability, improved chemical resistance, and lower surface tension because of the low polarizability and the strong electronegativity of the fluorine atom, its small van der Waals radius (1.32Å), and the strong C–F bond (485 kJ/mol). The important fluoropolymers for membrane operations are poly(vinylidene fluoride) (PVDF), poly(tetrafluoroethylene) (PTFE), poly(ethylene chlorotrifluoroethylene) (ECTFE), poly(chlorotrifluoroethylene) (PCTFE), polyvinyl fluoride (PVF), poly(fluorenyl ether), Hyflon® AD, Teflon® AF, and Cytop®. In the following sections, PVDF and Hyflon will be described in detail due to their unique properties for MD/MCr applications.

12.6.1 PVDF

PVDF is a hydrophobic polymer with excellent chemical resistance that can be dissolved in common organic solvents. As a result, PVDF membranes can be formed through phase inversion processes.

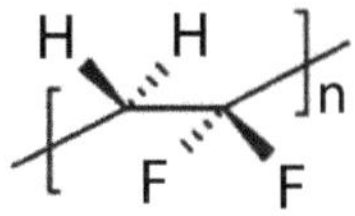

Figure 12.1 Chemical structure of PVDF.

There are mainly five mechanical crystalline forms of PVDF—namely, α, β, γ, δ, and ε—depending on the chain conformations as trans (T) or gauche (G) linkages. β and γ forms are polar. The β form is particularly important because it makes the greatest contribution to the piezoelectric/ferroelectric property of PVDF (Lovinger, 1983), while the γ form shows a lower piezoelectric coefficient but still among the largest ones obtained for polymers (Lopes et al., 2011). However, due to the fact that conformation of the α form (TGT$\overline{\text{G}}$) is more stable than that of the planar zigzag β form (all trans, TTT) (Yu et al., 2009) and the γ form (T3GT3$\overline{\text{G}}$), pure PVDF without any special treatment is usually dominated by the α form.

Sugihara et al. (1979), Bottino et al. (1980), and Oritz de Zfirate et al. (1995) have done extensive work on the preparation and characterization of PVDF flat-sheet ultrafiltration and microfiltration membranes using dry/wet phase inversion processes. Regarding the use of additives, Tomaszewska (1996) appears to be the first to fabricate flat-sheet PVDF membranes using lithium chloride (LiCl) as an additive and applied them for membrane contactor process. The additive LiCl increased the porosity and pore size of the fabricated membranes. This led to some flux enhancement. In a similar study, Khayet and Matsuura (2001) found that the pore size and porosity of PVDF membranes increased with increasing the concentrations of pure water as a non-solvent added into the dope solution. Also they concluded that the flux increases exponentially with the water content in the PVDF casting solution. Other research efforts have been directed toward the development of hollow-fiber PVDF membranes, rather than flat sheets. The hollow-fiber configuration has the advantage of compactness, since it generally allows dramatically higher ratios of membrane surface area to volume. Wang et al. (1999) reported the preparation and characterization of PVDF hollow fibers for ultrafiltration. They investigated the effects of different spinning parameters such as polymer concentration, air-gap, polyvinyl pyrrolidone (PVP) content and its molecular weight in the polymer dope, and the internal coagulant on the membrane structure and its permeation properties. Khayet et al. (2002) investigated the effect of ethylene glycol (EG) non-solvent in the PVDF dope solution on the membrane morphology and its ultrafiltration performance. Later, Wang et al. (2008) fabricated similar hollow fibers and tested them in the MD process. They observed a great increase in flux for the fibers fabricated using an EG non-solvent in the dope solution compared to the fibers fabricated out of the neat PVDF polymer solutions. Similar studies have been carried out on the fabrication and characterization of PVDF hollow fibers for other membrane contactor applications such as CO_2 and H_2S capture as well as removal of volatile organic compounds (VOCs) (Ren et al., 2006; Wang et al., 2002; Zhen et al., 2006).

Although good progress has been made in the fabrication of PVDF hollow fibers through the NIPS process recently, almost all of the fibers prepared in the literature suffer from the low surface porosity of the fibers at the outer layer. This is mainly due to the fact that water is commonly and preferably applied as the external coagulant due to economic and environmental considerations. Water is a strong non-solvent that induces a rapid phase inversion in the nascent fiber outer surface and consequently leads to the formation of a dense skin with a low surface porosity at the outer layer of the fibers. This makes it rather subtle to fabricate PVDF fibers with high outer surface porosity through the NIPS process. In contrast, it is quite straightforward to tailor the morphology of the inner surface by adopting a suitable composition of a bore fluid such as a mixture of solvent and non-solvent.

As reported earlier, extensive studies have been carried out to improve the surface porosity of PVDF membranes by using pore-forming additives in the casting solutions. The use of pore formers such as PVP, calcium carbonate ($CaCO_3$), LiCl, or tri-hydrated lithium perchlorate $LiClO_4 \cdot 3H_2O$ or EG allowed PVDF membranes with different morphologies to be manufactured. However, the films exhibited often macrovoids or cavities, affecting the mechanical stability of the membranes and skin with more or less open-end pores of the order of a few nanometers, yielding significant changes in the effective porosity or in the resistance to the LEP of water (LEP_w).

Another particular aspect to be considered is related to the effects of the pore-forming additive on the final membrane surface properties. Although most parts of pore formers are soluble in water, their removal from the polymer matrix is not easy. The surface free energy of the membrane is consequently affected, and the PVDF matrix loses its intrinsic hydrophobic character as a function of the enrichment in the hydrophilic domains dispersed onto the membrane surface. The use of hydrophilic pore former represents, therefore, a critical factor, especially if highly hydrophobic membrane surfaces are required (Gugliuzza and Drioli, 2007).

Bonyadi and Chung (2009) reported the fabrication of highly porous and macrovoid-free PVDF hollow-fiber membranes, with high outer surface porosity, by applying a two-phase flow consisting of a solvent and a dope solution in the air-gap region of spinning through NIPS. He et al. (2003) applied for the first time this approach to fabricate porous polysulfone fibers by using *N*-methylpyrrolidone (NMP) as the external liquid at the outside orifice of the spinneret. They observed considerable improvements in pure water and gas permeability of the fibers spun through this approach.

In Bonyadi and Chung's (2009) work, the dope solution and the *N*-methylpyrrolidone (NMP) solvent were co-discharged from the middle and outer channels of a triple orifice spinneret, respectively. Then, the two-phase flow went through an air-gap region and finally entered the coagulation bath. It was observed that the introduction of the two-phase flow greatly increased the outer surface porosity of the PVDF fibers and eliminated the formation of macrovoids in the cross section of the fibers as well. It was also found that the energy efficiency and the flux of the fibers spun through the solvent–dope solution co-extrusion were two to three times higher than the standard dry jet wet-spun fibers.

In order to improve the mechanical stability of the PVDF membranes, another possibility is to use the PVDF–Hyflon AD superhydrophobic membranes (Gugliuzza and Drioli, 2009). The excellent mechanical resistance of the fluorinated polymer Hyflon AD combined with its highly hydrophobic character makes this material a good candidate for the fabrication of PVDF–Hyflon composite membranes. These membranes can be tailored by using no cost-effective techniques such as dry–wet phase inversion and wet chemical treatment.

12.6.2 Hyflon

Amorphous glassy perfluoropolymers (PFPs) such as Teflon AF®, Hyflon AD® (Fig. 12.2), or Cytop® are known for their good film-forming properties and their high thermal and chemical stability. The latter is due to the very stable carbon–fluorine bond (485 kJ/mol) in comparison with the carbon–carbon bond (360 kJ/mol) (Jansen et al., 2011).

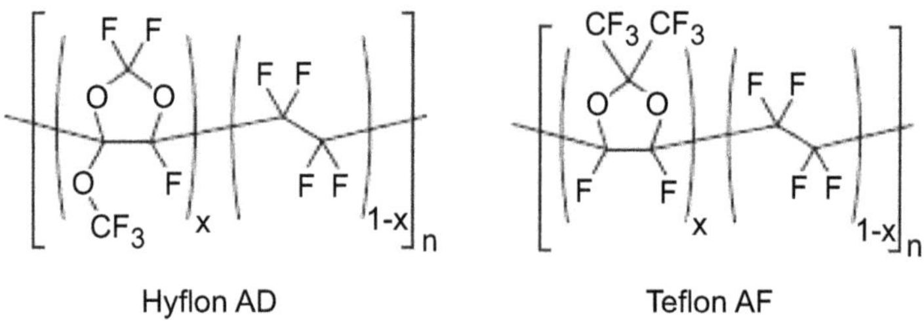

Figure 12.2 Molecular structure of Hyflon AD and Teflon glassy perfluoropolymers. Reprinted from Jansen et al. (2006b) with permission from Elsevier.

The bulky substituted dioxole monomers in the structure of Teflon AF and Hyflon AD reduce the chain mobility and obstruct potential crystal formation, and in combination with the low cohesive energy of the fully fluorinated molecules, this results in completely amorphous polymers with poor packing and a high fractional free volume (FFV). The perfluorinated structure also gives these polymers a low tendency toward swelling, insolubility in common organic solvents, and a strong hydrophobic character. All such properties make them interesting for application as membranes in "wet" gas treatment, where conventional polymers might suffer from plasticization by condensable species or from chemical attack in corrosive environment. In spite of the low swelling, these polymers are nevertheless remarkably permeable to some organic vapors because of the high interconnected free volume (FV); therefore, they have a certain potential for specific organic–organic separations (Chae et al., 2009).

A problem of perfluorinated polymers is their high price. Therefore, it is desirable to develop composite membranes with a thin layer of the perfluorinated polymer on a suitable porous support.

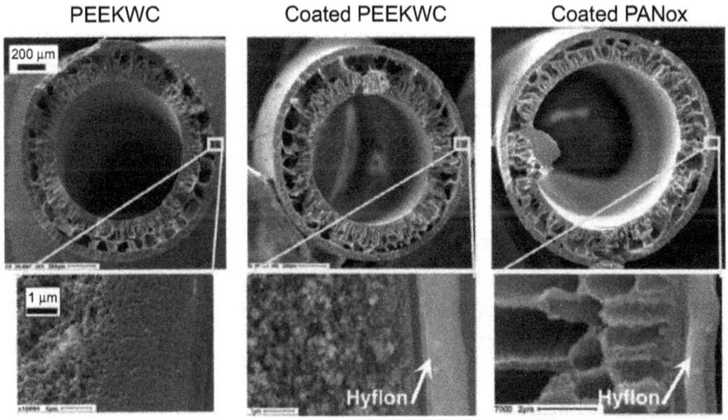

Figure 12.3 SEM images of a PEEKWC hollow-fiber ultrafiltration membrane before and after external coating with Hyflon AD60x (left and middle), and a PANox membrane after coating with Hyflon (right). The close-ups (bottom) clearly show the deposited Hyflon film on the external skin of the membranes. Reprinted from Jansen et al. (2006b) with permission from Elsevier.

An example can be found in the work of Jansen et al. (2006b), where composite membranes were prepared by dip coating of perfluorinated copolymer Hyfon® AD60x on a support of hollow-fiber ultrafiltration membrane (Fig. 12.3). Two types of polymers were used for the support: a glassy amorphous poly(ether ether ketone) and a crystalline partially oxidized acrylonitrile-vinylacetate copolymer. Hollow-fiber membranes were prepared by the dry–wet spinning method. The fibers perform well in pure gas permeation measurements, maintaining the typical permselectivity and high permeance of dense Hyflon. Jansen et al. (2011) found that application of a similar coating at the bore side of the hollow fiber results in a thicker film with low selectivity and is not suitable, possibly due to shrinkage and rupture of the film in the drying stage. Moreover, a careful selection of supporting fibers with proper characteristics (surface porosity, pore size distribution, surface roughness) and optimization of the coating procedure for reducing the skin thickness are desirable in order to maximize the gas permeance of the membranes and their stability under severe operating conditions.

Superhydrophobic fluorinated membranes were tailored by combining traditional dry–wet phase inversion and wet chemical treatment techniques by Zhen et al. (2006). In this work, PVDF and Hyflon AD60x were selected as raw polymers for their chemical and mechanical resistance as well as hydrophobic and solvophobic properties. In particular, PVDF membranes were coated with Hyflon AD.

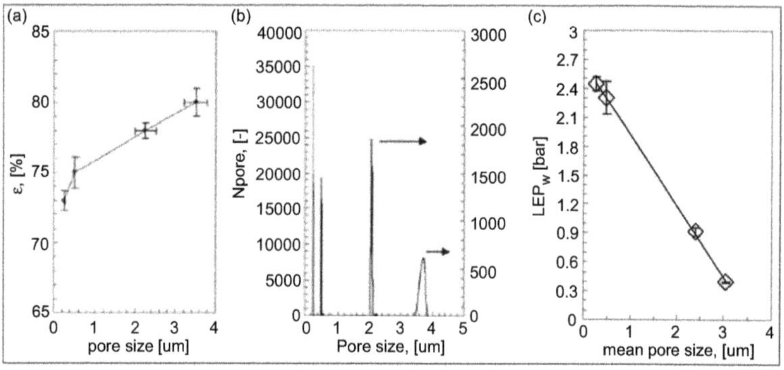

Figure 12.4 (a) Overall porosity versus mean pore size; (b) pore distribution versus mean pore size; (c) resistance to LEP_w. Reprinted from Zhen et al. (2006) with permission from John Wiley and Sons.

Membranes with high overall porosity (Fig. 12.4a), modulated pore size, and narrow pore distribution (Fig. 12.4b) were manufactured without using additional additives or modifiers. High resistance to liquid water entry pressure (LEP_w) (Fig. 12.4c), high mass transfer, and low surface free energy of the membrane surfaces were achieved. The combination of the two fluorinated polymers improved dramatically the mechanical resistance and the water repellence of the PVDF membranes. In fact, contact angle values ranging from 126±3° to 141±4° confirmed the superhydrophobic character of the membranes. The obtained well-controlled structure combined with aimed supra-molecular chemistry makes these porous layers ideal interfaces to be processed in membrane contactor devices.

12.7 New Materials for Membranes

Among the technological developments, fullerene, graphene membranes, carbon nanotubes (CNTs), and biomimic membranes are emerging as developed membranes with superior permeability, durability, and selectivity for water purification.

Fullerenes. Various studies found that C_{60} carbon fullerenes hinder the ability of bacteria and other microorganisms to accumulate on membranes. This attribute has made the researchers to believe that coating pipes and membranes with these nanoparticles may prove to be an effective strategy for addressing one of the major problems and costs of treating water.

Chaea et al. (2009) showed that the deposition of fullerene C_{60} on the surface of ceramic alumina microfiltration membranes decreased the membrane surface zeta potential, increased the hydrophobicity of the membrane surface, decreased microbial attachment, and inhibited microbial respiratory activity. Kim et al. (2015) fabricated polyvinylidene fluoride/fullerene nanoparticle (PVDF/C_{60}) composite microfiltration membranes. They found that membranes containing C_{60} have higher surface porosity and pore density than the pristine membrane (without C_{60}). The presence of numerous pores on the membrane caused weaker mechanical strength, but the water flux of the composite membranes increased in spite of their smaller size.

Graphene membrane. In water treatment, graphene might have interesting applications. Using classical molecular dynamics simulations, Cohen-Tanugi and Grossman (2012) reported the desalination performance of nanoporous graphene membranes as a function of pore size, chemical functionalization, and applied pressure. By simulating saltwater flow across a variety of pore diameters and chemical functionalizations, they established that desalination performance is most sensitive to pore size and pore chemistry. The examination of the structure of water in the pore vicinity revealed that the hydrophobic character of hydrogenated pores reduces the water flow by imposing additional conformational order on the system, even as the limited hydrogen bonding allows for greater salt rejection relative to hydroxylated pores. Overall, their results indicate that the water permeability of this material is several orders of magnitude higher than conventional RO membranes and that nanoporous graphene may have a valuable role to play for water purification. Applications of graphene membranes in water treatment can be found in (Han et al., 2013; Musico et al., 2014; Lee et al., 2013). The drawback of graphene is its very high thermal conductivity that needs to be reduced before considering its practical application in MD.

Nanotube membranes. The use of CNTs in water desalination is also emerging in lab-scale investigations. CNTs consist of rolled-up cylinder of graphene with nanoscale dimensions. Their exceptional mechanical strength, chemical resistance, and thermal properties are well known. The high transport of water molecules inside the CNTs together with their potential to change the water–membrane interaction (able to stop the permeation of water molecules while favoring the preferential transport of vapors through the pores) has encouraged their incorporation into the membrane matrix (Gethard et al., 2011; Dumée et al., 2011). Additionally, CNTs-based membranes provide excellent porosity and hydrophobicity, two main requisites for MD membranes. The initial studies demonstrating the potential of CNT membranes for desalination through MD have been provided by Dumée et al. (2010){Formatting Citation}. The application of CNTs-based membranes in MD has caused considerable increase in flux enhancement and salt rejection.

Biomimic membranes. Due to their high permeability and selectivity toward water molecules, biomimic membranes such as aquaporins (AQPs) have a great potential for desalination applications. AQPs have a unique hourglass architecture with a “pore opening” of 2.8Å;

the narrow pore prevents the passage of large molecules. Therefore, under the right conditions, AQP membranes form water channels allowing the passage of only water molecules and exclusion of all ions. The commercial application of these membranes is, however, still far away due to insufficient stability of the membranes, difficulties associated with commercial production, and limited rejection of salts exhibited by the existing membranes (Tang et al., 2013).

New fluoropolymers. Recently some new fluoropolymers have received attention for improving wetting resistance, thermal stability, chemical resistance, and mechanical strength of MD membranes. One of these is ECTFE, mainly provided by Solvay Specialty Chemicals under the trade name HALAR®. It has excellent hydrophobicity, mechanical stability, and solvent and chemical resistance properties. The main issue of ECTFE-based membrane is the necessity to fabricate the membranes at very high temperature. Similarly, Cytop® from Asahi Glass is another interesting example for MD applications. It is highly soluble in fluorinated solvents, and the solution viscosity is low: These characteristics make it an attractive candidate for membrane synthesis.

12.8 Conclusion

Membrane distillation is a relatively new process, investigated worldwide as a low-cost and energy-saving alternative with respect to conventional separation processes (such as distillation and RO). It is one of the few membrane operations based on a thermal process. Therefore, energy consumption is, in principle, the same as the traditional evaporative process. However, the required operating temperature is much lower than that of a conventional distillation column because it is not necessary to heat the process liquids above their boiling temperatures. In fact, the process can be conducted at temperatures typically below 70°C and driven by low temperature difference (20°C) of the hot and cold solutions. Therefore, low-grade waste and/or alternative energy sources such as solar and geothermal energy can be coupled with MD systems for a cost- and energy-efficient liquid separation system.

Membrane crystallization is conceived as an extension of MD, an alternative technology for producing crystals and pure water from supersaturated solutions. MCr can promote crystals nucleation and growth in a well-controlled pathway, starting from undersaturated solutions. In a membrane crystallizer, the membrane matrix acts as a selective gate for solvent evaporation, modulating the final degree and the rate for the generation of supersaturation. Hence, the possibility to act on the transmembrane flow rate, by changing the driving force of the process, allows to modulate the final properties of the crystals produced in terms of both structure (polymorphism) and morphology (habit, shape, size, and size distribution).

Despite the great potential of MD and MCr, they are still far from fulfilling all expectations. To overcome the existing barriers, new and better membranes are necessary for a further industrial exploitation of these technologies. For example, the investigation of the complex correlations between physicochemical properties of the membrane and MD/MCr performance confirms the need for a customized hardware, i.e., high porosity hydrophobic membranes with appropriate thickness and made by low-heat conductive polymers in order to reduce the amount of waste energy. New amorphous perfluoropolymers are becoming available, which allow improvement in membrane characteristics fundamental for the process, such as (i) *the long-term stability of the membrane* (which is important from an economic point of view); (ii) *high LEP* in order to prevent wetting of the membrane pores (high LEP may be achieved using membrane materials with high hydrophobicity); (iii) *process sustainability* (also through the use of "green" solvents to be used in membrane preparation, not dangerous for the environment and health, as tributyl O-acetyl citrate (ATBC) (Cui et al., 2013), PolarClean (Hassankiadeh et al., 2015).

Appendix

Abbreviations

AQP	aquaporin
ATBC	tributyl O-acetyl citrate
CF4	carbon tetrafluoride
CNT	carbon nanotube
DIPS	diffusion-induced phase separation

ECTFE	poly(ethylene chlorotrifluoroethylene)
EG	ethylene glycol
FAS	fluoroalkylsilane
FFV	fractional free volume
FV	free volume
LEP	liquid entry pressure
MC	membrane contactor
MCr	membrane crystallization
MD	membrane distillation
MED	multiple effect distillation
MF	microfiltration
MSF	multi-stage flash
NIPS	non-solvent-induced phase separation
NMP	*N*-methylpyrrolidone
OFCB	octafluorocyclobutane
PCTFE	poly(chlorotrifluoroethylene)
PDMS	polydimethylsiloxane
PE	polyethylene
PEEKWC	polyetheretherketone
PEG	polyethylene glycol
PFP	perfluoropolymer
PFS	1H,1H,2H,2H-perfluoro decyltriethoxysilane
PP	polypropylene
PTFE	polytetrafluoroethylene
PV	pervaporation
PVA	polyvinyl alcohol
PVDF	polyvinylidenefluoride
PVF	polyvinyl fluoride
PVP	polyvinyl pyrrolidone
RO	reverse osmosis
SMM	surface modifying macromolecule

TCS	trichloromethylsilane
TIPS	temperature-induced phase separation
UF	ultrafiltration
VOC	volatile organic compound
VTMS	vinyltrimethylsilicon

Symbols

R	pore size
$<r^{\alpha}>$	average pore size
T_g	glass transition temperature
T_m	melting point

Greek symbols

α	exponent depending on transport mechanism (for Knudsen diffusion α=1, and for viscous flux α=2)
γ_L	liquid surface tension
Δ	membrane thickness
E	membrane porosity
Θ	contact angle
T	membrane tortuosity

References

Albrecht, W., Hilke, R., Kneifel, K., Weigel, Th., and Peinemann, K.-V. (2005). Selection of microporous hydrophobic membranes for use in gas/liquid contactors: An experimental approach, *J. Membr. Sci.*, **263**, 66.

Alklaibi, A. M. and Lior, N. (2005). Membrane-distillation desalination: Status and potential, *Desalination*, **171**, 111.

Al-Obaidani, S., Curcio, E., Macedonio, F., Di Profio, G., Al-Hinai, H., and Drioli, E. (2008). Potential of membrane distillation in seawater desalination: Thermal efficiency, sensitivity study and cost estimation, *J. Membr. Sci.*, **323**, 85.

Alves, V. D. and Coelhoso, I. M. (2004). Effect of membrane characteristics on mass and heat transfer in osmotic evaporation process, *J. Membr. Sci.*, **228**, 159.

Bonyadi, S. and Chung, T. S. (2007). Flux enhancement in membrane distillation by fabrication of dual layer hydrophilic–hydrophobic hollow fiber membranes, *J. Membr. Sci.*, **306**, 134.

Bonyadi, S. and Chung, T.-S. (2009). Highly porous and macrovoid-free PVDF hollow fiber membranes for membrane distillation by a solvent-dope solution co-extrusion approach, *J. Membr. Sci.*, **331**, 66.

Bottino, A., Capannelli, G., and Munari, S. (1980). Permeation characteristics of poly(vinylene fluoride) membranes, *Polymer*, **21**, 21.

Brunel, L., Tingry, S., Servat, K., Cretin, M., Innocent, C., Jolivalt, C., and Rolland, M. (2006). Membrane contactors for glucose/O_2 biofuel cell, *Desalination*, **199**, 426.

Calabrò, V., Drioli, E., and Matera, F. (1991). Membrane distillation in textile wastewater treatment, *Desalination*, **83**, 209.

Chae, S. R., Wang, S., Hendren, Z. D., Wiesner, M. R., Watanabe, Y., and Gunsch, C. K. (2009). Effects of fullerene nanoparticles on Escherichia coli K12 respiratory activity in aqueous suspension and potential use for membrane biofouling control, *J. Membr. Sci.*, **329**, 68.

Charcosset, C. (2006). Membrane processes in biotechnology: An overview, *Biotechnol. Adv.*, **24**, 482.

Cheremisinoff, N. P. (1997). *Handbook of Engineering Polymeric Materials*, Dekker, New York.

Chouikh, R., Bouguecha, S., and Dhahbi, M. (2005). Modeling of a modified air gap distillation membrane for the desalination of seawater, *Desalination*, **181**, 257.

Cohen-Tanugi, D. and Grossman, J. C. (2012). Water desalination across nanoporous graphene, *Nano Lett.*, **12**, 3602.

Cui, Z., Drioli, E., and Lee, Y. M. (2014). Recent progress in fluoropolymers for membranes, *Prog. Polym. Sci.*, **39**, 164.

Cui, Z., Hassankiadeh, N. T., Lee, S. Y., Lee, J. M., Woo, K. T., Sanguineti, A., Arcella, V., Lee, Y. M., and Drioli, E. (2013). Poly(vinylidene fluoride) membrane preparation with an environmental diluent via thermally induced phase separation. Journal of membrane science, **444**, 223.

Drioli, E., Ali, A., and Macedonio, F. (2015). Membrane distillation: Recent developments and perspectives, *Desalination*, **356**, 56.

Drioli, E., Criscuoli, A., and Curcio, E. (2006). *Membrane Contactors: Fundamentals, Applications and Potentialities*, vol. 11, Elsevier, Amsterdam, Boston.

Drioli, E., Di Profio, G., and Curcio, E. (2012). Progress in membrane crystallization, *Curr. Opin. Chem. Eng.*, **1**, 178.

Dumée, L. F., Sears, K., Schütz, J., Finn, N., Huynh, C., Hawkins, S., Duke, M., and Gray, S. (2010). Characterization and evaluation of carbon nanotube bucky-paper membranes for direct contact membrane distillation, *J. Membr. Sci.*, **351**, 36.

Dumée, L., Campbell, J. L., Sears, K., Schütz, J., Finn, N., Duke, M., and Gray, S. (2011). The impact of hydrophobic coating on the performance of carbon nanotube bucky-paper membranes in membrane distillation, *DES*, **283**, 64.

El-Bourawi, M. S., Ding, Z., Ma, R., and Khayet, M. (2006). A framework for better understanding membrane distillation separation process, *J. Membr. Sci.*, **285**, 4.

Feng, C. S., Shi, B. L., Li, G. M., and Wu, Y. L. (2004a). Preliminary research on microporous membrane from F2.4 for membrane distillation, *Sep. Purif. Technol.*, **39**, 221.

Feng, C., Shi, B., Li, G., and Wu, Y. (2004b). Preparation and properties of microporous membrane from poly (vinylidene fluoride-co-tetrafluoro ethylene) (F2.4) for membrane distillation, *J. Membr. Sci.*, **237**, 15.

Gethard, K., Sae-Khow, O., and Mitra, S. (2011). Water desalination using carbon-nanotube-enhanced membrane distillation, *ACS Appl. Mater. Interface.*, **3**, 110.

Gugliuzza, A. and Drioli, E. (2007). PVDF and Hyflon AD membranes: Ideal interfaces for contactor applications, *J. Membr. Sci.*, **300**, 51.

Gugliuzza, A. and Drioli, E. (2009). New performance of hydrophobic fluorinated porous membranes exhibiting particulate-like morphology, *Desalination*, **240**, 14.

Gugliuzza, A., Aceto, M. C., Macedonio, F., and Drioli, E. (2008). Water droplets as template for next-generation self-assembled poly-(etheretherketone) with cardo membranes, *J. Phys. Chem. B*, **112**, 10483.

Han, Y., Xu, Z., and Gao, C. (2013). Ultrathin graphene nanofiltration membrane for water purification, *Adv. Funct. Mater.*, **23**, 3693.

Hassankiadeh, N. T., Cui, Z., Kim, J. H., Shin, D. W., Lee, S. Y., Sanguineti, A., Arcella, V., Lee, Y. M., and Drioli, E. (2015). Microporous poly(vinylidene fluoride) hollow fiber membranes fabricated with PolarClean as water-soluble green diluent and additives, *J. Membr. Sci.*, **479**, 204.

He, T., Mulder, M. H. V., and Wessling, M. (2003). Preparation of porous hollow fiber membranes with a triple-orifice spinneret, *J. Appl. Polym. Sci.*, **87**, 2151.

Hendren, Z. D., Brant, J., and Wiesner, M. R. (2009). Surface modification of nanostructured ceramic membranes for direct contact membrane distillation, *J. Membr. Sci.*, **331**, 1.

Ho, J. Y., Matsuura, T., and Santerre, J. P. (2000). The effect of fluorinated surface modifying macromolecules on the surface morphology of polyethersulfone membranes, *J. Biomater. Sci. Polym. Ed.*, **11**, 1085.

Iverson, S. B., Bhatia, V. K., Dam-Johansen, K., and Jonsson, G. (1997). Characterization of microporous membranes for use in membrane contactors, *J. Membr. Sci.*, **130**, 205.

Jansen, J. C., Buonomenna, M. G., Figoli, A., and Drioli, E. (2006a). Asymmetric membranes of modified poly(ether ether ketone) with an ultra-thin skin for gas and vapour separation, *J. Membr. Sci.*, **271**, 188.

Jansen, J. C., Friess, K., and Drioli, E. (2011). Organic vapour transport in glassy perfluoropolymer membranes: A simple semi-quantitative approach to analyze clustering phenomena by time lag measurements, *J. Membr. Sci.*, **367**, 141.

Jansen, J. C., Tasselli, F., Tocci, E., and Drioli, E. (2006b). High-flux composite perfluorinated gas separation membranes of Hyflon AD on a hollow fibre ultrafiltration membrane support, *Desalination*, **192**, 207.

Jin, Z., Yang, D. L., Zhang, S. H., and Jian, X. G. (2008). Hydrophobic modification of poly(phthalazinone ether sulfone ketone) hollow fiber membrane for vacuum membrane distillation, *J. Membr. Sci.*, **310**, 20.

Khayet, M. (2011). Membranes and theoretical modeling of membrane distillation: A review, *Adv. Colloid Interface Sci.*, **164**, 56.

Khayet, M. and Matsuura, T. (2001). Preparation and characterization of polyvinylidene fluoride membranes for membrane distillation, *Ind. Eng. Chem. Res.*, **40**, 5710.

Khayet, M., Feng, C. Y., and Matsuura, T. (2003). Morphological study of fluorinated asymmetric polyetherimide ultrafiltration membranes by surface modifying macromolecules, *J. Membr. Sci.*, **213**, 159.

Khayet, M., Feng, C., Khulbe, K. C., and Matsuura, T. (2002). Preparation and characterization of polyvinylidene fluoride hollow fiber membranes for ultrafiltration, *Polymer*, **43**, 3879.

Khayet, M., Mengual, J. I., and Matsuura, T. (2005). Porous hydrophobic/hydrophilic composite membranes: Application in desalination using direct contact membrane distillation, *J. Membr. Sci.*, **252**, 101.

Kim, K. H., Lee, J. S., Hong, H. P., Han, J. Y., Park, J. W., and Min, B. (2015). The effect of fullerene (C 60) nanoparticles on the surface of PVDF composite membrane, *Membr. Water Treatment*, **6**, 423.

Kong, Y., Lin, X., Wu, Y., Chen, J., and Xu, J. (1992). Plasma polymerization of octafluorocyclobutane and hydrophobic microporous composite membranes for membrane distillation, *J. Appl. Polym. Sci.*, **46**, 191.

Krajewski, S. R., Kujawski, W., Bukowska, M., Picard, Ć., and Larbot, A. (2006). Application of fluoroalkylsilanes (FAS) grafted ceramic membranes in membrane distillation process of NaCl solutions, *J. Membr. Sci.*, **281**, 253.

Larbot, A., Gazagnes, L., Krajewki, S., Bukowska, M., and Kujawski, W. (2004). Water desalination using ceramic membrane distillation, *Desalination*, **168**, 367.

Lee, J., Chae, H.-R., Won, Y. J., Lee, K., Lee, C.-H., Lee, H. H., Kim, I.-C., and Lee, J. (2013). Graphene oxide nanoplatelets composite membrane with hydrophilic and antifouling properties for wastewater treatment, *J. Memb. Sci.*, **448**, 223.

Li, B. and Sirkar, K. K. (2005). Novel membrane and device for vacuum membrane distillation-based desalination process, *J. Membr. Sci.*, **257**, 60.

Lopes, A., Costa, C., Tavares, C., Neves, I., and Lanceros-Mendez, S. (2011). Nucleation of the electroactive γ phase and enhancement of the optical transparency in low filler content poly(vinylidene)/clay nanocomposites, *J. Phys. Chem. C*, **115**, 18076.

Lovinger, A. J. (1983). Ferroelectric polymers, *Science*, **220**, 1115.

Membrana GmbH. (2006). Membrane contactors improve production yields, *Membr. Technol*, **4**, 5.

Mourgues, A., Heng, N., Belleville, M. P., Paolucci-Jeanjean, D., and Sanchez, J. (2010). Study of a new membrane evaporator with a hydrophobic metallic membrane, *J. Membr. Sci.*, **355**, 112.

Musico, Y. L. F., Santos, C. M., Dalida, M. L. P., and Rodrigues, D. F. (2014). Surface modification of membrane filters using graphene and graphene oxide-based nanomaterials for bacterial inactivation and removal, *ACS Sustain. Chem. Eng.*, **2**, 1559.

Ortiz de Zfirate, J. M., Pefia, L., and Mengual, J. I. (1995). Characterization of membrane distillation membranes prepared by phase inversion, *Desalination*, **100**, 139.

Peng, P., Fane, A. G., and Li, X. (2005). Desalination by membrane distillation adopting a hydrophilic membrane, *Desalination*, **173**, 45.

Pham, V. A., Santerre, J. P., Matsuura, T., and Narbaitz, R. M. (1999). Application of surface modifying macromolecules in polyethersulfone membranes: Influence on PES surface chemistry and physical properties, *J. Appl. Polym. Sci.*, **73**, 1363.

Qtaishat, M., Khayet, M., and Matsuura, T. (2009). Guidelines for preparation of higher flux hydrophobic/hydrophilic composite membranes for membrane distillation, *J. Membr. Sci.*, **329**, 193.

Quist-Jensen, C. A., Macedonio, F., Conidi, C., Cassano, A., Aljlil, S., Alharbi, O. A., and Drioli, E. (2016). Direct contact membrane distillation for the concentration of clarified orange juice, *J. Food Eng.*, **187**, 37.

Ren, J., Wang, R., Zhang, H. Y., Li, Z., Liang, D. T., and Tay, J. H. (2006). Effect of PVDF dope rheology on the structure of hollow fiber membranes used for CO_2 capture, *J. Membr. Sci.*, **281**, 334.

Simone, S., Figoli, A., Carnevale, M. C., Rosselli, A., and Drioli, E. (2010). Preparation of hallow fibre membranes from PVDF/PVP blends and their application in VMD, *J. Membr. Sci.*, **364**, 219.

Song, L., Li, B., Sirkar, K. K., and Gilron, J. L. (2007). Direct contact membrane distillation-based desalination: Novel membranes, devices, larger-scale studies, and a model, *Ind. Eng. Chem. Res.*, **46**, 2307.

Sugihara, M., Fujimoto, M., and Uragami, T. (1979). Effect of casting solvent on permeation characteristics of polyvinylidene fluoride membranes, *Polym. Prepr. Am. Chem. Soc. Div. Polym. Chem.*, **20**, 999.

Suk, D. E., Matsuura, T., Park, H. B., and Lee, Y. M. (2006). Synthesis of a new type of surface-modifying macromolecules (nSMM) and characterization and testing of nSMM blended membranes for membrane distillation, *J. Membr. Sci.*, **277**, 177.

Suk, D. E., Matsuura, T., Park, H. B., and Lee, Y. M. (2010). Development of novel surface modified phase inversion membranes having hydrophobic surface-modifying macromolecule (nSMM) for vacuum membrane distillation, *Desalination*, **261**, 300.

Tang, C. Y., Zhao, Y., Wang, R., Hélix-Nielsen, C., and Fane, A. G. (2013). Desalination by biomimetic aquaporin membranes: Review of status and prospects, *Desalination*, **308**, 34.

Tomaszewska, M. (1996). Preparation and properties of flat-sheet membranes from poly(vinylidene fluoride) formembrane distillation, *Desalination*, **104**, 1.

Wang, D., Li, K., and Teo, W. K. (1999). Preparation and characterization of polyvinylidene fluoride (PVDF) hollow fiber membranes, *J. Membr. Sci.*, **163**, 211.

Wang, D., Teo, W. K., and Li, K. (2002). Removal of H_2S to ultra-low concentrations using an asymmetric hollow fibre membrane module, *Sep. Purif. Technol.*, **27**, 33.

Wang, K. Y., Chung, T. S., and Gryta, M. (2008). Hydrophobic PVDF hollow fiber membranes with narrow pore size distribution and ultra-thin skin for the freshwater production through membrane distillation, *Chem. Eng. Sci.*, **63**, 2587.

Wu, Y., Kong, Y., Lin, X., Liu, W., and Xu, J. (1992). Surface-modified hydrophilic membranes in membrane distillation, *J. Membr. Sci.*, **72**, 189.

Yu, S., Zheng, W., Yu, W., Zhang, Y., Jiang, Q., and Zhao, Z. (2009). Formation mechanism of β-phase in PVDF/CNT composite prepared by the sonication method, *Macromolecules*, **42**, 8870.

Zhen, H., Jang, S. M. J., Teo, W. K., and Li, K. (2006). Modified silicone–PVDF composite hollow fiber membrane preparation and its application in VOC separation, *J. Appl. Polym. Sci.*, **99**, 2497.

Chapter 13

Nanostructured Membranes for Biosensor Development

Francesca Militano, Teresa Poerio, Rosalinda Mazzei, and Lidietta Giorno
Institute on Membrane Technology of the National Research Council of Italy (ITM-CNR), Via P. Bucci, Cubo 17C, 87036 Rende (CS), Italy
f.militano@itm.cnr.it

The increasing attention toward safety and the continuous need for rapid and simple monitoring techniques for controlling specific molecules have motivated a rapid growth of biosensor technology, in terms of application areas and design of device architecture. This chapter provides an analysis of the use of nanostructured membranes in biosensor design, including several aspects related to different membrane materials, properties, and functions as well as applications in different areas.

13.1 Introduction

Since the development of the first biosensor in 1962 (Clark and Lyons, 1962), biosensor technology has been intensively studied

Functional Nanostructured Membranes
Edited by Enrico Drioli, Lidietta Giorno, and Annarosa Gugliuzza

ISBN 978-981-4774-79-6 (Hardcover), 978-1-351-13511-5 (eBook)
www.panstanford.com

and increasingly used in different fields, ranging from medicine to food quality and safety control and environmental monitoring. Great efforts have been devoted to improving the sensitivity, selectivity, stability, and portability of biosensors leading to the development of even more sophisticated and complex devices. The integration or combination of biosensors with nanostructured materials is playing an important role in such advancements (Jianronga et al., 2004; Tuantranont, 2013). In particular, nanostructured membranes are becoming increasingly attractive due to their unique properties. They exhibit a large ratio of surface area to volume, which permits the immobilization of an enhanced quantity of bioreceptor units, facilitates rapid solution exchange, allows selective transport and analyte separation/concentration, thus having the potential to provide higher sensitivity and selectivity, faster response time, and lower detection limits. Also a large number of materials can be used for their fabrication ranging from inorganic and organic synthetic materials to natural biocompatible polymers. Moreover, they can be incorporated into miniaturized systems that integrate multiple functions for the real-time and cost-effective detection of analytes.

This chapter discusses the use of nanostructured membranes in biosensor design. Before analyzing membrane-based biosensors, the basic concepts of a biosensor and a general classification are given. The chapter then explains the importance and advantages of using membranes in the construction of such devices. The first part identifies the membrane functions and properties as well as the different membrane types and fabrication methods and explains the incorporation or integration of these membranes into miniaturized systems. The next part focuses on the techniques for immobilizing the bioreceptor that plays a crucial role in determining the performance of the biosensor. The different techniques are examined in detail, including the advantage and disadvantage of such methods. The preparation of membrane-based biomimetic receptors by molecular imprinting technology is also considered. After this, transduction methods and systems commonly combined with membranes for signal transduction and detection are explained. In the last part, a summary of membrane-based biosensor applications in different areas is given.

13.2 Basic Concepts and Classification of Biosensors

A biosensor is an analytical device capable of providing specific quantitative or semi-quantitative analytical information, using a biological sensing element (bioreceptor) intimately associated with or integrated within a physicochemical transducer (Thévenota et al., 2001). The basic concept of operation of a biosensor is illustrated in Fig. 13.1. The bioreceptor, which is an immobilized biological element (i.e., enzyme, antibody, DNA probe, etc.), recognizes a specific analyte through a biocatalytic reaction or a specific binding, and the transducer converts the result of this recognition into a usable signal, which can be quantified. Thus, the bioreceptor is responsible for the selectivity of the sensor; it is very specific to the analyte to which it is sensitive.

Biosensors can be classified into two ways according to either the basic principles of signal transduction or the bioreceptor (Scheme 13.1).

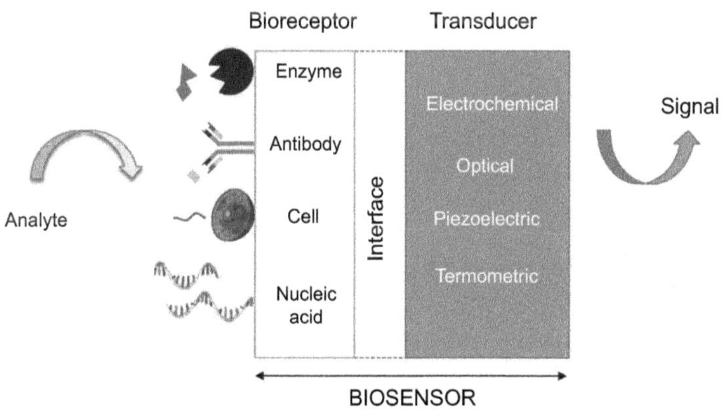

Figure 13.1 A schematic representation of a biosensor and working principle.

According to the recognition properties of the bio-element, the two biosensor groups are catalytic biosensors and affinity biosensors. Catalytic biosensors are based on biological catalysts such as enzymes, cells, microorganisms that promote a specific reaction with the analyte producing the species to be detected. Otherwise, affinity biosensors are based on biomolecules able to selectively and reversibly bind specific ligands (i.e., antibodies/antigens, nucleic acids) and the binding event is monitored (Leech, 1994).

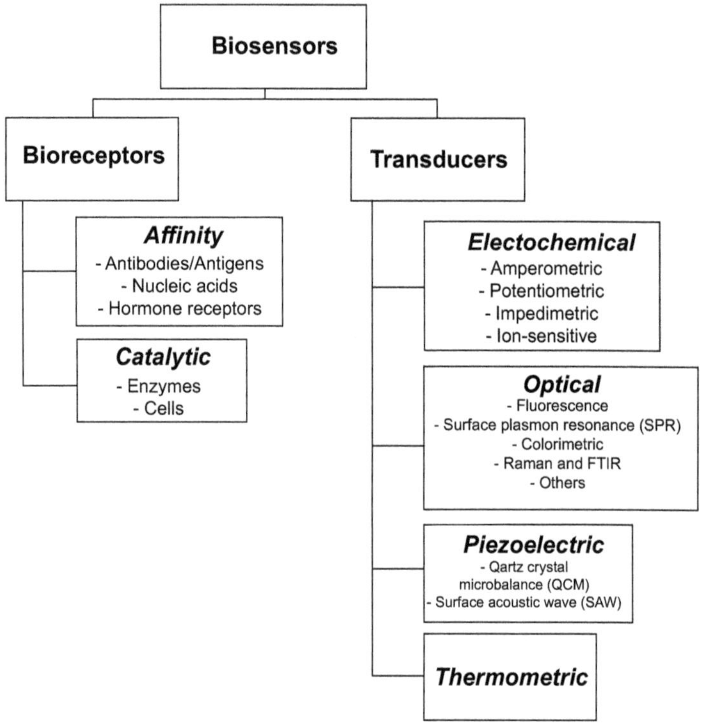

Scheme 13.1 Schematic of biosensors classification.

Depending on the transduction method, biosensors can also be classified into four main classes: (i) electrochemical, (ii) optical, (iii) piezoelectric, and (iv) thermometric. Each of these classes contains many different subclasses generating a great number of possible transduction methods. For example, the electrochemical biosensors, based on the parameter measured, can be further classified as amperometric, potentiometric, and impedimetric. Transduction mechanisms will be explored in detail in the last part of the chapter.

13.2.1 Catalytic Biosensors

Catalytic biosensors are mostly represented by enzymatic biosensors. Enzymes were the first biological components used in biosensors and remain the most commonly used bio-recognition elements due to their specific binding capabilities as well as their catalytic activity. An enzyme-catalyzed reaction may generate a variety of measurable products, including protons, electrons, light, and heat, which can be directly determined using one of the transduction methods mentioned above.

The most largely studied and commercially available enzymatic biosensors are those for measuring glucose in blood samples using glucose oxidase or glucose dehydrogenase for the monitoring of diabetes (Yoo and Lee, 2010). The enzyme urease has also been widely used for urea monitoring for both medical and environmental applications (Barhoumi et al., 2006).

The performance of enzymatic biosensors can be influenced by several factors, including enzyme loading, pH, temperature, and, in some cases, the use of cofactor (inorganic ions or coenzymes) required for enzyme activity. Besides, the method of immobilization and the thickness of enzyme layer on the sensor surface may have a great impact on the biosensor performance.

13.2.2 Affinity Biosensors

Affinity biosensors involve mainly immunosensors and DNA sensors. Immunosensors make use of specific interactions between an antibody and an antigen. Antibodies are complex biomolecules with molecular recognition properties exhibiting very specific binding capabilities for specific structures (antigens). These properties of antibody are crucial to their usefulness in immunosensors where only the specific analyte of interest, the antigen, fits into the antibody binding site. The formation of antigen–antibody complex can be monitored by different transduction mechanisms based, for example, either on the generation of an optical or electrochemical signal or a mass change on the sensor surface. Immunosensors can be designed to detect both antibodies and antigens in biological fluids and also in other media, including food and the natural environment (Ricci et al., 2007).

The sensitivity and stability of an immunosensor directly depend on the type of antibody used for the immunosensor design, the quantity of immobilized molecules, and the method of immobilization that affects antibody activity and orientation on the sensing interface.

DNA sensors are based on the specific recognition process of nucleic acids consisting in the hybridization of two single strands having complementary base sequences. The hybridization occurs between a known DNA sequence, called probe, and an unknown counterpart, the target. The duplex formation can be detected by using a hybridization indicator (i.e., a fluorescent label) or by following other changes occurred during the binding event. The probes can be DNA fragments synthesized by DNA amplification or oligonucleotide sequences having no natural analogue, named aptamers.

The immobilization of DNA probe is the crucial step for proper operation of the biosensor. DNA sensors were used for monitoring DNA—ligand interactions, for the detection of DNA damage, determination of drug in blood, for toxins, etc. (Diculescu et al., 2005).

13.3 Nanostructured Membranes for Biosensing

Nanostructured membranes of different materials, porosity, and geometry have been largely used in several medical and biological applications, including separation and sorting of biomolecules, analysis of single molecules, sensing of ions and proteins, as well as immunoisolation and drug delivery (Adiga et al., 2009). In biosensing application, the membrane technologies offer multiple advantages. The high ratio of surface area to volume is one of the most important properties of nanostructured membranes. The increased pore surface can interact with biomolecules trapped into the pores or flowing through the membrane providing a large and often highly reactive surface area, which enables more effective capture and detection of molecules on the sensing surface. In addition, compared with other nanostructures, such as nanotubes and nanowires, membranes allow to separate biomolecules based on their size, shape, charge, and interactions with nanopores, increasing the selectivity of the sensor. They can be manufactured using various techniques with high control in pore size, pore-size distribution, pore location, nano-architecture, nano-patterns, and the overall physical and chemical properties. Besides, surface modification techniques can be used to improve biocompatibility, anti-biofouling capability, to promote/control interaction with biomolecules.

Moreover, with the increasing interest in microfluidic and lab-on-a-chip systems, membranes have also been easily integrated into microfabricated platforms for the "real-time" and cost-effective detection of target analyte (Prasad et al., 2010).

13.3.1 Membrane Properties and Functions

Membranes may have different functions in biosensing systems, acting or not as integral part of the transduction mechanism. They can be used as a barrier structure between the sensor system (bioreceptor and transducer) and its environment. In this case, by exploiting the transport properties of membranes, the interfering species can be excluded from the sensor surface allowing the only access of the analyte to the bioreceptor and protecting the bioreceptor from the environment. Requirements such as permeability and selectivity become important to permit an adequate flow of the analyte to the transducer and to prevent the flow of other active species, respectively (Davies et al., 1992). Permselective membranes are generally used to design the front and interface of the biosensor. They are used to coat the electrode surface in the form of a single membrane layer or multilayered membranes with different transport properties. The mechanism of selective transport can be based on size-exclusion, charge-exclusion, or both. The different species can be discriminated based on their molecular size by using a size-exclusion membrane that allows small molecules to pass through while preventing larger molecules to reach the sensor surface. Equally, charged interferences can be eliminated by using a membrane that has a charge similar to the species to be discriminated. Additional selectivity can be achieved by the use of multilayered systems that combine the properties of different membranes. Moreover, the permeability of such membrane can be tuned to meet specific needs. Thus, the selectivity can be achieved by taking advantage of analyte properties (i.e., charge, size, shape) and the selection of a proper membrane.

In implantable biosensors, the protective function of the membrane plays a crucial role in the overall sensitivity: both inner and outer membranes are used, to either immobilize the bioreceptor or eliminate interferences and to prevent biofouling or to govern analyte diffusion, respectively (Vaddirajua et al., 2010).

Other biosensing systems involve the use of porous membranes for retaining the analyte and then increasing its local concentration, to improve the overall sensitivity (Lee and Yanavich, 2004).

Another important role of membrane in biosensing systems is related to their use as immobilization matrix for the bioreceptor. The bioreceptor (i.e., enzyme, antibody, etc.), which ensures the molecular recognition of the analyte, can be immobilized in/on the membrane/surface and the membrane can be used as a tool for bringing it in the immediate vicinity of the transducer, or even the bioreceptor can be embedded in a layer coating the active tip of the transducer (Coulet, 1992). The binding efficiency is mainly dependent on the surface properties of the membrane such as hydrophobicity, roughness, or the presence of reactive functional groups.

13.3.2 Membrane Types and Fabrication

Given the main membrane functions in biosensor applications, two aspects should be considered in membrane design: the physical structure and the chemical composition. The physical structure is more critical for filtering applications, and it includes parameters such as surface area and roughness, membrane thickness, pore size, and size distribution, which are tightly correlated to the method used for membrane fabrication. Based on their pore size, membranes are classified as nonporous, microporous (pore diameter ≤ 2 nm), mesoporous (2–50 nm), and macroporous (50–500 nm) (Ulbricht, 2006). Based on their structure, they can be classified as symmetric (with the same structure and transport properties over the entire cross section) and asymmetric (with structure and transport properties varying over the membrane cross section; they can be integrally asymmetric or composite structures). They can be prepared in different configurations, such as flat-sheet, hollow fiber, tubular, etc. The chemical composition is more critical for application as support. Based on their intrinsic materials, membranes can be broadly classified as inorganic, organic, and composite.

13.3.2.1 Inorganic membranes

Inorganic membranes are made from oxides, ceramics, and metals. With respect to polymeric membranes, they are versatile, can be used at high temperatures, and are highly resistant to chemical attack.

At present, among the inorganic materials, nanoporous membranes with well-ordered pores such as anodic alumina and nanoporous silica receive the major attention for use in biosensor applications.

Nanoporous alumina membranes, also known as anodic aluminum oxide, with flat shape, have been extensively used due to their excellent biocompatibility, non-conductivity, ease of functionalization, and well-established fabrication process resulting in well-defined nanopores, small pore size, and high pore density. Pore size from 20 nm to 200 nm can be achieved via anodization etching process. Figure 13.2 shows the SEM image of a porous alumina membrane prepared via anodization, with an average pore diameter of 60 nm. The membrane was developed for application in label-free immunosensors (Alvarez et al., 2009). Alumina membranes are also commercially available. They have been used, for example, for nucleic acid detection by modifying the nanopores with covalently bound DNA probe and monitoring the variation in ionic conductivity in nanopores due to DNA hybridization determining the pore blocking to ionic flow (Vlassiouk, 2005).

For silicon and silicon nitride surfaces, highly ordered and cylindrical arrays of nanopores have been fabricated by means of micro- and nanofabrication technology. Several contributions in the literature are devoted to the development of porous silicon membranes by using lithography techniques for biosensor applications. For example, Desai et al. (2000) proposed a novel nanoporous micromachined silicon membrane with selective permeability and low biofouling for implantable biosensor application prepared by using photolithography and deposition/selective removal of sacrificial layers. The novel membrane showed uniform pore sizes and stability in biological environments, with the capability of simultaneous exclusion of albumin and diffusion of glucose. Other approaches that allow a high degree of pore-size control are focused ion beam (FIB) and electron beam lithography (EBL) techniques, which have been used for the fabrication of silicon and silicon nitride membranes (Striemer et al., 2007; Tong et al., 2004). These processes can precisely control nanopore size as well as easily integrate with microdevices. Figure 13.3 shows the SEM image of a silicon nitride (SiN) membrane with uniform cylindrical nanopores and pore diameter as small as 25 nm, fabricated using FIB etching for selective size-exclusion-based separations.

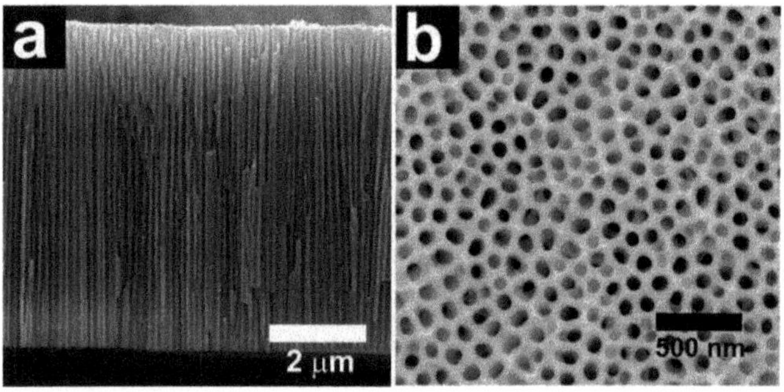

Figure 13.2 Cross-sectional view (a) and plan-view image (b) of porous alumina membrane prepared via anodization, with an average pore diameter of 60 nm. Reprinted from Alvarez et al. (2009) with permission from American Chemical Society.

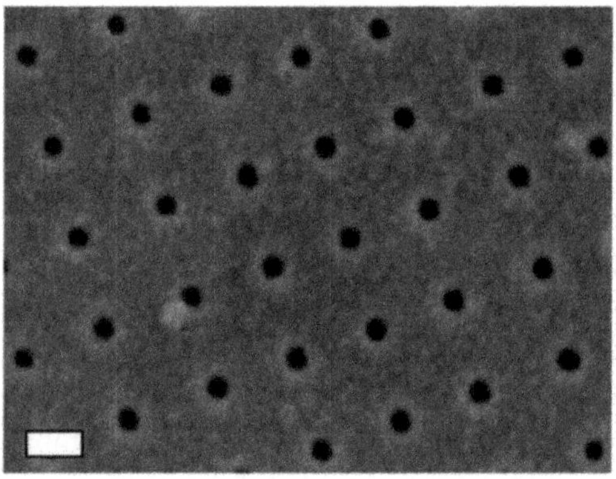

Figure 13.3 Silicon nitride membrane with uniform and circular nanopores of around 25 nm diameter. Reprinted from Tong et al. (2004) with permission from American Chemical Society.

In addition to aluminum oxide and silicon membranes, gold and silver membranes have been used in several cases (Wigginton and Vikesland, 2010; Zhanga et al., 2014).

13.3.2.2 Organic membranes

Polymeric membranes are most commonly used in biosensor applications compared to inorganic membranes. Since the development of the first biosensors (Clark and Lyons, 1962), they have played an important role and continue to contribute to the advancement of even more complex and efficient systems.

Polymeric membranes have been used for various functions: structure support, inner and outer protective barrier, for the immobilization of the bio-recognition element, acting in some cases as integral part of the sensing process. The chemical structure of the commonly used polymeric materials for membrane preparation is illustrated in Fig. 13.4.

The two earliest and most commonly used membranes are Nafion (Wirguin, 1996) and cellulose acetate (Maines et al., 1997). Nafion is a fluorinated negatively charged polymer with both hydrophobic and hydrophilic properties, good biocompatibility; it is chemically inert, stable with temperature, and limits the diffusion of small neutral or negatively charged interfering molecules. Due to the advantageous properties described, it has been extensively used in biosensors as coating material to reduce biofouling, also in vivo applications (Wisniewski and Reichert, 2000). Most of the applications of Nafion membranes as outer protective barrier are referred to the construction of glucose biosensors (Mignani et al., 2009; Ren et al., 2012; Yang et al., 2009). But recently, Nafion outer protective membranes have also been used in DNA-based biosensors showing anti-fouling effect in different matrices (Ambrozy et al., 2013).

Membranes of natural origin, such as cellulose and its derivatives, are very suitable for biomedical applications. Cellulose acetate membranes are commonly used for protective function and for enhancing biocompatibility, for example, in electrochemical biosensors (Maines et al., 1997). Thanks to its strong adsorption capacity, nitrocellulose membranes have been used as functional membranes in different biosensing systems, for example for the realization of an enzyme-linked immunostrip biosensor for the detection of *Escherichia coli* (Park et al., 2008). Nitrocellulose membrane was also used in combination with magnetic bead for the detection of two different influenza A viruses (Hong et al., 2011), and again as support for antibody immobilization for the development of *E. coli* amperometric biosensors (Theegala et al., 2008). Other membranes used in biosensors were made with polyvinylidene fluoride (PVDF), polycarbonate (PC), polypropylene (PP), polyethylene terephthalate (PET), polyethersulfone (PES), and polydimethylsiloxane (PDMS). Many methods can be used to manufacture polymeric membranes. The main preparation processes include phase separation of polymer solutions (NIPS, TIPS, PIPS), extrusion and stretching of polymer films, and track-etching of polymer films (Ulbricht, 2011).

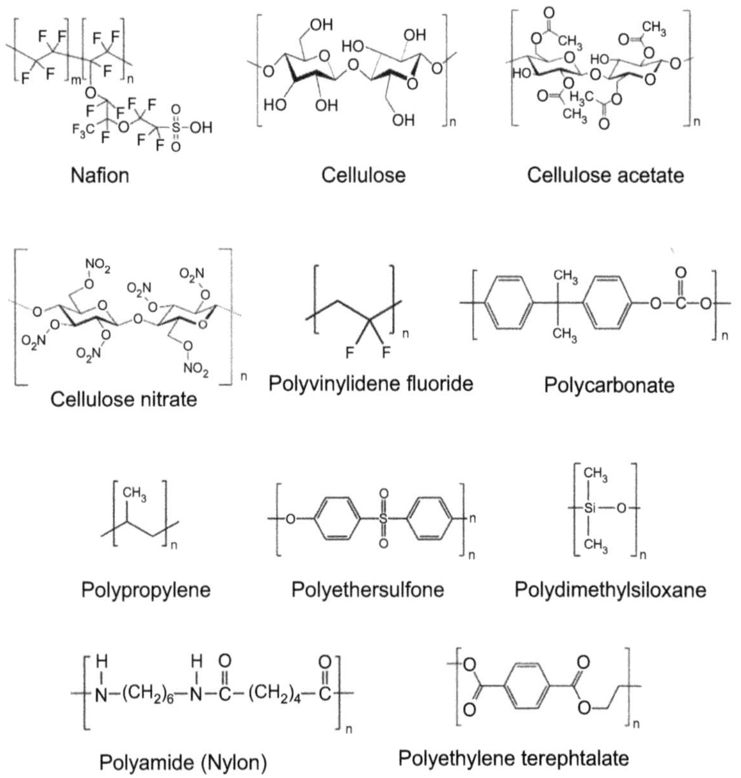

Figure 13.4 Chemical structure of commonly used polymeric membrane materials.

Recently, electrospinning is emerging as a powerful technique for the fabrication of nanofiber membranes with high specific surface area and high porosity, making it increasingly attractive for biosensor application. Figure 13.5 shows the SEM image of a typical polyamidic nanofibrous membrane prepared by electrospinning. This type of membrane has been used for the development of biocatalytic membranes, tyrosinase-biosensor for the detection of phenolic compounds (Arecchia et al., 2010), and for the development of glucose biosensor (Scampicchio et al., 2010). Electrospun polylactic acid (PLA) nanofiber membranes have been also used as substrates for biosensor assemblies (Lia et al., 2006).

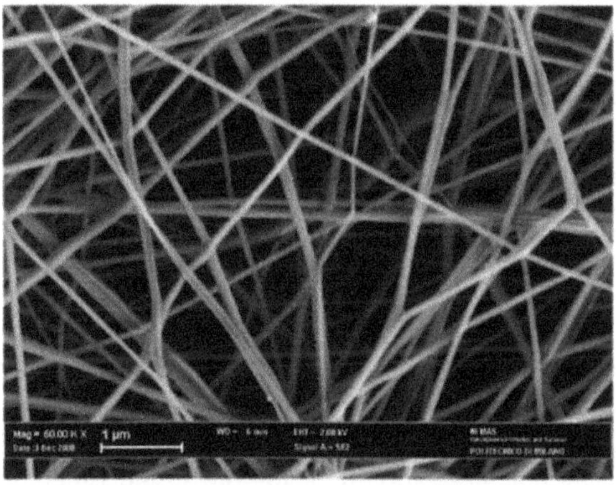

Figure 13.5 SEM image of a typical mesh of randomly oriented nanofibers made of nylon-6. Reprinted from Scampicchio et al. (2012) with permission from John Wiley and Sons.

On the other hand, polymeric porous membranes with relatively ordered pore arrays as well as pores close to cylindrical or conical geometry have been also prepared by ion-track technology. Figure 13.6 shows the SEM image of a track-etched polycarbonate membrane used for DNA detection (Kececi et al., 2015).

13.3.2.3 Composite membranes

Composite membranes are intended membranes containing different functional materials/layers, obtained by combining, for example, a polymer with a ceramic as well as nanoparticles, hydrogels, etc., to obtain membranes with improved stability, biocompatibility, selectivity, separation properties, and, in some cases, with multifunctional capabilities.

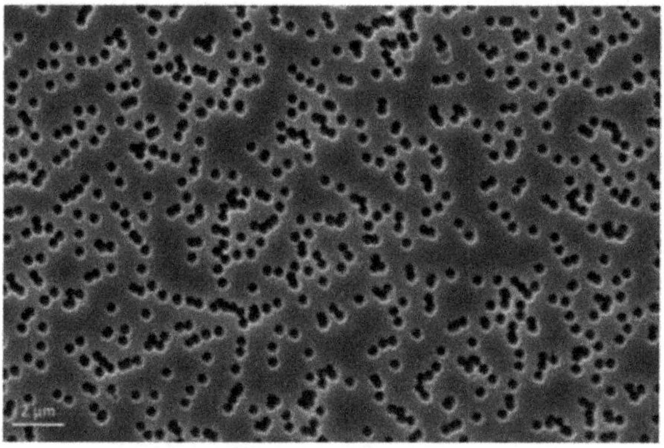

Figure 13.6 SEM image of multipore PC membrane surface. Reprinted from Kececi et al. (2015) with permission from Elsevier.

Composite membranes are prepared by starting with a base support membrane (organic or inorganic membrane) that defines the shape of the final membrane. The main types of composite membranes are thin-film, pore-filled, and pore-surface functionalized membranes. Thin-film composite membranes are prepared both by coating a polymer on the support membrane or by interfacial polymerization by using the support membrane as a functional interface where the reaction takes place. The pore-filled composite membranes can be prepared via graft copolymerization from the base polymer, via in situ polymerization, or via cross-linking of pre-synthesized polymers (Ulbricht, 2006).

The layer-by-layer (LBL) assembly is one of the methods that is increasingly used to prepare thin-film composite membranes for biosensor applications. This technique is based on the alternate electrostatic adsorption of polyelectrolytes, which leads to controllable ultrathin layers (Fig. 13.7). This technique allows the fabrication of tailored sensing membranes for optimum performance with regard to controlled thickness, structural morphology, and/or bioreceptor immobilization as a vast range of functional groups can be incorporated within the structure of the film (Smuleac et al., 2006). LBL composite membranes have been used, for example, for the construction of amperometric glucose biosensors (Tipnis et al., 2007).

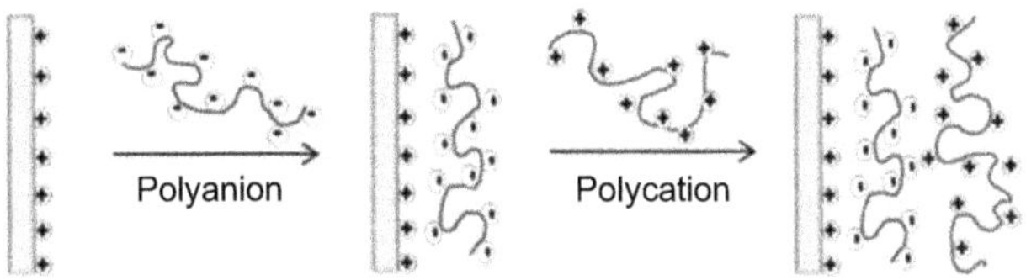

Figure 13.7 Schematic representation of layer-by-layer deposition process starting with a positively charged membrane.

The combination of polymeric functional hydrogels and porous membranes is also commonly performed for the preparation of composites of different types (pore-filled, thin-film, or pore-surface functionalized) with improved or novel functionality, such as improved binding capacity and biocompatibility or stimuli-responsive properties (Yang et al., 2011). Molecularly imprinted polymer (MIP) membranes are one type of thin-film or pore-filled composite membranes used in biosensors. The characteristics of these membranes are well illustrated in Section 13.5.

13.3.3 Membrane Integration with Microfluidic Systems

One of the emerging needs in the biosensing field is the real-time and cost-effective detection of target analytes. The rapid development in the field of microfluidic systems has allowed the construction of even more sophisticated miniaturized bioanalytical devices that aim to integrate mixing, reaction, separation, and detection functions into miniaturized chips, for facilitating the automation of sample analysis, reducing the time and reagent consumption and limiting costs. Hence, a lot of the present research in the biosensing area focuses on the integration of these functions on microfluidic devices for real-time measurements. One of the innovative approaches adopted has been the integration of nanostructured nanoporous membranes into these systems. Often, the nanoporous membrane has been incorporated for biomolecular separation and sample pre-concentration.

The sample concentration before analysis offers multiple advantages to the sensor system: It enables detection of species present in very low concentration, such as DNA or proteins from biological systems, which can be significant for clinical diagnosis and improve the detection sensitivity and signal-to-noise ratio, enabling the use of more different detection techniques.

Membrane-based sample pre-concentration has been obtained by means of three main techniques: filtration, evaporation, and concentration polarization (Song and Singh, 2006). Song et al. (2004) performed an electrophoretic protein concentration in a microfluidic chip by using a nanoporous polymer membrane, increasing the local protein concentration up to 10^4-fold in less than 2 min.

Over the years, considerable attention has been devoted to ion-selective membranes for the development of new membrane-based microfluidic systems that integrate multiple functionalities comprising pumping, manipulation, pre-concentration, and detection of biomolecules. It has been demonstrated that the ion concentration polarization phenomenon on ion-exchange membranes can be employed for separation, concentration, and detection of analytes on integrated membrane point-of-care diagnostic unit (Slouka et al., 2014).

Nanostructured membranes have also been used for the development of microfluidic immunoassays and immunosensors (Hartwell and Grudpan, 2010). Due to the porous structure, they have been extensively used in solid-state immunoassay, offering several advantages such as a greater surface area for the analysis, with respect to nonporous plastic surfaces. Moreover, they can be easily adapted for flow-injection immunoassays playing different roles: They have been used as support for antibody immobilization, retaining the target analyte (Hamid et al., 1999) or they have been placed at the electrode, participating in detection event (Nguyen et al., 2009), or even they have been used as unit for analyte extraction and pre-concentration, improving the sensitivity of the analysis (Tudorache and Emnéus, 2005). Nitrocellulose and PVDF commercial membranes are commonly used in microfluidic assays (Kim and Herr, 2013). Electrospun nanofibrous membranes have been used for the development of microfluidic immunoassays for HIV (Yang et al., 2008). Alumina nanoporous membranes have been integrated into a PDMS-based microfluidic immunosensor and have been used for antibody immobilization and detection of food pathogens (Tan et al., 2011).

Several approaches have been used for integrating membranes into microfluidic chips. Commercial membranes comprising polycarbonate, nitrocellulose, and PVDF have been directly inserted and physically clamped into microfluidic chips (Gao et al., 2001). Membranes have also been prepared in situ, for example, by condensation, interfacial polymerization (Hisamoto et al., 2003), and electrospinning (Yang et al., 2008). Alternatively, the membranes have been prepared during the fabrication process of the chip, or even the membrane materials themselves (i.e., PDMS and polyamide membranes) have been used to fabricate the chip (Jong et al., 2006).

13.4 Bioreceptor Immobilization

The immobilization of the bioreceptor is one of the critical steps for the development of an efficient biosensor. The sensitivity, selectivity, and stability of the biosensor are strongly influenced by the method chosen to link the bioreceptor to the sensor surface. Indeed, it directly affects the bioreceptor orientation, loading, mobility, stability, structure, and biological activity.

The most used immobilization strategies are adsorption, entrapment, covalent binding, affinity, crosslinking, or a combination of them, also making use of spacers and linkers to help minimize steric hindrances between the biomolecule and the surface. The selection of a technique depends on the nature of the bioreceptor, the nature of the membrane (i.e., chemical composition, available functional groups), and the final application. Regarding the surface, a large number of modification techniques are available for enhancing the membrane-binding capability. The different techniques are illustrated in Chapter 14.

The aforementioned immobilization techniques can be grouped into two main classes: random and oriented immobilization (Scheme 13.2). The random immobilization is simple and generally inexpensive; it is not a controlled process and in some cases may lead to reduction in bioreceptor activity and conformational changes. The oriented immobilization is more specific; it generally provides a homogeneous and reproducible bioreceptor layer with a properly oriented and enhanced activity. In some cases, it may require the functionalization of the target molecule or the derivatization of the membrane surface or both.

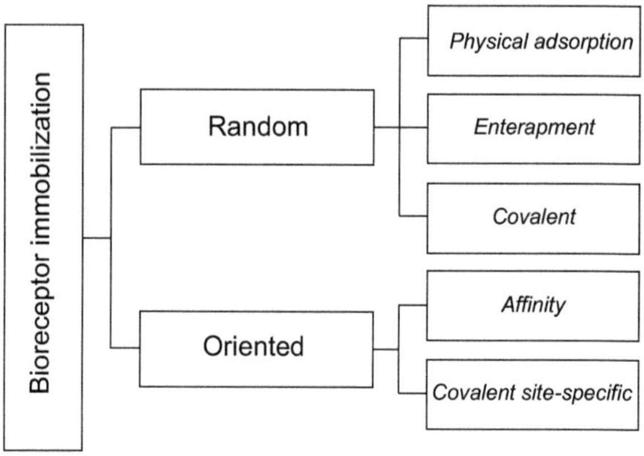

Scheme 13.2 Classification of bioreceptor immobilization techniques.

13.4.1 Random Immobilization

13.4.1.1 Physical adsorption

Physical adsorption is the most simple and inexpensive method for immobilizing bioreceptors. Proteins adsorb on the membrane surface via weak intermolecular forces, including hydrogen bonds, hydrophobic, electrostatic, and van der Waals interactions or a combination of these. The technique consists of the static incubation of the protein solution with the membrane or the continuous flow of the solution over the membrane surface. The resulting protein layer is generally heterogeneous and randomly oriented, which sometimes may result in low reproducibility and reduction in protein activity. Indeed, proteins can form multiple contacts, in different orientations, with the surface, resulting, in some cases, in conformational changes and reduction in accessibility to the functional binding sites (Nakanishi et al., 2008). Both immobilization density and the extent of conformational change may depend on protein (i.e., protein size, concentration, stability, charge, etc.), surface type, and chemistry (i.e., hydrophilicity or hydrophobicity), and the immobilization conditions (i.e., temperature, pH, and ionic strength of the solution).

Despite the disadvantages related to weak protein coverage, surface orientation, and reproducibility, this immobilization technique may be useful for certain applications. The adsorption of proteins is commonly practiced in polymeric membranes such as nitrocellulose and PVDF. Due to their strong adsorption capacity, these membranes are frequently used in Western blotting and microfluidic assays (Kim and Herr, 2013). The adsorption of biomolecules to nitrocellulose membranes is attributed to both charge–charge interactions and hydrogen interactions. Recently, wax-patterned nitrocellulose membranes have been used for the development of paper microfluidic devices for immunoassay applications (Lu et al., 2010).

13.4.1.2 Entrapment

Entrapment is a physical immobilization method based on the passive trapping of the bioreceptor into the membrane pores, involving some advantages such as the ability to entrap large amounts of biomolecule with enhanced thermal and chemical stability and the ease of preparation, without the need of modification of the biological element. This technique is more useful when cells or cellular organelles are used as bio-recognition elements (D'Souza, 2001). Moreover, enzyme entrapment has also been widely used for the construction of biosensors showing increasing operational and storage stability (Sassolas, 2012). Some limitations of this method for a practical use are the large diffusional barrier and the possible leakage of biomolecules through the membrane pores.

13.4.1.3 Covalent immobilization

Covalent immobilization of the bioreceptor is performed when high surface coverage and long-term stability of the protein layer are required. The covalent bond is generally formed between the reactive functional groups, already present or properly introduced, of the membrane surface and the functional groups of exterior amino acid residues of proteins, resulting in an irreversible and stable binding. Ideally, the reaction should be performed in mild condition (physiological condition, i.e., aqueous buffer and neutral pH) in order to avoid protein denaturation during the immobilization process. The surface coverage depends on both the number of available functional groups on the membrane surface (Vitola et al., 2015) and the protein properties (Militano et al., 2016). The reactive groups of amino acids that can be involved are primary amine (N-terminus and ε-amino group of Lysine), thiol (cysteine), carboxylic acid (C-terminus, glutamic acid, and aspartic acid), hydroxyl (serine and threonine), phenol (tyrosine), and imidazole (histidine).

Based on the type of functional group involved, several bioconjugation techniques have been developed for the covalent immobilization of biomolecules (Hermanson, 1996). The most commonly used are *amine chemistry, thiol chemistry, epoxy chemistry,* and *carboxyl chemistry.*

These chemical couplings are often random, and they may occur in multiple binding involving simultaneously many amino acid residues and restricting the conformational flexibility of the protein. Protein activity often decreases, mostly due to the linkage with amino acid residues located on the active site of the protein. Although this technique is not a controlled immobilization, it still can result in a quite homogeneous orientation of the proteins on the surface, which may be useful for many applications. Moreover, in some cases, it has been demonstrated that the multipoint covalent immobilization may contribute to protein stabilization (Blanco et al., 1989).

13.4.2 Oriented Immobilization

13.4.2.1 Affinity immobilization

Affinity immobilization is a type of bio-specific adsorption that exploits the selectivity of specific interactions that exist in nature. It is based on complementary interactions between biomolecules such as antibodies and antigens, nucleic acids and nucleic acid-binding proteins, avidin and biotin, etc., giving a relatively strong, highly specific, and oriented protein immobilization. The procedure is based on the pre-bio/functionalization of the membrane with the affinity capture reagent, which is then used for the site-specific immobilization of the target protein (the bioreceptor). This procedure, sometimes, may require protein modification/engineering for conjugation with affinity pair or tag.

One advantage/disadvantage of affinity immobilization is the possibility of surface regeneration by using chemical treatment, pH change, ionic strength, or heat treatment. For some applications, this can be considered an advantage as it enables the surface reuse. For other applications, it may represent a limitation, as the oriented affinity immobilization can be followed by a covalent linkage.

The most common affinity interactions are avidin/streptavidin-biotin, protein A/G-antibody, polyhistidine-metal ions, DNA-mediated and aptamers.

The avidin/streptavidin-biotin is one of the strongest non-covalent affinity bindings, almost insensitive to pH, temperature, organic solvents, and other denaturtating agents (Smith et al., 2005). Avidin is a tetrameric glycoprotein that can bind up to four molecules of biotin. Biotin is a small molecule normally used in conjugation with the target protein. Due to its small size, it does not affect protein functionality or conformation. Alternatively to avidin or streptavidin, engineered proteins such as neutravidin are also available with high specificity of binding to biotin, but lower stability to pH changes. Avidin and streptavidin have been attached to different surfaces by either physical adsorption or covalent binding (Anzai et al., 2000) for the subsequent immobilization of biotinylated proteins.

Protein A and protein G are widely used for antibody (IgG)-oriented immobilization. They have been extensively used for application in affinity chromatography for antibody purification (Zou et al., 2001), but their use in immunosensor design is increasing (Makaraviciute and Ramanaviciene, 2013). These proteins specifically interact with the Fc constant region of IgGs; thus, the Fab variable region remains well accessible for antigen binding. Protein G has three IgG-binding domains, while protein A has five IgG-binding domains. With respect to protein A, protein G has a wider antibody binding range of IgG class. Both proteins have been attached to membranes by physical adsorption and covalent binding. The protein A/G-antibody complex can be dissociated by using acidic or basic treatments and the surface can be reused for further applications.

Other affinity interactions used for the site-specific immobilization of the bioreceptor are based on protein engineering with affinity tag to the C- or N-terminus, such as polyhistidine, to have a segment that specifically chelates metal ions (i.e., Ni^{2+}) (Redeker et al., 2013). The interaction is reversible upon the addition of competitive ligands or metal chelators.

13.4.2.2 Site-specific covalent immobilization

Site-specific covalent immobilization allows the attachment of proteins in a stable, definite, and controlled manner on the sensor surface. The process is based on the introduction of a single chemical group or a sequence in a specific location of the protein, which will then react with a complementary group previously introduced into the surface. Ideally, the functional group should not cross-react with endogenous amino acids, and its position should not affect the conformation of the target-binding site (Redeker et al., 2013). The reaction between the functional groups should be chemoselective and should work under physiological conditions (i.e., aqueous buffer and neutral pH) to avoid protein denaturation. This type of modification can be obtained by means of both chemical and enzymatic reactions (Duckworth et al., 2006). The most used chemical reactions are the so-called "click" reactions such as copper-catalyzed azide–alkyne cycloadditions, Staudinger ligations, Diels–Alder cycloadditions, thiol–ene additions, and oxime formation (Lutz, 2008). These reactions are highly selective and, in general, work in mild conditions by using water as a reaction medium.

13.5 Molecularly Imprinted Recognition Sites in Membranes

Although the high specificity and selectivity of natural receptors, sometimes their low chemical and physical stability, together with their high costs, limit their use especially in harsh environments. An alternative approach involves the use of biomimetic receptor systems capable of binding target molecules with affinities and specificities comparable to natural receptors. One technique that is being increasingly adopted is molecular imprinting in synthetic polymers. The process consists of the generation of cavities, complementary in size, shape, and functionality to the analyte, in the three-dimensional polymeric matrix during its formation. This is achieved by the copolymerization of a monomer and a crosslinker in the presence of the target analyte (the imprint molecule), which acts as a molecular template. The functional monomers initially form a complex with the imprint molecule (via covalent or non-covalent interactions), which is subsequently polymerized with the cross-linker into a rigid polymer to maintain the position of the functional groups. The subsequent removal of the imprint molecule from the polymeric matrix reveals the cavity that possesses the molecular memory of the analyte, and it is capable of rebinding the analyte with high specificity. Therefore, MIPs are synthetic recognition elements that mimic the recognition processes found in nature. They have the advantages of higher stability, facilitating their application in extreme environments, and they are cheap to produce and can be stored in the dry state at room temperature for long periods.

These polymeric materials have been used as recognition element in biosensing, instead of natural biomolecules (i.e., antibodies), in various detection formats. The binding of the target molecule or some properties of the analyte may cause variations in physicochemical parameters of the system that can be used for detection. They have been combined with different types of transducers such as optical, electrochemical, etc. Interfacing with the transducer can be accomplished by either an in situ synthesis at the transducer surface or the surface can be coated with a preformed polymer (Haupt and Mosbach, 2000).

The molecular imprinting technology has been explored to prepare membranes with specific molecular recognition sites in its matrix and then a pre-determined affinity for a variety of molecules. These membranes are named "molecular imprinted membranes" (MIMs) and are able to selectively recognize target molecules from solutions by simple static adsorption or permeation through membrane devices (Algieri et al., 20014). They combine the selectivity of MIPs with mechanical integrity and separation properties of membranes.

An MIM can be either composed of or contain an MIP. Three main strategies are used for the preparation of MIMs: (i) preparation of membranes from previously synthesized MIPs, (ii) simultaneous formation of MIP structure and membrane morphology by either in situ crosslinking polymerization (Yoshikawa et al., 1995) or polymer solution phase inversion (Piletsky et al., 1990) in the presence of the template, and (iii) preparation of MIPs on or in support membranes with suited morphology (composite membranes). Generally, cellulose acetate, nylon, polyvinylidene-difluoride (PVDF) are used as supporting membranes due to their chemical and thermal stability and mechanical strength.

Two major mechanisms for selective transport of the target molecule are distinguished: (i) the facilitated permeation mediated by the affinity binding and desorption to neighbored MIP sites (carrier-mediated transport), while the permeation of other solutes is hindered by the micropore structure of the membrane, and (ii) the retarded permeation due to affinity binding to the MIP sites, while the permeation of other solutes is faster through the membrane; in this case, the membrane is generally macroporous, and since selectivity is based on specific adsorption, these MIMs are considered membrane adsorbers (Ulbricht, 2004).

MIMs have been mostly studied for separation purpose (high selective separation systems), but they are being increasingly attractive also for biosensing applications. The binding of the template molecule may change the barrier properties of the MIM, such as the membrane permeability that may either increase or decrease ("gate effect"), which can be used as a transducer in a sensor system (Haupt and Mosbach, 2000; Hattori et al., 2004).

MIMs have been used for the detection of small molecules such as atrazine (Piletsky et al., 1995), 2-deoxyadenosine (Scorrano et al., 2015). One of the general problems of the imprinting technology is the random distribution and uneven accessibility of receptor sites. Also it is difficult to obtain a high yield of imprinting sites contemporary maintaining the separation efficiency of the membrane.

13.6 Recognition–Transduction Mechanisms and Systems

Transduction is the process by which the biochemical interaction between the bioreceptor and the target molecule is converted into observable output signal, whose intensity is proportional to the analyte concentration. The bio-recognition processes, affinity recognition and catalytic recognition, offer different methods of detection (Scheme 13.1). The most used detection methods in membrane-based biosensors are electrochemical and optical. Each detection mechanism differs in the design of the device, method of fabrication, and application area. Also an increasing interest is devoted to miniaturization of the systems and improvement in portability.

13.6.1 Electrochemical Detection

In electrochemical biosensors, the biochemical interaction between the immobilized bioreceptor and the target molecule results in the production or consumption of ions or electrons, affecting the electrical conditions of the system, for example the membrane electrical resistance (McGraw et al., 2012) or the membrane potential. Thus, the membrane can be directly involved in the transduction process, or in other systems, it may act as physical support or barrier structure.

Based on the type of signal measured, the electrochemical detection techniques can be further classified into amperometric, potentiometric, and impedimetric.

13.6.1.1 Amperometric

Amperometric methods are based on the measurement of the current resulting from the electrochemical oxidation or reduction of an electroactive species, by maintaining a constant potential at a working electrode (generally made of platinum, gold, or carbon) with respect to a reference electrode (generally made of Ag/AgCl) in an electrochemical cell. The current is then related to the concentration of the analyte present. Amperometric biosensors make use of redox enzymes (enzymes that catalyze reactions that produce or consume electrons), such as oxidases, dehydrogenases, and peroxidases that either consume oxygen, or produce hydrogen peroxide, or produce (indirectly) the reduced form of NAD(P), during the course of the catalytic reaction with the substrate of interest (Prodromidis and Karayannis, 2001). The first enzyme-based biosensor was developed in 1962 by Clark and Lyons for glucose monitoring, by entrapping the enzyme glucose oxidase (GOx) over an oxygen electrode via a semipermeable dialysis membrane. Today, most in vivo and ex vivo glucose biosensors present a negatively charged inner cellulose membrane to decrease the interfering effect of electroactive species, such as ascorbate or ureate, and an outer barrier membrane to control the substrate diffusion and to avoid the access of high molecular weight interferences to the electrode (Cordeiro et al., 2016; Thévenota et al., 2001). The membrane may also act as a matrix for enzyme immobilization to the electrode.

Enzyme-immobilized membranes have been largely used for the construction of other amperometric biosensors. For example, the enzyme urease was immobilized on acrylonitrile (AN) copolymer membrane for urea detection. The enzyme-immobilized membrane was attached to a Pt working electrode with the non-selective side facing the surface of the electrode. Then it was placed in an electrochemical cell and a potential was applied to the working electrode, and the electrochemical current was awaited to become stationary. The addition of the urea substrate into the cell resulted in a current change, which was recorded (Gabrovska et al., 2011).

In another example, choline oxidase (ChOx) was immobilized by entrapment in a hybrid mesoporous alumina membrane for choline detection. The sensor system consisted of three electrodes—an Ag/AgCl reference electrode, a Pt auxiliary electrode, and a Pt working electrode coated with the enzyme-immobilized membrane. After injection of the substrate, the resulting current from the generation of hydrogen peroxide was measured (Shimomura et al., 2009).

In labeled detection systems, such as sandwich-type immunosensors, enzyme-labeled antibodies are generally used to obtain an electrochemical response after target antigen binding to the capture antibody. For example, Hamid et al. (1999) developed a flow-through immunosensor system equipped with an immunofiltration membrane where a sandwich immunoassay was employed to electrochemically detect *E. coli* cells. The amperometric immunosensor consisted of an antibody-coupled membrane resting on a carbon working electrode, an Ag/AgCl reference electrode, and a carbon counter electrode. The solution containing the *E. coli* cells was allowed to flow through the membrane; then a peroxidase-conjugated second antibody was captured on the membrane surface and the enzyme activity was measured electrochemically.

13.6.1.2 Potentiometric

Potentiometric methods consist of the measurement of a potential difference between two electrodes, a working and a reference electrode, which accumulates during the recognition process in an electrochemical cell when zero or no significant current flows through the electrode, i.e., in equilibrium conditions (Grieshaber et al., 2008). The dominant part of potentiometric biosensors consists of ion-selective electrodes (ISEs) that are based on ion-selective membranes used with ion-exchanger or ionophore (plasticized PVC is the most popular membrane material). In an electrochemical cell, the ion-selective membrane separates an inner reference solution and an outer sample solution of different ionic activities and the potential developed at the membrane is the result of either an ion-exchange process or an ion transport process occurring at each interface between the membrane and the solution. The potential developed is proportional to the logarithms of the concentration of the active species, measured in relation to a reference electrode. The membrane establishes the preference with which the sensor responds to the analyte even in the presence of other interfering ions, thus giving the selectivity of the sensor. A typical potentiometric biosensor, such as for urea detection, employs an enzyme (urease) that catalyzes the transformation of the neutral analyte (urea) into an ionic product (NH_4^+) that the electrode (ISE) is designed to sense (Mascini and Guilbault, 1977).

More recent biosensor designs are based on the modulation of ion fluxes at the membrane/sample interface caused by a bio-recognition event that takes place at or near the membrane surface (Pawlak and Bakker, 2014).

Recently, Ozdemir et al. (2013) developed a potentiometric biosensor for the detection of antigen–antibody interaction at an ion-selective membrane surface for point-of-care (POC) diagnostics. A PVC cation-responsive membrane electrode was prepared, and a steady-state flux of a quaternary ammonium marker ion across the membrane was established. The mechanism of action of the sensor was based on the disturbance of the marker ion flux as a result of the target molecule binding on the membrane surface. As the membrane was sensitive to the marker ion that leached out from an internal solution to the sample side where the antigen was present, the binding of the antigen to the antibody (immobilized on the membrane surface) disturbed the flux of the marker ion, resulting in an increase in its surface concentration and, hence, in the measured potential.

Pawlak et al. (2015) developed a new electrochemical sensor for Concanavalin A based on blocking the surface of plasticized PVC membranes modified with D-mannose. The interaction of D-mannose with Concanavalin A at the surface perturbed the flux of a marker ion for which the ion-selective membrane was responsive, and this resulted in a change in the electrochemical signal.

Shishkanova et al. (2007) developed a potentiometric biosensor based on PVC membrane ion-selective electrode modified with single-stranded oligonucleotides for the detection of hybridization events. The hybridization process between the complementary single-stranded (ss) oligonucleotides induced an ion concentration redistribution at the phase boundary PVC membrane/solution that was detected by potentiometric methods.

13.6.1.3 Impedimetric

Impedimetric methods are based on the measurement of the impedance change caused by binding of targets to bioreceptors immobilized on the electrode surface. Impedimetric detection is mostly used for affinity biosensors to monitor immunological binding events such as antibody–antigen binding or hybridization processes. When the bio-recognition event occurs, the electrode is coated with a blocking layer, resulting in an increase in the electron transfer resistance (Prodromidis, 2010). The most popular method is electrochemical impedance spectroscopy (EIS).

Membranes with well-defined nanopores such as aluminum-anodized membranes are commonly used for immobilizing bioreceptors to construct impedance biosensors. These membranes can be fabricated with nanopore sizes comparable to the size of small biomolecules such as DNA, so they can be used to detect DNA molecules by monitoring the ion conductivity change in the nanopores. The sensing mechanism basically relies on the blocking of the nanochannels of a membrane-modified electrode upon the formation of an affinity complex at the nanopore membrane.

For example, nanoporous Al_2O_3 membrane with covalently linked DNA has been used to detect target DNA by monitoring the increase in impedance at the electrode upon DNA hybridization, which resulted from blocking the channels to ionic flow. A platinum working electrode was placed in close contact with the tested side of the membrane, while a screen counter electrode was in contact with the opposite side and the $Fe(CN)_6^{4-/3-}$ and $Ru(NH_3)_6^{2+/3+}$ redox pairs were used to probe the efficiency of ion blockage (Vlassiouk et al., 2005). In another similar example, DNA hybridization was again used to create ion blockages in alumina membranes. As before, the hybridization of the probe and target DNA took place in the pores of the membranes, and EIS was used to quantify the DNA. In addition to the hybridization process, the DNA probe was extended by Taq DNA polymerase as long as the target DNA strand and this led to a substantial enhancement of the ion blockage, thus increasing the sensitivity of the sensor (Wang et al., 2009).

In another study, nanoporous alumina membranes with 100 nm pore sizes were used to develop a membrane-based impedimetric biosensor for pathogenic bacteria detection. The measurement system consisted of the antibody-coupled membrane, which divided the reaction space into an upper and a lower chamber, and the electrodes. The binding of target bacteria on the surface of the antibody-coupled membrane affected the flow of ions through the membrane in a concentration-dependent manner, which was monitored by impedance spectra and analyzed by normalized impedance change (Joung et al., 2013).

Bacterial cells have also been detected in a similar manner. The bacteria were captured by the antibody-immobilized membrane, resulting in the blockage of pores, and thus the flow of the electrolyte, leading to an increase in the impedance of the sensor (Tan et al., 2012).

In a different example, conducting polymer electrodes were fabricated by chemical deposition of polypyrrole on the surface of a microporous polycarbonate membrane to detect Salmonella-infecting phage. Salmonella host cells were absorbed onto the surface of the film and reacted with infecting bacteriophage. The bacteriophage-mediated host cell lysis resulted in electrochemical changes in the supporting conducting polymer film, which was detected using impedance spectroscopy (Dadarwal et al., 2009).

13.6.2 Optical Detection

Optical methods are widely used in biosensors as they are characterized by a high sensitivity and the possibility of using various spectroscopic techniques. Most of the optical detections are based on the measurement of luminescence, fluorescence, color changes, by the measurement of absorbance, reflectance, or fluorescence emissions. Moreover, the use of methods such as surface-enhanced Raman spectroscopy and interferometry is also common (Wigginton and Vikesland, 2010).

In most cases, the detection mechanism involves the use of label molecules such as dyes or fluorescent markers that emit light at specific wavelengths, or enzymes that catalyze colorimetric reactions. For example, liposomes with entrapped dye were conjugated to DNA probes and were used as labels to optically detect pathogen mRNA in a sandwich-hybridization assay format. The DNA probes were immobilized on a polyetersulfone membrane and used for hybridization with the target sequences. Then the target sequences hybridized to a second liposome-conjugated DNA probe and the amount of liposomes captured in the detection zone were quantified with a reflectometer (Baeumner et al., 2004). In a different system, horseradish peroxidase (HRP) enzyme was used as label for the colorimetric detection of *E. coli*. A capture antibody was immobilized on the detection zone (nitrocellulose membrane) of an immunostrip; then the HRP-conjugated second antibody was used to bind to the captured pathogen, and the addition of the tetramethylbenzidene substrate allowed the colorimetric reaction to take place and the optical detection (Park et al., 2008).

Another way to produce a visual color change is through the attachment of gold or carbon nanoparticles to sensing molecules (Blazkova et al., 2011; Chua et al., 2011).

Fluorescent dyes are also commonly used for optical detection. For example, liposomes filled with fluorescent dyes were used as labels to antibodies for the detection of antigens. In this case, an integrated microfluidic system was developed combining sample pre-concentration and liposome-based fluorescent detection and signal amplification. The fluorescent dye-liposomes conjugated with the antibody were mixed with the antigens to form the antigen–liposome complexes, and then they were electrokinetically concentrated at a nanoporous membrane by applying a potential across the membrane. The concentrated complexes were eluted to a detection region where captured liposomes were lysed to release fluorescent dye molecules that were then quantified using image processing (Connelly et al., 2012).

In a different example, antibodies were conjugated with CdSe/ZnS quantum dots (QDs) that were used as fluorescent label for the detection of *Salmonella typhi*. The biosensor design was based on the traditional immunoreaction between antibodies and antigens. A capture antibody was immobilized on porous PC membrane, and then the membrane with the captured bacteria was incubated with antibody-conjugated QDs for the formation of sandwich complex and fluorescent detection (Jackeraya et al., 2011).

Due to the moderate transparency in the UV and IR regions, nanoporous alumina membranes were used for label-free detection of DNA hybridization by optical and IR adsorption. Single-stranded DNA (ss-DNA) were covalently immobilized inside the pores of the membrane that was used for both target DNA separation and direct detection after hybridization (Vlassiouk et al., 2004).

13.7 Applications of Nanostructured Membranes in Biosensors

The increased surface area, together with the properties of selective barrier and the possibility of integration with both electrochemical and optical detection methods, allowed the use of nanostructured membranes in numerous biosensing applications, such as glucose detection, nucleic acid detection, as well as bacteria and virus detection for either diagnostics or environmental and food analysis. The integration of nanostructured membranes in biosensor design has different advantages in terms of sensitivity, response time, and lower detection limits. Table 13.1 summarizes various applications of membranes for biosensing, comprising membrane sensor types, detection methods, limits, and time of detection.

Table 13.1 Some applications of membrane biosensors

Biosensor application	Analyte	Membrane biosensor	Detection method	Detection limit	Detection time	Ref.
Bacteria detection	*E.coli* DNA sequence	Alumina membrane with immobilized DNA probe	Impedimetric	0.5 nM	-	Wanga et al. (2009)
	B. anthracis mRNA sequence	Polyethersulfone membrane with immobilized DNA probe	Optical (dye labels)	1 nM	4 h	Baeumner et al. (2004)
	Salmonella typhi	Polycarbonate membrane with immobilized antibody	Optical (fluorescent labels)	100 cells/ml	2.5 h	Jackeraya et al. (2011)
	S. aureus	Alumina membrane with immobilized antibody integrated in microfluidic chip	Impedimetric	10^2 CFU/ml	2 h	Tan et al. (2011)
	E. coli cells	Nylon immunofilter membrane with immobilized anti-*E. coli* antibody	Amperometric (enzyme-labeled)	100 cells/ml	30 min	Hamid et al. (1999)
			Optical (fluorescent labels)	2.3 CFU/ml	30 min	Liu et al. (2007)
		Alumina membrane with immobilized antibody	Impedimetric	22 CFU/ml	90 min	Cheng et al. (2011)
Virus detection	Hepatitis B antigen	Nylon immunofilter membrane with immobilized antibody	Optical (fluorescent labels)	5 ng/mL	30 min	Liu et al. (2007)

Biosensor application	Analyte	Membrane biosensor	Detection method	Detection limit	Detection time	Ref.
Virus detection	Dengue virus RNA	Polyethersulfone membrane strip with immobilized capture DNA probes	Optical (dye labels)	50 molecules for the serotype 2	25 min	Zaytseva et al. (2004)
	West Nile viral proteins and particles	Alumina membrane with immobilized antibody	Impedimetric	4 pg/ml proteins, 2 viral particles per 100 ml	30 min	Nguyen et al. (2009)
Cancer biomarker detection	CA15-3 glycoprotein	Nanoporous alumina filter membrane with immobilized anti CA15-3 antibody	Voltammetric	52 U/ml	-	Muñiz and Merkoç (2011)
Cholesterol detection	Cholesterol	Silicic sol–gel membrane with immobilized cholesterol oxidase	Amperometric	1.2×10^{-7} M	-	Li et al. (2003)
		Cholesterol oxidase/ polypyrrole membrane	Amperometric	6×10^{-7} M	-	Li and Gu (2006)
Glucose detection	Glucose	Nylon nanofibrous membrane with immobilized glucose oxidase	Amperometric	6×10^{-6} M	-	Scampicchio et al. (2010)
		Nanoporous ZrO_2/Chitosan composite film with immobilized glucose oxidase and covered with Nafion film	Amperometric	1.0×10^{-5} M	10 s response time	Yanga et al. (2004)

13.8 Conclusion

Nanostructured membranes are increasingly used for the development of biosensors in a wide range of applications, including diagnostics as well as environmental and food analysis, as demonstrated by the literature data of the last few years. The main advantages of using membranes to develop biosensors are the high ratio of surface area to volume (which, compared to flat surfaces, enables the immobilization of enhanced quantity of bioreceptor, which results in an increased surface reaction and capturing area) and the ability to selectively separate molecules based on different mechanisms (i.e., on their size, shape, charge, and interactions with pores). Therefore, the membranes have been directly involved in the transduction process or/and they have been used as physical support or barrier structure, and even they have been used to filter or concentrate the analyte. Their physical structure and chemical composition may be very different, but they may accomplish one or combine all of the mentioned functions. The combination of these membranes with realizable detection techniques such as electrochemical and optical detection methods allowed the development of detection systems with the advantages of rapid response, high sensitivity, and low detection limits.

Finally, the integration of nanostructured membranes in miniaturized systems is expected to strongly contribute to the development of novel detection systems with enhanced and multiple functionality, by providing devices that are small, portable, disposable, low cost, and versatile.

Thanks to the increased attention toward safety (food, water, air, environment, etc.), it is expected that biosensors will diffuse more and more also in uncommon areas and among traditional users. To implement this popularization, simple detection systems (such as optical signal) are needed.

Acknowledgments

The authors gratefully acknowledge for the financial support to the project Nanomax (N-CHEM): "Nanomax-integrable sensors for pathological biomarkers diagnosis" from Programma Nazionale della Ricerca 2011-2013 (Progetto Bandiera).

Appendix

Abbreviations

AAO	anodic aluminum oxide
AN	acrylonitrile
ChOx	choline oxidase
DNA	deoxyribonucleic acid
EBL	electron beam lithography
EIS	electrochemical impedance spectroscopy
FIB	focused ion beam
LBL	layer-by-layer
GOx	glucose oxidase
HRP	horseradish peroxidase
IgG	immunoglobulin G
IR	infrared
ISEs	ion-selective electrodes
MIMs	molecular imprinted membranes
MIPs	molecularly imprinted polymers
mRNA	messanger ribonucleic acid
NAD(P)	nicotinamide adenine dinucleotide (phosphate)
NC	nitrocellulose
NIPS	non-solvent-induced phase separation
SEM	scanning electron microscopy
SiN	silicon nitride
PC	polycarbonate
PDMS	polydimethylsiloxane
PES	polyethersulfone
PET	polyethylene terephthalate
PIPS	polymerization-induced phase separation
PLA	polylactic acid
POC	point-of-care
PP	polypropylene
PVC	polyvinyl chloride
PVDF	polyvinylidene fluoride
QDs	quantum dots
ss-DNA	single-stranded DNA
TIPS	thermally induced phase separation
UV	ultraviolet–visible

Notations

CFU	colony forming units
M	molar (mol/L)
U	units

References

Adiga, S. P., Jin, C., Curtiss, L. A., Monteiro-Riviere, N. A., and Narayan, R. J. (2009). Nanoporous membranes for medical and biological applications, *WIREs Nanomed. Nanobiotechnol.*, **1**, pp. 568–581.

Algieri, C., Drioli, E., Guzzo, L., and Donat, L. (2014). Bio-mimetic sensors based on molecularly imprinted membranes, *Sensors*, **14**, pp. 13863–13912.

Alvarez, S. D., Li, C. P., Chiang, C. E., Schuller, I. K., and Sailor, M. J. (2009). A label-free porous alumina interferometric immunosensor, *ACS Nano*, **3**, pp. 3301–3307.

Ambrozy, A., Hlavata, L., and Labuda, J. (2013). Protective membranes at electrochemical biosensors, *Acta Chim. Slov.*, **6**, pp. 35–41.

Anzai, J., Hoshi, T., and Osa, T. (2000). Avidin–biotin mediated biosensors, in: *Biosensors and Their Applications*, Yang, V. C. and Ngo, T. T. (eds.) (Springer, US), pp. 35–46.

Arecchi, A., Scampicchio, M., Druschb, S., and Mannino, S. (2010). Nanofibrous membrane-based tyrosinase-biosensor for the detection of phenolic compounds, *Anal. Chim. Acta*, **659**, pp. 133–136.

Arecchi, A., Scampicchio, M., Brenna, O. V., and Mannino, S. (2010). Biocatalytic nylon nanofibrous membranes, *Anal. Bioanal. Chem.*, **398**, pp. 3097–3103.

Baeumner, A. J., Leonard, B., McElwee, J., and Montagna, R. A. (2004). A rapid biosensor for viable B-anthracis spores, *Anal. Bioanal. Chem.*, **380**, pp. 15–23.

Barhoumia, H., Maarefa, A., Rammaha, M., Marteletb, C., Jaffrezicb, N., Moustyc, C., Viald, S., and Forano, C. (2006). Urea biosensor based on Zn_3Al–urease layered double hydroxides nanohybrid coated on insulated silicon structures, *Mater. Sci. Eng. C*, **26**, pp. 328–333.

Blanco, R. M., Calvete, J. J., and Guisán, J. M. (1989). Immobilization–stabilization of enzymes: Variables that control the intensity of the trypsin (amine)-agarose (aldehyde) multi-point covalent attachment, *Enzyme Microb. Technol.*, **11**, pp. 353–359.

Blazkova, M., Javurkova, B., Fukal, L., and Rauch, P. (2011). Immunochromatographic strip test for detection of genus cronobacter, *Biosens. Bioelectron.*, **26**, pp. 2828–2834.

Cheng, M. S., Lau, S. H., Chow, V. T., and Toh, C. S. (2011). Membrane-based electrochemical nanobiosensor for *Escherichia coli* detection and analysis of cells viability, *Environ. Sci. Technol.*, **45**, pp. 6453–6459.

Chua, A. L., Yean, C. Y., Ravichandran, M., Lim, B., and Lalitha, P. (2011). A rapid DNA biosensor for the molecular diagnosis of infectious disease, *Biosens. Bioelectron.*, **26**, pp. 3825–3831.

Clark, L. C. and Lyons, C. (1962). Electrode systems for monitoring in cardiovascular surgery, *Ann. N Y Acad. Sci.*, **102**, pp. 29–45.

Cordeiro, C. A., de Vriesa, M. G., Cremersa, T. I. F. H., and Westerink, B. H. C. (2016). The role of surface availability in membrane-induced selectivity for amperometric enzyme-based biosensors, *Sensor Actuat. B Chem.*, **223**, pp. 679–688.

Connelly, J. T., Kondapalli, S., Skoupi, M., Parker, J. S. L., Kirby, B. J., and Baeumner, A. J. (2012). Micro-total analysis system for virus detection: Microfluidic pre-concentration coupled to liposome-based detection, *Anal. Bioanal. Chem.*, **402**, pp. 315–323.

Coulet, P. R. (1992). Polymeric membranes and coupled enzymes in the design of biosensors, *J. Membr. Sci.*, **68**, pp. 217–228.

Dadarwal, R., Namvar, A., Thomas, D. F., Hall, J. C., and Warriner, K. (2009). Organic conducting polymer electrode-based sensors for detection of salmonella-infecting bacteriophages, *Mater. Sci. Eng. C*, **29**, pp. 761–765.

Davies, M. L., Hamilton, C. J., Murphy, S. M., and Tighe, B. J. (1992). Polymer membranes in clinical sensors, *Biomaterials*, **13**, pp. 971–978.

Desai, T. A., Hansfordb, D. J., Leonia, L., Essenpreisc, M., and Ferrari, M. (2000). Nanoporous anti-fouling silicon membranes for biosensor applications, *Biosens. Bioelecron.*, **15**, pp. 453–462.

Diculescu, V. C., Paquim, A. M. C., and Brett, A. M. O. (2005). Electrochemical DNA sensors for detection of DNA damage, *Sensors*, **5**, pp. 377–393.

D'Souza, S. F. (2001). Immobilization and stabilization of biomaterials for biosensor applications, *Appl. Biochem. Biotechnol.*, **96**, pp. 225–238.

Duckworth, B. P., Xu, J., Taton, T. A., Guo, A., and Distefano, M. D. (2006). Site-specific, covalent attachment of proteins to a solid surface, *Bioconjugate Chem.*, **17**, pp. 967–974.

Gabrovska, K., Ivanov, J., Vasileva, I., Dimova, N., and Godjevargova, T. (2011). Immobilization of urease on nanostructured polymer membrane and preparation of urea amperometric biosensor, *Intern. J. Biol. Macromol.*, **48**, pp. 620–626.

Gao, J., Xu, J., Locascio, L. E., and Lee, C. S. (2001). Integrated microfluidic system enabling protein digestion, peptide separation, and protein identification, *Anal. Chem.*, **73**, pp. 2648–2655.

Grieshaber, D., MacKenzie, R., Vörös, J., and Reimhult, E. (2008). Electrochemical biosensors: Sensor principles and architectures, *Sensors*, **8**, pp. 1400–1458.

Hamid, I. A., Ivnitski, D., Atanasov, P., and Wilkins, E. (1999). Flow-through immunofiltration assay system for rapid detection of *E. coli* O157:H7, *Biosens. Bioelectron.*, **14**, pp. 309–316.

Hamid, I. A., Ivnitski, D., Atanasov, P., and Wilkins, E. (1999). Highly sensitive flow-injection immunoassay system for rapid detection of bacteria, *Anal. Chim. Acta*, **399**, pp. 99–108.

Hartwell, S. K. and Grudpan, K. (2010). Flow based immuno/bioassay and trends in micro-immuno/biosensors, microfluidic immunoassays, *Microchim. Acta*, **169**, pp. 201–220.

Hattoria, K., Hiwataria, M., Iiyamab, C., Yoshimib, Y., Kohoria, F., Sakaia, K., and Piletskyc, S. A. (2004). Gate effect of theophylline-imprinted polymers grafted to the cellulose by living radical polymerization, *J. Membr. Sci.*, **233**, pp. 169–173.

Haupt, K. and Mosbach, K. (2000). Molecularly imprinted polymers and their use in biomimetic sensors, *Chem. Rev.*, **100**, pp. 2495–2504.

Hermanson, G. T. (1996). *Bioconjugate Techniques*, 2nd Ed. (Academic Press, San Diego).

Hisamoto, H., Shimizu, Y., Uchiyama, K., Tokeshi, M., Kikutani, Y., Hibara, A., and Kitamori, T. (2003). Chemicofunctional membrane for integrated chemical processes on a microchip, *Anal. Chem.*, **75**, pp. 350–354.

Hong, H. B., Krause, H. J., Song, K. B., Choi, C. J., Chung, M. A., Son, S. W., and Offenhaeusser, A. (2011). Detection of two different influenza A viruses using a nitrocellulose membrane and a magnetic biosensor, *J. Immunol. Methods*, **365**, pp. 95–100.

Jackeraya, R., Abida, C. K. V. Z., Singha, G., Jaina, S., Chattopadhyayaa, S., Saprab, S., Shrivastavc, T. G., and Singh, H. (2011). Selective capturing and detection of *Salmonella typhi* on polycarbonate membrane using bioconjugated quantum dots, *Talanta*, **84**, pp. 952–962.

Jianronga, C., Yuqing, M., Nongyueb, H., Xiaohuaa, W., and Sijiaoa, L. (2004). Nanotechnology and biosensors, *Biotechnol. Adv.*, **22**, pp. 505–518.

Jong, J., Lammertink, R. G. H., and Wesslinga, M. (2006). Membranes and microfluidics: A review, *Lab. Chip.*, **6**, pp. 1125–1139.

Joung, C. K., Kima, H. N., Lima, M. C., Jeonb, T. J., Kima, H. Y., and Kim, Y. R. (2013). A nanoporous membrane-based impedimetric immunosensor for label-free detection of pathogenic bacteria in whole milk, *Biosens. Bioelectron.*, **44**, pp. 210–215.

Kececi, K., Sanb, N., and Kayab, D. (2015). Nanopore detection of double stranded DNA using a track-etched polycarbonate membrane, *Talanta*, **144**, pp. 268–274.

Kim, D. and Herr, A. E. (2013). Protein immobilization techniques for microfluidic assays, *Biomicrofluidics*, **7**, pp. 1–47.

Lee, G. U. and Yanavich, C. (2004). Nanoporous membrane immunosensor, US Patent 6,676,904.

Leech, D. (1994). Affinity biosensors, *Chem. Soc. Rev.*, **24**, pp. 205–216.

Li, J., Peng, T., and Pen, Y. (2003). A cholesterol biosensor based on entrapment of cholesterol oxidase in a silicic sol-gel matrix at a Prussian blue modified electrode, *Electroanalysis*, **15**, pp. 1031–1037.

Li, J. P. and Gu, H. N. (2006). A selective cholesterol biosensor based on composite film modified electrode for amperometric detection, *J. Chin. Chem. Soc.*, **53**, pp. 575–582.

Lia, D., Freya, M. W., and Baeumnerb, A. J. (2006). Electrospun polylactic acid nanofiber membranes as substrates for biosensor assemblies, *J. Membr. Sci.*, **279**, pp. 354–363.

Liu, Y., Brandon, R., Cate, M., Peng, X., Stony, R., and Johnson, M. (2007). Detection of pathogens using luminescent CdSe/ZnS dendron nanocrystals and a porous membrane immunofilter, *Anal. Chem.*, **79**, pp. 8796–8802.

Lu, Y., Shi, W., Qin, J., and Lin, B. (2010). Fabrication and characterization of paper-based microfluidics prepared in nitrocellulose membrane by wax printing, *Anal. Chem.*, **82**, pp. 329–335.

Lutz, J. F. (2008). Copper-free azide–alkyne cycloadditions: New insights and perspectives, *Angew. Chem. Int. Edit.*, **47**, pp. 2182–2184.

Maines, A., Ashworth, D., and Vadgama, P. (1997). Diffusion restricting outer membranes for greatly extended linearity measurements with glucose oxidase enzyme electrode, *Anal. Chim. Acta*, **323**, pp. 223–231.

Mascini, M. and Guilbault, G. G. (1977). Urease coupled ammonia electrode for urea determination in blood serum, *Anal. Chem.*, **49**, pp. 795–798.

Makaraviciute, A. and Ramanaviciene, A. (2013). Site-directed antibody immobilization techniques for immunosensors, *Biosens. Bioelectron.*, **50**, pp. 460–471.

McGraw, S. K., Alocilja, E., Senecal, K., and Seneca, A. (2012). A resistance based biosensor that utilizes conductive microfibers for microbial pathogen detection, *OJAB*, **1**, pp. 24554–24562.

Mignani, A., Scavetta, E., Guadagnini, L., and Tinelli, D. (2009). Comparative study of protective membranes for glucose biosensors based on electrodeposited hydrotalcites, *Sensor Actuat. B Chem.*, **136**, pp. 196–202.

Militano, F., Poerio, T., Mazzei, R., Piacentini, E., Gugliuzza, A., and Giorno, L. (2016). Influence of protein bulk properties on membrane surface coverage during immobilization, *Colloids Surf. B*, **143**, pp. 309–317.

Muñiz, A. E. and Merkoç, A. (2011). A nanochannel/nanoparticle-based filtering and sensing platform for direct detection of a cancer biomarker in blood, *Small*, **7**, pp. 675–682.

Nakanishi, K., Sakiyama, T., Kumada, Y., Imamura, K., and Imanaka, H. (2008). Recent advances in controlled immobilization of proteins onto the surface of the solid substrate and its possible application to proteomics, *Curr. Proteomics*, **5**, pp. 161–175.

Nguyen, B. T. T., Koh, G., Lim, H. S., Chua, A. J. S., Ng, M. M. L., and Toh, C. S. (2009). Membrane-based electrochemical nanobiosensor for the detection of virus, *Anal. Chem.*, **81**, pp. 7226–7234.

Ozdemir, M. S., Marczak, M., Bohets, H., Bonroy, K., Roymans, D., Stuyver, L., Vanhoutte, K., Pawla, M., and Bakker, E. (2013). A label-free potentiometric sensor principle for the detection of antibody–antigen interactions, *Anal. Chem.*, **85**, pp. 4770–4776.

Park, S., Kim, H., Paek, S. H., Hong, J. W., and Kim, Y. K. (2008). Enzyme-linked immuno-strip biosensor to detect *Escherichia coli* O157: H7, *Ultramicroscopy*, **108**, pp. 1348–1351.

Pawlak, M. and Bakker, E. (2014). Chemical modification of polymer ion-selective membrane electrode surfaces, *Electroanalysis*, **46**, pp. 1121–1131.

Pawlak, M., Mistlberger, G., and Bakker, E. (2015). Concanavalin A electrochemical sensor based on the surface blocking principle at an ion-selective polymeric membrane, *Microchim. Acta*, **182**, pp. 129–137.

Piletsky, S. A., Dubey, I. Y., Fedoryak, D. M., and Kukhar, V. P. (1990). Substrate-selective polymeric membranes: Selective transfer of nucleic acids components, *Biopolym. Cell.*, **6**, pp. 55–58.

Piletsky, S. A., Piletskaya, E. V., Elgersma, A. V., Yano, K., and Karube, I. (1995). Atrazine sensing by molecularly imprinted membranes, *Biosens. Bioelectron.*, **10**, pp. 959–964.

Prasad, S., Yadav, Y., Bothara, M., Kunduru, V., and Muthukumar, S. (2010). Nanoporous membrane-based microfluidic biosensors, in: *Microfluidic Devices in Nanotechnology*, Kumar, C. S. (ed.) (John Wiley & Sons, Inc., Hoboken, NJ, USA), pp. 47–90.

Prodromidis, M. I. (2010). Impedimetric immunosensors: A review, *Electrochim. Acta*, **55**, pp. 4227–4233.

Prodromidis, M. I. and Karayannis, M. I. (2001). Enzyme-based amperometric biosensors for food analysis, *Electroanalysis*, **14**, pp. 241–261.

Redeker, E. S., Ta, D. T., Cortens, D., Billen, B., Guedens, W., and Adriaensens, P. (2013). Protein engineering for directed immobilization, *Bioconjugate Chem.*, **24**, pp. 1761–1777.

Ren, J., Shi, W., Li, K., and Ma, Z. (2012). Ultrasensitive platinum nanocubes enhanced amperometric glucose biosensor based on chitosan and nafion film, *Sensor Actuat. B Chem.*, **163**, pp. 115–120.

Ricci, F., Volpe, G., Micheli, L., and Palleschi, G. (2007). A review on novel developments and applications of immunosensors in food analysis, *Anal. Chim. Acta*, **605**, pp. 111–129.

Sassolas, A., Blum, L. J., and Leca-Bouvier, B. D. (2012). Immobilization strategies to develop enzymatic biosensors, *Biotechnol. Adv.*, **30**, pp. 489–511.

Scampicchio, M., Arecchi, A., Lawrenceb, N. S., and Manninoa, S. (2010). Nylon nanofibrous membrane for mediated glucose biosensing, *Sensor Actuat. B Chem.*, **145**, pp. 394–397.

Scampicchio, M., Bulbarello, A., Arecchi, A., Cosio, M. S., Benedetti, S., and Mannino, S. (2012). Electrospun nonwoven nanofibrous membranes for sensors and biosensors, *Electroanalysis*, **24**, pp. 719–725.

Scorrano, S., Mergola, L., Di Bello, M. P., Lazzoi, M. R., Vasapollo, G., and Del Sole, R. (2015). Molecularly imprinted composite membranes for selective detection of 2-deoxyadenosine in urine samples, *Int. J. Mol. Sci.*, **16**, pp. 13746–13759.

Shimomuraa, T., Itohb, T., Sumiyaa, T., Mizukamib, F., and Ono, M. (2009). Amperometric determination of choline with enzyme immobilized in a hybrid mesoporous membrane, *Talanta*, **78**, pp. 217–220.

Shishkanova, T. V., Volf, R., Krondak, M., and Král, V. (2007). Functionalization of PVC membrane with ss oligonucleotides for a potentiometric biosensor, *Biosens. Bioelectron.*, **22**, pp. 2712–2717.

Slouka, Z., Senapati, S., and Chang, H. C. (2014). Microfluidic systems with ion-selective membranes, *Annu. Rev. Anal. Chem.*, **7**, pp. 317–335.

Smith, C. L., Milea, J. S., and Nguyen, G. H. (2005). Immobilization of nucleic acids using biotin-strept(avidin) systems, in: *Immobilisation of DNA on Chips II 261*, Wittmann, C. (ed.) (Springer Berlin Heidelberg), pp. 63–90.

Smuleac, V., Butterfield, D. A., and Bhattacharyya, D. (2006). Layer-by-layer-assembled microfiltration membranes for biomolecule immobilization and enzymatic catalysis, *Langmuir*, **22**, pp. 10118–10124.

Song, S. and Singh, A. K. (2006). On-chip sample preconcentration for integrated microfluidic analysis, *Anal. Bioanal. Chem.*, **384**, pp. 41–43.

Song, S., Singh, A. K., and Kirby, B. J. (2004). Electrophoretic concentration of proteins at laser-patterned nanoporous membranes in microchips, *Anal. Chem.*, **76**, pp. 4589–4592.

Striemer, C. C., Gaborski, T. R., McGrath, J. L., and Fauchet, P. M. (2007). Charge- and size-based separation of macromolecules using ultrathin silicon membrane, *Nature*, **445**, pp. 749–753.

Tan, F. (2012). *Foodborne Pathogens Detection with Nanoporous Anodic Aluminum Oxide Membrane Based Biosensor* (Hong Kong Polytechnic University, Hong Kong, China).

Tan, F., Leungb, P. H. M., Liua, Z., Zhangc, Y., Xiaoa, L., Yea, W., Zhange, X., Yic, L., and Yang, M. (2011). A PDMS microfluidic impedance immunosensor for *E. coli* O157:H7 and *Staphylococcus aureus* detection via antibody-immobilized nanoporous membrane, *Sensor Actuat. B Chem.*, **159**, pp. 328–335.

Theegala, C. S., Small, D. D., and Monroe, W. T. (2008). Oxygen electrode-based single antibody amperometric biosensor for qualitative detection of *E. coli* and bacteria in water, *J. Environ. Sci. Heal A*, **43**, pp. 478–487.

Thévenota, D. R., Tothb, K., Durstc, R. A., and Wilsond, G. S. (2001). Electrochemical biosensors: Recommended definitions and classifications, *Biosens. Bioelectron.*, **16**, pp. 121–131.

Tipnis, R., Vaddiraju, S. S., Jain, F., Burgess, D. J., and Papadimitrakopoulos, F. (2007). Layer-by-layer assembled semipermeable membrane for amperometric glucose sensors, *J. Diabetes Sci. Technol.*, **1**, pp. 193–200.

Tong, H. D., Jansen, H. V., Gadgil, V. J., Bostan, C. G., Berenschot, E., Rijn, C. J. M., and Elwenspoek, M. (2004). Silicon nitride nanosieve membrane, *Nano Lett.*, **4**, pp. 283–287.

Tuantranont, A. (2013). *Applications of Nanomaterials in Sensors and Diagnostics*, 1st Ed. (Springer Heidelberg).

Tudorache, M. and Emnéus, J. (2005). Selective immuno-supported liquid membrane (ISLM) extraction enrichment and analysis of 2,4,6-trichlorophenol, *J. Membr. Sci.*, **256**, pp. 143–149.

Ulbricht, M. (2004). Membrane separations using molecularly imprinted polymers, *J. Chromatogr. B*, **804**, pp. 113–125.

Ulbricht, M. (2006). Advanced functional membranes, *Polymer*, **47**, pp. 2217–2262.

Ulbricht, M. (2011). Separation membranes, in: *Porous Polymers*, Silverstein, M. S., Cameron, N. R., and Hillmyer, M. A. (eds.) (John Wiley & Sons), pp. 277–321.

Vaddirajua, S., Tomazosb, I., Burgessc, D. J., Jaind, F. C., and Papadimitrakopoulos, F. (2010). Emerging synergy between nanotechnology and implantable biosensors: A review, *Biosens. Bioelectron.*, **25**, pp. 1553–1565.

Vitola, G., Mazzei, R., Fontananova, E., and Giorno, L. (2015). PVDF membrane biofunctionalization by chemical grafting, *J. Membr. Sci.*, **476**, pp. 483–489.

Vlassiouk, I., Krasnoslobodtsev, A., Smirnov, S., and Germann, M. (2004). "Direct" detection and separation of DNA using nanoporous alumina filters, *Langmuir*, **20**, pp. 9913–9915.

Vlassiouk, I., Takmakov, P., and Smirnov, S. (2005). Sensing DNA hybridization via ionic conductance through a nanoporous electrode, *Langmuir*, **21**, pp. 4776–4778.

Wanga, L., Liub, Q., Hub, Z., Zhangb, Y., Wub, C., Yangc, M., and Wang, P. (2009). A novel electrochemical biosensor based on dynamic polymerase-extending hybridization for *E. coli* O157:H7 DNA detection, *Talanta*, **78**, pp. 647–652.

Wigginton, K. R. and Vikesland, P. J. (2010). Gold-coated polycarbonate membrane filter for pathogen concentration and sers-based detection, *Analyst*, **135**, pp. 1320–1326.

Wirguin, C. H. (1996). Recent advances in perfluorinated ionomer membranes: Structure, properties and applications, *J. Membr. Sci.*, **120**, pp. 1–33.

Wisniewski, N. and Reichert, M. (2000). Methods for reducing biosensor membrane biofouling, *Colloids Surf. B*, **18**, pp. 197–219.

Yang, D., Niu, X., Liu, Y., Wang, Y., Gu, X., Song, L., Zhao, R., Ma, L., Shao, Y., and Jiang, X. (2008). Electrospun nanofibrous membranes: A novel solid substrate for microfluidic immunoassays for HIV, *Adv. Mater.*, **20**, pp. 4770–4775.

Yang, L., Ren, X., Tang, F., and Zhang, L. (2009). A practical glucose biosensor based on Fe_3O_4 nanoparticles and chitosan/nafion composite film, *Biosens. Bioelectron.*, **25**, pp. 889–895.

Yang, Q., Adrus, N., Tomickiab, F., and Ulbricht, M. (2011). Composites of functional polymeric hydrogels and porous membranes, *J. Mater. Chem.*, **21**, pp. 2783–2811.

Yanga, Y., Yanga, H., Yanga, M., Liua, Y., Shena, G., and Yua, R. (2004). Amperometric glucose biosensor based on a surface treated nanoporous ZrO_2/chitosan composite film as immobilization matrix, *Anal. Chim. Acta*, **525**, pp. 213–220.

Yoo, E. H. and Lee, S. Y. (2010). Glucose biosensors: An overview of use in clinical practice, *Sensors*, **10**, pp. 4558–4576.

Yoshikawa, M., Izumi, J., Kitao, T., Koya, S., and Sakamoto, S. (1995). Molecularly imprinted polymeric membranes for optical resolution, *J. Membr. Sci.*, **108**, pp. 171–175.

Zaytseva, N. V., Montagna, R. A., Lee, E. M., and Baeumner, A. J. (2004). Multi-analyte single-membrane biosensor for the serotype-specific detection of Dengue virus, *Anal. Bioanal. Chem.*, **380**, pp. 46–53.

Zhanga, L., Wanga, Y., Chenb, M., Luoa, Y., Denga, K., Chena, D., and Fua, W. (2014). A new system for the amplification of biological signals: RecA and complimentary single strand DNA probes on a leaky surface acoustic wave biosensor, *Biosens. Bioelectron.*, **60**, pp. 259–264.

Zou, H., Luo, Q., and Zhou, D. (2001). Affinity membrane chromatography for the analysis and purification of proteins, *J. Biochem. Bioph. Meth.*, **49**, pp. 199–240.

Chapter 14

Nanoengineered Membranes in Biotechnology

Giuseppe Vitola, Rosalinda Mazzei, and Lidietta Giorno
Institute on Membrane Technology of the National Research Council of Italy (ITM-CNR), Via P. Bucci, Cubo 17C, 87036 Rende (CS), Italy
g.vitola@itm.cnr.it

This chapter provides basic knowledge of the production, characterization, and applications of nanoengineered membranes in biotechnological sectors. First, it describes the nanoscale methodologies for membrane synthesis and the functionalization techniques that can be applied to engineering existing membranes. Afterward, it discusses the analytical techniques that can be used to characterize the nanoengineered membranes and, finally, presents many examples of their applications in different biotechnological fields.

14.1 Introduction

Membrane and membrane processes play an important role in a wide range of applications either directly or indirectly. Different

Functional Nanostructured Membranes
Edited by Enrico Drioli, Lidietta Giorno, and Annarosa Gugliuzza

ISBN 978-981-4774-79-6 (Hardcover), 978-1-351-13511-5 (eBook)
www.panstanford.com

applications require different membranes in terms of structures and functions. Membranes are widely used in biotechnology with applications ranging from bioproduct purification and laboratory filtration to use as component of bioreactors, biosensors, and biomedical devices. Nanoengineered membranes, in addition to the function of selective barrier, exhibit advanced and tailored properties, provided by functional parts, of nano size, introduced with precise control in the membrane structure.

The interesting properties of high specific surface area and the possibility to combine separation with other advanced functions make membranes optimal candidate for the development of hybrid systems. Biomolecule–membrane hybrid systems are examples of nanoengineered membranes. A biomolecule is any molecule that is present in living organisms. Examples of biomolecules are large macromolecules such as proteins and nucleic acids, as well as small molecules such as metabolites. Biomolecules are nanoscale materials with intrinsic structure and properties.

Membranes nanoengineered by means of biomolecules (biofunctionalized membranes) have been employed in several biotechnological applications such as catalysis (membrane-based bioreactors), analysis (biosensors), separations (affinity membranes), and for the development of artificial organs. In bioreactors, the membranes can be used just as separation units downstream of a traditional reactor containing the biocatalyst, as in membrane bioreactors (MBRs), or can be exploited either as site for the biocatalyst confinement or separation unit, as in biocatalytic membrane reactors (BMRs).

This chapter mainly deals with BMRs, which are typical examples of devices using nanoengineered membranes. In the following paragraph, the membrane-engineering methodologies required for BMR development are analyzed. In particular, first the chemical surface modification methods for creating reactive sites on the membrane are reported. These sites can be exploited in some types of immobilization of the biocatalyst on membranes. Afterward, the techniques for biocatalyst immobilization and the characterization methods for examining the chemical modifications and/or the presence of biocatalysts on the membrane are discussed. In the last section, some applications of nanoengineered membranes in the biotechnology sectors, either in laboratory processes or at the industrial level, are presented.

14.2 Nanoengineered Membrane Production

Nanoengineering methodologies are valuable tools for designing appropriate membranes, as demanded interfacial properties and advanced functions can seldom be achieved by bulk modification of the membrane-forming material without disadvantages in membrane production. Several membrane-forming materials have been investigated to introduce binding sites for the immobilization of functional molecules. Most of these materials are polymers because a large variety of membrane structures can be realized with this group of materials. Applications of polymeric membranes such as cellulose (Tyagi et al., 2009), regenerated cellulose (RC) (Chen et al., 2007), cellulose acetate (CA) (Brady and Jordaan, 2009), polyacrylonitrile (PAN) (Godjevargova et al., 2004), polysulfone (PS) (Ulbricht and Riedel, 1998), nylon (Smuleac et al., 2006), polyvinylidene fluoride (PVDF) (Jolivalt et al., 2000), polypropylene (PP) (Becker et al., 2002), and polyethersulfone (PES) (Fang et al., 2009) have been reported. Also inorganic membranes made of silica, alumina, and titanium have been used for this purpose (Ranieri et al., 2016). The chemical functionalization strategies applied can be divided into five wide groups: coating, self-assembly, chemical treatment, plasma treatments, and surface graft polymerization. In addition, other promising and more recent approaches that permit production of well-defined nanometer-scale feature membranes are molecular imprinting, lithography, 3D printing, and anodization. Nanoporous membranes with uniform pore shape and pore-size distribution have a lot of applications, such as in catalysis, selective molecule separation, optics, electronics, purification, biosensing, molecule detection, and so on.

14.2.1 Coating

Coating is a simple but effective method for membrane surface modification, and it has been widely adopted to tailor the surface properties of reverse osmosis, nanofiltration, ultrafiltration, and microfiltration membranes (Louie et al., 2006; Kim and Lee, 2006; Musale et al., 1999; Wang et al., 2008). It is a modification method based on the addition of an extra layer on top of the native membrane surface. The materials are physically deposited on the membrane surface via one of the following mechanisms:

- Adsorption–adhesion, in which the functional layer is only physically attached on the base polymer; the binding strength can be increased via multiple interactions between functional groups in the macromolecular layer and on the solid surface
- Interpenetration by mixing between the added functional material and the base polymer in an interphase
- Entanglement of an added material layer on the pore structure of the membrane

Owing to the coating of these functional materials on the membrane, the surface property of the membrane could completely mutate, e.g., from hydrophobic or non-biocompatible to hydrophilic and biocompatible. However, this method cannot gain a stable surface for the materials absorbed on the membrane surface, which run away easily due to the relatively weak interaction (Kang and Cao, 2014).

14.2.2 Self-Assembly

Self-assembly is a relatively new technique for membrane surface modification. The technique involves self-assembled monolayers (SAMs) and layer-by-layer (LbL) assembly, which enable the fabrication of "engineerable" highly selective membrane top layers. SAMs are highly ordered molecular assemblies formed by the adsorption of functionalized alkanes onto a solid surface (Biebuyck et al., 1996). The thickness of SAM is commonly 2–3 nm and can be tuned with an accuracy of about 0.1 nm by varying the number of carbon atoms in the alkyl chain. It leads to a structure that is at, or close to, thermodynamic equilibrium and thus tends to self-healing/defect rejection and leads to a closely packed, well-ordered, and stable configuration on the surface (Ruckenstein and Li, 2005). The LbL technique is extremely versatile in that uniform thin films with a wide range of properties can be prepared onto various substrates, including membranes (Lee et al., 2007). The LbL process is simple and can be performed in mild conditions using environmentally friendly reagents. It has become a primary tool for thin-film fabrication. The basic process involves dipping a charged (e.g., cationic) substrate into a dilute aqueous solution of an anionic polyelectrolyte and allowing the polymer to adsorb and reverse the charge of the substrate surface. The negatively charged coated substrate is rinsed and dipped into a solution of cationic polyelectrolyte, which adsorbs and re-creates a positively charged surface. Sequential adsorptions of anionic and cationic polyelectrolytes allow the construction of multilayer films (Chen and McCarthy, 1997). The unique advantages of the method are that a wide variety of materials, both organic and inorganic, can be incorporated into LbL thin films and that the film architecture is largely controlled by the deposition sequence.

14.2.3 Chemical Treatment

Wet chemical treatment is one of the most common methods for membrane functionalization. Usually, the modifications allow retaining mechanical properties of the membrane, while changing the interfacial properties. In wet chemical surface modification, a membrane is treated with liquid reagents to generate reactive functional groups on the surface. This approach to surface modification can be conducted in most laboratories because it does not require expensive or special equipment. This method is also more capable of penetrating into the membrane pores compared to plasma and other energy source surface modification techniques (Goddard and Hotchkiss, 2007). Chemical treatments include oxidation, addition, substitution, and hydrolysis. Oxidative treatment is the dominant means for chemically modified reactions, which generally involve dry or wet oxidation. In the case of dry oxidation, a gaseous oxidation agent, such as oxygen, ozone, or carbon dioxide, is often used. Oxygen-containing groups such as hydroxyl and carboxyl are introduced onto the membrane surface by flame treatment or corona discharge in these gases. These groups can also be introduced onto a membrane by wet oxidation. Wet chemical oxidation involves the use of nitric acid, sulfuric acid, phosphoric acid, alone or in combination with hydrogen peroxide, sodium hypochlorite, permanganate, chromate or dichromate of potassium, transition metal nitrates, etc. (Xu et al., 2009).

14.2.4 Plasma Treatment

A very surface-selective functionalization method is excitation with plasma (Denes and Manolache, 2004). Plasma modifications exhibit multifunctional chemistries such as oxidation, degradation, changes in the carbon/hydrogen ratio, which cause crosslinking and structural changes occurring in the surface layer without substantially changing the bulk structure. Thus, these chemistries occurring only at the contact point between plasma and surface of the solid materials result in changed physicochemical properties of the surface. Plasma treatment for the surface modification of membranes includes plasma sputtering, etching, implantation, and spraying.

14.2.5 Surface Graft Polymerization

Surface functionalization by grafting methods can be usually divided into two classes: "surface graft polymerization (SGP)-to" and "SGP-from" processes (Jochems et al., 2011). In the "SGP-to" process, incorporation of preformed polymer chains bringing reactive groups can be tailor-made. The "SGP-from" method involves the incorporation and polymerization of monomers using an initiator and active species on the membrane surface. In this case, the polymerization of monomers goes from the membrane surface toward the outside bulk phase. Different methods have been used for the generation of reactive groups in the "SGP-from"; they can be classified as chemical, radiation, photochemical, and plasma-induced. The main advantage of SGP is that the membrane surface can be modified or tailored to acquire the desired properties by the choice of suitable grafting monomers, but at the same time the bulk properties can be maintained. In addition, another important advantage of the SGP methodology compared with the physical modification methods such as coating is the long-term chemical stability of the modified surface, ensured by the covalent attachment of polymer chains onto the membrane surface, which avoids detaching.

14.2.6 Molecular Imprinting

Molecular imprinting is a technique for designing artificial receptors with a predetermined selectivity and specificity for a given analyte, which can be used as ideal materials in various fields (Vasapollo et al., 2011). Molecular imprinting of polymers has been explored to prepare membranes with a predetermined affinity for a variety of molecules (He et al., 2009). Molecularly imprinted membranes (MIMs) combine the characteristics of membrane separation technology with the highly selective recognition of the molecular imprinting technique.

Several methods have been reported in the literature for preparing MIMs: bulk polymerization, which consists of simultaneous formation of imprinted sites and membrane structure; surface imprinting, an approach in which a molecularly imprinted polymer (MIP) is prepared on or in a preformed support membrane; and physical mixing, which consists of embedding MIP particles into a porous polymer matrix (Ulbricht, 2004). In all these approaches, a polymer solution containing a template is used to form an MIP.

14.2.7 Lithography

Lithography is an extremely complex tool based on the concept of "imprinting" an original template version onto mass output (Landis, 2013). Lithography techniques are essential for creating new devices, new functionalities, and exploring new application fields. Nanolithography is a fabrication technique for nanostructuring with resolution between the size of an individual atom and approximately 100 nm. It is the branch of nanotechnology that allows the study and fabrication of nanostructured devices in single or multiple patterning steps. At the beginning, nanolithography was carried out using techniques such as optical exposure, masking, and etching; nowadays, more advanced techniques are also employed, such as exposure to X-rays, high-energy UV light, and electron beams. Nanolithography is extensively applied, among others, in the fabrication of field-effect transistors (FET), semiconductor integrated circuits, and nanoelectromechanical systems (NEMS) (Landis, 2013). One of the most investigated nanolithographic techniques for polymer membrane nanoengineering or preparation is photolithography. Based on the chemistry of the photo-reactive group, controlled elimination or addition reactions are possible, also including dimerization reactions leading to crosslinking of macromolecule chains. Figure 14.1 shows a scheme of the photolithographic process applied for producing nanostructured membrane. The substrate surface is coated with a photosensitive solution, for example, by spin-coating. The substrate can then be baked to ensure complete evaporation of the solvent, and the solid coating is irradiated through an optical mask bearing the desired micropatterns. After the irradiation, the substrate is treated with a developing solution, which removes the nonirradiated portion of the photosensitive compound. Afterward, a precursor polymer solution is casted on the developed pattern, and the membrane with a well-defined pore size and shape is obtained after curing.

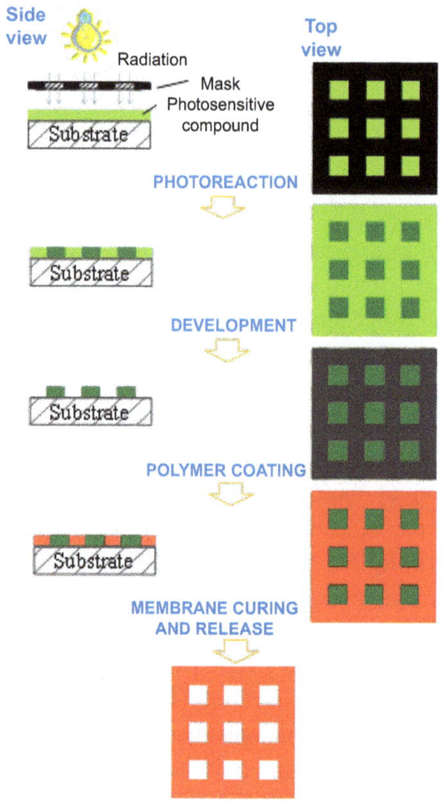

Figure 14.1 Example of a photolithographic process.

Lithography processes have high throughput and good control of pore diameter and are easy to integrate into microfluidic devices, which is important, for example, in lab-on-a-chip applications (De Jong et al., 2006).

For example, using direct photolithography followed by reactive ion etching, Chen et al. (2011) produced polydimethylsiloxane (PDMS) microfiltration membranes containing a large array of closely packed through holes with various pore size and shape (Fig. 14.2).

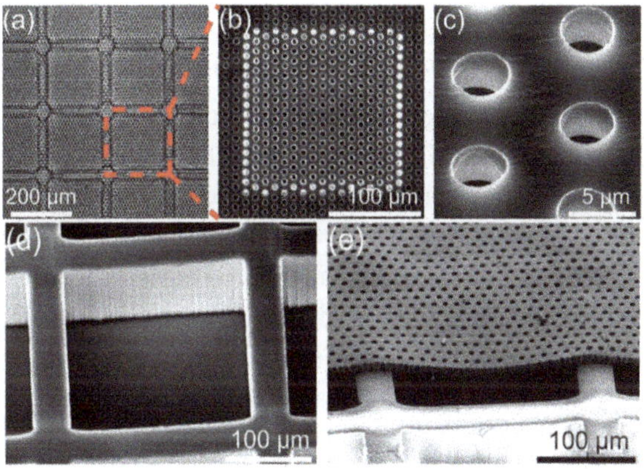

Figure 14.2 Bright-field (a) and SEM (b–e) images of a PDMS microfiltration membrane, bonded to a PDMS support structure (d). The arrays of through holes of different diameters were generated using the photolithographic surface micromachining method. Reprinted from Chen et al. (2011) with permission from Royal Society of Chemistry.

14.2.8 3D Printing

The conception of 3D printing dates back to 1986 when Charles Hull patented a system for obtaining 3D objects (Hull, 1986). In the subsequent years, 3D printing technology found industrial applications in fields that require rapid prototyping of 3D models such as automotive and aerospace industries (Gross et al., 2014). Several 3D printing systems have been developed, including fused deposition, selective heat or laser sintering, photo-polymerization, and thin-film lamination. A 3D printer works by dividing the file representing the 3D object into a sequence of 2D horizontal cross sections. In this way, it allows the 3D object to be printed, starting at the base, in consecutive layers of the desired material.

Nowadays, 3D printing enables the creation of porous media characterized by a combination of fine precision, scalability, and versatility, at low cost (Fee, 2014). It is expected that this approach will gain more and more importance in a large number of applications where intimate fluid–solid contact is required.

In the field of modifying surface materials, inkjet printing is an interesting technology, which has been used to create multidimensional structures in a lot of applications (e.g., electronic circuits). This technology has emerged as a promising material-processing technique because it makes it possible to rapidly and precisely deposit functional materials on a substrate surface. Appling this technique, Gao et al. (2016) prepared the so-called "multifunctional mosaic membrane," which, thanks to its unique nanostructure (consisting of both positively and negatively charged domains), can transport dissolved salts more rapidly than similarly sized neutral solutes and/or solvents. The multifunctional mosaic membranes were prepared by printing inks that contained polyelectrolytes on a polycarbonate track-etched membrane with pores 30 nm in diameter.

14.2.9 Anodization

Anodization is the electrolytic passivation of chemical species by applying an appropriate voltage of current in an electrolytic solution. It is used to cover the surface of various materials (such as aluminum, titanium, magnesium, zinc, niobium, tantalum, and others) by a nanoporous oxide layer (Edwards, 1997). This process has been intensively utilized in the industry because it increases resistance to corrosion and wear and provides better adhesion for paint primers and glues than bare metal does. The anodizing process is most commonly used to treat aluminum parts. By means of controlled anodic oxidation, it is possible to produce a thin layer of dense aluminum oxide, followed by an ordered array of smaller nanopores (Alwitt and Xu, 1993). Aluminum anodic oxide (AAO) membranes with high pore densities can be produced by this process by treating the aluminum substrate in an aqueous solution of acidic electrolytes, such as oxalic, sulfuric, chromic, and phosphoric acids (Stroeve and

Ileri, 2011). The resulting membranes consist of a well-defined pore structure, and the pore diameter depends on the type of electrolyte used and the anode voltage applied. Their self-ordered pore arrangement is important to maximize permeation and molecular flux. Yi et al. (2013) provide an example of AAO membranes with controlled morphology (honeycomb-like pore structure, shown in Fig. 14.3) for application in biology, catalytic chemistry, and medical science.

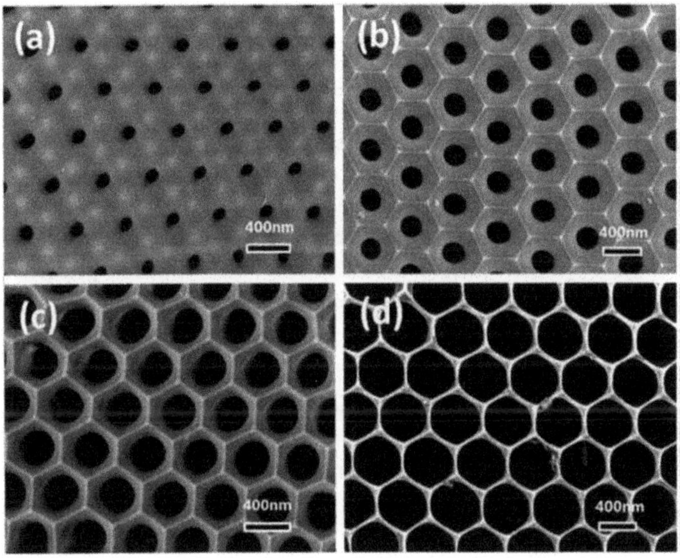

Figure 14.3 Aluminum anodic oxide (AAO) membranes. Reprinted from Yi et al. (2013) with permission from Elsevier.

14.3 Immobilization of Functional Molecules on Membranes

Attachment of groups or molecules depends on the final application of the membrane. Since in biotechnological applications the immobilization procedure concerns biomolecules, further discussions are referred to those molecules only. The techniques for biomolecule immobilization on membrane can be roughly divided into physical, where weak interactions between the support and biomolecules exist, and chemical, wherein covalent bonds are formed with the biomolecules. The different types of interactions are presented in Table 14.1.

Table 14.1 Biomolecule immobilization methods on membranes

Method	Features	Drawbacks
Entrapment	It can be carried out by cross-flow filtration of the biomolecule solution through an asymmetric membrane so that the biomolecules are entrapped in the pores.	Leakages and blockage of membrane pore due to the biomolecules
Gelification	It is based on one of the main drawbacks of membrane processes: concentration polarization phenomena and subsequent protein precipitation upon reaching a critical concentration at the membrane interface	Reduction in the catalytic efficiency, due to mass transport limitations and possible preferential pathways in the enzyme gel layer
Adsorption	It can be performed on unmodified or modified membranes. Interactions between the biomolecule and the membrane happen through weak and nonspecific bonds, promoted by van der Waals forces, hydrogen bonds, or hydrophobic interaction. The method is cheap and very simple.	Leakages
Covalent linking	It is the most stable biomolecule immobilization technique. It exploits active groups present on the membrane such as aldehyde, amine, epoxide, anhydride, and carboxylic acid. The main advantage is that the covalent bond between the biomolecule and the membrane is irreversible; therefore, no leakage of biomolecules occurs.	Loss of active properties of immobilized biomolecules

Method	Features	Drawbacks
Crosslinking	It is an immobilization technique performed by intermolecular crosslinking of the enzymes to either other protein molecules or functional groups present in an insoluble support.	Loss of active properties of immobilized biomolecules
Ionic binding	This technique involves interaction of a charged biomolecule with the membrane or groups introduced on the membrane that has opposite charge.	Leakage
Site specific	It is performed by introduction of a spacer molecule. Strategies are based on gene fusion, onsite direct mutagenesis introducing unique cystein molecules on biomolecule and on biotin–avidin strategy. This technique retains the activity of the biomolecule along with the benefits of traditional covalent bonding.	Very expensive

Besides these traditional methods, nanoparticle-mediated immobilization is a new promising trend in biocatalyst immobilization. Owing to the specific physical, chemical, optical, and electrical properties of nanoparticles, they appear in a lot of novel biotechnological applications. The use of nanoparticles as carriers for immobilization of biocatalyst on membranes can provide several advantages such as reversible recovery of the biocatalyst when necessary (i.e., during membrane-cleaning steps or because of biomolecule denaturation), increase in the immobilization surface/area, and avoidance of biocatalyst interaction with the membrane. An example of biocatalyst recovery was reported by Gebreyohannes et al. (2015) where the biocatalyst was covalently immobilized on the surface of magnetic responsive nanoparticles, which were successively deposited on or removed from the membrane surface, thanks to the action of a magnetic field. In the work of Vitola et al. (2016), in order to improve the biocatalytic performances of enzymes immobilized on a highly hydrophobic membrane (PVDF), colloidal particles (microgels) were employed as carriers for covalent immobilization of enzymes maintained in hydrated microenvironment on/in the membrane (Fig. 14.4).

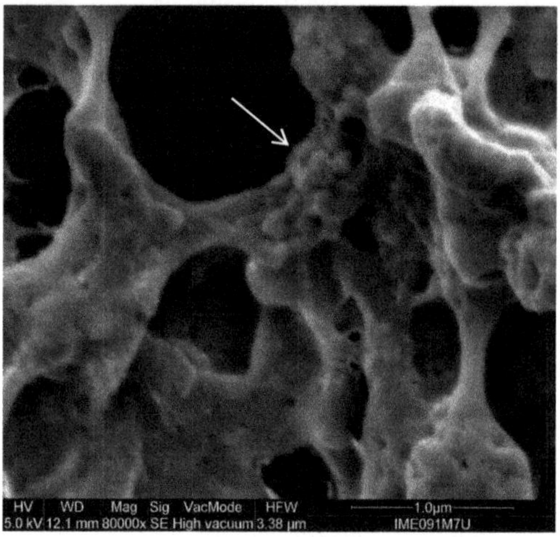

Figure 14.4 Cross-sectional SEM image of a PVDF membrane; the arrow points out the immobilized hydrogels. Reprinted from Vitola et al. (2016) with permission from John Wiley and Sons.

14.4 Surface Characterization of Modified Membrane and Evidences of Biomolecule Immobilization

Several analytical techniques can be employed to investigate the effective modification of the membrane and the subsequent biomolecule immobilization. Some techniques are appropriate for analysis of the changes in membrane chemistry, while some other techniques can point out the presence of immobilized biomolecules. The type of technique adopted for membrane characterization before and after biomolecule immobilization depends on the type of modification pursued with the relative anticipated nature, on the degree of accuracy required, and on the resources available. The surface chemistry of membrane must be investigated to confirm the success of the applied protocol and to understand the chemical changes that alter the membrane properties. The most used characterization techniques are reported in the following paragraphs.

14.4.1 Fourier Transform Infrared Spectroscopy

Fourier transform infrared (FTIR) spectroscopy is a technique used to obtain an infrared spectrum of absorption or emission of the sample. Some of the infrared radiations are absorbed by the sample, and some are passed through (transmitted). The resulting spectrum plotted in absorbance (or transmittance) versus wavenumber represents the molecular fingerprint of the sample, with absorption peaks that correspond to the frequencies of vibrations between the bonds of atoms making up the material. In attenuated total reflectance (ATR) FTIR, the incident infrared beam first passes through a ZnSe, Ge, or diamond crystal, improving the surface sensitivity of the technique.

FTIR can provide information about the chemistry of the sample indicating, for example, the presence of new functional groups as a consequence of functionalization (Vitola et al., 2015). It can also be used to confirm the presence of bounded biomolecule on the membrane when their characteristic absorption peeks are not overlapped by the absorption peeks of the membrane material (Belfer et al., 2000).

14.4.2 X-Ray Photoelectron Spectroscopy

X-ray photoelectron spectroscopy (XPS) analysis is a spectroscopic technique that provides quantitative and chemical state information from the top 0 to 10 nm of the material being studied. XPS is typically accomplished by exciting a sample surface with low-energy X-rays causing photoelectrons to be emitted from the sample surface. An electron energy analyzer is used to measure the energy of the emitted photoelectrons. From the binding energy and intensity of a photoelectron peak, the elemental identity, chemical state, and quantity of a detected element can be determined.

XPS is the most widely used surface analysis technique because of its relative simplicity in use and data interpretation. Using XPS investigation, the presence of functional groups introduced on the membrane surface by functionalization procedure can be validated (Ross et al., 2000).

14.4.3 Time of Flight Secondary Ion Mass Spectroscopy

Time of flight secondary ion mass spectroscopy (TOF-SIMS) analyzes the composition of solid surfaces by sputtering the surface of the sample with a focused primary ion beam and collecting and analyzing the ejected secondary ions. The secondary ions are separated according to their mass-to-charge ratio in a mass spectrometer. The resulting plot is of signal intensity versus mass-to-charge ratio (m/z). TOF-SIMS can provide detailed information on composition, structure, orientation, and spatial distribution of molecules on the surface (Goddard and Hotchkiss, 2007). The sampling depth of this technique is from 1 to 2 nm, and the detection limit is of the order of parts per billion; therefore, it is more sensitive of XPS. TOF-SIMS can be used to integrate XPS results or to differentiate samples that have similar XPS spectra. TOF-SIMS is generally considered a qualitative technique, because ionization of surface molecules is dependent on the sample matrix.

14.4.4 Scanning Electron Microscopy

Scanning electron microscopy (SEM) is one of the more widely available tools for surface analysis. It provides structural information on the surface topography of materials at the nanometer-scale resolution. The SEM image of a sample is produced by scanning the sample with a focused beam of electrons, which causes emission of secondary electrons and X-rays. The secondary electrons are collected using a detector to generate the surface image. Non-conducting materials have to be sputtered or coated before SEM analysis. In membrane science, SEM analysis provides important information about the topography and thickness of the membrane surface (Torras and Garcia-Valls, 2004). After functionalization, SEM can be used to investigate whether the native membrane topography has been modified.

14.4.5 Field Emission Scanning Electron Microscopy

Field emission SEM (FESEM) provides topographical and elemental information with virtually unlimited depth of field. FESEM produces clearer and less electrostatically distorted images than conventional SEM. It is used to visualize small morphological details on the membrane surface (Zhao et al., 2013).

14.4.6 Transmission Electron Microscopy

Transmission electron microscopy (TEM) operates on the same basic principles of a light microscope but transmits electrons, instead of light, through a sample. Compared to a light microscope, TEM provides a much higher resolution and allows visualization of details as small as a single column of atoms. The high energetic electrons, produced using a tungsten filament in a vacuum chamber, are shot through the sample having a very thin thickness (50–500 nm). The electrons transmitted through the sample hit a fluorescent screen, a film, or a sensor such as a charge-coupled device (CCD), producing an image.

TEM finds application particularly in biomolecule immobilization to understand the biomolecule distribution on the membrane (Vitola et al., 2016). In this aim, the biomolecules have to be labeled.

14.4.7 Circular Dichroism Spectroscopy

Circular dichroism (CD) is a spectroscopic technique widely used for the examination of conformation and stability of proteins and peptide structures in different media and environmental conditions (Mohamad et al., 2015). CD spectroscopy is useful for investigating the secondary structures of proteins because the peptide bond is asymmetric and influences the spectra of these molecules in the far ultraviolet regions. In fact, in the regions below 250 nm, their spectra, which are dominated by the n–η* and η–η* transitions of amide groups and are influenced by the geometries of the polypeptide backbones, reflect the different types of secondary structures (Whitmore and Wallace, 2008).

The near-UV CD spectrum of proteins provides information on the tertiary structure. The tertiary structure of proteins is characterized by η0– η* transition observation of CD in the region between 250 and 300 nm. However, these signals are due to the absorption, dipole orientation, and the nature of the surrounding environment of the aromatic amino acids. These signals cannot be assigned to any particular 3D structure because the theoretical treatment of near-UV CD spectra is not sufficiently advanced yet; by the way, important progress has been made in assigning features of the spectrum to particular residues by sequential removal of aromatic side chains by site-directed mutagenesis (Kelly et al., 2005).

CD spectroscopy is non-destructive and quite easy to operate and requires a small quantity of sample and minimal data collection. Several methods have been developed to process the various contributions arising from the different kinds of secondary structures present in each protein providing information on the overall structure of that protein. Exploiting reference spectra derived from proteins of determined structure, several different empirical algorithms have been developed (Whitmore and Wallace, 2008).

CD spectroscopy is very useful in enzyme immobilization studies as it provides structural information regarding protein–ligand and protein–protein interactions, as well as structural compositions of proteins (Mohamad et al., 2015).

14.4.8 Atomic Force Microscopy

Atomic force microscopy (AFM) is a versatile and powerful microscopy technology for studying samples at the atomic scale. The AFM probe consists of a tip attached to a flexible cantilever that can be moved in three dimensions by means of piezoelectric scanners. The changes in surface height are measured by a laser beam localized at the top of the cantilever and reflected onto a photodiode generating a surface topographical map from which surface roughness values can be calculated (Vrijenhoek et al., 2001).

In addition, AFM can produce 2D chemical affinity maps by modifying the cantilever tip with specific functional groups. Attaching an interacting molecule to the AFM tip and the other molecule to the sample surface, the molecular binding forces can be quantified. This opportunity, for example, can be used for quantifying available active sites in immobilized enzymes or measuring the force of antibody–antigen interactions.

14.4.9 Thermogravimetric Analysis

Thermogravimetric analysis (TGA) is a method of thermal analysis in which a physical property of a substance is measured in a controlled atmosphere as a function of increasing temperature (with constant heating rate), or as a function of time (with constant temperature and/or constant mass loss).

TGA analysis provides information about the thermal stability of the substance, oxidative stability, quantitative determination of components, reaction rates, and lifetime of a product (Mohamad et al., 2015). TGA makes it possible to evaluate the amount of functional groups attached on the membrane or understand the amount and new structure of the immobilized biomolecules.

14.4.10 Differential Scanning Calorimetry

Differential scanning calorimetry (DSC) is a thermoanalytical technique that measures the amount of energy absorbed or released by a sample when it is heated or cooled, providing quantitative and qualitative data on endothermic (heat absorption) and exothermic (heat evolution) processes. In the biomolecule immobilization field, the technique can provide information about thermodynamic effects of immobilization, from multipoint covalent attachment to simple absorption (Battistel and Rialdi, 2006).

14.4.11 Water Contact Angle

The contact angle is defined as the angle formed by the intersection of the liquid–solid interface and the liquid–vapor interface (geometrically acquired by applying a tangent line from the contact point along the liquid–vapor interface in the droplet profile). This angle is determined by both properties of the solid and the liquid and the interaction and repulsion forces between the liquid and the solid and by the three-phase interface properties (gas, liquid, and solid). Those interactions are described by cohesion and adhesion forces, which are intermolecular forces. The balance between the cohesive forces of similar molecules such as between the liquid molecules (i.e., hydrogen bonds and van der Waals forces) and the adhesive forces between dissimilar molecules such as between the liquid and solid molecules (i.e., mechanical and electrostatic forces) will determine the contact angle created in the solid and liquid interface. Water contact angle is a method to measure the surface wettability of a material by measuring how much a droplet of water spreads on its surface. Conventionally, a wetted (hydrophilic) surface forms a contact angle with the water drop, which is smaller than 90°. A non-wetted (hydrophobic) surface forms a contact angle with the water drop between 90° and 180° (Yuan and Lee, 2013).

Contact angle values can be calculated from Young's equation (Eq. 14.1). The contact angle of a liquid drop on an ideal solid surface is defined by the mechanical equilibrium of the drop under the action of three interfacial tensions:

$$\gamma_{\mathrm{lv}} \cos \theta_{\mathrm{Y}} = \gamma_{\mathrm{sv}} - \gamma_{\mathrm{sl}} \tag{14.1}$$

where γ_{lv}, γ_{sv}, and γ_{sl} represent the liquid–vapor, solid–vapor, and solid–liquid interfacial tensions, respectively, and θ_{Y} is Young's contact angle.

Considering Eq. 14.1, the parameters γ_{lv}, γ_{sv}, and γ_{sl} determine exactly the contact angle θ_{Y}, but experimentally observed contact angles often have different values with respect to θ_{Y}. This difference is due to the surface roughness and/or heterogeneity of the surface in contact with the liquid. It is known as the "contact angle hysteresis." This phenomenon can be understood by looking what happens in a droplet if a small-enough amount of liquid is added or removed. In both cases, the droplet will not move. However, in the first case, the contact angle will increase (advancing contact angle), and in the latter case, it will decrease (receding contact angle). Therefore, the contact angle hysteresis can be defined as the difference of the advancing (maximal) contact angle and the receding (minimal) contact angle (Eral et al., 2013). However, on ideal solid surfaces, there is no contact angle hysteresis, and the experimentally observed contact angle perfectly coincides with Young's contact angle (Yuan and Lee, 2013).

Surfaces having chemical differences, but identical with respect to other surface effects, show differences in the observed contact angle. For example, after a membrane functionalization treatment, the water contact angle provides information about the change in surface hydrophilicity as a consequence of the nature of functional groups introduced.

14.4.12 Zeta Potential

The zeta potential is described as the difference of electrical potential between a dispersion medium and a double layer of liquid ions attached to a charged solid surface. The liquid double layer in contact with the charged solid surface consists of tightly bound liquid phase ions of opposite charge forming the fixed layer, and loosely bound liquid phase ions of opposite charge forming the mobile layer. The zeta potential is particularly studied in the colloidal chemistry, because it is a fundamental indicator of the stability of colloidal dispersion, but it is also an important parameter to describe solid surfaces and the interaction with the surrounding liquid. Zeta potential can be used to determine surface isoelectric points and quantify a change in surface ionizable groups after membrane surface modification and biomolecule immobilization (Katoa et al., 2003). In addition, important information can be obtained by measuring the initial zeta potential value of the biomolecules in the different chemical conditions (ionic strength and pH) before immobilization.

14.4.13 Water Permeability

Pure water permeability is a characteristic property of micro- and ultrafiltration membranes (Strathmann et al., 2006). The hydrodynamic permeability is obtained by the slope of the plotted membrane flux determined at different operative pressures. The flux through a membrane is determined by Darcy's law (Eq. 14.2):

$$J_\mathrm{v} = L_\mathrm{p}\,\Delta p \tag{14.2}$$

where J_v is the membrane flux, L_p is the hydrodynamic solvent permeability, and Δp is the applied pressure.

The modification of the membrane surface or the immobilization of biomolecules alters the membrane permeability. In fact, the different steps of modification can modify the hydrophilicity of the membrane, increasing or decreasing its permeability; while the biomolecule immobilization, which is a surface or entrapment phenomenon, can change either the surface wettability of the membrane or the pore conformation, leading to a decrease in permeability.

14.4.14 Colorimetric Tests

14.4.14.1 Determination of functional groups

The surface chemistry of membranes can also be investigated by colorimetric tests. These tests fall into the analytical chemistry or biochemistry field. Qualitative and quantitative assays can be performed by means of reagents that develop or change color in the presence of appropriate chemical groups (see Table 14.2). Fluorescent dyes can also be used for the quantification of functional groups; they can complex with functional groups in an established stoichiometric ratio and can be desorbed and quantified afterward (Goddard and Hotchkiss, 2007). In addition, fluorescent dyes can be used for direct labeling and visualization of biomolecules (in situ) (Bindman and van der Donk, 2013).

Table 14.2 Colorimetric tests for the analysis of functional groups

Functional groups	Reagent	Reference
Carboxylic acid	Bromocresol green	Blaih et al. (2000)
Hydroxyl	Dansyl chloride	Naassner et al. (2002)
Carbony	2,4-dinitrophenyl-hydrazine	Burfield and Law (1979)
Primary and secondary amino	Ninhydrin	Sun et al. (2006)

However, colorimetric tests are more appropriate as qualitative test only, because in quantitative determinations, these can suffer from low sensibility due to the limited amount of groups that have to be determined or their poor accessibility.

14.4.14.2 Protein assay

Four commonly used colorimetric methods for the determination of protein concentration in solution are Biuret reaction, Lowry method, Bradford assay, and bicinchoninic acid (BCA) assay (Sapan et al., 1999). Each assay has advantages and disadvantages relative to sensitivity, ease of performance, acceptance in the literature, and accuracy. The amount of proteins immobilized on a membrane is calculated from mass balance between the amount of protein present in the solution before and after immobilization by using Eq. 14.3.

$$C_iV_i = C_fV_f + C_{ws}V_{ws} + m \quad (14.3)$$

where m indicates the immobilized protein mass in the membrane, C and V represent the concentration and volume, respectively; the subscripts i, f, and ws indicate the initial, final, and washing solutions, respectively.

14.4.15 Activity Assay

Activity assays are laboratory methods for assessing the biological activity of biomolecules, such as enzymes. The biological activities of immobilized biomolecules can also be treated as the evidence of immobilization (Caldwell and Raushel, 1991). However, after immobilization, biomolecules often suffer from decrease in biological activity. But this is not a general rule; in fact there are cases in which an improved biological activity has been showed after immobilization (Mazzei et al., 2009). On the other hand, immobilization often improves the stability of the immobilized biomolecule compared to the free one.

14.5 Nanoengineered Membrane Applications in Biotechnology

Biotechnology is the application of scientific and engineering principles to the processing of materials by living systems and organisms to develop or make new products (Gavrilescu and Chisti, 2005). Among others, some fields that involve biotechnology include genetic engineering, metabolic engineering, hybridoma technology, bioelectronics, nanobiotechnology, protein engineering, transgenic animals and plants, tissue and organ engineering, immunological assays, genomics and proteomics, bioseparations, and bioreactor technologies. Modern biotechnology has been divided into four main branches:

1. **Green biotechnology**: It is the technology applied to agricultural processes.

2. **Red biotechnology**: It is the technology used in medical applications.
3. **Blue biotechnology**: It is the term used to describe aquatic and marine applications of biotechnology.
4. **White biotechnology**: It is the technology used in industrial processes.

Membrane processes are employed extensively throughout the production, purification, and formulation of biotechnology products. Biotechnological applications involving nanoengineered membranes basically cover each of the aforementioned branches; some examples of these applications are discussed in the following paragraphs. A special focus is devoted to biocatalytic membranes, which are the most important example of nanoengineered membranes applied in the field of biotechnology.

14.5.1 Membrane Bioreactors

Membranes can be associated or functionalized with catalyst of biological origin such as enzymes, cells, and microorganisms to develop biohybrid systems. On the basis of membrane role they can be divided in MBRs or BMRs. In the first case, a chemical reaction carried out by a biocatalyst is implemented by a membrane operation. In the second case, the membrane simultaneously acts as a support for the biocatalyst and as a separation unit (Fig. 14.5). Nanoengineering of artificial membranes by biocatalyst immobilization is one of the most promising technologies for production, processing, and manufacturing operations. BMRs operate concentration and separation processes without the use of heat, lowering energy consumption and operating costs (Giorno and Drioli, 2000). Owing to the outstanding selectivity and the high efficiency of the biological catalyst, the products and side-products are of excellent quality. In general, biocatalysts are poorly stable for industrial applications, which are often hindered by the lack of long-term stability, recovery, and reuse of the biocatalysts (Jochems et al., 2011). In addition, problems can also occur, such as low amount of biocatalyst in the active form, as well as kinetic and/or thermodynamic inhibition of the biomolecules by reactants and products. Compartmentalizing the biocatalyst within the membrane overcomes these drawbacks.

In fact, immobilization of the biocatalyst can enhance its stability, making easy recovery and reuse. BMRs allow continuous and selective removal of products from the reaction site, increasing the conversion of product-inhibited or thermodynamically unfavorable reactions (Giorno et al., 2003).

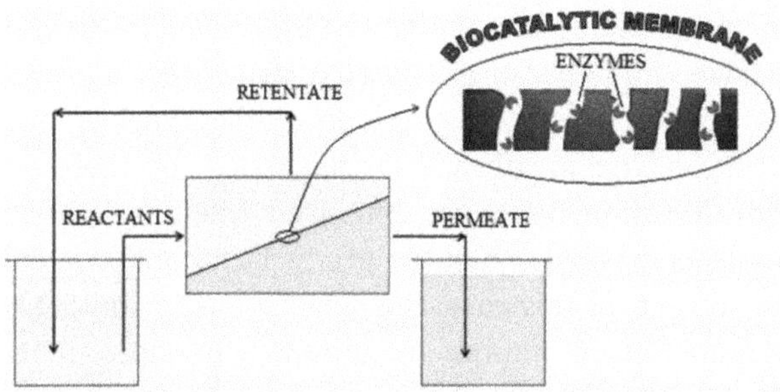

Figure 14.5 Scheme of a BMR.

Reaction rate and selectivity may be influenced by membrane properties, because immobilization of the biocatalyst entails the interaction of two species: the biocatalyst and the membrane. The physicochemical properties of the membranes can influence catalyst structure and conformation. The physical–chemical properties of the membrane material, as well as the nanoarchitecture of the membrane where the biocatalyst is immobilized, have great influence on biocatalyst performance. In fact, they dictate the chemical interactions acting on the biomolecule, the steric hindrance, the local pH, ionic strength, and mass and heat transfer.

In order to select the membrane to be used in BMRs, the size of the biocatalyst, substrates, and products must also be considered. Besides, membranes are required to have some specific properties such as adequate permselectivity and mechanical and chemical stability.

Membranes specifically designed to answer these bioreactor needs should also enable the site-specific immobilization of the biocatalyst to minimize undesired changes in membrane morphology and average pore size, and retain maximum biocatalyst activity. In addition, the membrane fabrication process should provide good control over the average pore diameter and produce a narrow distribution for pore diameter.

BMRs often employ polymeric membranes since they are relatively cheap, are suitable to work under mild conditions of temperature and pressure, and can be prepared in a large variety of nanostructures. However, BMRs using inorganic membranes have also been reported (Giorno and Drioli, 2000). The membrane conformations can be flat sheet and assembled in plate-and-frame or spiral-wound modules, or tubular and assembled in tube modules; they can have symmetric or asymmetric structure.

BMRs can be classified considering the type of biocatalyst immobilization, the membrane conformation, or their operation mode. Ultrafiltration BMRs are employed when the substrate has higher molecular weight than the product and both are soluble in the same solvent. In this case, the substrate is in contact with the enzyme immobilized on the membrane surface, and it cannot permeate through the membrane, whereas the product can pass through it. In this case, on the basis of the Le Chatelier principle, the reaction equilibrium is shifted toward the reaction product and the total high reaction yield can be obtained. When the substrate and the product can both permeate through the membrane, it is important to define the optimal "residence time" to ensure that the substrate is in contact with the enzyme. It is converted, and the product is recovered downstream, together with the unconverted substrate (Álvarez-Hornos et al., 2011; Mazzei et al., 2009). In this case, the reaction yield is governed by the reaction equilibrium.

When the substrate and the product have different solubility (e.g., an ester and its hydrolysis products), a biphasic membrane reactor can be used. In this type of BMR, the enzyme-loaded membrane separates two immiscible phases and is able to carry out the bioconversion at the interface allowing, for example, the bioconversion of the substrate contained in the organic phase and its continuous extraction in the aqueous phase.

The BMR concept dates back to the 1950s, when protease was first immobilized on diazotized polystyrene. Afterward, technological advancements in the development of membrane materials and modules allowed the investigation of several applications involving BMRs, most of which fall in the biotechnological field. For example, BMRs have been used for the production of pharmaceuticals, in biomedical applications, food applications, control of fouling in waste water treatment, and so on.

14.5.2 Nanoengineered Membranes in Pharmaceutical Applications

BMRs have been used for the production of amino acids, antibiotics, vitamins, anticancer drugs, and so on. These devices mainly use enzymes or cells immobilized on the membrane. In this field, the use of a BMR improves the yield and purity of the downstream with respect to the compound of interest, enabling cost reduction in downstream purification. In addition, BMRs show higher conversion rates when a product inhibition or a reversible reaction occurs (Gebreyohannes et al., 2013).

For example, BMRs have been reported for the hydrolysis of oleuropein with immobilized β-glucosidase on a polysulfone capillary membrane (Mazzei et al., 2009). The nanoengineered membrane, having asymmetric structure, was prepared by entrapping the β-glucosidase enzyme into the pores by cross-flow ultrafiltration. The biocatalytic membrane allowed the control of oleuropein feeding and aglycon removal from the reaction environment in a single unit operation. Kinetic performance of immobilized β-glucosidase was studied and compared with that of free enzyme. Results showed that intrinsic kinetics of β-glucosidase immobilized in the BMR was not negatively affected compared to the free enzyme. On the contrary, it was slightly improved; in fact, a lower K_m value that indicates higher β-glucosidase–oleuropein affinity was obtained. This shows that the lower catalytic activity of immobilized enzyme, commonly observed in the literature, is not a general rule and can be avoided by choosing an appropriate immobilization technique, membrane material, and optimizing process parameters.

Oleuropein is the major phenolic compound in olive fruits and leaves. It has well-known hypotensive effect and protection against induced arrhythmias. The hydrolysis product of oleuropein, "aglycon," is an antimicrobial agent for olive trees and has anti-inflammatory properties. Oleuropein is abundant in olive leaves and mill wastewater. The production of aglycon from oleuropein has glucose as a side product; therefore, this kind of BMR represents also a strategy to valorize wastes coming from the olive industry.

BMRs are also employed in studies that concern the production of optically pure enantiomers. The pharmacological activity of chiral bioactive compounds depends on their enantiomeric shape;

therefore, in most cases, it is mandatory to produce molecules with high enantiomeric purity. An example of BMR application in this field is the production of (S)-naproxen by the optical resolution of racemic naproxen methyl ester (Giorno et al., 2007). This reaction was carried out using a multiphase membrane reactor in two separate phase configurations. The naproxen methyl ester was dissolved in the organic phase (isooctane), while an aqueous phase extracted the hydrophilic product. A biocatalytic membrane was prepared immobilizing lipase by entrapment on a 10 kDa polyamide membrane, which was used to also separate the two phases. The nanoengineered biofunctionalized membrane could operate the hydrolysis at the organic/aqueous interface, achieving an enantiomeric excess of 97% or 100% using methyl ester or butyl ester, respectively, with no decay of activity over a period of 2 weeks of continuous operation for both cases.

BMRs with immobilized lipase have also been developed using asymmetric ceramic hollow-fiber membranes (Ranieri et al., 2016). Ceramic membranes are more resistant to solvents, so they might be more suitable for biocatalytic membranes operating in organic solvent.

An example of a multiphase BMR process operated at the large scale is the one developed by Tanabe Seyaku (Japan) and Sepracor Inc. (USA) for producing (2R,3S)-methylmethoxyphenyl-glycidate ester, a chiral intermediate for the preparation of diltiazem, a drug used in the treatment of hypertension and angina (Lopez and Matson, 1997). In this case, an asymmetric hollow-fiber polyacrylonitrile membrane with entrapped lipase was used. The plant consisted of two banks with 12 membrane modules, each one with an active membrane area per module of 60 m^2. It could produce 53 kg/year-m^2 of 99% enantiomeric excess ester. The plant was operated for a few years.

A business in which a membrane process provides an efficient and sustainable solution is the production of L-lactic acid. It has lots of applications in the pharmaceutical, food, textile, leather, and chemical industries. Nowadays, the production of L-lactic acid at the industrial level is carried out by membrane-based hybrid reactor system, in a continuous batch fermentation way (Ghaffar et al., 2014). In fact, compared to conventional processes, this process offers advantages in terms of L-lactic acid productivity.

Another advantage of the continuous batch over batch culture is that the process can be run for a long period of time. The membrane reactor system consists of a traditional fermenter combined with an ultrafiltration unit. During the bioreactor's operation, its volume is maintained by adding fresh fermentation broth at the same rate as it permeates through the membrane. The permeated solution contains L-lactic acid and other small molecules such as residual sugars, color, nutrients, and other organic acids. Afterward, L-lactic acid can be purified by membrane-based operations.

14.5.3 Nanoengineered Membranes in Biomedical Applications

Nanoengineered membranes can mimic natural filtration systems, so they have been developed for use in bioartificial organs, tissues, implantable drug-delivery systems, and other novel nano-enabled medical devices (Lee, 2006; Warkiani et al., 2013). Millions of patients are sustained by organ substitution. In this field, the use of nanoengineered membranes can perform toxin separation, and metabolic organ function is an attractive challenge. For example, membrane bioartificial pancreas, in which porcine islets of Langerhans were segregated in the shell side of a hollow-fiber module, was investigated as an alternative approach to the transplantation of the whole pancreas in patients with insulin-dependent diabetes (Giorno and Drioli, 2000; Stamatialis et al., 2008). The islets produce insulin and regulate the patient's blood-glucose levels, while the membrane allows the transport of molecules from the segregated cells to the blood and vice versa.

One of the most successful membrane processes is hemodialysis or hemofiltration, which has been an important life-sustaining technology. This therapy replaces the solute clearance function of glomerulus. Its limitation is due to the lack of the metabolic and endocrinologic functions of the tubular cells, which in patients with end-stage renal disease or acute renal failure cause suboptimal clinical outcomes. In these cases, kidney transplantation is the best medical care, but the lack of organs in comparison with the number of patients leads to the use of hemofiltration. The combination of hemofiltration with living human cells could be a promising solution. In fact, the cells could provide functions such as reabsorption of glucose, amino acids, bicarbonate, electrolytes and water, secretion of toxins and xenobiotics, production of 1,25-dihydroxy vitamin D3, and immunomodulatory functions. Polysulfone/polyvinylpyrrolidone (PSF/PVP) is the membrane material of choice in bioartificial kidney applications and is generally used as a substrate for kidney epithelial cells. So far commercial hemofiltration/hemodialysis cartridges with hollow-fiber membranes consisting of PES/PVP have been used as bioreactor units of bioartificial kidneys in animal experiments and clinical trials (Humes et al., 2002). One of the main challenges in the development of a bioartificial kidney (BAK) is purification of blood from toxins while minimizing leakage of albumin and other important macromolecules and ions. These challenges can be pursued by exploiting nanoengineering techniques that enable replication in the polymer membrane of the uniform, elongated, slit-shaped structures of glomerulus (Humes et al., 2014). Silicone-nanopore membranes have been fabricated with this goal by using bulk and surface micromachining techniques derived from microelectromechanical systems technology (Fissel et al., 2009).

Also nanoporous AAO membranes have been investigated for potential application in hemodialysis (Huang et al., 2007). The uniform pore size, high porosity, and high hydraulic conductivity make these membranes advantageous for this application because a precise control over molecular transport is possible.

BMRs with housed hepatocytes can be used as temporary liver support for maintaining organ function until transplantation or for facilitating liver regeneration with the aim of preventing severe complications and improving survival of patients with acute liver failure (Giorno et al., 2003; Morelli et al., 2016). Several bioreactor designs have been explored for bioartificial liver devices. The semipermeable hollow-fiber MBR is the one most frequently used in clinical investigation. Bioartificial liver devices have been developed using either animal cells (pig hepatocytes) or human cells (Drioli and Giorno, 2010; Ren and Irudayam, 2015). Circe Biomedical Inc. developed a hollow fiber–based model with porcine hepatocytes: the HepatAssist™ liver support system. The Hybrid Organ GmbH Company (Germany) also developed a biohybrid liver device based on a hollow-fiber reactor with porcine hepatocytes (Giorno and Drioli, 2013). Vital Therapies Inc. (USA) proposed the first human-cell-based bioartificial liver system designed to increase the liver's ability to regenerate and recover—the so-called Extracorporeal Liver Assist Device (ELAD®). It is based on a hollow-fiber bioreactor that uses human hepatoblastoma cell line C3A. The device is designed to be used continuously for up to 10 days (Giorno and Drioli, 2010; Morelli, 2016).

Membrane oxygenators can be used as extracorporeal blood oxygenation systems. Since the widespread success of single and double lung transplantation in the early 1990s, demand for donor lungs has steadily outgrown the supply. Lung diseases are among the main causes of morbidity and mortality in the world. In fact, four of the leading causes of global death are chronic obstructive pulmonary disease, acute respiratory tract infections, lung cancer, and tuberculosis, while a fifth, asthma, causes enormous global morbidity (Schluger and Koppaka, 2014). Membrane oxygenators are composed of hydrophobic hollow-fiber membranes (usually polypropylene) that are wrapped into specific configuration. Blood and oxygen are flushed counter current through the shell and the lumen, respectively. The membrane's micropores, which are not wetted, enable respiratory gas diffusion.

Clearly, these membranes, besides permitting exchange of gaseous oxygen and carbon dioxide, have to prevent leakage of blood and its components from passing from one side to the other. Gas exchange can be optimized by maximizing membrane permeability and reducing membrane thickness.

Nanoengineered membranes have also been proposed for the development of drug-delivery systems (Adiga et al., 2009). In fact, nanoporous membranes are excellent devices for controlling the transport rate of drugs. The main goals of controlled delivery systems include protecting drugs that have a short half-life in vivo, minimizing the possibility of improper dosing, reaching a continuous drug release over extended time periods, and local delivery of drugs. For example, Orosz et al. (2004) employed nanoporous alumina membranes for releasing ophthalmic drugs, including endostatin catalase and vitamin C. Desai et al. (2006) developed a small subcutaneous nanoporous implant named NanoGATE, incorporating a microfabricated silicon nanoporous membrane (Fig. 14.6) engineered to slowly release the encapsulated drug (e.g., Interferon-α for the treatment of chronic hepatitis C) at an optimal rate. The nanopore membrane consists of arrays of parallel rectangular channels, which, in their smallest aspect, range from 7 nm to 50 nm.

In addition, thanks to the well-controlled pore size, pore shape, porosity, thickness, and the possibility to manipulate the surface chemical properties to suit the application by modification methods, nanoengineered membranes are also attractive systems for making capsules that can be used in the controlled release and delivery of drugs (Piacentini et al., 2010).

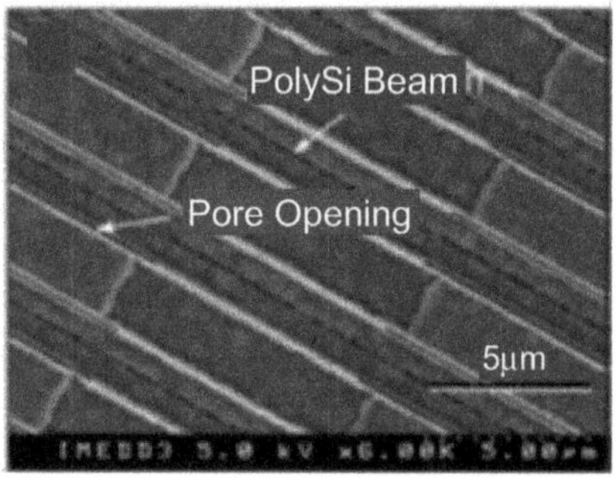

Figure 14.6 6000X top SEM view of membrane showing details of pore and anchor structures. Reprinted from Desai (2006) with permission from Springer.

In the biomedical field, nanoengineered membranes also find application in tissue engineering. They are optimal support for cell culture because they ensure efficient mass transfer through the cell-cultured scaffold under aseptic conditions in bioreactors designed for tissue engineering (often used in a hollow-fiber configuration). To date, the main areas of tissue-engineered replacements in clinical trials and applications are skin, cardiovascular organs, bone, and cartilage (Morelli et al., 2016). The functionality of the grown tissue is determined by the scaffold design (Bernhardt et al., 2008; Stamatialis et al., 2008). In fact, chemical–physical properties, shape, size, and micropattern of supporting materials influence cells behavior.

14.5.4 Nanoengineered Membranes in Food Applications

Membrane technologies have several advantages over other separation approaches in food processing, such as the possibility to preserve or improve flavor, color, and texture, as well as the opportunity to eliminate, minimize, valorize, and recycle wastes (Drioli and Giorno, 2010). By integrating nano-functionalized membranes and biocatalytic reactors, the efficiency of traditional membrane operations can be improved and innovative production and processing systems developed. Reduction in fluid viscosity, control of membrane fouling, improvement in product stability, and minimal waste production are among the main benefits of BMRs, but also new food products with improved texture and shelf-life can be produced (Giorno et al., 2009).

Reducing the viscosity of juices by hydrolyzing pectins is one of the investigated applications of BMR in the food sector. Pectins are linear polymers of D-galacturonic acid. The viscosity of fruit juices is principally due to the interaction between pectins and sugars (e.g., galactose and arabinose). Pectinases (able to hydrolyze the polygalacturonic chain) and pectinesterases (able to hydrolyze the ester bonds of pectins to produce pectic acids) are predominantly used to reduce the viscosity of fruit juices. Usually, at the industrial level, depectinization is carried out in batch and then the depectinized juice is used as feed for a tank that supplies juice to the membrane module. Using the pectinases immobilized on a polysulfone membrane (spiral-wound module), the permeate flux can be increased by about 30% with respect to that obtained when the enzyme is freely suspended in solution (Giorno et al., 1998).

BMR applications have been reported in must treatment (Drioli and Giorno, 2010; Peuchot et al., 2010). Must is the pressed juice of grapes. Polyphenolic compounds present in the must confer the characteristic taste and color to wine. In order to stabilize the must of white and rosè wines, the enzyme laccase is used to oxidize polyphenols in solution (Durán et al., 2002). During the maturation process, a secondary fermentation named malolactic fermentation occurs, which converts malic acid into lactic acid. This fermentation is very important for the organoleptic properties of wine because lactic acid increases the pH of wine from 3.4 to 4.2, preventing precipitation of proteins in the mouth. Therefore, controlling this reaction is extremely important. Cells of *Leuconostic oenos* immobilized by cross-flow filtration through polypropylene capillary membrane have been used to carry out malolactic fermentation in white wine (Giorno and Drioli, 2000). The production of lactic acid, which influences the taste of the wine, can be tuned acting on amount of cells and residence time of wine within the membrane.

BMRs have been used on a large scale in the dairy industry to hydrolyze either lactose (in whole milk) or cheese whey (Harju et al., 2012). The production of milk with low lactose content is useful for people who are lactose intolerant, while production of hydrolyzed cheese whey is important for economic and environmental reasons. Cheese whey is a polluting product, consisting of water (~93%), lactose (~5%), proteins (~0.7%), and salts. These compounds can be recovered and reused obtaining benefits either in terms of process cost effectiveness or waste reduction. In fact, whey proteins have excellent functional properties and can be recovered by ultrafiltration; then the lactose present in the permeate can be converted to glucose and galactose syrup using immobilized β-galactosidase. In the system proposed by Novalin et al. (2005), β-galactosidase was used in combination with a hollow fiber in order to continuously hydrolyze lactose in skim milk. The reactor consists of an ultrafiltration module (10 kDa cut-off, 4.9 m^2 membrane area). A continuous flow of the feed is applied through the tubes, and the high molecular mass enzyme remained outside, on the shell side. The temperature of the feed was tuned, and β-galactosidase solution was circulated in a closed loop. Microbial growth in the β-galactosidase solution was controlled using an UV irradiation module and a sterile filtration unit included in the circulation line. The driving force for this system is lactose diffusion, which mainly depends on the concentration gradient, the temperature and flow rates of the substrate, and the β-galactosidase solution. In comparison to a batch process, in this system enzymes retain activity for a longer time and it is not released into the product.

Membrane processes for separation and fractionation of milk components such as fat globules, serum proteins, reduction in bacteria and spores in skim milk, concentration of casein micelles (for cheese manufacturing) have been reported in the literature (Gésan-Guiziou, 2014; Schroën et al., 2010). A common challenge for the membranes used in these processes is to reduce flux decline (fouling) and improve selectivity. Therefore, nanoengineered membranes having uniform pore sizes and pore shapes are optimal candidate for these applications. BMRs with immobilized lipases and esterases represent a sustainable alternative for the treatment of oils and fats by replacing traditional chemical processes with biotechnological ones (Giorno et al., 2003; Pugazhenthi and Kumar, 2004). Biocatalytic systems using immobilized lipase on membranes have been used for hydrolysis of triglycerides in olive oil, palm oil, and various other vegetable oils, for improving oil quality and producing foods, home, and health-care ingredients (Gupta et al., 2013).

14.5.5 Nanoengineered Membranes in Chromatography

Membrane chromatography is a combination of membrane filtration with another well-established separation technique: affinity chromatography. Membrane chromatography can be considered an alternative to conventional packet-bed chromatography columns, for many chromatographic purification operations, including ion-exchange, hydrophobic, reversed-phase, and affinity chromatography (Charcosset, 2006).

In membrane chromatography, the main advantage is that dissolved molecules are carried to the adsorptive sites in the membranes by the convective flow, which can significantly increase the throughput of the process. This involves in the absence of pore diffusion, which is the major transport resistance in traditional chromatography. This advantage can be particularly important for purification of large biomolecules and viruses that can have significant diffusion limitations in conventional chromatographic media. The challenge in the production of chromatographic membranes has been to achieve binding capacities that are competitive with beads. Membrane chromatography usually employs micro- or macro-porous membranes, which contain tentacle ligands in the inner pore surface. The ligands provide highly selective separations through adsorption-binding interactions. Acting on membrane chemistry, pore size, membrane thickness, and number of membrane layers, improved binding capacity and flow distribution have been achieved (Van Reis and Zydney, 2007).

The preparation of the basic membrane, its activation, and the coupling of ligands to the activated membrane are the three fundamental steps usually involved in the preparation of adsorptive membranes (Zeng and Ruckenstein, 1999). The basic membranes must have micro- or macro-porous structure for rapid flow, available reactive groups for further coupling of functional ligands, chemical–physical stability to harsh conditions, and hydrophilic surface for minimizing nonspecific binding. A variety of base membrane materials have been engineered to create chromatographic membranes (Zeng and Ruckenstein, 1999). Methods such as coating or graft polymerization have been used to create the reactive groups in the membrane structure. However, membranes specifically prepared for affinity binding also exist (Klein, 2000). In this case, two functional monomers are in situ copolymerized. One functional monomer provides the reactive groups; the other one functions as a crosslinker, ensuring mechanical strength.

After the membranes are activated, different ligands can be coupled to them under mild conditions. Purification procedures using chromatographic membranes have been reported for several bioprocess purposes, such as primary protein capture, purification of large biomolecules (e.g., nucleic acids, proteins, antibodies, and viruses), and polishing of bioprocess streams (Boi et al., 2008; Orr et al., 2013). Affinity appears to be the most prevalent chemistry for experimental chromatographic membranes. Common affinity ligands include protein A for purification of immunoglobulins (IgGs), immunoaffinity ligands, low-molecular-mass ligands (Cibacron Blue, histidine, tryptophan), and other ligands (peptide, Cu^{2+}) (Ghosh, 2002).

14.5.6 Other Applications of Nanoengineered Membranes in Biotechnology

14.5.6.1 Nanoengineered membranes in biosensing

Biosensors, according to the IUPAC definition, are "devices that use specific biochemical reactions mediated by isolated enzymes, immunosystems, tissues, organelles or whole cells to detect chemical compounds usually by electrical, thermal or optical signals." A biosensor consists of two combined elements: (i) a bioreceptor, which is an immobilized biological element that recognizes the analyte and (ii) a transducer, which is used to convert the biochemical signal resulting from the interaction of the analyte with the bioreceptor into a physical signal. Membranes are optimal support for bioreceptor immobilization because they can either participate in the sensing mechanisms or immobilize the component responsible for sensing the analyte. There are many advantages leading to the use of membranes as components of sensor devices. For example, the membrane can perform the function of a barrier between the sensor system and its environment, allowing selective access (e.g., of the analyte only) to the receptor or/and protecting the receptor from disturbing influences of the environment. In addition if the transducer is a separate chemical species, the membrane, used either as support for the immobilization of the receptor or tool for bringing it close to the detector, can allow to integrate the sensing system. For further details, see Chapter 13 "Nanostructured Membranes for Biosensor Development."

14.5.6.2 Nanoengineered membranes in wastewater treatment

During the last few years, membrane technologies have increasingly gained attention in the field of wastewater treatment (Judd, 2010; Strathmann et al., 2006). Wastewater comes from domestic, municipal, industrial, and agricultural discharges. Wastewaters have to be treated to minimize their environmental release and to promote water reuse and recycling. However, in the treatment of industrial wastewater, the water recovery is not necessarily the most important objective. In fact, often very toxic or valuable compounds have to be removed to avoid environmental contamination. Membrane processes present a lot of advantages, which make them a valuable alternative to other wastewater treatment techniques such as the activated sludge process. The combination of a membrane operation with a biological treatment is a technology called MBR, which finds wide application in wastewater treatment. To date, the mainly used MBRs are the one in which bacteria are suspended and the membrane acts as a selective barrier (Ramakrishna et al., 2011). In fact, the retention of bacteria, viruses, suspended matter, and most soluble compounds carried out by means of the membrane provides high-quality effluents, already suitable for some reuse without the need of additional treatment. The bacteria can be suspended in the bioreactor chamber or immobilized on the ultrafiltration membranes where they form a biofilm (Yang et al., 2006). In the latter case, we can speak more precisely of BMRs. Biosolid separation is the most investigated type of membrane application for wastewater treatment. The MBR processes are configured with the membranes placed either outside or inside a tank. The immersed configuration (Fig. 14.7a) is the most frequent adopted because it requires a lower transmembrane pressure (TMP) than the external configuration (Fig. 14.7b). In fact, in the submerged process, the permeate is withdrawn with a suction pump at TMP 0.05–0.5 bar (Drioli and Fontananova, 2010).

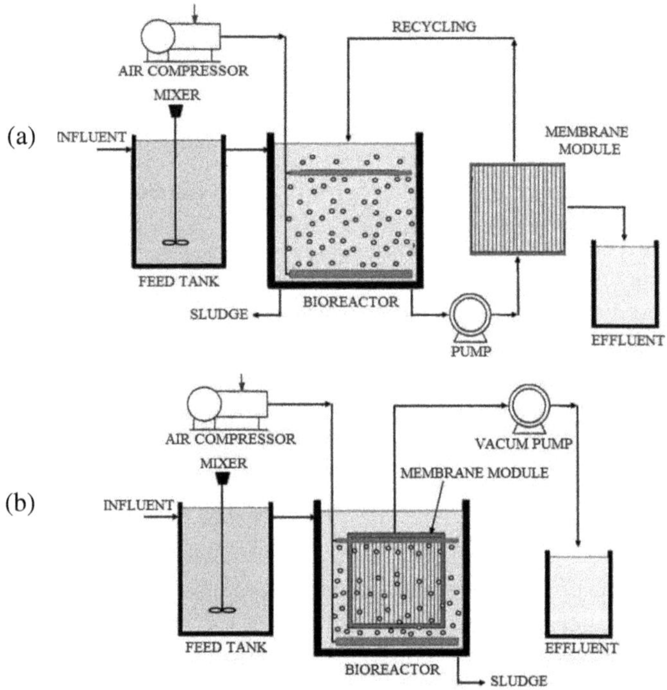

Figure 14.7 Configuration of a membrane bioreactor with a submerged membrane module (a) and an external membrane module (b).

The main difficulty in the application of MBR in wastewater treatment is membrane fouling (Yeon et al., 2009). It increases the aeration demand and in turn, coupled with the reduced duration of fouled membranes, increases operating cost. Nowadays, it is clear that MBRs have a great future in the treatment of wastewater (especially where a highly treated water quality is required and space is limited) and major efforts are devoted to contrast membrane fouling (Judd, 2016). Low fouling membranes can be produced by engineering strategies or modifying the chemical and physical properties of the surface such as roughness and hydrophobicity. Grafting methods have been shown to be an effective way of coating the surface of membranes to reduce absorption of bacteria, proteins, and other organic compounds. Kochan et al. (2009) modified polyethersulfone hollow-fiber ultrafiltration membranes by using the LbL technique for wastewater treatment. The polyelectrolytes used were poly(ethylenimine), poly(diallyldimethyl-ammonium chloride), and poly(styrene sulfonate) to obtain up to three layers. The total organic carbon rejection was higher for the modified membranes as compared to the virgin membranes. Byun et al. (2013) modified poly(vinylidene)-fluoride flat and hollow-fiber membranes to improve antifouling characteristics, via thermal crosslinking of acrylic acid. The method enabled the formation of a crosslinked layer and, consequently, hydrophilization of membranes at the surface as well as inside the pores. Wardrip et al. (2016) generated patterned polymer coatings on ultrafiltration membranes by a lithographic technique.

Organophosphate pesticides (OPs) are a class of dangerous polluting compounds that have motivated many detoxification efforts in the last few years. These toxic compounds are employed in the agriculture industry (Ghanem and Raushel, 2005). Owing to the high costs or environmental concerns related to the physical and chemical methods for the destruction of OPs, enzymatic detoxification has generated great interest (Le Jeune and Russell, 1996). The use of phosphotriesterase enzymes is hindered by the poor stability of these enzymes. Vitola et al. (2016) developed biocatalytic membrane systems that can work in liquid and vapor phases by employing an extremely stable phosphotriesterase enzyme extracted from an extremophile archaea (Merone et al., 2010). The use of this enzyme shows promise for the development of BMRs able to decontaminate OPs, and this aim is currently being pursued.

14.5.6.3 Nanoengineered membranes in carbon dioxide (CO_2) separation

The emission mitigation pledges aim at reducing emission of CO_2 and other greenhouse gases. It is a well-accepted fact that the increased CO_2 concentration in the atmosphere, which started since the beginning of industrial revolution, is the cause of greenhouse effect (Rosenberg et al., 2016).

Currently, CO_2 capture is dominated by amine-based technologies (Ducher et al., 2015), but these are really energy intensive and less attractive from the environmental point of view due to emissions of volatile solvents (Ramdin et al., 2012).

Membranes can be used to capture CO_2 during power generation. The streams can be pumped through the feed side of the membrane, while the strip side of the membrane is an alkaline aqueous stream for the entrapment of CO_2. For these membranes, it is very important to be highly permeable, selective, and thermally and chemically resistant.

Nanoengineering by carbonic anhydrase, an enzyme that catalyzes the conversion of CO_2 into bicarbonate, improves CO_2 absorption and separation. Carbozyme Inc. first proposed this approach (Kawatra et al., 2011). The main problem with this technology is solvent management, since water evaporates even at low temperatures.

This problem can be overcome by replacing volatile solvents with non-volatile solvents, such as ionic liquids, which are compounds having negligible vapor pressure and a lot of interesting properties (Ramdin, 2012). Neves et al. (2012) studied an innovative strategy for capturing CO_2 from flue gas streams. They used a supported liquid membrane with ionic liquids immobilized inside the porous structure comprising a bovine carbonic anhydrase that enhances the selective transport of CO_2.

Abdelrahima et al. (2017) recently developed a thermostable biomimetic membrane system using supported ionic liquid membranes and a thermophilic enzyme (SspCA). This enzyme manages to improve the selective transport of CO_2 at high temperatures through the supported ionic liquid membrane.

14.6 Conclusion

For a long time, membranes have been employed in operations related to biotechnology processes, such as micro- and ultrafiltration. Other applications of membranes in biotechnology, including MBRs, membrane biosensors, and so on, have been reported more recently. These operations are carried out by nanoengineered membranes. In this chapter, methodologies for membrane nanoengineering have been discussed, illustrating the techniques that can be used for membrane chemical functionalization and biomolecule immobilization. In addition, a special section has been devoted to the description of the characterization techniques applied to validate modification protocols. Afterward, some examples of processes exploiting nanoengineered membranes, at either laboratory or industrial scale, have been discussed. The nanoengineered membranes, in particular biocatalytic membranes, are very suitable for application in biotechnology processes because they possess advantages characteristics of high efficiency, high selectivity and permeability for the bioconversion and transport of specific components, low energy requirements, environmental compatibility, easy control and scale-up. However, this technology is still at an emerging stage of industrial development, because of some technological difficulties related to rate-limiting aspects such as the lifespan, cost, and microbial contamination of the biocatalyst. The integration of advanced technologies, including membrane engineering, molecular modeling, molecular biology and genetics, will contribute to overcoming these problems. The field is expected to progress and contribute to innovative production, processing, and treatment strategies beyond the current technological solutions.

Appendix

Young's equation

$$\gamma_{\mathrm{lv}} \cos \theta_{\mathrm{Y}} = \gamma_{\mathrm{sv}} - \gamma_{\mathrm{sl}} \tag{A.1}$$

Darcy's law

$$J_{\mathrm{v}} = L_{\mathrm{p}}\, \Delta \mathrm{p} \tag{A.2}$$

Mass balance equation

$$C_{\mathrm{i}} V_{\mathrm{i}} = C_{\mathrm{f}} V_{\mathrm{f}} + C_{\mathrm{ws}} V_{\mathrm{ws}} + m \tag{A.3}$$

References

Abdelrahim, M. Y. M., Martins, C. F., Neves, L. A., Capasso, C., Supuran, C. T., Coelhoso, I. M., and Barboiu, M. (2017). Supported ionic liquid membranes immobilized with carbonic anhydrases for CO_2 transport at high temperatures. *J. Membr. Sci.*, **528**, pp. 225–230.

Adiga, S. P., Jin, C., Curtiss, L. A., Monteiro-Riviere, N. A., and Narayan, R. J. (2009). Nanoporous membranes for medical and biological applications. *Wiley Interdiscip. Rev. Nanomed. Nanobiotechnol.*, **1**, pp. 568–581.

Álvarez-Hornos, F. J., Volckaert, D., Heynderickx, P. M., and Van Langenhove, H. (2011). Performance of a composite membrane bioreactor for the removal of ethyl acetate from waste air. *Bioresour. Technol.*, **102**, pp. 8893–8898.

Alwitt, R. S. and Xu, J. (1993). Stresses in sulfuric acid anodized coatings on aluminum. *J. Electrochem. Soc.*, **140**, pp. 1241–1246.

Battistel, E. and Rialdi, G. (2006). Characterization of immobilized enzymes by microcalorimetry, in: *Immobilization of Enzymes and Cells*, Guisan, J. M. (ed.) (Humana Press, Totowa), pp. 295–310.

Becker, M., Provart, N., Lehmann, I., Ulbricht, M., and Hicke, H.-G. (2002). Polymerization of Glucans by enzymatically active membranes. *Biotechnol. Prog.*, **18**, pp. 964–968.

Belfer, S., Fainchtain, R., Purinson, Y., and Kedem, O. (2000). Surface characterization by FTIR-ATR spectroscopy of polyethersulfone membranes-unmodified, modified and protein fouled. *J. Membr. Sci.*, **172**, pp. 113–124.

Bernhardt, A., Lode, A., Boxberger, S., Pompe, W., and Gelinsky, M. (2008). Mineralised collagen—an artificial, extracellular bone matrix—improves osteogenic differentiation of bone marrow stromal cells. *J. Mater. Sci. Mater. Med.*, **19**, pp. 269–275.

Bernkop-Schnürch, A., Schwarz, V., and Steininger, S. (1999). Polymers with thiol groups: A new generation of mucoadhesive polymer? *Pharm. Res.*, **16**, pp. 876–881.

Biebuyck, H. A., Gerber, C., Delamarche, E., and Michel, B. (1996). Golden interfaces: The surface of self-assembled monolayers. *Adv. Mater.*, **8**, pp. 719–729.

Bindman, N. A. and van der Donk, W. A. (2013). A general method for fluorescent labeling of the N-termini of lanthipeptides and its application to visualize their cellular localization. *J. Am. Chem. Soc.*, **135**, pp. 10362–10371.

Blaih, S. M., Abdine, H. H., El-yazbi, F. A., and Shaalan, R. A. (2000). Spectrophotometric determination of enalapril maleate and ramipril in dosage forms. *Spectrosc. Lett.*, **33**, pp. 91–102.

Boi, C., Dimartino, S., and Sarti, G. C. (2008). Performance of a new protein A affinity membrane for the primary recovery of antibodies. *Biotechnol. Progr.*, **24**, pp. 640–647.

Brady, D. and Jordaan, J. (2009). Advances in enzyme immobilization. *Biotechnol. Lett.*, **31**, pp. 1639–1650.

Byuna, Y. J., Kim, J. H., and Kim, S. S. (2013). Surface modification of PVDF membranes for water treatment via hydrophilic thermal cross-linking method. *Desalin. Water Treat.*, **51**, pp. 5371–5378.

Burfield, D. R. and Law, K. S. (1979). Determination of carbonyl groups in polymers by reaction with 2,4-dinitrophenylhydrazine. *Polymer*, **20**, pp. 620–626.

Caldwell, S. R. and Raushel, F. M. (1991). Detoxification of organophosphate pesticides using a nylon based immobilized phosphotriesterase from *Pseudomonas dirninuta*. *Appl. Biochem. Biotech.*, **31**, pp. 59–73.

Charcosset, C. (2006). Membrane processes in biotechnology: An overview. *Biotechnol. Adv.*, **24**, pp. 482–492.

Chen, C., Chen, C., Huang, C., and Liu, Y. (2007). Simultaneous purification and immobilization of penicillin G acylase using bifunctional membrane. *J. Membr. Sci.*, **298**, pp. 24–29.

Chen, W., Lama, R. H. W., and Fu, J. (2011). Photolithographic surface micromachining of polydimethylsiloxane (PDMS). *Lab. Chip.*, **12**, pp. 391–395.

Chen, W. and McCarthy, T. J. (1997). Layer-by-layer deposition: A tool for polymer surface modification. *Macromolecules*, **30**, pp. 78–86.

De Jong, J., Lammertink, R. G. H., and Wessling, M. (2006). Membranes and microfluidics: A review. *Lab. Chip.*, **6**, pp. 1125–1139.

Denes, F. S. and Manolache, S. (2004). Macromolecular plasma-chemistry: An emerging field of polymer science. *Prog. Polym. Sci.*, **29**, pp. 815–885.

Desai, T. A., Sharma, S., Walczak, R. J., Boiarski, A., Cohen, M., Shapiro, J., West, T., Melnik, K., Cosentino, C., Sinha, P. M., and Ferrari, M. (2006). Nanoporous implants for controlled drug delivery, in: *BioMEMS and Biomedical Nanotechnology*, Desai, T. A. and Bhatia, S. (eds.) (Therapeutic Micro/Nanotechnology, Springer; US), pp. 263–286.

Drioli, E. and Fontananova, E. (2010). Membrane reactors, in: *Ullmann's Encyclopedia of Industrial Chemistry* (Wiley-VCH Verlag GmbH & Co. KGaA, Weinheim, Germany), pp. 393–412.

Drioli, E. and Giorno, L. (2010). *Comprehensive Membrane Science and Engineering* (Newnes).

Dutcher, B., Fan, M., and Russell, A. G. (2015). Amine-based CO_2 capture technology development from the beginning of 2013: A review. *ACS Appl. Mater. Interf.*, **7**, pp. 2137–2148.

Durán, N., Rosa, M. A., D'Annibale, A., and Gianfreda, L. (2002). Applications of laccases and tyrosinases (phenoloxidases) immobilized on different supports: A review. *Enzyme Microb. Tech.*, **31**, pp. 907–931.

Edwards, J. (1997). Coating and surface treatment systems for metals: A comprehensive guide to selection (ASM International Material Park, United States).

Eral, H. B., Mannetje, D. J. C. M., and Oh, J. M. (2013). Contact angle hysteresis: A review of fundamentals and applications. *Colloid. Polym. Sci.*, **291**, pp. 247–260.

Fang, B., Ling, Q., Zhao, W., Ma, Y., Bai, P., Wei, Q., Li, H., and Zhao, C. (2009). Modification of polyethersulfone membrane by grafting bovine serum albumin on the surface of polyethersulfone/poly(acrylonitrile-co-acrylic acid) blended membrane. *J. Membr. Sci.*, **329**, pp. 46–55.

Fee, C., Nawada, S., and Dimartino, S. (2014). 3D printed porous media columns with fine control of column packing morphology. *J. Chromatogr. A*, **1333**, pp. 18–24.

Fissell, W. H., Dubnisheva, A., Eldridge, A. N., Fleischman, A. J., Zydney, A. L., and Roy, S. (2009). High-performance silicon nanopore hemofiltration membranes. *J. Membr. Sci.*, **326**, pp. 58–63.

Gebreyohannes, A. Y., Bilad, M. R., Verbiest, T., Courtin, C. M., Dornez, E., Giorno, L., and Vankelecom, I. F. (2015). Nanoscale tuning of enzyme localization for enhanced reactor performance in a novel magnetic-responsive biocatalytic membrane reactor. *J. Membr. Sci.*, **487**, pp. 209–220.

Gebreyohannes, A. Y., Mazzei, R., Curcio, E., Poerio, T., Drioli, E., and Giorno, L. (2013). Study on the in situ enzymatic self-cleansing of microfiltration membrane for valorization of olive mill wastewater. *Ind. Eng. Chem. Res.*, **52**, pp. 10396–10405.

Gésan-Guiziou, G. (2014). Integrated membrane operations in whey processing, in: *Integrated Membrane Operations: In the Food Production,* Cassano, A. and Drioli, E. (eds.) (De Gruyter, Berlin), pp. 133–146.

Gao, P., Hunter, A., Summe, M. J., and Phillip, W. A. (2016). A method for the efficient fabrication of multifunctional mosaic membranes by inkjet printing. *Appl. Mater. Interf.*, DOI: 10.1021/acsami.6b06048.

Gavrilescu, M. and Chisti, Y. (2005). Biotechnology: A sustainable alternative for chemical industry. *Biotechnol. Adv.*, **23**, pp. 471–499.

Ghaffar, T., Irshad, M., Anwar, Z., Aqil, T., Zulifqar, Z., Tariq, A., Kamran, M., Ehsan, N., and Mehmood, S. (2014). Recent trends in lactic acid biotechnology: A brief review on production to purification. *J. Radiat. Res.*, **7**, pp. 222–229.

Ghanem, E. and Raushel, F. M. (2005). Detoxification of organophosphate nerve agents by bacterial phosphotriesterase. *Toxicol. Appl. Pharmacol.*, **207**, pp. 459–470.

Giorno, L., Donato, L., Todisco, S., and Drioli, E. (1998). Study of fouling phenomena in apple juice clarification by enzyme membrane reactor. *Sep. Sci. Technol.*, **33**, pp. 739–756.

Giorno, L. and Drioli, E. (2000). Biocatalytic membrane reactors: Applications and perspectives. *Trends Biotechnol.*, **18**, pp. 339–349.

Giorno, L., De Bartolo, L., and Drioli, E. (2003). Membrane bioreactors for biotechnology and medical application, in: *New Insights into Membrane Science and Technology: Polymeric and Biofunctional Membranes*, Bhattacharyya, D. and Butterfield, D. A. (eds.) (Elsevier Science, Amsterdam), pp. 187–217.

Giorno, L., D'Amore, E., Mazzei, R., Piacentini, E., Zhang, J., Drioli, E., Cassano, R., and Picci, N. (2007). An innovative approach to improve the performance of a two separate phase enzyme membrane reactor by immobilizing lipase in presence of emulsion. *J. Membr. Sci.*, **295**, pp. 95–101.

Giorno, L., Mazzei, R., and Drioli, E. (2009). Biochemical membrane reactors in industrial processes, in: *Membrane Operations: Innovative Separations and Transformations*, Drioli, E. and Giorno, L. (eds.) (Wiley-VCH Verlag GmbH & Co. KGaA, Weinheim, Germany).

Giorno, L. and Drioli, E. (2010). Biocatalytic membrane reactors, in: *Encyclopedia of Industrial Biotechnology: Bioprocess, Bioseparation, and Cell Technology*, Flickinger, M. C. (eds.) (John Wiley & Sons, Inc.), pp. 1–23.

Giorno, L. and Drioli, E. (2013). Biocatalytic membrane reactors, in: *Upstream Industrial Biotechnology*, Flickinger, M. C. (ed.) (John Wiley & Sons).

Goddard, J. M. and Hotchkiss, J. H. (2007). Polymer surface modification for the attachment of bioactive compounds. *Prog. Polym. Sci.*, **32**, pp. 698–725.

Godjevargova, T., Dayal, R., and Marinov, I. (2004). Simultaneous covalent immobilization of glucose oxidase and catalase onto chemically modified acrylonitrile copolymer membranes. *J. Appl. Polym. Sci.*, **91**, pp. 4057–4063.

Ghosh, R. (2002). Protein separation using membrane chromatography: Opportunities and challenges. *J. Chromatogr. A*, **95**, pp. 13–27.

Gross, B. C., Erkal, J. L., Lockwood, S. Y., Chen, C., and Spence, D. M. (2014). Evaluation of 3D printing and its potential impact on biotechnology and the chemical sciences. *Anal. Chem.*, **86**, pp. 3240–3253.

Gupta, S., Bhattacharya, A., and Murthy, C. N. (2013). Tune to immobilized lipases on polymer membranes: Techniques, factors and prospects. *Biocatal. Agri. Biotechnol.*, **2**, pp. 171–190.

Harju, M., Kallioinen, H., and Tossavainen, O. (2012). Lactose hydrolysis and other conversions in dairy products: Technological aspects. *Int. Dairy J.*, **22**, pp. 104–109.

He, D., Susanto, H., and Ulbricht, M. (2009). Photo-irradiation for preparation, modification and stimulation of polymeric membranes. *Prog. Polym. Sci.*, **34**, pp. 62–98.

Huang, Z., Zhang, W., Yu, J., and Gao, D. (2007). Nanoporous alumina membranes for enhancing hemodialysis. *J. Med. Devices*, **1**, pp. 79–83.

Hull, C. W. (1986). Apparatus for production of three-dimensional objects by stereolithography. U.S. Patent 4,575, 330.

Humes, H. D., Fissell, W. H., Weitzel, W. F., Buffington, D. A., Westover, A. J., MacKay, S. M., and Gutierrez, J. M. (2002). Metabolic replacement of kidney function in uremic animals with a bioartificial kidney containing human cells. *Am. J. Kidney Dis.*, **39**, pp. 1078–1087.

Humes, H. D., Buffington, D., Westover, A. J., Roy, S., and Fissell, W. H. (2014). The bioartificial kidney: Current status and future promise. *Pediatr. Nephrol.*, **29**, pp. 343–351.

Jochems, P., Satyawali, Y., Dielsa, L., and Dejonghe, W. (2011). Enzyme immobilization on/in polymeric membranes: Status, challenges and perspectives in biocatalytic membrane reactors (BMRs). *Green Chem.*, **13**, pp. 1609–1623.

Jolivalt, C., Brenon, S., Caminade, E., Mougin, C., and Ponti, M. (2000). Immobilization of laccase from *Trametes versicolor* on a modified PVDF microfiltration membrane: Characterization of the grafted support and application in removing a phenylurea pesticide in wastewater. *J. Membr. Sci.*, **180**, pp. 103–113.

Judd, S. (2010). *The MBR Book: Principles and Applications of Membrane Bioreactors for Water and Wastewater Treatment* (Elsevier, Great Britain).

Judd, S. J. (2016). The status of industrial and municipal effluent treatment with membrane bioreactor technology. *Chem. Eng. J.*, **305**, pp. 37–45.

Kang, G. and Cao, Y. (2014). Application and modification of polyvinylidene fluoride (PVDF) membranes: A review. *J. Membr. Sci.*, **463**, pp. 145–165.

Katoa, K., Uchidab, E., Kangc, E., Uyamaa, Y., and Ikada, Y. (2003). Polymer surface with graft chains. *Prog. Polym. Sci.*, **28**, pp. 209–259.

Kawatra, S. K., Eisele, T. C., and Simmons, J. J. (2011). U.S. Patent Application No. 13/079,459.

Kelly, S. M., Jess, T. J., and Price, N. C. (2005). How to study proteins by circular dichroism. *Biochim et Biophys Acta*, **1751**, pp. 119–139.

Kim, I. C. and Lee, K. H. (2006). Dyeing process waste water treatment using fouling resistant nanofiltration and reverse osmosis membranes. *Desalination*, **192**, pp. 246–251.

Klein, E. (2000). Affinity membranes: A 10-year review. *J. Membr. Sci.*, **179**, pp. 1–27.

Kochan, J., Wintgens, T., Wong, J. E., and Melin, T. (2009). Polyelectrolyte-modified polyethersulfone ultrafiltration membranes for wastewater treatment applications. *Desalin. Water Treat.*, **9**, pp. 175–180.

Landis, S. (2013). Nano lithography (Wiley-ISTE).

Le Jeune, K. E. and Russell, A. J. (1996). Covalent binding of a nerve agent hydrolyzing enzyme within polyurethane foams. *Biotechnol. Bioeng.*, **51**, pp. 450–457.

Lee, D., Gemici, Z., Rubner, M. F., and Cohen, R. E. (2007). Multilayers of oppositely charged SiO_2 nanoparticles: Effect of surface charge on multilayer assembly. *Langmuir*, **23**, pp. 8833–8837.

Lee, L. J. (2006). Polymer nanoengineering for biomedical applications. *Ann. Biomed. Eng.*, **34**, pp. 75–88.

Lopez, J. L. and Matson, S. L. (1997). A multiphase/extractive enzyme membrane reactor for production of diltiazem chiral intermediate, *J. Membr. Sci.*, **125**, pp. 189–211.

Louie, J. S., Pinnau, I., Ciobanu, I., Ishida, K. P., Ng, A., and Einhard, M. R. (2006). Effects of polyether-polyamide block copolymer coating on performance and fouling of reverse osmosis membranes. *J. Membr. Sci.*, **280**, pp. 762–770.

Mazzei, R., Giorno, L., Piacentini, E., Mazzuca, S., and Drioli, E. (2009). Kinetic study of a biocatalytic membrane reactor containing immobilizer β-glucosidase for the hydrolysis of oleuropein. *J. Membr. Sci.*, **339**, pp. 215–223.

Merone, L., Mandrich, L., Porzio, E., Rossi, M., Müller, S., Reiter, G., Worek, F., and Manco, G. (2010). Improving the promiscuous nerve agent hydrolase activity of a thermostable archaeal lactonase. *Bioresour. Technol.*, **101**, pp. 9204–9212.

Morelli, S. (2014). Biohybrid artificial liver (BAL) systems, in: *Encyclopedia of Membranes* (Springer Berlin Heidelberg), pp. 1–3.

Morelli, S., Salerno, S., Ahmed, H. M., Piscioneri, A., and De Bartolo, L. (2016). Recent strategies combining biomaterials and stem cells for bone, liver and skin regeneration. *Curr. Stem Cell Res. Ther.*, **11**, pp. 676–691.

Mohamad, N. R., Marzuki, N. H. C., Buang, N. A., Huyop, F., and Wahab, R. A. (2015). An overview of technologies for immobilization of enzymes and surface analysis techniques. *Biotechnol. Biotec. Eq.*, **29**, pp. 205–220.

Musale, D. A., Kumar, A., and Pleizier, G. (1999). Formation and characterization of poly(acrylonitrile)/chitosan composite ultrafiltration membranes. *J. Membr. Sci.*, **154**, pp. 163–173.

Naassner, M., Mergler, M., Wolf, K., and Schuphan, I. (2002). Determination of the xenoestrogens 4-nonylphenol and bisphenol A by high-performance liquid chromatography and fluorescence detection after derivatisation with dansyl chloride. *J. Chromatogr. A*, **945**, pp. 133–138.

Neves, L. A., Afonso, C., Coelhoso, I. M., and Crespo, J. G. (2012). Integrated CO_2 capture and enzymatic bioconversion in supported ionic liquid membranes. *Sep. Purif. Technol.*, **97**, pp. 34–41.

Novalin, S., Neuhaus, W., and Kulbe, K. D. (2005). A new innovative process to produce lactose-reduced skim milk. *J. Biotechnol.*, **119**, pp. 212–218.

Orosz, K. E., Gupta, S., Hassink, M., Abdel-Rahman, M., Moldovan, L., Davidorf, F. H., and Moldovan, N. I. (2004). Delivery of antiangiogenic and antioxidant drugs of ophthalmic interest through a nanoporous inorganic filter. *Mol. Vis.*, **10**, pp. 555–565.

Orr, V., Zhong, L., Moo-Young, M., and Chou, C. P. (2013). Recent advances in bioprocessing application of membrane chromatography. *Biotechnol. Adv.*, **31**, pp. 450–465.

Peuchot, M. M. (2010). New applications for membrane technologies in enology, in: *Membrane Technology: Membranes for Food Applications*, Peinemann, K.-V., Pereira Nunes, S., and Giorno, L. (eds.) (Wiley-VCH Verlag GmbH & Co. KGaA, Weinheim, Germany.), pp. 119–127.

Piacentini, E., Drioli, E., and Giorno, L. (2011). Preparation of stimulus responsive multiple emulsions by membrane emulsification using con A as biochemical sensor. *Biotechnol. Bioeng.*, **108**, pp. 913–923.

Pugazhenthi, G. and Kumar, A. (2004). Enzyme membrane reactor for hydrolysis of olive oil using lipase immobilized on modified PMMA composite membrane. *J. Membr. Sci.*, **228**, pp. 187–197.

Ramakrishna, S., Ma, Z., and Matsuura, T. (2011). Polymer membranes in biotechnology: Preparation, functionalization and application (Imperial College Press, London).

Ramdin, M., de Loos, T. W., and Vlugt, T. J. (2012). State-of-the-art of CO_2 capture with ionic liquids. *Ind. Eng. Chem. Res.*, **51**, pp. 8149–8177.

Ranieri, G., Mazzei, R., Wu, Z., Li, K., and Giorno, L. (2016). Use of a ceramic membrane to improve the performance of two-separate-phase biocatalytic membrane reactor. *Molecules*, **21**, pp. 345–359.

Ren, S. and Irudayam, J. I. (2015). Bioartificial liver device based on induced pluripotent stem cell-derived hepatocytes. *J. Stem Cell Res. Ther.*, **5**, pp. 263.

Rosenberg, N. J., Easterling III, W. E., Crosson, P. R., and Darmstadter, J. (2016). *Greenhouse Warming: Abatement and Adaptation* (Routledge, Oxon).

Ross, G. J., Watts, J. F., Hill, M. P., and Morrissey, P. (2000). Surface modification of poly(vinylidene fluoride) by alkaline treatment 1. The degradation mechanism. *Polymer*, **41**, pp. 1685–1696.

Ruckenstein, E. and Li, Z. F. (2005). Surface modification and functionalization through the self-assembled monolayer and graft polymerization. *Adv. Colloid. Interf.*, **113**, pp. 43–63.

Sapan, C. V., Lundblad, R. L., and Price, N. C. (1999). Colorimetric protein assay techniques. *Biotechnol. Appl. Biochem.*, **29**, pp. 99–108.

Schluger, N. W. and Koppaka, R. (2014). Lung disease in a global context. A call for public health action. *Ann. Am. Thorac. Soc.*, **11**, pp. 407–416.

Schroën, K., Van Dinther, A. M. C., Bogale, S., Vollebregt, M., Brans, G., and Boom, R. M. (2010). Membrane processes for dairy fractionation, in: *Membrane Technology: Membranes for Food Applications*, Peinemann, K.-V., Pereira Nunes, S., and Giorno, L. (eds.) (Wiley-VCH Verlag GmbH & Co. KGaA, Weinheim, Germany), pp. 25–43.

Smuleac, V., Butterfield, D. A., and Bhattacharyya, D. (2006). Layer-by-layer-assembled microfiltration membranes for biomolecule immobilization and enzymatic catalysis. *Langmuir*, **22**, pp. 10118–10124.

Stamatialis, D. F., Papenburg, B. J., Girones, M., Saiful, S., Bettahalli, S. N. M., Schmitmeier, S., and Wessling, M. (2008). Medical applications of membranes: Drug delivery, artificial organs and tissue engineering. *J. Membr. Sci.*, **308**, pp. 1–34.

Strathmann, H., Giorno, L., and Drioli, E. (2006). *An Introduction to Membrane Science and Technology* (CNR-Servizio Pubblicazioni, Rome).

Stroeve, P. and Ileri, N. (2011). Biotechnical and other applications of nanoporous membranes. *Trends Biotechnol.*, **879**, pp. 259–266.

Sun, S. W., Lin, Y. C., Weng, Y. M., and Chen, M. J. (2006). Efficiency improvements on ninhydrin method for amino acid quantification. *J. Food Comp. Anal.*, **19**, pp. 112–117.

Torras, C. and Garcia-Valls, R. (2004). Quantification of membrane morphology by interpretation of scanning electron microscopy images. *J. Membr. Sci.*, **233**, pp. 119–127.

Tyagi, C., Tomar, L. K., and Singh, H. (2009). Surface modification of cellulose filter paper by glycidyl methacrylate grafting for biomolecule immobilization: Influence of grafting parameters and urease immobilization. *J. Appl. Polym. Sci.*, **111**, pp. 1381–1390.

Ulbricht, M. and Riedel, M. (1998). Ultrafiltration membrane surfaces with grafted polymer tentacles: Preparation, characterization and application for covalent protein binding. *Biomaterials*, **19**, pp. 1229–1237.

Ulbricht, M. (2004). Membrane separations using molecularly imprinted polymers. *J. Chromatogr. B*, **804**, pp. 113–125.

Vasapollo, G., Del Sole, R., Mergola, L., Lazzoi, M. R., Scardino, A., Scorrano, S., and Mele, G. (2011). Molecularly imprinted polymers: Present and future prospective. *Int. J. Mol. Sci.*, **12**, pp. 5908–5945.

Van Reis, R. and Zydney, A. (2007). Bioprocess membrane technology. *J. Membr. Sci.*, **297**, pp. 16–50.

Vitola, G., Büning, D., Schumacher, J., Mazzei, R., Giorno, L., and Ulbricht, M. (2016). Development of a novel immobilization method by using microgels to keep enzyme in hydrated microenvironment in porous hydrophobic membranes. *Macromol. Biosci.*, DOI: 10.1002/mabi.201600381.

Vitola, G., Mazzei, R., Fontananova, E., and Giorno, L. (2015). PVDF membrane biofunctionalization by chemical grafting. *J. Membr. Sci.*, **476**, pp. 483–489.

Vitola, G., Mazzei, R., Fontananova, E., Porzio, E., Manco, G., Gaeta, S. N., and Giorno, L. (2016). Polymeric biocatalytic membranes with immobilized thermostable phosphotriesterase. *J. Membr. Sci.*, **516**, pp. 144–151.

Vrijenhoek, E. M., Hongb, S., and Elimelech, M. (2001). Influence of membrane surface properties on initial rate of colloidal fouling of reverse osmosis and nanofiltration membranes. *J. Membr. Sci.*, **188**, pp. 115–128.

Wang, X. Y., Chen, C., Liu, H. L., and Ma, J. (2008). Preparation and characterization of PAA/PVDF membrane-immobilized Pd/Fe nanoparticles for dechlorination of trichloroacetic acid. *Water Res.*, **42**, pp. 4656–4664.

Wardrip, N. C., Dsouza, M., Urgun-Demirtas, M., Snyder, S. W., Gilbert, J. A., and Arnusch, C. J. (2016). Printing-assisted surface modifications of patterned ultrafiltration membranes. *ACS Appl. Mater. Inter.*, **8**, pp. 30271–30280.

Warkiani, M. E., Bhagat, A. A. S., Khoo, B. L., Han, J., Lim, C. T., Gong, H. Q., and Fane, A. G. (2013). Isoporous micro/nanoengineered membranes. *ACS Nano*, **7**, pp. 1882–1904.

Whitmore, L. and Wallace, B. A. (2008). Protein secondary structure analyses from circular dichroism spectroscopy: Methods and reference databases. *Biopolymers*, **89**, pp. 392–400.

Xu, Z., Huang, X., and Wan, L. (2009). *Surface Engineering of Polymer Membranes* (Springer Science + Business Media).

Yang, W., Cicek, N., and Ilg, J. (2006). State-of-the-art of membrane bioreactors: Worldwide research and commercial applications in North America. *J. Membr. Sci.*, **270**, pp. 201–211.

Yeon, K.-M., Cheong, W.-S., Oh, H.-S., Lee, W.-N., Hwang, B.-K., Lee, C.-H., Beyenal, H., and Lewandowski, Z. (2009). Quorum sensing: A new biofouling control paradigm in a membrane bioreactor for advanced wastewater treatment. *Environ. Sci. Technol.*, **43**, pp. 380–385.

Yi, C., Zhiyuan, L., Yi, L., and Xing, H. (2013). Hydrothermal synthesis of aluminum oxy-hydroxide nanorod and nanotube arrays. *Electrochim Acta*, **93**, pp. 241– 247.

Yuan, Y. and Lee, T. R. (2013). Contact angle and wetting properties, in: *Surface Science Techniques*, Bracco, G. and Holst, B. (eds.) (Springer-Verlag Berlin Heidelberg), pp. 3–34.

Zeng, X. and Ruckenstein, E. (1999). Membrane chromatography: Preparation and applications to protein separation. *Biotechnol. Progr.*, **15**, pp. 1003–1019.

Zhao, C., Xue, J., Ran, F., and Sun, S. (2013). Modification of polyethersulfone membranes: A review of methods. *Prog. Mater. Sci.*, **58**, pp. 76–150.

Chapter 15

Encapsulation of Bioactive Compounds in Nano- and Micro-Matrices, Emulsions, and Capsules Produced by Membrane Technology

Emma Piacentini and Lidietta Giorno
Institute on Membrane Technology of the National Research Council of Italy (ITM-CNR), Via P. Bucci, Cubo 17C, 87036 Rende (CS), Italy
e.piacentini@itm.cnr.it

This chapter describes the manufacturing of encapsulating materials (nano- and micro-matrices, emulsions, and capsules) by membrane technology. It highlights the parameters (architecture and physical structure, particle size, particle size distribution, surface properties) controlling the encapsulating material properties (cell uptake, drug release, encapsulated efficiency, drug loading, stability).

The application of membrane technology in the production of particulate materials for drug delivery and biocatalysis will also be discussed.

Functional Nanostructured Membranes
Edited by Enrico Drioli, Lidietta Giorno, and Annarosa Gugliuzza

ISBN 978-981-4774-79-6 (Hardcover), 978-1-351-13511-5 (eBook)
www.panstanford.com

15.1 Introduction

Encapsulation is a rapidly expanding technology that has gained importance in the field of drug delivery as well as in biocatalytic process development with potential application in pharmaceutical, food, and cosmetic industries. The method is employed mainly to protect bioactive compounds (drugs, nutraceuticals, enzymes, antioxidants, flavors, etc.) from adverse environment and for their release over longer periods of time or at certain locations in the body (Gouin, 2004). Encapsulation is a process by which particles or droplets containing bioactive compounds are surrounded by a wall material also called shell, coating, or membrane. The bioactive compounds can be confined in a core (capsules) or uniformly dispersed in matrix systems (matrices) or in a liquid dispersed in another liquid where it is immiscible (emulsions) (Fig. 15.1).

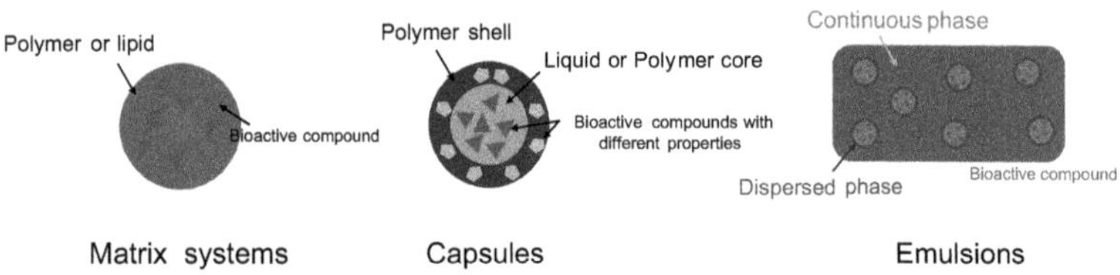

Figure 15.1 Encapsulating materials.

The architecture of encapsulating materials depends on the materials and methods used in their preparation. Particle size and size distribution are important characteristics of encapsulating materials due to their direct impact on in-vivo distribution, biological fate, toxicity, and targeting ability as well as drug loading, drug release, and stability. Based on the size, particles ranging in size from 10 to 1000 nm and with size more than 1000 nm are referred to as nanoparticles and microparticles, respectively. Adequate size must be designed to determine the successful application of particle-based products. The surface of encapsulating materials can be modified to alter the functionality and/or improve the colloidal stability. Particles' surface can be designed by introducing various signal-responsive units to respond to different physical and chemical signals to produce delivery devices for targeted drug release triggered (Table 15.1).

Table 15.1 Effecting parameters of encapsulating materials

Effecting parameter	Encapsulating material properties				
	Cell uptake	Drug release	Encapsulated efficiency	Drug loading	Stability
Architecture and physical structure		x	x	x	x
Particle size	x	x	x	x	x
Particle size distribution	x	x	x	x	x
Surface properties	x	x			x

Advanced technologies for manufacturing encapsulating materials are needed to turn the challenges into opportunities of drug delivery and biocatalytic reactions. Membrane technology has been demonstrated as a promising method for the preparation of micro- and nanomatrices, capsules, and emulsions. The method permits to assist particle formation at micro- and nanoscale, governing and fine-tuning particle size, size distribution, and architecture. Membrane emulsification has been used to fabricate simple or multiple liquid droplets that will then be solidified by solvent diffusion, solvent evaporation, and coacervation, depending on the chemical nature of materials, active ingredients, and target particle properties. They have also paved the way for encapsulating hydrophilic and hydrophobic active ingredients in complex nanocapsules with precise hierarchy.

15.2 Manufacturing of Encapsulating Materials by Membrane Technology

Membrane technology can be used to produce micro- and nanoparticles in different ways according to the phases and membrane properties, which dictate the membrane role and droplet mechanism formation (Fig. 15.2).

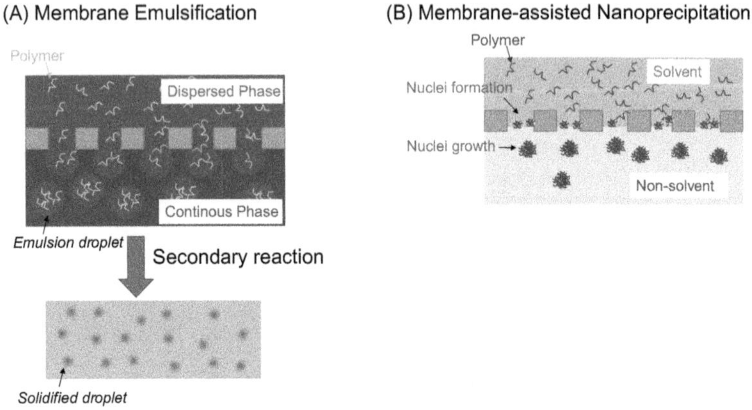

Figure 15.2 Particle production by membrane technology.

When the two phases are immiscible and a high interfacial tension exists between the membrane surface and the droplet-forming phase, membrane emulsification based on a drop-by-drop mechanism occurs (direct membrane-emulsification process; Fig. 15.2A). Here, one phase (referred to as dispersed phase) can be dispersed into the other (referred to as continuous phase) by forming droplets at the membrane pore opening, where the two phases meet, as a result of the pressure gradient driving force and the interfacial tension existing between the dispersed phase and the membrane wetted by the continuous phase. The method is used to produce oil-in-water (O/W), water-in-oil (W/O), or multiple (O/W/O and W/O/W) emulsions. The droplets size is mainly controlled by the pore size of membrane. A linear correlation between droplet and pore size is generally observed with a coefficient between 2 and 10, depending on the physical–chemical properties of the phases. Droplets that are two or three times the pore size of the membrane are usually produced at the fluid-dynamic optimized conditions.

When the two phases are miscible and meet each other at the pore exit while one phase is passing through the membrane under a pressure gradient, the two phases mix on or near the surface of the membrane, which is wetted by both phases (membrane-assisted nanoprecipitation; Fig. 15.2B). The polymer is contained in the phase permeating through the membrane; however, droplets are not formed at the pore level because of the absence of the interfacial tension between the two liquids and the membrane. The polymeric nanoparticles are formed by precipitation due to solvent displacement.

15.2.1 Control of Architecture and Physical Structure

Membrane emulsification is a suitable method for producing oil-in-water (O/W) or water-in-oil (W/O) emulsions with a uniform particle size distribution and controlled mean particle size as well as multiple suspensions and emulsions, such as solid-in-oil-in-water (S/O/W), oil-in-water-in-oil (O/W/O), and water-in-oil-in-water (W/O/W). The integration of membrane-emulsification technology with appropriate secondary reactions allows extension to the production of micro-/nanocarriers with different architecture and physical structure (Table 15.2). If a preformed polymer is dissolved in the dispersed phase, the emulsification is followed by a solidification step such as solvent evaporation/diffusion and gelification to produce micro-nanocapsules and micro-nanobeads, respectively. If polymers (usually water-soluble polymers) with opposite charge are dissolved in the continuous phase, core–shell matrices are obtained by complex coacervation. If the dispersed phase is a lipid, solid at room temperature, solid–lipid micro-nanocapsules or spheres are obtained by cooling the emulsion below the melting point. If the dispersed phase contains a polymeric monomer solution, micro-nanocapsules or spheres are produced by suspension polymerization or polycondensation or mini-emulsion polymerization. Hydrophilic bioactive compounds can be encapsulated in micro-nanobeads based on hydrophilic polymer (alginate, chitosan, agarose, gelatin, PVA) or in multicore-matrix particles produced starting from W/O/W or S/O/W emulsions. Hydrophobic bioactive compounds can be encapsulated in micro-nanospheres, solid–lipid micro-nanospheres, liquid-core/polymer-shell capsules, polymer-core/polymer-shell capsules, and multicore-matrix particles (produced starting from W/O/W or S/O/W emulsions). Coencapsulation of hydrophilic and hydrophobic bioactive compounds can also be assessed in multicore-matrix particles. Alternatively, bioactive compounds can be attached by chemical bonds to the surface of preformed particles. Table 15.2 illustrates some examples of formulation based on the methods mentioned above.

Table 15.2 Micro/nanoparticles produced by membrane technology

Encapsulating material type		Method to produce	Wall material	References
Micro- and nanomatrix systems	Beads	An aqueous solution containing the polymer is injected through the membrane pores followed by cross-linking reaction (by internal/external gelation, cooling, heating).	Alginate	Liu (2003); You (2001)
			Chitosan	Wang (2005); Akamatsu (2010a,b); Lv (2009); Wu (2008)
			Agarose	Zhou (2007, 2008, 2009)
			Hemoglobin	Lai (2015a,b)
			PVA	Piacentini (2017)
	Spheres	An organic solution containing the polymer is injected through the membrane pores followed by solvent removal.	PCL	Imbrogno (2014a,b)
			PLGA	Ito (2011); Doan (2011)
			PLA	Ito (2004); Zhang (2014)
			PES	Piacentini (2013)
	Solid–lipid spheres	A lipid, heated to its melting point, is injected through the membrane pores followed by cooling the emulsion below that temperature.	Gelucire 44/14 or Compritol 888	D'oria (2009)
			Trimyristin	Kukizaki (2009)

Encapsulating material type		Method to produce	Wall material	References
Micro- and nanocapsules	Hollow beads	Calcium alginate microspheres as core particles are produced by membrane emulsification and then coated with chitosan. Using tripolyphosphate, cross-linking of the chitosan shell layer and solubilization and removal of the alginate core are achieved at the same time.	Alginate/chitosan	Akamatsu (2010b)
	Liquid-core/ polymer-shell capsules	A mixture of polymer (i.e., PLA), a volatile water-immiscible good solvent for the polymer (i.e., DCM), and non-volatile water-immiscible nonsolvent (i.e., hydrocarbon) are emulsified with a nonsolvent phase (water or a water–alcohol mixture) by membrane emulsification followed by solvent (i.e., DCM) diffusion.	PLLA	Sawalha (2008)
			Eudragit FS 30D	Wagdare (2011)
		A mixture of monomer and the initiator are permeated trough the membrane followed by one-stage suspension polymerization or two-stage suspension polymerization.	Poly(styrene-co-N,N-dimethylaminoethyl methacrylate) [P(St-DMAEMA)]	Ma (2003)
			Poly(MMA-co-EGDMA-co-2-EHA)	Omi (2000)

(*Continued*)

Table 15.2 (*Continued*)

Encapsulating material type		Method to produce	Wall material	References
Micro- and nanocapsules	Liquid-core/ polymer-shell capsules	A hydrophobic monomer dissolved in organic solvent is injected through the membrane into an aqueous phase followed by interfacial polycondensation (by addition of a hydrophilic monomer).	Poly(N-isopropylacrylamide) (PNIPAM)	Chu (2002)
		An oil core is injected through the membrane, while the continuous phase contains a monomer (polymer precursor) followed by in situ polymerization.	Melamine-formaldehyde	Pan (2012)
		An oil core is injected through the membrane while the continuous phase contains a polycationic polymer (i.e., protein from animal and vegetal source) followed by complex coacervation (by mixing a polyanionic polymer (i.e., polysaccaraide).	Gelatin–Gum Arabic	Piacentini (2013)

Encapsulating material type		Method to produce	Wall material	References
Micro- and nanocapsules	Polymer-core/polymer-shell capsules	A hydrophobic monomer dissolved in organic solvent together with a polymer is injected through the membrane into an aqueous phase followed by interfacial polymerization and solvent removal.	PLGA-PE2CA (poly(ethyl 2-cyanoacrylate)	Lee (2009)
	Composite core/shell particles	An organic solution containing the polymer is injected through the membrane pores followed by solvent removal and layer-by-layer.	PAH/PLGA/silica particles	Ito (2010)
	Multicore-matrix particles	A W_1/O emulsion, in which the oil phase is a polymeric solution (or a lipid, heated up its melting point) is permeated through the membrane followed by solvent removal (or cooling the emulsion below that temperature).	PCL	Imbrogno (2015)
			PLA	Liu (2005a,b)
			PELA	Wei (2008)
			PLGA	Qi (2014); Zhang (2014)
			Tripalmitin	Kukizaki (2007)
			Lipid/PLGA	Ma (2014a,b)
	Multicore-composite matrix particles	The oil phase containing a polymeric solution (or a monor) and the inorganic particles is permeated through the membrane followed by solvent removal (suspension polymerization).	QDs + PSMA	Wang (2013)
			CdSe/ZnS QDs + PS	Han (2015)
			TiO_2–Acrylic monomer	Takeda (2009)

15.2.2 Control of Particle Size and Particle Size Distribution by Membrane Emulsification and Membrane-Assisted Nanoprecipitation

The droplet size distribution in membrane emulsification depends on numerous parameters, which can be divided into the following:

- Membrane parameters (surface energy, porosity, pore distance, pore size distribution, pore morphology)
- Process parameters (axial velocity, which determines the shear stress on the membrane surface, transmembrane flux)
- Formulation parameters (viscosity and density of the dispersed and continuous phase, type and concentration of emulsifier, interfacial tension)

One of the most important parameters related to the membrane is the surface energy at the pore border where the dispersing phase meets the continuous phase. For a fine control of particle size distribution, membrane surfaces unwetted by the dispersed phase are needed in such a way that under the acting interfacial forces, the dispersing phase assumes the shape of droplet while appearing at the pore mouth. On the contrary, low interfacial tension promotes adhesion and displacement of the dispersing phase on the membrane surface, and no controlled dispersion of one phase into another is obtained (Piacentini et al., 2014b). Usually, a hydrophilic membrane ($\theta < 90°$) is used for the production of O/W emulsion, while a hydrophobic membrane ($\theta > 90°$) is used for the production of W/O emulsion. This means that low dispersed phase flux values are usually obtained resulting in low productivity. Alternatively, using membranes with asymmetric wettability, i.e., wetted by the dispersing phase through the thickness while non-wetted at the surface, both high productivity and uniform droplet size distribution can be promoted (Piacentini et al., 2014b).

A low porosity and sufficiently distanced pores are also required to prevent coalescence of droplets at neighboring pores (Kobayashi et al., 2002). The particle size increases linearly with the membrane pore size (Katoh et al., 1996), while the uniformity of droplets size reflects the uniform membranes pore size distribution. A linear correlation between the particle size (d_d) and the membrane pore size (d_p) is observed when using porous membranes (while for mesoporous and microporous membranes, the behavior is not always clear). A minimum value of the correlation coefficient around three can be obtained when the optimized fluid-dynamic conditions and interfacial tension are achieved.

Concerning the process parameters, the particle size decreases with an increase in the shear stress because of (i) fast droplets detachment from the membrane pore and (ii) prevention of the coalescence of droplets at the pore level (Katoh et al., 1996). The influence is greater for low-wall shear stress range values and more effective for submicron membrane pore sizes. Particle size increases with increase in the dispersed phase flux; however, for high value of dispersed phase flux, also a decrease in particle size uniformity is observed since droplets coalesce before droplets detach from the pore (Piacentini, 2016). Particle size is smaller when the emulsifier rapidly adsorbs at the newly formed interface due to the influence of emulsifier on the interfacial tension. Particle size also decreases as a function of emulsifier concentration as the interfacial tension had decreased to a constant (Schröder et al., 1998).

Few examples are reported in the literature about the production of nanoparticles by membrane-assisted nanoprecipitation. The particle size has been demonstrated to be related to the solvent/antisolvent ratio and to the membrane pore size. Larger particles are produced when the solvent/antisolvent ratio is increased as a result of solvent displacement decrease (Othman et al., 2016b). In membrane-assisted nanoprecipitation, particle size is sensitive to the membrane pore size; however, particles significantly smaller in size than the pore size of the membrane are produced (Charcosset et al., 2005; Othman et al., 2016b). The dispersed phase flux and wall shear stress do not show a significant effect on the nanoparticles size (Charcosset et al., 2005; Othman et al., 2016b).

15.2.3 Control of Particle Surface Properties

The layer-by-layer (LbL) technique has attracted particular interest, largely because of the ability to readily tailor particle surface properties. It allows the introduction of multiple functionalities, thus providing opportunities to engineer a new class of materials with advanced structure and function. The surfaces of the quantum dot (QD)-embedded polystyrene (PS) microbeads prepared by membrane emulsification were functionalized to generate carboxyl groups, which are suitable for covalent immobilization of antibodies (Han, 2015). The particles' surface was first modified with poly(allylamine hydrochloride) (PAH) polycations/polyacrylic acid (PAA) polyanion layers using the LbL deposition method. After the deposition of PAA, the antibody was covalently attached onto the particles via an (*N*-ethyl-N′-(3-(dimethylamino)propyl)carbodiimide/*N*-hydroxysuccinimide) (EDC/NHS)-mediated amidation reaction. Multilayer oil-in-water (M-O/W) emulsions can be prepared by electrostatic LbL deposition of polyions onto the surface of oil droplets produced by membrane emulsification by using charged emulsifier or biopolymer (Gudipati, 2010; Nazir, 2012).

Another strategy used for particle surface functionalization is the introduction of functional groups by chemical reaction for covalent conjugation. Nanoclusters composed of magnetite nanoparticles prepared by emulsifying a mixture of magnetite nanoparticles and hexane in an aqueous surfactant solution by membrane emulsification followed by solvent removal were coated with silica particles by ammonia-catalyzed hydrolysis and condensation of silica precursor tetraethyl orthosilicate (TEOS) (Chang, 2012). The silica coating provides silanol groups for surface functionalization of nanoclusters with various organic ligands. Lipid nanoparticles for target organ delivery were obtained by covalent conjugating, on the surface, an M-cell selective molecular signature, ulex europaeus agglutinin-1 (UEA-1), and a toll-like receptor (TLR)-agonist adjuvant monophosphoryl lipid (AMPL) to intimately interact with the mucosal dendritic cells (T. Ma, 2014). The functionalization with UEA-1 allowed the almost exclusive adhesion to M-cells, leading to specific absorption and continuous retention in the target organ (gut), while the TLR-agonist MPL played a crucial role in up-regulating the phagocytosis of mucosal dendritic cells. Molecularly imprinted polymers are also used to introduce specific functionality of the particle surface. Particles are synthesized by polymerization in the presence of a template molecule that is extracted afterward, thus leaving nanocavities in the polymer network, which are complementary in shape, size, and chemical functionality to the template molecule.

Molecularly imprinted poly(ethyleneglycol dimethacrylate-co-methacrylic acid) (P(MAA-co-EGDMA)) particles are produced by membrane emulsification using chloramphenicol as a template, methacrylic acid (MAA) as a functional monomer, ethylene glycol dimethacrylate (EGDMA) as a crosslinker and ethyl acetate as a porogen (Kou, 2012). Molecularly imprinted P(MMA-co-EGDMA) particles are produced by membrane emulsification using bisphenol as a template, methylmethacrylate (MMA) as a functional monomer, ethylene glycol dimethacrylate (EGDMA) as a cross-linker, and n-octane as a porogen (Wolska, 2014). The prepared microspheres were used as highly selective sorbents for the removal of chloramphenicol and bisphenol from aqueous solutions, respectively.

Surface functionalization of particles can also be carried out simultaneously with particle formation. The surface of nanoparticles was embedded with one layer of phospholipids by using PLGA–lipid as the dispersed phase by membrane emulsification (Ma, 2014). There were strong interactions between the lipids hydrophobic tails and PLGA blocks, and they re-arrange at the oil/water interface during the solvent extraction.

15.3 Applications

15.3.1 Encapsulation for Drug Delivery

Membrane emulsification has been used to develop micro- and nanoparticles for application in different fields. The major area of application is in the pharmaceutical field for the production of drug-delivery systems (Piacentini 2014; Piacentini, 2017). Representative examples of the application of membrane emulsification for the preparation of encapsulating materials for drug delivery will be described in this section.

Micro- and nanoparticles containing bioactive materials with a uniform size have a great potential as carriers in drug-delivery systems, since the release rate and loading efficiency of drug can be constant in each particle. Lipophilic drug-delivery systems are designed by membrane emulsification to improve the handling, water-dispersibility, chemical stability, and oral bioavailability of many types of lipophilic bioactive agents such as anti-inflammatory and anticancer drugs. Flurbiprofen-loaded nanoparticles prepared by using membranes and a spray-drying technique demonstrated an improved bioavailability of a poorly water-soluble flurbiprofen (Oh, 2013). By using the same technique, silymarin-loaded nanoemulsion was also formulated with improved hepatoprotective effects compared with silymarin powder and the commercial product as well as itraconazole-loaded solid nanoemulsion with improved oral bioavailability of the drug in in-vivo experiments (Yang, 2013). Many anticancer drugs are lipophilic molecules, and the use of smart, targeted micro- and nanoparticles by membrane-emulsification technology offers the possibility to reach some attractive features needed in this medical area. Doxorubicin-loaded W/O/W microemulsions produced by membrane-emulsification technique showed a sustained release profile, prolonged the plasma drug concentration after intravenous administration to rats, and gave about 17-fold higher area under the drug concentration–time curve compared to free doxorubicin solution (Pradhan, 2014). Paclitaxel-porous quaternized chitosan nanoparticles with pH responsiveness prepared by membrane emulsification demonstrated improved therapeutic efficacy in non-small-cell lung cancer after oral administration (Lv, 2011). Docetaxel-loaded poly(lactide)-D-α-tocopheryl polyethylene glycol 1000 succinate (PLA-TPGS) nanoparticles produced by using premix membrane-emulsification technique demonstrated superior pharmacokinetic and anticancer effects compared to the commercially available Taxotere (Yu, 2013). W/o/w emulsions have also been prepared by membrane emulsification to transport and deliver anticancer drugs for hepatocellular carcinoma (Higashi, 1995, 2000). Microdroplets accumulated only in hepatocellular carcinoma tissue and remained in the tissue for more than 3 weeks affecting tumor cells. The clinical tests showed a cancer volume decreased to a quarter of its initial size.

Vitamin E (Charcosset et al., 2005), spironolactone (Blouza, 2006), and rapamycin-loaded PCL nanoparticles (Othman et al., 2016a) were produced by membrane-assisted nanoprecipitation. High encapsulation efficiency of vitamin E in PCL nanocapsules was achieved for formulation at laboratory scale (98% ± 1.68) and at pilot scale (97.65% ± 2.06) (Charcosset et al., 2005). The scale-up at the pilot scale for spironolactone-loaded PCL nanocapsules has been successfully assessed with a slight increase in size and a slight reduction in drug encapsulation efficiency (Blouza, 2006). The encapsulation efficiency of rapamycin was ≥98.8%, due to very low aqueous solubility of the drug and strong hydrophobic interactions with PCL (Othman et al., 2016a). Rapamycin-loaded PCL nanoparticles were significantly smaller and more uniform than rapamycin nanocrystals prepared under the same conditions (without PCL as a matrix system), revealing the important role of PCL in inhibiting particle growth and agglomeration during nanoprecipitation (Othman et al., 2016a).

Hydrophilic systems for encapsulating, protecting, and releasing hydrophilic bioactive components, such as protein and peptides, and small drugs have also been produced by membrane emulsification. The suitability of membrane emulsification to design protein/peptide-loaded particles has been demonstrated in many works as a result of the mild operative conditions required for emulsion production (Ma, 2014). Applications are reported for insulin-delivery systems (Liu, 2006; Nan, 2014; Wei, 2010; Toorisaka, 2003) and vaccine adjuvants (Wang, 2015; Ma, 2014).

PLA/PLGA microcapsules loaded with recombinant human insulin, with higher encapsulation efficiency (91.82%) than the conventional mechanical stirring method, were obtained by combining membrane emulsification and a double emulsion–evaporation method (Liu, 2006). Insulin oral delivery systems were also based on chitosan-coated alginate particles (Nan, 2014), and hollow quaternized chitosan (Wei, 2010) with pH-sensitive properties was successfully prepared by membrane emulsification. The colloidosomes achieved a high encapsulation efficiency of 97.6% and a pH-sensitive drug release (slow in simulated gastric fluid and rapid in simulated intestinal fluid). Furthermore, animal testing confirmed the strong hypoglycemic effects of insulin-loaded colloidosomes even up to 6 h, suggesting colloidosomes as efficient oral insulin-delivery vehicles.

PLGA–lipid nanoparticles produced by membrane emulsification and the double emulsion solvent evaporation method displayed promising properties as oral vaccine delivery systems (Ma, 2014). Surface particle properties such as hydrophilic/hydrophobic nature or charge surface have been demonstrated to influence the immune responses induced by particulate used as vaccine adjuvants (Chen, 2014; Liu, 2014, 2013).

15.3.2 Encapsulation for Biocatalytic Process Development

Membrane emulsification is emerging as a more promising emulsification method for producing biocatalytic particles. The entrapment of enzymes in polymer networks has been realized with various hydrogels consisting of crosslinked natural polymers such as chitosan or gelatin and synthetic polymers such as polyvinyl alcohol (PVA). Lysozyme-encapsulated chitosan microspheres were produced by using membrane-emulsification technique followed by crosslinking with glutaraldehyde (Akamatsu et al., 2012). The method enabled to control the size, monodispersity, and enzyme-loading amount, and to maintain half of its activity during the preparation procedure. Gelatin microspheres were prepared by using the same procedure to encapsulate yeast alcohol dehydrogenase (Li et al., 2016). The preparation conditions such as water/oil volume ratio, emulsifier concentration, transmembrane pressure, and stirring rate have been demonstrated to significantly affect the diameter size and the uniformity of the microspheres. The encapsulated enzyme demonstrated enhanced pH stability, storage stability, and operational stability. Lipase from *Candida rugosa* has been immobilized by using different immobilization methods (adsorption, crosslinking, and entrapment) on glutaraldehyde crosslinked PVA microspheres prepared by membrane emulsification (Piacentini et al., 2017). Each of the immobilization methods used was successfully applied for the production of enzyme-loaded microspheres. The entrapment method was demonstrated to be the most suitable method for lipase immobilization, enabling an improved operational stability and promoting the intrinsic optimal interaction between enzyme and substrate. Lipase (that is an interfacial enzyme) is distributed at the W/O interface during emulsion production, exhibiting its open form (in which the catalytic site is accessible to the substrate). This allowed achieving an oriented immobilization of the enzyme, which is the basis of bio-specific interaction between the enzyme and the substrate. In the case of the adsorption and crosslinking methods, the interaction of the enzyme with the carrier increased the rigidity of the enzyme molecules, probably limiting the interaction with the substrate, which was hampered further by the hydrophilic properties of the carrier. In a previous work, lipase from *Candida rugosa* was distributed at the interface of stable oil-in-water emulsions for the kinetic resolution of racemic naproxen methyl ester mixture by using membrane emulsification (Giorno et al., 2008).

15.4 Conclusion

Membrane technology offers great potential in the manufacturing of simple and multiple emulsions with tuned size and uniform size distribution. Emulsion droplets are used as template in the production of structured solid particles by combining membrane technology with the process of polymerization, gel formation, crystallization, and self-assembly. The main advantage offered by the use of membrane-based technology for particles manufacturing is the control of particle size and size uniformity, which is of great importance in all the applications especially for drug delivery. The mild operative conditions applied extend the application of membrane emulsification toward the encapsulation of shear-sensitive compounds such as proteins (antibody or enzyme) and cells.

The application of membrane technology for particle manufacturing is expected to implement the development of bio-functionalized micro-nanostructured devices with improved functions, sensitivity, and stability in important applications such as biosensors, biomolecule-responsive drug-delivery systems, and biocatalysis.

Abbreviations

O/W	oil-in-water
W/O	water-in-oil
S/O/W	solid-in-oil-in-water
O/W/O	oil-in-water-in-oil
W/O/W	water-in-oil-in-water
PVA	polyvinyl alcohol
PCL	polycaprolacton
PLGA	poly(lactic-co-glycolic acid)

PLA	polylactic acid
PES	polyethersulfone
PLLA	poly-L-lactic acid
PNIPAM	poly(*N*-isopropylacrylamide)
P(MAA-co-EGDMA)	poly(ethyleneglycol dimethacrylate-co-methacrylic acid)
PE2CA	poly(ethyl 2-cyanoacrylate)
PAH	poly(allylamine hydrochloride)
PAA	polyacrylic acid
QDs	quantum dot
PSMA	polystyrene–maleic anhydride
PS	polystyrene
LbL	layer-by-layer

References

Akamatsu, K., Ikeuchi, Y., Nakao, A., and Nakao, S. (2012). Size-controlled and monodisperse enzyme-encapsulated chitosan microspheres developed by the SPG membrane emulsification technique. *J. Colloid. Interface Sci.*, **371**, pp. 46–51.

Akamatsu, K., Chen, W., Suzuki, Y., Ito, T., Nakao, A., Sugawara, T., et al. (2010a). Preparation of monodisperse chitosan microcapsules with hollow structures using the SPG membrane emulsification technique. *Langmuir*, **26**, pp. 14854–14860.

Akamatsu, K., Kaneko, D., Sugawara, T., Kikuchi, R., and Nakao, S. (2010b). Three preparation methods for monodispersed chitosan microspheres using the Shirasu porous glass membrane emulsification technique and mechanisms of microsphere formation. *Ind. Eng. Chem. Res.*, **49**, pp. 3236–3241.

Blouza, I. L., Charcosset, C., Sfar, S., and Fessi, H. (2006). Preparation and characterization of spironolactone-loaded nanocapsules for paediatric use. *Int. J. Pharm.*, **325**, pp. 124–131.

Chang, E. P. and Hatton, T. A. (2012). Membrane emulsification and solvent pervaporation processes for the continuous synthesis of functional magnetic and Janus nanobeads. *Langmuir*, **28**, pp. 9748–9758.

Charcosset, C. and Fessi, H. (2005). Preparation of nanoparticles with a membrane contactor. *J. Membr. Sci.*, **266**, pp. 115–120.

Chen, X. M., Liu, Y. Y., Wang, L. Y., Liu, Y., Zhang, W., Fan, B., et al. (2014). Enhanced humoral and cell mediated immune responses generated by cationic polymer-coated PLA microspheres with adsorbed HBsAg. *Mol. Pharm.*, **11**, pp. 1772–1784.

Chu, L. Y., Park, S. H., Yamaguchi, T., and Nakao, S. (2002). Preparation of micron-sized monodispersed thermoresponsive core–shell microcapsules. *Langmuir*, **18**, pp. 1856–1864.

Doan, T. V. P., Couet, W., and Olivier, J. C. (2011). Formulation and in vitro characterization of inhalable rifampicin-loaded PLGA microspheres for sustained lung delivery. *Int. J. Pharm.*, **414**, pp. 112–117.

D'oria, C., Charcosset, C., Barresi, A. A., and Fessi, H. (2009). Preparation of solid lipid particles by membrane emulsification: Influence of process parameters. *Colloid. Surf. A*, **338**, pp. 114–118.

Giorno, L., Piacentini, E., Mazzei, R., and Drioli, E. (2008). Membrane emulsification as a novel method to distribute phase-transfer biocatalysts at the oil/water interface in bioorganic reactions. *J. Membr. Sci.*, **317**, pp. 19–25.

Gouin, S. (2004). Microencapsulation: Industrial appraisal of existing technologies and trends. *Trends Food Sci. Technol.*, **15** (7–8), pp. 330–347.

Gudipati, V., Sandra, S., McClements, D. J., and Decker, E. A. (2010). Oxidative stability and in vitro digestibility of fish oil-in-water emulsions containing multilayered membranes. *J. Agric. Food Chem.*, **58**, pp. 8093–8099.

Han, S. W., Jang, E., and Koh, W. G. (2015). Microfluidic-based multiplex immunoassay system integrated with an array of QD-encoded microbeads. *Sens. Actuators B*, **209**, pp. 242–251.

Higashi, S. and Setoguchi, T. (2000). Hepatic arterial injection chemotherapy for hepatocellular carcinoma with epirubicin aqueous solution as numerous vesicles in iodinated poppy-seed oil microdroplets: Clinical application of water-in-oil-in-water emulsion prepared using a membrane emulsification technique. *Adv. Drug Deliv. Rev.*, **45**, pp. 57–64.

Higashi, S., Shimizu, M., Nakashima, T., Iwata, K., Uchiyaemotherapyma, F., Tateno, S., et al. (1995). Arterial injection chemotherapy for hepatocellular carcinoma using monodispersed poppy-seed oil micro droplets containing fine aqueous vesicles of epirubicin. *Cancer*, **75**, pp. 1245–1254.

Imbrogno, A., Dragosavac, M. M., Piacentini, E., Vladisavljević, G. T., Holdich, R. G., and Giorno, L. (2015). Polycaprolactone multicore-matrix particle for the simultaneous encapsulation of hydrophilic and hydrophobic compounds produced by membrane emulsification and solvent diffusion processes. *Colloid. Surf. B*, **135**, pp. 116–125.

Imbrogno, A., Piacentini, E., Drioli, E., and Giorno, L. (2014a). Micro and nano polycaprolactone particles preparation by pulsed back-and-forward cross-flow batch membrane emulsification for parenteral administration. *Int. J. Pharm.*, **477**, pp. 344–350.

Imbrogno, A., Piacentini, E., Drioli, E., and Giorno, L. (2014b). Preparation of uniform plycaprolactone microparticles by membrane emulsification/solvent diffusion process. *J. Membr. Sci.*, **467**, pp. 262–268.

Ito, F., Kanakubo, Y., and Murakami, Y. (2011). Rapid preparation of monodisperse biodegradable polymer nanospheres using a membrane emulsification technique under low gas pressure. *J. Polym. Res.*, **18**, pp. 2077–2085.

Ito, F. and Makino, K. (2004a). Preparation and properties of monodispersed rifampicin-loaded poly(lactide-co-glycolide) microspheres. *Colloids Surf. B*, **39**(1–2), pp. 17–21.

Ito, F., Uchida, Y., and Murakami, Y. (2010). Facile technique for preparing organic–inorganic composite particles: Monodisperse poly(lactide-co-glycolide) (PLGA) particles having silica nanoparticles on the surface. *Colloids Surf. A*, **361**, pp. 109–117.

Katoh, R., Asano, Y., Furuya, A., Sotoyama, K., and Tomita, M. (1996). Preparation of food emulsions using a membrane emulsification system. *J. Membr. Sci.*, **113**, pp. 131–135.

Khayata, N., Abdelwahed, W., Chehna, M. F., Charcosset, C., and Fessi, H. (2012). Preparation of vitamin E loaded nanocapsules by the nanoprecipitation method: From laboratory scale to large scale using a membrane contactor. *Int. J. Pharm.*, **423**, pp. 419–427.

Kobayashi, I., Yasuno, M., Iwamoto, S., Shono, A., Satoh, K., and Nakajima, M. (2002). Microscopic observation of emulsion droplet formation from a polycarbonate membrane. *Colloids Surf. A*, **207**, pp. 185–196.

Kou, X., Li, Q., Lei, J., Geng, L., Deng, H., Zhang, G., et al. (2012). Preparation of molecularly imprinted nanospheres by premix membrane emulsification technique. *J. Membr. Sci.*, **417–418**, pp. 87–95.

Kukizaki, M. and Goto, M. (2007). Preparation and evaluation of uniformly sized solid lipid microcapsules using membrane emulsification. *Colloids Surf. A*, **293**, pp. 87–94.

Kukizaki, M. (2009). Relation between salt rejection and electrokinetic properties on Shirasu porous glass (SPG) membranes with nano-order uniform pores. *Sep. Sci. Technol.*, **69**, pp. 87–96.

Lai, Y. T., Sato, M., Ohta, S., Akamatsu, K., Nakao, S., Sakai, Y., et al. (2015a). Preparation of uniform sized hemoglobin–albumin microspheres as oxygen carriers by Shirasu porous glass membrane emulsification technique. *Colloids Surf. B*, **127**, pp. 1–7.

Lai, Y. T., Sato, M., Ohta, S., Akamatsu, K., Nakao, S., Sakai, Y., et al. (2015b). Size-dependent interaction of cells and hemoglobin–albumin based oxygen carriers prepared using the SPG membrane emulsification technique. *Biotechnol. Prog.*, **31**, pp. 1676–1684.

Lee, S. H., Baek, H. H., Kim, J. H., and Choi, S. W. (2009). Core–shell poly(d. l-lactide-co-glycolide)/poly(ethyl 2-cyanoacrylate) microparticles with doxorubicin to reduce initial burst release. *Macromol. Res.*, **17**, pp. 1010–1014.

Li, J., Ma, J., Jiang, Y., Jiang, T., Wang, Y., Chen, Y., et al. (2016). Immobilizing enzymes in regular-sized gelatin microspheres through a membrane emulsification method. *J. Mater. Sci.*, **51**, pp. 6357–6369.

Liu, R., Huang, S. S., Wan, Y. H., Ma, G. H., and Su, Z. G. (2006). Preparation of insulin-loaded PLA/PLGA microcapsules by a novel membrane emulsification method and its release in vitro. *Colloids Surf. B*, **51**, pp. 30–38.

Liu, R., Ma, G. H., Meng, F. T., and Su, Z. G. (2005a). Preparation of uniform-sized PLA microcapsules by combining Shirasu porous glass membrane emulsification technique and multiple emulsion-solvent evaporation method. *J. Controlled Release*, **103**, pp. 31–43.

Liu, R., Ma, G. H., Wan, Y. H., and Su, Z. G. (2005b). Influence of process parameters on the size distribution of PLA microcapsules prepared by combining membrane emulsification technique and double emulsion-solvent evaporation method. *Colloids Surf. B*, **45**, pp. 144–153.

Liu, X. D., Bao, D. C., Xue, W. M., Xiong, Y., Yu, W. T., Yu, X. J., et al. (2003). Preparation of uniform calcium alginate gel beads by membrane emulsification coupled with internal gelation. *J. Appl. Polym. Sci.*, **87**, pp. 848–852.

Liu, Y., Yin, Y., Wang, L. Y., Zhang, W., Chen, X., Yang, X., et al. (2013). Engineering biomaterial-associated complement activation to improve vaccine efficacy. *Biomacromolecules*, **14**, pp. 3321–3328.

Liu, Y. Y., Chen, X. M., Wang, L. Y., Yang, T., Yuan, Q., and Ma, G. (2014). Surface charge of PLA microparticles in regulation of antigen loading, macrophage phagocytosis and activation, and immune effects in vitro. *Particuology*, **17**, pp. 74–80.

Lv, P. P., Wei, W., Gong, F. L., Zhang, Y. L., Zhao, H. Y., Lei, J. D., et al. (2009). Preparation of uniformly sized chitosan nanospheres by a premix membrane emulsification technique. *Ind. Eng. Chem. Res.*, **48**, pp. 8819–8828.

Lv, P. P., Wei, W., Yue, H., Yang, T. Y., Wang, L. Y., and Ma, G. H. (2011) Porous quaternized chitosan nanoparticles containing paclitaxel nanocrystals improved therapeutic efficacy in non-small-cell lung cancer after oral administration. *Biomacromolecules*, **12**(12), pp. 4230–4239.

Ma, G. H., Su, Z. G., Omi, S., Sundberg, D., and Stubbs, J. (2003). Microencapsulation of oil with poly(styrene-N,N-dimethylaminoethyl methacrylate) by SPG emulsification technique: Effects of conversion and composition of oil phase. *J. Colloid. Interface Sci.*, **266**, pp. 282–294.

Ma, T., Wang, L., Yang, T., Ma, G., and Wang, S. (2014). Homogeneous PLGA-lipid nanoparticle as a promising oral vaccine delivery system for ovalbumin. *Asian J. Pharm. Sci.*, **9**, pp. 129–136.

Ma, T., Wang, L., Yang, T., Ma, G., and Wang, S. (2014b). M-cell targeted polymeric lipid nanoparticles containing a toll-like receptor agonist to boost oral immunity. *Int. J. Pharm.*, **473**, pp. 296–303.

Ma, T., Wang, L., Yang, T., Wang, D., Ma, G., and Wang, S. (2014a). PLGA-lipid liposphere as a promising platform for oral delivery of proteins. *Colloids Surf. B*, **117**, pp. 512–519.

Ma, G. (2014). Microencapsulation of protein drugs for drug delivery: Strategy, preparation, and applications. *J. Controlled Release*, **193**, pp. 324–340.

Nan, F., Wu, J., Qi, F., Fan, Q., Ma, G., and Ngai, T. (2014). Preparation of uniform-sized colloidosomes based on chitosan-coated alginate particles and its application for oral insulin delivery. *J. Mater. Chem. B*, **2**, pp. 7403–7409.

Nazir, H., Wang, L., Lian, G., Zhu, S., Zhang, Y., Liu, Y., et al. (2012). Multilayered silicone oil droplets of narrow size distribution: Preparation and improved deposition on hair. *Colloids Surf. B*, **100**, pp. 42–49.

Oh, D. H., Din, F. U., Kim, D. W., Kim, J. O., Yong, C. S., and Choi, H. G. (2013). Flurbiprofen-loaded nanoparticles prepared with polyvinylpyrrolidone using Shirasu porous glass membranes and a spray-drying technique: Nano-sized formation and improved bioavailability. *J. Microencapsulation*, **30**, pp. 674–680.

Omi, S., Ma, G. H., and Nagai, M. (2000). Membrane emulsification: A versatile tool for the synthesis of polymeric microspheres. *Macromol. Symp.*, **151**, pp. 319–330.

Othman, R., Vladisavljević, G. T., Nagy, Z. K., and Holdich, R. G. (2016a). Encapsulation and controlled release of rapamycin from polycaprolactone nanoparticles prepared by membrane micromixing combined with antisolvent precipitation. *Langmuir*, DOI: 10.1021/acs.langmuir.6b03178.

Othman, R., Vladisavljević, G. T., Shahmohamadi, H., Nagy, Z. K., and Holdich, R. G. (2016b). Formation of size-tuneable biodegradable polymeric nanoparticles by solvent displacement method using micro-engineered membranes fabricated by laser drilling and electroforming. *Chem. Eng. J.*, **304**, pp. 703–713.

Pan, X., York, D., Preece, J. A., and Zhang, Z. (2012). Size and strength distributions of melamine formaldehyde microcapsules prepared by membrane emulsification. *Powder Technol.*, **227**, pp. 43–50.

Piacentini, E., Drioli, E., and Giorno, L. (2014a). Membrane emulsification technology: Twenty-five years of inventions and research through patent survey. *J. Membr. Sci.*, **468**, pp. 410–422.

Piacentini, E., Imbrogno, A., Drioli, E., and Giorno, L. (2014b). Membranes with tailored wettability properties for the generation of uniform emulsion droplets with high efficiency. *J. Membr. Sci.*, **459**, pp. 96–103.

Piacentini, E., Poerio, T., Bazzarelli, F., and Giorno, L. (2016). Microencapsulation by membrane emulsification of biophenols recovered from olive mill wastewater. *Membranes*, **6**, pp. 1–11.

Piacentini, E., Shanthana Lakshmi, D., Figoli, A., Drioli, E., and Giorno, L. (2013). Polymeric microspheres preparation by membrane emulsification-phase separation induced process. *J. Membr. Sci.*, **448**, pp. 190–197.

Piacentini, E., Dragosavac, M., and Giorno, L. (2017). Pharmaceutical particles design by membrane emulsification: Preparation methods and applications in drug delivery. *Curr. Pharm. Design*, **23**, DOI 10.2174/1381612823666161117160940.

Piacentini, E., Yan, M., and Giorno, L. (2017). Development of enzyme-loaded PVA microspheres by membrane emulsification. *J. Membr. Sci.*, **524**, pp. 79–86.

Pradhan, R., Kim, Y. I., Jeong, J. H., Choi, H. G., Yong, C. S., and Kim, J. O. (2014). Fabrication characterization and pharmacokinetic evaluation of doxorubicin-loaded water-in-oil-in-water microemulsions using a membrane emulsification technique. *Chem. Pharm. Bull.*, **62**(9), pp. 875–882.

Qi, F., Wu, J., Hao, D., Yang, T., Ren, Y., Ma, G., et al. (2014). Comparative studies on the influences of primary emulsion preparation on properties of uniform-sized exenatide-loaded PLGA microspheres. *Pharm. Res.*, **31**, pp. 1566–1574.

Sawalha, H., Fan, Y., Schroën, K., and Boom, R. (2008). Preparation of hollow polylactide microcapsules through premix membrane emulsification: Effects of nonsolvent properties. *J. Membr. Sci.*, **325**, pp. 665–671.

Schröder, V., Beherend, O., and Schubert, H. (1998). Effect of dynamic interfacial tension on the emulsification process using microporous ceramic membranes. *J. Colloid. Interface Sci.*, **202**, pp. 334–340.

Takeda, M., Tanabe, E., Iwaki, T., Yabuki, A., and Okuyama, K. (2009). Importance of dispersibility of TiO_2 in preparation of TiO_2-dispersed microspheres by Shirasu porous glass (SPG) membrane emulsification. *Adv. Powder Technol.*, **20**, pp. 361–365.

Wagdare, N. A., Marcelis, A. T. M., Boom, R. M., and van Rijn, C. J. M. (2011). Porous microcapsule formation with microsieve emulsification. *J. Colloid. Interface Sci.*, **355**, pp. 453–457.

Wang, G., Leng, Y., Dou, H., Wang, L., Li, W., Wang, X. et al. (2013). Highly efficient preparation of multiscaled quantum dot barcodes for multiplexed hepatitis B detection. *ASC Nano*, **7**, pp. 471–481.

Wang, L. Y., Ma, G. H., and Su, Z. G. (2005). Preparation of uniform sized chitosan microspheres by membrane emulsification technique and application as a carrier of protein drug. *J. Controlled Release*, **106**, pp. 62–75.

Wang, L., Yang, T., and Ma, G. (2015). Particle design of membrane emulsification for protein drug and vaccine delivery. *Curr. Pharm. Design*, **21**(19), pp. 2563–2598.

Wei, Q., Wei, W., Tian, R., Wang, L. Y., Su, Z. G., and Ma, G. H. (2008). Preparation of uniform-sized PELA microspheres with high encapsulation efficiency of antigen by premix membrane emulsification. *J. Colloid. Interface Sci.*, **323**, pp. 267–273.

Wei, W., Ma, G. H., Wang, L. Y., Wu, J., and Su, Z. G. (2010). Hollow quaternized chitosan microspheres increase the therapeutic effect of orally administered insulin. *Acta Biomaterialia*, **6**(1), pp. 205–209.

Wolska, J. and Bryjak, M. (2014). Removal of bisphenol A from aqueous solution by molecularly imprinted polymers. *Sep. Sci. Technol.*, **49**, pp. 1643–1653.

Wu, J., Wei, W., Wang, L. Y., Su, Z. G., and Ma, G. H. (2008). Preparation of uniform-sized pH-sensitive quaternized chitosan microsphere by combining membrane emulsification technique and thermal-gelation method. *Colloids Surf. B*, **63**, pp. 164–175.

Yang, K. Y., Hwang, D. H., Yousaf, A. M., Kim, D. W., Shin, Y. J., Bae, O. N., et al. (2013). Silymarin-loaded solid nanoparticles provide excellent hepatic protection: Physicochemical characterization and in vivo evaluation. *Int. J. Nanomedicine*, **8**, pp. 3333–3343.

You, J. O., Park, S. B., Park, H. Y., Haam, S., Chung, C. H., and Kim, W. S. (2001). Preparation of regular sized Ca-alginate microspheres using membrane emulsification method. *J. Microencapsulation*, **18**, pp. 521–532.

Yu, Y., Tan, S., Zhaom, S., Zhuang, X., Song, Q., Wang, Y., et al. (2013). Antitumor activity of docetaxel-loaded polymeric nanoparticles fabricated by Shirasu porous glass membrane-emulsification technique. *Int. J. Nanomed.*, **8**, pp. 2641–2652.

Zhang, W., Wang, L., Liu, Y., Chen, X., Li, J., Yang, T., et al. (2014). Comparison of PLA microparticles and alum as adjuvants for H5N1 influenza split vaccine: Adjuvanticity evaluation and preliminary action mode analysis. *Pharm. Res.*, **31**, pp. 1015–1031.

Zhang, W., Wang, L., Liu, Y., Chen, X., Li, Q., Jia, J., et al. (2014). Immune responses to vaccines involving a combined antigene-nanoparticle mixture and nanoparticle-encapsulated antigen formulation. *Biomaterials*, **35**, pp. 6086–6097.

Zhou, Q. Z., Ma, G. H., and Su, Z. G. (2009). Effect of membrane parameters on the size and uniformity in preparing agarose beads by premix membrane emulsification. *J. Membr. Sci.*, **326**, pp. 694–700.

Zhou, Q. Z., Wang, L. Y., Ma, G. H., and Su, Z. G. (2008). Multi-stage premix membrane emulsification for preparation of agarose microbeads with uniform size. *J. Membr. Sci.*, **322**, pp. 98–104.

Zhou, Q. Z., Wang, L. Y., Ma, G. H., and Su, Z. G. (2007). Preparation of uniform-sized agarose beads by microporous membrane emulsification technique. *J. Colloid. Interface Sci.*, **311**, pp. 118–127.

Chapter 16

Nanostructured Membranes for Engineering Organs and Tissues

Sabrina Morelli, Simona Salerno, Antonella Piscioneri, and Loredana De Bartolo
Institute on Membrane Technology of the National Research Council of Italy (ITM-CNR), Via P. Bucci, Cubo 17C, 87036 Rende (CS), Italy
l.debartolo@itm.cnr.it; loredana.debartolo@cnr.it

In order to mimic natural tissues, nanotechnology and nanomaterials have become the center of interest in the field of organ and tissue repair and regeneration. Nanostructured membranes have important properties, including controlled composition, pore size and geometry, morphology, mechanical properties, and surface functionality, which make them suitable for creating novel tissue-engineered constructs. Research results have shown that nanostructured membranes can be tuned to control cell behaviors at multiple levels, including adhesion, migration, proliferation, signaling, genetic expression, and stem cell fate. This chapter reports the recent advances in the realization of bioengineered tissue substitutes by using nanostructured membrane systems.

Functional Nanostructured Membranes
Edited by Enrico Drioli, Lidietta Giorno, and Annarosa Gugliuzza

ISBN 978-981-4774-79-6 (Hardcover), 978-1-351-13511-5 (eBook)
www.panstanford.com

16.1 Introduction

Natural tissues possess certain properties from hierarchical architectures that are precisely controlled from the nanoscale to the macroscale, and the extracellular matrix (ECM) modulates cell behavior in terms of spreading, migration, proliferation, differentiation, and metabolism. When damage occurs, the tissue homeostasis is compromised, which in turn determines cell death and tissue disruption. Tissue-engineered strategies aim at establishing a proper cellular microenvironment in which new tissue and organ regeneration is promoted by providing topographical, chemical, and bio-functional cues that enable dynamic cell interaction. The treatment of diseases or injuries by means of a tissue-engineered approach requires the development of sophisticated tissue analogues able to mimic the in vivo complex hierarchical tissue disposition and functions. The development of biomaterials with a feature complexity, which mimic the native ECM, offers powerful tools to regenerative medicine not only for the ultimate goal of clinical translational application, but also for the study and comprehension of a step-wise analysis of cell behavior, both in physiological and pathological context (Chiono et al., 2015; Khan et al., 2015). The choice of the appropriate biomaterial, which presents biophysical and biochemical signals at the cellular interface, able to evoke a cellular organization according to a tissue-like cytoarchitecture is of primary importance for the realization of a reliable in vitro platform, where a defined parameter setting can be established giving rise to a well-controlled microenvironment (Ai et al., 2014; Salerno and De Bartolo, 2017).

Many efforts have been made to design biomaterials capable of replicating the natural physical, chemical, and topographical cues that cells receive in vivo, which are necessary to produce a specific tissue or organ. Different materials—natural and synthetic, resorbable and non-biodegradable—have been proposed to support cells and promote their differentiation and proliferation toward the formation of a new tissue. Structural features, surface chemistry, topography, and roughness exert influence on cell migration and polarization as a consequence of a different cytoskeleton organization. Nano- and micro-topographical cues can be incorporated in the biomaterial structure by shaping its morphology accordingly. It is noteworthy that nanostructured materials have great properties, such as nano-scale topography, large surface area to volume ratios, enhanced mechanical properties, and surface functionality, which offer the opportunity to engineer constructs that can directly influence the behavior and fate of cells, tissues, and organs (Cao and Wang, 2011; Wang, 2002; Sato and Webster, 2004; Emerich and Thanos, 2003; Chen et al., 2016; Krishna et al., 2016). The intrinsic utility of these materials for biomedical applications is due to their ability to match characteristic length scales of natural physiologic sub-cellular structures in engineered constructs with nanoscale features that directly impact cell function. Designing biomaterials to stimulate specific cellular activities provides a basis for numerous high-impact applications in regenerative medicine, including bioactive materials that direct and guide tissue regeneration. Embedded nanostructures can ultimately direct cell behavior by virtue of interactions such as modulated protein adsorption, receptor–ligand binding, and contact guidance.

Over the last decade, different types of nanostructured materials with properties that more closely fulfill the requirement in the course of recovery of native tissues were designed, synthesized, and utilized. In this scenario, nanostructured polymeric membranes have been successfully used for the reconstruction of tissue analogues because of the high control of cell microenvironment at the molecular level (Morelli et al., 2015a; Messina et al., 2015; Morelli et al., 2016a; Salerno et al., 2013a; Salerno et al., 2017a; Morelli et al., 2017b; Salerno and De Bartolo, 2017). These artificial systems compartmentalize cells providing a wide surface area with nanostructured complexes for cell adhesion and ensuring a continuous and selective transport of nutrients and metabolites to and from cells. Membrane systems can create a biomimetic environment with highly selective and specific physicochemical, morphological, and transport properties. Notably, tailor-made membranes (organic, functionalized with specific biomolecules, in flat and hollow-fiber configurations), designed and operated according to well-defined engineering criteria, can provide adequate transport of oxygen, nutrients, and catabolites throughout the cellular compartment, and supply appropriate biomechanical stimuli of the developing tissue (Morelli et al., 2009; De Bartolo et al., 2009; Curcio et al., 2012; Salerno et al., 2013b).

In tissue-engineered constructs, the surface and transport properties of the membranes play an important role in the promotion of cell adhesion, proliferation, and viability (De Bartolo et al., 2007a; De Bartolo et al., 2008; Morelli and Piscioneri, 2013). The membrane surface properties such as chemical composition, hydrophilicity/hydrophobicity, charge, free energy, and roughness affect the cell adhesion through the modulation of proteins secreted by cells or contained in the physiological liquids. To this purpose, novel strategies aimed at improving cell-biomaterial interactions have been developed in the last decades. These approaches include modification of surface chemistry through grafting of functional groups, immobilization of molecules leaving the bulk properties unaltered, coating with ECM proteins, and topographical pattern.

In the present chapter, we summarize the nanostructured membrane systems currently used in liver, bone, neuronal, and skin tissue engineering. We also present and discuss the most recent research achievements.

Although more tasks and challenges need to be addressed and resolved in order to translate them into commercialized products, nanostructured membrane systems are very promising candidates in the development of new tissue-engineered constructs in the future.

16.2 Advances in Membrane Bioartificial Liver

The suitability of an in vitro liver tissue with a complete ultra-architecture remains a critical issue in the development of a hepatic bioartificial organ and tissue engineering. Hepatocyte cultures represent not only a source for cell therapy and liver supplementation, but they are a cell model system for studying liver metabolic pathways and disease, for drug screening test, and for therapeutic and diagnostic applications (Salerno and De Bartolo, 2017). With the aim to fabricate biological substitutes, liver tissue engineering strategies are applied for the realization of hepatocyte constructs by mimicking as much as possible the in vivo three-dimensional microenvironment and arrangement. After the isolation, hepatocytes lose their integration with their natural microenvironment, constituted by the highly organized architecture of the extracellular matrix and by non-parenchymal cells. One of the most used approaches to realize hepatocyte constructs is based on the seeding of cells on biomaterial scaffolds with specific biochemical and physical cues that provide the necessary microenvironment for hepatic adhesion, differentiation, proliferation, and survival. Nanostructured polymeric membranes, with suitable properties of biocompatibility, biostability, and selectivity in transport phenomena, offer interesting opportunities for the design of bioartificial liver (Drioli and De Bartolo, 2006; Morelli et al., 2010; Salerno and De Bartolo, 2017; Salerno et al., 2013a; Salerno et al., 2017a). In these systems, primary human hepatocytes reconstituted many of the in vivo liver features and exhibited specific functional integrity and differentiated state. Biocompatible polymeric membranes modulate the adhesion, proliferation, and differentiation of cells by controlling the mass transfer of molecules and by providing the physical and chemical instructive signals to the cells (De Bartolo et al., 2005; De Bartolo et al., 2006; De Bartolo et al., 2007b; De Bartolo et al., 2009; Morelli et al., 2007; Piscioneri et al., 2011; Salerno et al., 2009; Salerno et al., 2013b). Some examples of our recent developments that allowed us to identify the membranes able to support the differentiated functions of primary human hepatocytes are reported in the following section (Salerno et al., 2011; Salerno et al., 2013c; Ahmed et al., 2017).

16.2.1 Two-Dimensional Human Liver Organotypic Membrane Systems

Both non-degradable and biodegradable membranes have been employed for the in vitro creation of liver tissue. A synthetic membrane constituted of a polymeric blend of modified polyetheretherketone (PEEK-WC) and polyurethane (PU) exhibited properties of biocompatibility, thermal and mechanical resistance, elasticity, permeability, selectivity, and well-defined geometry, which were crucial for supporting the long-term maintenance of primary human hepatocytes (De Bartolo et al., 2005). On the other hand, a biodegradable chitosan (CHT) membrane was able to promote the expansion and functional differentiation of embryonic liver cells (Piscioneri et al., 2011).

A liver organotypic co-culture system by using non-degradable and degradable membranes with primary human hepatocytes and endothelial cells was realized with the aim of obtaining an in vitro physiological model for drug testing and studying metabolic diseases as well as for therapeutic applications (Salerno et al., 2011). The organotypic co-culture of parenchymal and non-parenchymal cells represents an additional strategy to improve culture maintenance of primary hepatocyte phenotype with the final ambition of reproducing complex vascularized networks and integrated membrane bioartificial systems. It is well known that cell–cell communication between multiple cell types in vivo is essential to maintain differentiated cell functions. Heterotypic cell interactions between neighboring cells are of fundamental importance in physiological and pathophysiological liver processes as well as in the organ development, differentiation and regeneration processes (Olson et al., 1990; Michalopoulos and De Frances, 1997). Organotypic co-culture membrane systems were realized by seeding primary human hepatocytes on a confluent monolayer of human endothelial cells by using PEEK-WC-PU and CHT membranes. The CHT membrane has a lowly porous structure and a thickness of 5.5 ± 0.23 μm. Differently, the PEEK-WC-PU membrane has a very porous surface with a mean pore size of 0.1 μm and a thickness of 26.5 ± 1.0 μm. This membrane has a moderate wettability with a water contact angle (WCA) of 77.7°, while the CHT membrane displays a more hydrophilic character with a WCA of 53.8° and is biodegradable with a dissolution degree of 69.7% after 48 h. Membranes exhibit different mechanical properties; in fact, CHT membrane shows an average Young's modulus (E) and a tensile strength (R_m) significantly higher with respect to the PEEK-WC-PU membrane, as a result of its mechanical stiffness property.

In the realization of the organotypic layered co-culture systems, the early main membrane interactions were established by the endothelial cells. Both membranes supported proper endothelial cells attachment and growth, preserving their specific cobblestone shape and favoring the formation of ring- and tube-like structures that increased with time, and whose size varied on the different substrates (Fig. 16.1). The formation of tube- or capillary-like structures has been observed in specialized conditions, including the culture on ECM components suggesting that the tension-dependent biochemical interaction between endothelial cells and matrix proteins or biomaterials may regulate capillary development. In this dynamic process, which implies a migration and the rearrangement of cells besides the matrix remodeling coupled to the proteolytic activity of cells, the mechanical properties of the matrices and substrates are essential to promote capillary-like structure formation. Membrane properties such as the mechanical stiffness and porosity of CHT and PEEK-WC-PU membranes, respectively, not only improved cell adhesion but also induced both the formation of the tube-like structures by endothelial cells and the secretion of ECM proteins by hepatocytes and endothelial cells, which furthermore increased in the organotypic co-culture systems. In these systems, ring-like structures delineated by endothelial cells were observed from the first days of culture. Successively, numerous heterotypic cell–cell interactions and a homogeneous distribution of cells in tight contact were observed. An increase in the morphological complexity was achieved with time, and a self-assembly of cells gave rise to the loss of the original layered distribution. Many tube-like structures appeared bordered with elongated or twisted cells in a network with closely associated hepatocytes. The mutual cell heterotypic interactions help both hepatocytes in the maintenance of their phenotypic morphology and in the improvement of their specific functions and endothelial cells in the formation of capillary-like structures (Fig. 16.1). These results confirm that in the co-culture systems, hepatocytes and endothelial cells undergo both homotypic and heterotypic cell–cell interactions besides the interactions with the substrate for their adhesion. Furthermore, the variation in the heterotypic interface between hepatocytes and endothelial cells interacting with the membrane influences the performance of functional biofabricated tissue constructs. Indeed, it has been demonstrated that the hepatocellular construct changes varying the cell type interacting with the membrane and the heterotypic interface area between cells (Salerno et al., 2013c). In the biofabrication of layered co-culture membrane systems, the sequential cell seeding on membrane and the establishment of heterotypic cell–cell interactions have a strong influence on the performance of the tissue constructs. In particular, it has been observed that when primary human hepatocytes are seeded on a confluent layer of endothelial cells on PEEK-WC-PU membrane, the morphological and functional behavior of hepatocytes improves. In this hepatocellular construct, non-parenchymal cells occupy a substrate surface area dedicated to hepatocyte adhesion, providing the necessary important cues to induce hepatocellular adhesion and functions. This strategy permits a cellular organization that favors heterotypic interactions, improving the interfacial and contact points between endothelial cells and hepatocytes, which affect the stabilization and the enhancement of cell functionality in terms of albumin and urea synthesis, drug biotransformation as well as ECM protein secretion.

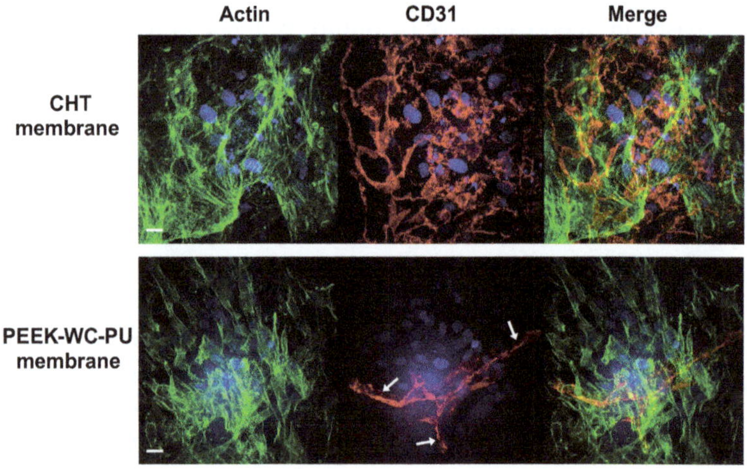

Figure 16.1 Confocal laser micrographs of primary human hepatocytes and endothelial cells in the organotypic co-culture systems on CHT and PEEK-WC-PU membranes after 13 days of culture. Cells were stained for the cytoskeleton protein actin (green) and nuclei (blue); endothelial cells were identified for the staining of the cell adhesion receptor CD31 (red). White arrows indicate the tube-like structures. Scale bar 20 μm. Reprinted from Salerno et al. (2011), with permission from Elsevier.

16.2.2 Three-Dimensional Human Liver Organotypic Membrane Systems

Hollow-fiber membranes of polyethersulfone (PES) modified with coating of fibronectin have been successfully employed for the development of a human organotypic systems, which were realized by a sequential seeding of primary human sinusoidal endothelial cells, stellate cells, and hepatocytes, in order to mimic the liver layers, in static and dynamic culture conditions (Ahmed et al., 2017).

The system consists of two bundles of PES hollow-fiber membranes cross-assembled at a distance of 250 μm. One bundle of fibers feeds fresh medium that flows inside the lumen, and the other bundle removes catabolites and cell products from cell compartment. Human liver cells are cultured in the extracapillary space formed by the network of hollow fibers.

The PES hollow-fiber membranes have an internal diameter and wall thickness of 300 ± 40 μm and 100 ± 25 μm, respectively, and a mean pore size of 0.2 μm. They have a hydrophilic character (WCA = 43 ± 3.5°) and hydraulic permeance of 15.2 L/m^2h mbar. The hollow-fiber membranes provided a wide surface area for the adhesion of cells and promoted the formation of tissue-like structures ranging in size from 250 μm to more than 1 mm, thanks to their physicochemical properties. Cells organized themselves in the liver hollow-fiber membrane system, producing a complex cytoarchitecture with the presence of tube-like structures with a lumen in an anastomosis network formed by sinusoidal endothelial cells. Immunohistochemical studies confirmed the tissue-like arrangement of parenchymal and non-parenchymal liver cell populations in the organotypic system where the tube-like structures surrounded hepatocytes and stellate cells. Fibers appeared completely covered by multilayers of cells that, in some points, formed big aggregates that reached 1 mm in size as a result of cell self-assembling and growth. The formation of aggregates around and between fibers is due to the contribution of stellate cells that facilitate the aggregation (Thomas et al., 2005), by both the physical contact between the cell types and the chemotaxis mechanism through paracrine factors.

The organotypic system was maintained functionally active in the hollow-fiber membrane bioreactor, which allowed a continuous perfusion of cells and the selective mass transfer of molecules to and from cell compartment, creating a physiologically relevant microenvironment. Urea and albumin synthesis as well as the diazepam biotransformation functions were performed at high levels up to 28 days and were significantly higher in the organotypic hollow-fiber membrane system with respect to the hepatocyte monoculture. Metabolites formed by the phase I reactions catalyzed by P450 cytochrome, nordiazepam, temazepam, and oxazepam were detected, with temazepam being the major metabolite. The oxygen uptake rate data underlined the efficiency of the bioreactor to provide an adequate oxygen supply to the cells. Cells displayed rates of oxygen consumptions at levels that implied the maintenance of the oxygen concentration at around 20% in the atmosphere without any oxygen limitation. Additionally, the establishment of an organotypic culture in the crossed hollow-fiber membrane is correlated with the maintenance of the metabolic activity of cells over an extended period of time. The membrane bioreactor created a physiologically relevant microenvironment for cells since it ensures a continuous perfusion of nutrients and metabolites that permeate through the porous wall of feeding fibers from the lumen to the extracapillary space where cells are cultured. In the meantime, catabolites and specific cellular products are removed from the extracapillary space through the removing fibers. Therefore, the microporous modified PES hollow-fiber membranes, because of their structural and physicochemical properties, ensured a non-hindered mass transfer of molecules to and from cell compartment. Furthermore, the crossed configuration allowed a more efficient packing of hollow fibers (Curcio et al., 2012). Data highlight that human liver cells well preserved the biotransformation activity in the hollow-fiber membrane bioreactor.

16.3 Nanostructured Membranes for Bone Tissue Engineering

Bone defects caused by trauma, osteoporosis, injuries, and congenital malformations are serious worldwide problems and are one of the main causes of morbidity and disability in aged patients. Current treatments of large bone defects are based on the use of autologous grafts, which represent the gold standard of bone substitution, and on the transplantation of allograft, but they have disadvantages such as donor site morbidity, available quantities, and host immune rejection that limited their clinical potential (Karp et al., 2006). Therefore, innovative strategies are required to develop effective therapies for bone regeneration. Bone tissue engineering scaffolds are new emerging approaches for bone regeneration compared to the traditional orthopedic implants and contain a complex mixture of molecules in three-dimensional pattern (Remya et al.,

2013; Shalumon et al., 2013; Lee et al., 2013). Cell differentiation, proliferation, and growth can be achieved simultaneously to meet the requirements of bone regeneration. Temporary frameworks to support bone regeneration and controlled release of growth factors to regulate bone formation are also desired properties for the ideal scaffold. It also needs to have macro- and microporosity, biodegradable or biocompatible properties, and good mechanical strength for safe handling, as well as keeping functionality under physical conditions in vivo (Sun et al., 2013; Jin et al., 2013). However, most current conventional tissue scaffolds cannot meet these ideal conditions and suffer from limitations in terms of insufficient fulfillment for requirements of mechanical strength, cell growth promotion, and growth factor release (Bose et al., 2012; Li et al., 2013). Current strategies are focused on the employment of various kinds of constructs that differ in their constituting elements such as scaffolds, cells, and a combination of them. Compared with conventional materials, nanostructured materials present properties in the aspect of overcoming the limitations of current materials. For instance, the presence of nanotubes or nanocrystals in composite materials can improve the mechanical properties for bone tissue engineering materials. Cellular responses to nanomaterials, such as cell attachment, proliferation, and differentiation, can be regulated by the presence of nanostructures. Nanostructured surfaces with chemical modifications have shown increased surface energy and wettability for specific purposes, such as enhancement for cell response (Gu et al., 2013). Many studies have also shown that nanostructured surfaces promote inorganic phase mineralization and enhance in vitro osteogenesis (McMahon et al., 2013). Much research effort has been made to develop nanostructured bone tissue engineering materials.

The natural bone is a composite material with both inorganic and organic phases. Mimicking the nanostructure of bones in which the inorganic component interacts with the organic phase of ECM, composite scaffolds constituted of both organic and inorganic phases have been employed for the creation of bone substitutes. Polymeric ceramic composites combine the intrinsic properties of each component and optimize the physicochemical and biological properties that the hard tissues need. In particular, polymer–ceramic composite materials can increase the integration between graft materials and native bone tissue. Bioresorbable composite scaffolds made of biodegradable polymers and calcium phosphates are suitable for bone repair and regeneration because they do not need to be removed by surgery after bone healing. Hydroxyapatite (HA) is the most investigated ceramic material for creating bone tissue scaffolds as it is the major inorganic component of natural bone. Several biodegradable polymers have been combined with HA for bone tissue engineering such as poly-L-glycolic acid-HA, poly-L-lactic acid-HA (PLLA-HA), poly(3-hydroxybutyrate)-nano-hydroxyapatite, cellulose-HA, polycaprolactone (PCL-HA), etc. PCL is a biocompatible and biodegradable polymer that, thanks to its excellent mechanical properties, flexibility, easy processability, is one of the most attractive polymers with potential applications in tissue engineering (Morelli et al., 2015a). However, pure PCL has disadvantages such as hydrophobic character, low stiffness and cell affinity that limited its clinical applications. A strategy to overcome these limitations is to combine PCL polymer with inorganic particles, such as HA, biphasic calcium phosphate, and bioactive glasses.

In an effort to develop more tunable bone grafts, we have recently reported on the development of a nanostructured composite membrane scaffold by using biodegradable polyester (PCL) and nanoparticles of HA (nHA) in order to obtain bone substitutes with structural similarity to the mineral phase of bone and osteoconductive and bone-binding properties (Morelli et al., 2013). Composite membrane scaffolds of PCL and nHA were prepared by phase-inversion technique from a starting polymeric solution mixed with a bio-ceramic dispersion. The developed scaffolds exhibited high porosity (61.5 ± 0.04%) and low degradability properties and a Young's modulus of 32.3 ± 8.8 N/mm^2. The performance of the developed PCL-nHA membrane scaffolds was assessed in terms of their ability to improve growth and differentiation of hMSCs in osteoblast-like cells. In particular, the osteogenic differentiation of hMSCs was investigated in order to evaluate the expression of osteoblast-specific marker proteins, osteocalcin and osteopontin. Osteocalcin is an ECM protein, related to bone formation, and expressed during the post-proliferative period that reaches its maximum expression during mineralization and accumulates in the mineralized bone. Osteopontin is an extracellular structural protein and is one of the earlier markers of osteoblastic differentiation.

The in vitro experiments demonstrated that cells on PCL-nHA membrane scaffolds express both osteocalcin and osteopontin, which are representative of their osteogenic differentiation and it also suggests that cells mineralization took place on PCL-nHA membrane scaffolds. This study provides strong evidence that PCL-nHA membrane scaffold represents a suitable system for bone tissue regeneration. Such composite membrane system could provide insights for the creation of a controllable osteoinductive microenvironment based on the specific properties of the involved scaffold (e.g., basic composition, surface chemistry, architecture), and on the proliferation functions and differentiation of defined cellular systems.

Nanofibrous scaffolds, due to the architectural, functional, and morphological similarities to the collagen fibrils in bone, are one of the most suitable materials for bone tissue engineering (Smith and Ma, 2004). Woo et al. (2003) designed a three-dimensional polylactic acid (PLA) nanofibrous scaffold to improve protein absorption and enhance cell attachment on the scaffold. Furthermore, cell shape and morphology and cell spreading can also be affected by the presence of the nanofibrous scaffolds (Tuzlakoglu et al., 2005). PLA and its copolymer poly-L-glycolic acid are bioresorbable materials that are approved for clinical use, but these polymers alone usually have the adverse clinical effect that the degraded acidic monomers may cause inflammatory or allergic reactions. Incorporating osteoconductive fillers into the PLA matrix is a useful way to improve clinical performance. The introduction of HA into PLA not only neutralizes the degraded acidic products of PLA, but also makes the implant more osteoconductive, thus improving tissue compatibility. To date, a large number of reports indicate that scaffolds composed of nHA and PLA could fulfill many of the requirements to use as substrates for bone tissue engineering. Recently, HA-mineralized PLA nanofiber sheets created by immersing electrospun nanofiber sheets in body fluid solution were found to interact potently with the hADSCs and to promote osteogenic differentiation and, therefore, are suitable for bone regeneration (Kung et al., 2014).

Bone scaffolds when implanted into the natural bone undergo a remodeling process. This process that ensures the in vivo skeleton integrity involves the communication between the bone-building cells, osteoblasts, and the bone-resorbing cells, osteoclasts. Hence, for a successful bone scaffold, it is important not only to consider the interaction with the osteoblasts but also to investigate the interactions with osteoclasts that play a critical role in the conversion of the bone graft into native bone. Therefore, the bioengineered scaffolds should be considered in this context, evaluating the interaction between cells and scaffolds during the tissue formation and remodeling.

Previous studies have focused on the biocompatibility and osteoconductivity of nanostructured PLA-HA scaffolds with osteoblasts or bone stromal cells (Kutikov and Song, 2013; Lao et al., 2011). Only a few papers report on the co-culture of osteoblasts and osteoclasts by using other biomaterials (e.g., composite xerogel, polystyrene, silk fibroin, chitosan and PLLA films, poly-L-glycolic acid scaffolds) up to 2 weeks (Jones et al., 2009; Tortelli et al., 2009). Our strategy was to apply the knowledge of natural bone tissue remodeling in a bioengineered composite scaffold constituted of PLA-nHA-fiber mat combining hMSCs with osteoclasts in order to mimic the bone environment feature based on the cross-talk between cells and scaffold. Recently, we reported, for the first time, on the development of novel nanostructured PLA-nHA fiber scaffolds that were used with osteoblasts and osteoclast progenitor cells in order to mimic the natural bone tissue remodeling process (Morelli et al., 2015b). In particular, taking cues from nature, we studied the differentiation of hMSCs from bone marrow into osteoblasts and the osteoclastogenesis in the PLA and PLA-nHA scaffolds up to 46 days of culture. Three-dimensional PLA and composite PLA-nHA scaffolds with 20 and 50 wt.% of nHA prepared by electrospinning technique were used to induce the differentiation of hMSCs into osteoblasts and to trigger osteoclastogenesis from human monocytes. The PLA-nHA fiber scaffolds provided physical and chemical cues at micro- and nano-level to the cells. They are highly porous with the required pore interconnectivity, which is important for the diffusion of nutrients and molecules and for cell adhesion giving a large ratio of surface area to volume (Fig. 16.2). The developed scaffolds have different fiber diameters due to the different nHA concentrations in the polymer solution that affect also the roughness and mechanical strength. Fibers exhibited nanopores and nHA aggregates dispersed over the surface. The Young modulus values of the developed scaffolds are in line with those reported for the trabecular bone in the human mandible ranging from 3.5 to 125.6 MPa (Misch et al.,

1999). HMSCs adhered and proliferated especially on composite scaffolds where a higher density of cells was found. PLA composite scaffolds not only elicit the growth of hMSC-derived osteoblasts but also the viability of monocyte-derived osteoclasts and co-cultured cells (Fig. 16.3). The presence of nHA stimulated the growth of cells in homotypic and heterotypic culture. Moreover, co-cultured cells in the PLA-nHA scaffolds were able to cross-talk and to cooperate in the remodeling process as pointed out by the expression of both receptor activator of nuclear factor NF-kB and its ligand, which are specific markers of osteoclasts and osteoblasts, respectively. Monocytes-derived osteoclasts could express tartrate-resistant acid phosphatase activity and cathepsin K, which have a crucial role in bone remodeling and in the resorptive process, respectively. The osteoclast differentiation was also supported by the configuration of actin filaments, which are organized around the cellular membrane as the ruffled border. The typical ring-like distribution of actin was evident in osteoclasts. Overall, the combination of the composite PLA-nHA scaffolds with co-cultured hMSCs-derived osteoblasts and osteoclasts stimulated the cell viability, differentiation, and functions giving new insights on how to produce a suitable microenvironment for bone tissue engineering (Morelli et al., 2015b).

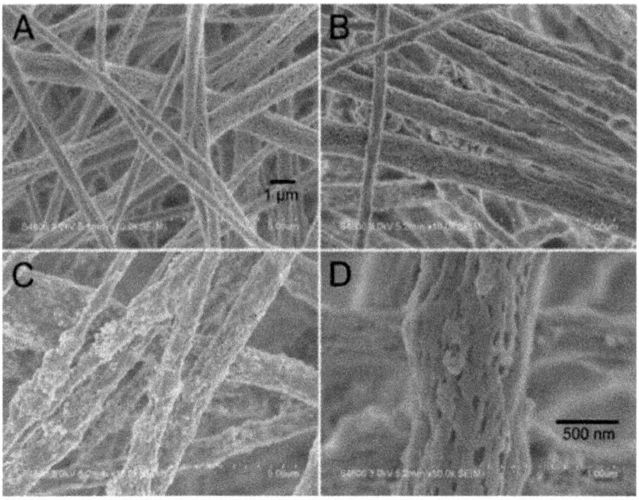

Figure 16.2 Scaffold properties: SEM images of (A) bare PLA fibers and PLA-nHA composite fibers with (B and D) 20 and (C) 50 wt.% nHA. Reprinted from Morelli et al. (2015b), with permission from Elsevier.

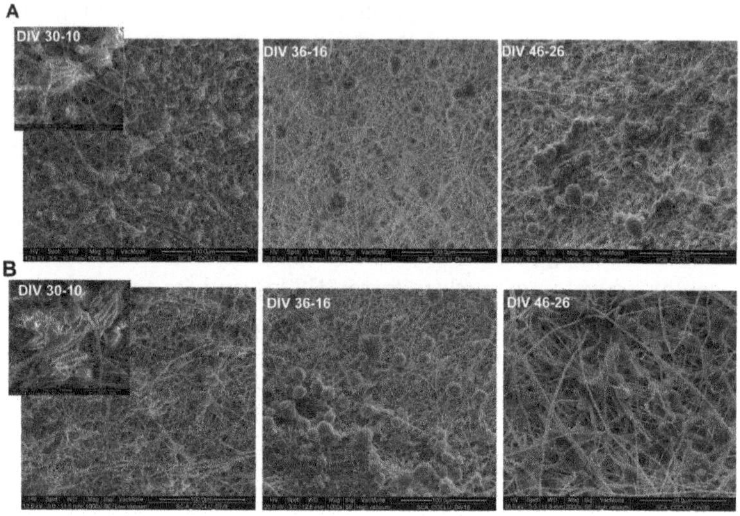

Figure 16.3 SEM images of hMSC-derived osteoblasts and monocyte-derived osteoclasts co-cultured for 46–26 days in (A) PLA-nHA-20 and (B) PLA-nHA-50 fiber mat scaffolds. Reprinted from Morelli et al. (2015b), with permission from Elsevier.

16.4 Membrane Systems with Nanostructural Features for Neuro-Engineering Applications

Recent advances in nanotechnology, biomaterials, and tissue engineering offer new strategies for the development of in vitro advanced devices that enhance the neuronal growth and differentiation by mimicking specific features of the in vivo environment (Zhang et al., 2005). Notably, polymeric membranes can promote the formation of membrane–neuron hybrid systems that, thanks to their selective structural, physicochemical, mechanical, and transport properties, are able to drive neurite outgrowth and branching, network connectivity, and synaptic plasticity, leading to the successful in vitro reconstruction of neuronal tissue. Neuronal biohybrid membrane systems are widely used as in vitro brain tissue models for pharmacological screening (Giusi et al., 2009; Di Vito et al., 2014; Pucci et al., 2013; Morelli et al., 2014), and as investigational platforms for neurodegenerative diseases (Piscioneri et al., 2015; Mele et al., 2016; Morelli et al., 2016b).

In the past decade, nanotechnology applications in the nervous system have often involved the study and use of novel nanomaterials to improve the diagnosis and therapy of neurological diseases (Fabbro et al., 2012; Garbayo et al., 2014; Das et al., 2016). The following section reports some examples of nanostructured membrane systems that have been engineered to support and guide neuronal outgrowth and differentiation in order to enhance regeneration in the nervous system.

16.4.1 Membrane-Based Approaches for Nerve Repair/Regeneration

The peripheral nerve fibers, unlike those of the central nervous system, can regenerate after injury. Due to this property, surgeons are able to achieve a good functional recovery in patients who have suffered an injury to one or more nerve trunks through the realignment and suturing of the two stumps of the resected nerve. The resected fibers can thus regenerate along the graft and reach the distal nerve stump that will lead to their peripheral nerves. Although a segment of autologous sensory nerve may be an excellent material to make nerve grafts, this surgical technique has inevitable disadvantages, including, among many others, limited supply of available nerve autografts and certain donor site morbidity. Xenografts and allografts have been evaluated as alternatives to autografts but have poor successful rate and problems of immune rejection (Evans et al., 1998). Thus, a wide range of biological materials (e.g., muscles, vessels) (Tos et al., 2007; Meek et al., 2004) and non-degradable materials (e.g., silicone tubes) have frequently been used for nerve regeneration (Wang-Bennett and Coker, 1990). Artificial materials have the disadvantages of engendering chronic foreign body reaction due to scar tissue formation, inflexibility, and lack of stability. To overcome these drawbacks, recent research has been focused on the production of biodegradable artificial nerve guides, which degrade within a reasonable period and manifest only mild foreign body reactions. Biodegradable polymers are advantageous because of their flexibility and biocompatibility; degradation behavior, porosity, and mechanical strength can be altered by changing their chemical properties. In this context, taking into account the encouraging results in the field of peripheral nerve regeneration, the investigation was focused on the development of novel biodegradable membranes supporting neuronal differentiation, aiming toward a tissue-engineered construct for nerve repair. Matching these main features, different membranes made up of PCL, polyurethane (PU), and of biosynthetic blend between PCL and PU (PCL-PU) were developed in flat configuration using the phase-inversion techniques (Morelli et al., 2015a).

The performance of the developed biomaterials was assessed in terms of their ability to improve cell adhesion, viability, and differentiation, as well as for their ability to allow axonal growth of neuronal cells, with special attention to the formation of synapse connections between the neurons, thus serving as a valuable tool in neuronal tissue engineering. All the membranes have been fully characterized in terms of structural and mechanical properties. PU membrane exhibits a homogenous and very smooth surface. The presence of repetitive microstructures is clearly visible on PCL surface, and this patterned distribution was revealed on PCL-PU membrane as well. This quite singular morphology is typical of the PCL membranes, and it is in fact conferred by the presence of this polymer in the casting solution. Because natural ECM is a fully hydrated gel, wettability is a key consideration factor; accordingly, different substrates were subjected to further physicochemical characterizations. All prepared biodegradable membranes have hydrophilic character as demonstrated by dynamic contact angle measurements. Synthetic biodegradable polymer-made membranes, namely PU, PCL, as well as PCL-PU blend, exhibit a similar wettability with advancing contact angle values ranging from 72° to 77°. Materials for the fabrication of nerve guides are required to have suitable mechanical properties, such as elasticity/stiffness, for the bearing of stresses during the surgical procedure (handling and suturing) and the implantation time (movements of the patient). Nerve guides should be flexible, tough, and capable of bending without kinking (Ciardelli and Chiono, 2006).

In order to serve in a peripheral nerve graft, biomaterials must function as growth substrates for the neuronal cells and must guide regenerating neurons. Therefore, we tested the capacity of generated biodegradable membranes to support and improve the adhesion, proliferation, and differentiation of neuronal cells.

In order to analyze how the membrane properties can affect cell morphology, adhesion, and proliferation, we performed quantitative measurements of the area of different membrane substrate covered by cells. This analysis demonstrated that the cellular growth increased with time and the cells displayed a different morphological behavior when seeded on different substrates. After 3 days of culture, cell adhesion on PCL-PU (20%), PCL (19%), and PU (18%) membranes considerably increased. After a period of 6 days, the highest value was reached on PCL-PU membranes representing, therefore, the most covered surface with a percentage of 50%. The same trend was observed in terms of the acquired neuronal morphology. Indeed, further, cell morphological features were also investigated by performing morphometric analysis in terms of neurite length of the cells. After 3 days of culture, cells seeded on PCL, PU, and PCL-PU membranes showed an intense neurite sprouting as evidenced by the average process length of 98 μm. Toward the end of the culture period, the longest neuritic processes were obtained on PCL-PU membranes (105 μm). The neuronal differentiation was confirmed by the expression and distribution of specific neuronal markers βIII-tubulin and synaptophysin. The βIII-tubulin at day in vitro (DIV) 6 was widely distributed, and the cells on the biodegradable membranes of PCL, PU, and PCL-PU appeared connected to each other through the extension of their neuritis, giving rise to well-developed neural networks (Fig. 16.4c). A more intense and diffuse distribution of synaptophysin was found in neurons cultured on the PCL-PU membranes (Fig. 16.4) compared to the other substrates. The highest expression of the synaptic protein synaptophysin on PCL-PU membranes suggests how this specific substrate is able to enhance neuronal differentiation, therefore acting as a tissue analogue in which cells reached maturation in terms of functionality and morphology. The successful differentiation of the cells toward a neuronal phenotype was also pointed out by the expression pattern profile of a specific neuronal protein: growth-associated protein (GAP 43), which is an axonal protein involved in neuronal outgrowth and synaptic plasticity of developing and regenerating neurons. GAP 43 distribution and expression were particularly enhanced in cells seeded on PU and PCL-PU membranes, suggesting that these substrates well supported the axonal elongation as well as, having therefore, the ability to support and promote neuronal outgrowth and synaptic plasticity in terms of axonal elongation.

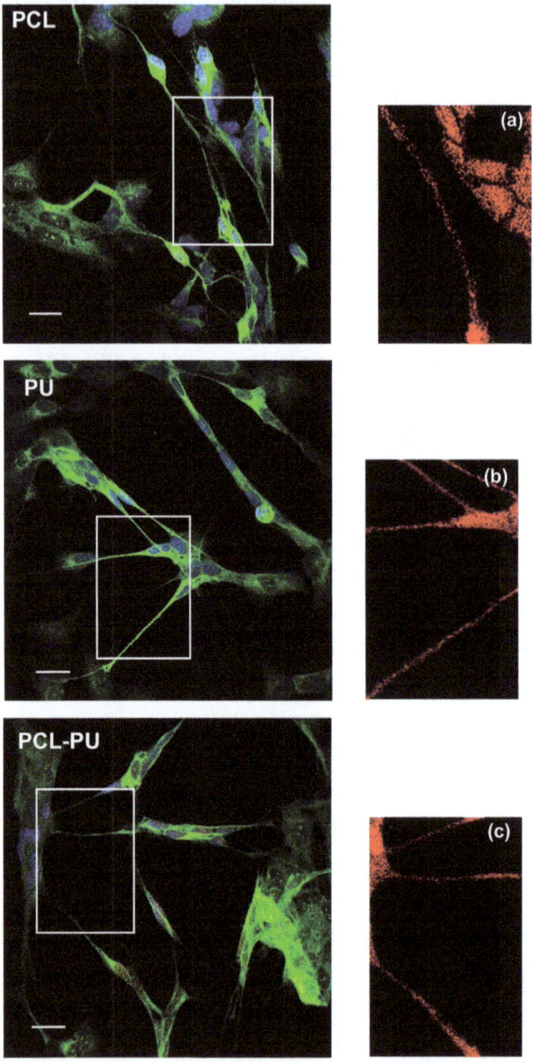

Figure 16.4 Confocal laser micrographs of neuronal cells after 6 days of culture on different substrates. Cells were stained for βIII-tubulin (green), synaptophysin (red), and nuclei (blue). (a,b,c) Magnification panels of synaptophysin distribution (red) on PCL, PU, and PCL-PU membranes, respectively. Scale bar 20 μm. Reprinted from Morelli et al. (2015a), with permission from John Wiley and Sons.

16.4.2 Innovative Tool for Neuronal Differentiation and Orientation: A Microtube Array Membrane Bioreactor

An important challenge in neuronal tissue engineering is to create innovative tools capable of promoting cellular response in terms of neuronal differentiation and neurite orientation, which may be used as investigational platforms for studying neurobiological events and neurodegenerative disorders. Specific cellular responses in a biomaterial-based system are significantly improved when associated and supported by a membrane technology approach as it happens in a membrane bioreactor (De Bartolo et al., 2009; Morelli et al., 2010; Curcio et al., 2012; Salerno et al., 2013; Morelli et al., 2016b). This kind of device offers a well-controlled microenvironment that can trigger, besides cell differentiation, the maintenance of specific functions that are often lost at early stage in two-dimensional traditional culture system. This system within its surrounding provides a wide range of physiological cues, which empowers their reliability as investigational platform (Giusi et al., 2009; Di Vito et al., 2014). Their experimental parameters can be tuned and controlled according to the needs of the cellular construct that has to be created, thus providing a neuronal interface quite similar to the in vivo situation. The entire set of membrane properties (e.g., roughness, physicochemical, permeability, pore size, porosity) as well as the fluid dynamic conditions of the membrane device can be tuned in order to favor the neurite outgrowth and elongation, giving rise to a rich arborized neuronal network (Morelli et al., 2016b; Piscioneri et al., 2015; Morelli et al., 2015a; Morelli et al., 2012a; Morelli et al., 2012b). Based on these considerations, we aimed at developing a pioneering neuronal membrane bioreactor able to accomplish a dual purpose (Morelli et al., 2017a). The peculiar component of the device, is an array of poly-L-lactic acid microtubes (PLLA-MTA) that are highly aligned and packed to promote cell polarization and orientation. The bioreactor design based on the use of PLLA-MTA membrane offers a continuous perfusion to the cells with oxygenated medium and removal of catabolites, avoiding profile concentration and shear stress that often occur in the traditional plastic plate culture systems or in rotating wall vessel (Di Loreto et al., 2006; Lin et al., 2004; Morabito et al., 2015; Valmikinathan et al., 2011). The PLLA-MTA membrane bioreactor supplied the neuronal cells with a balanced and uniform amount of nutrients and an appropriate oxygenation across the entire cell compartment volume. Moreover, this ideal surrounding is due to the combined synergism of the membrane geometrical construction and configuration within the device, according to a continuous one-by-one hollow-fiber array structured along a single layer, together with the intrinsic structural features of the PLLA-MTA membranes.

Membrane characterization revealed an overall membrane porosity of 46% and a moderate surface wettability with dynamic water contact values in advancing of 87.3° ± 3° and in receding of 50.1° ± 2.5°, as expression of apolar and polar domains of the surface, respectively. Once the membrane features and the fluid dynamic conditions were optimized, the differentiation of neuronal cells within the PLLA-MTA membrane bioreactor was investigated. The overall aim of the study was to establish if the developed device represents a satisfactory apparatus for the in vitro creation of a three-dimensional neuronal construct. Thus, neurons were seeded onto the surface of the aligned PLLA-MTA membranes and maintained in culture within the bioreactor up to 13 days, which is the timeframe to investigate the ability of the membrane bioreactor to support neuronal growth and differentiation. The evaluation of morphological features, as the analysis of the expression level of specific neuronal markers, in combination with metabolic activity assessment helped to identify the in vitro neuronal differentiation. Cells acquired and maintained their differentiated neuro-cytoarchitecture up to the investigated 13 days of culture in the membrane bioreactor, which then provided an appropriate in vitro microenvironment to create a neural tissue-like structure that mimics the in vivo neuronal tissue. In addition, membranes could guide the orientation of the new emerging neurites, which were elongated and aligned in parallel and in the same direction of the membrane structures. The investigation showed a uniform distribution of neurons above the PLLA-MTA membranes in the bioreactor with a high degree of adhesion. The acquired neuronal phenotype was evaluated and confirmed by the expression and distribution of neuronal cytoskeleton marker protein, βIII-tubulin, synaptic vesicles marker, synaptophysin, and axonal marker, GAP-43. Neuronal processes were aligned in a parallel manner to the long axis of the membranes. To this aim, a quantitative analysis was performed in order to determine the neurite outgrowth orientation of neurons cultured into the membrane bioreactor. The evaluation consisted in measuring the angle between the membrane long axis and neurite growth, assessing also the frequency of each angle. After 6 days of culture in PLLA-MTA membrane bioreactor, the highest frequency of neurite orientation (25%) was measured between 5° and 15°. After 13 days of culture, the orientation of neurons along membranes was maintained with a high degree of frequency (23%) within 15°. These findings suggest that the membrane bioreactor not only promotes neuritogenesis but also guides the neurite extension by enhancing their orientation and confirming an aligned elongation of neuritis process with membranes. Consistently, both the glucose consumption and the brain-derived neurotrophic factor release were significantly higher in bioreactor than in the static system, which corroborates the hypothesis that the dynamic system offers optimal culture conditions able to sustain specific metabolic activity besides the acquisition of specific morphological features, which is more related to the intrinsic membrane properties. The high-performing device built up in this study modulated and enhanced neuronal outgrowth, thanks to a synergistic action of the PLLA-MTA membrane properties and the uniform dynamic bioreactor microenvironment. The PLLA-MTA membrane bioreactor, besides enhancing acquisition of neuronal phenotype, guided the neurons into a defined aligned orientation according to the direction of the microtubes. Overall, the bioreactor accomplished two main achievements: it promotes long-term growth and differentiation of neuronal cells and orients neurite alignment generating a neuronal tissue-like construct.

16.5 Bioengineered Skin Substitutes by Using Nanostructured Membranes

Advances in tissue engineering led to the realization of human bioengineered skin substitutes for in vivo skin regeneration to restore functional and esthetical integrity in burn patients, in ulcers and necrosis from disease, and to develop in vitro skin models for toxicological test of a wide range of drugs, cosmetics, and pollutants. Notably, in vitro skin models are useful tools alternative to animal experimentation for testing the adsorption, accumulation, effects, and/or toxicity of chemical compounds. Human bioengineered skin substitutes give a rapid, accurate, and highly specific evaluation of chemical effects and toxicity on humans, considering the differences in enzyme pattern and expression between human and animals. Through the skin, many drugs are administered by percutaneous adsorption, and a high amount of drug metabolites are secreted in sweating solution.

Nanostructured biodegradable membranes of CHT, PCL, and a blend of CHT-PCL were developed by phase-inversion technique for the regeneration of human bioartificial skin equivalents and for the creation of different skin layers. In these systems, polymeric membranes, thanks to their different and specific properties, satisfied distinct requirements and fulfilled different applications in skin tissue engineering (Salerno et al., 2016; Salerno et al., 2017b). The use of biodegradable membranes represents a requisite factor for the development of implantable bioartificial constructs that should be resorbed by the surrounding tissues without their surgical removal. The different biodegradation characteristics exhibited by the developed membranes accomplish skin repair: the CHT membrane degradation time, of approximately 25 days, is suitable for healing acute wounds; a degradation rate of around 8 weeks, as displayed by CHT-PCL membrane, could be considered for repairing in chronic events. Since the membranes must play a role of selective barrier at the nanoscale level, they were developed in order to have surfaces with mean pore diameter of 26, 24, and 37 nm on CHT, PCL, and CHT-PCL, respectively. Membranes effectively display surfaces with different roughness, which is an important parameter in the cell adhesion (Fig. 16.5). CHT membranes exhibit a uniform roughness with values of Ra and RMS of 3.97 ± 0.4 and 5.12 ± 0.5 nm, respectively. PCL and CHT-PCL membranes display a surface with nano- and microstructures that confer Ra roughness values of 4.61 ± 1.1 and 4.87 ± 1.4 nm, respectively. The presence of the homogeneous microstructures is quite singular and typical for membranes made by PCL; indeed the presence of this polymer in the casting solution of CHT-PCL blend conferred the same morphology to the membrane surface.

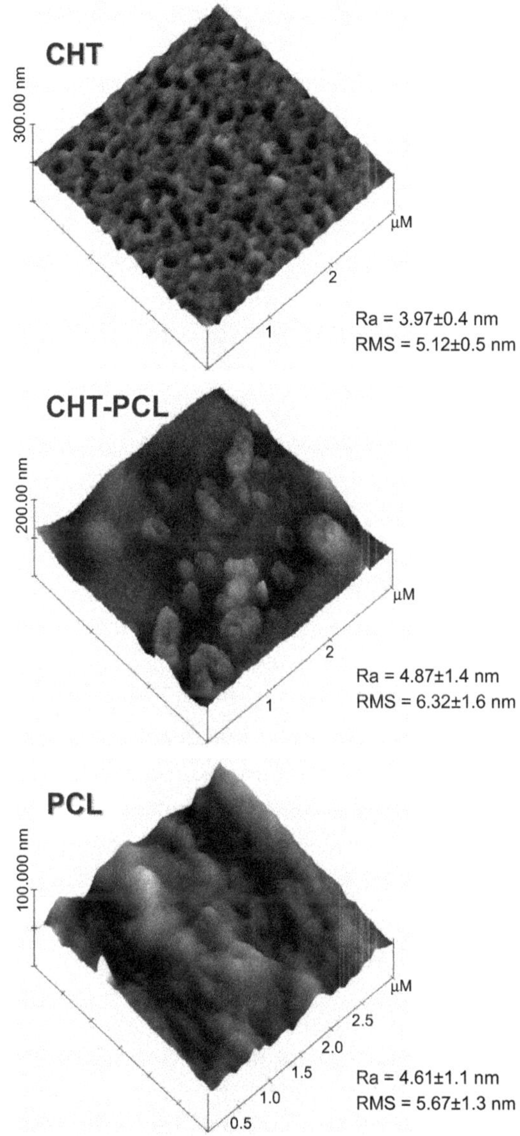

Figure 16.5 AFM images of chitosan (CHT), chitosan-polycaprolactone (CHT-PCL), and polycaprolactone (PCL) membranes, and relative roughness parameters R_a (average absolute distance from average flat surface), and RMS (root mean square from average flat surface). The values are the means of 30 measurements per sample ± SD. Reprinted from Salerno et al. (2016), with permission from Elsevier.

The developed membranes display different hydrophilic character as demonstrated by WCA values, ranging from 53.8 ± 2° for CHT to 77.6 ± 3° for PCL, while the CHT-PCL membrane consistently reveals a moderate wettability represented by an intermediate value of 61.8 ± 2.5°. The mechanical properties change in relation to the dry or wet membrane conditions. In particular, CHT and CHT-PCL membranes show high mechanical stiffness properties when dry, exhibiting tensile modulus (E) of 2831 ± 304 and 1727 ± 302 MPa, respectively. When wet, due to swelling, these values are drastically reduced to 5.7 ± 2.6 and 2.9 ± 1.4 MPa, respectively, and the percentages of elongation at break, ε, are 10-fold increased. Differently, the PCL dry membrane, due to the poor swelling process, appears more stable, exhibiting in dry condition a tensile modulus (E) of 443 ± 34 MPa, which decreased by only 22% when wet.

Biohybrid membrane epidermal, dermal, and dermal/epidermal substitutes were realized by using human keratinocytes and skin derived stem cells (SSCs) alone or in co-culture on CHT, PCL, and CHT-PCL membranes. These membranes played a crucial role in the guidance of the regeneration of different skin layers, modulating the cell proliferation and differentiation. In particular, these membranes strongly influenced the cell morpho-functional behavior of epidermal cells, regulating their terminal differentiation in distinct and specific epidermal strata (Salerno et al., 2016), and the differentiation and transdifferentiation of dermal mesenchymal stem cells in dermal fibroblasts and epidermal keratinocytes (Salerno et al., 2017b).

16.5.1 Epidermal Membrane Systems

Artificial human epidermis models commercially available and currently used for testing phototoxicity, corrosivity, irritancy, and drug penetration consist of human keratinocytes cultured on matrix acting as artificial dermis (i.e., polycarbonate filters in SkinEthic®, a type I bovine collagen matrix in Episkin®, and a silicone gel in Epiderm®) (Netzlaff et al., 2005). Even though these models currently are the closest to the human epidermis, with an extensively proven accuracy and reproducibility, they represent the entire epidermal sheet.

The in vivo epidermis is made up of self-renewing keratinocytes programmed to be replaced continuously from the basal lamina to the outermost layers. Therefore, cells acquire a migratory phenotype in a differentiation process for which well-defined cell layers or strata of increased differentiation and different morphology are recognizable from the bottom to the top. In vitro models of isolated epidermal strata could represent a powerful and essential tool for the understanding of the interactions with external agents along the epidermis layers.

For the first time, specific and distinct human epidermal strata were engineered by using nanostructured membranes and keratinocytes (Salerno et al., 2016). A high differentiation state of human keratinocytes, with morphological and functional behavior like the cells in the *lucidum* and *corneum* strata of natural epidermis, was triggered by a CHT membrane. On this membrane, human keratinocytes grew in monolayer and exhibited a flattened morphology with a high degree of spreading, and low cell proliferation. Interestingly, a high expression of cytokeratin 1 was found, as a specific marker of the suprabasal levels of epidermis, despite the quite missing expression of cytokeratin 18, as a marker of cell proliferation. The differentiated flattened shape that keratinocytes acquired on CHT membrane was driven by the surface hydrophilicity of the membrane and its intrinsic chemical structure, which is similar to the glycosaminoglycan matrix of the dermis (Fig. 16.6).

The CHT-PCL membrane, which displays a moderate wettability and surface microstructures, induced both flattened and more cuboidal cell morphology. Human keratinocytes on this membrane showed a lower cell spreading and maintained a proliferation activity, expressing, however, both cytokeratin 1 and cytokeratin 18 in the early days of culture, as a result of their incomplete terminally differentiated state. This morpho-functional behavior is more similar to that belonging to keratinocytes in *granulosum* and *lucidum* layers of the in vivo epidermis, underneath the outermost *corneum* layer.

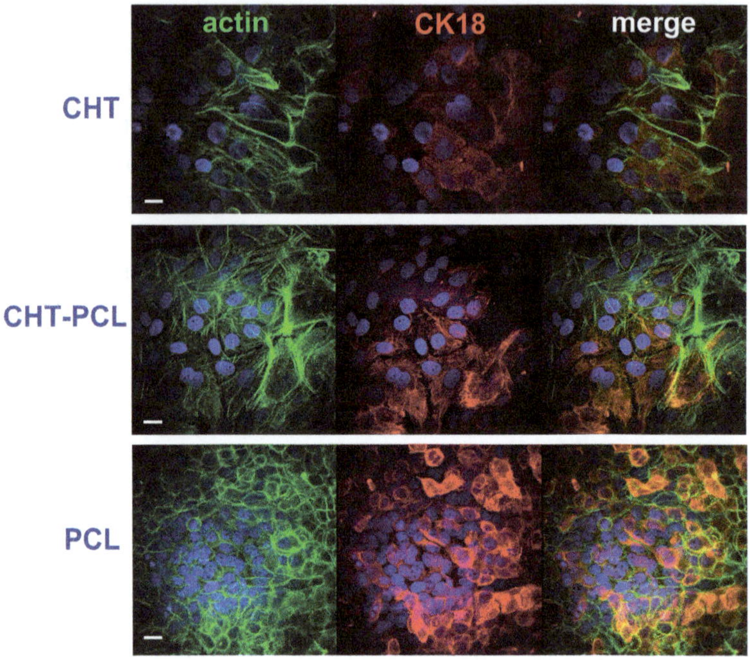

Figure 16.6 Confocal laser micrographs of human keratinocytes after 7 days of culture on biodegradable membranes of chitosan (CHT), chitosan-polycaprolactone (CHT-PCL), and poly-caprolactone (PCL). Cells are stained for actin (green), CK18 (red), and counterstained for nuclei (blue). Scale bar 20 μm. Reprinted from Salerno et al. (2016), with permission from Elsevier.

The PCL membrane, thanks to mechanical properties similar to those exhibited by the natural dermis, a moderate surface wettability and roughness, limited the cell spreading and favored a morphological behavior more similar to the cells of the *basal lamina*. On the PCL membrane, human keratinocytes were maintained in a round shape and in proliferating activity, like the cells in the *basal lamina* of the in vivo epidermis. Cells proliferated without reaching their terminal differentiation up to 14 days of culture, as evidenced also by their lower expression of cytokeratin 1. Over time, the sustained proliferation engendered the stratification of multilayered cells so that after 21 days, a complete proliferative and migratory differentiating bioartificial epidermal substitute was created on the PCL membrane.

16.5.2 Dermal and Dermal/Epidermal Membrane Systems

In the dermal membrane systems, SSCs adhered on the membranes and exhibited the typical fibroblasts-like spindle-shaped morphology with different spreading degree (2252 ± 781 μm^2 on CHT, 1600 ± 480 μm^2 on CHT-PCL, and 818 ± 219 μm^2 on PCL). In all the systems, a strong secretion and deposition of collagen I was observed. In the skin, it represents one of the primary collagens that provide structure and integrity as well as both strength and elasticity to the tissue. Collagen I is also involved in the stimulation and modulation of cell proliferation, differentiation, and migration in skin regeneration and in wound-healing processes (Midwood et al., 2004). The fibronectin secretion and deposition were dependent on the substrates on which SSCs were grown. In particular, SSCs secreted and deposited fibronectin on PCL and on CHT-PCL, but not on CHT membrane. This might be due to a reduced cell proliferation, which is related to the different spreading degree. Membranes with a more hydrophilic character, such as CHT and CHT-PCL, favored high spreading of the cells owing to the strong adhesion and interaction with the polar groups of chitosan polymer, which are protonated amino groups at physiological pH, and the anionic functional groups on the cellular membrane. The high spreading is inversely related to the cell proliferation potential, in contrast with the cell rounding. Furthermore, both fibronectin deposition and cell proliferation are strongly influenced not only by surface wettability and hydrophilicity but also by the mechanical properties of the substrates. It is well known that on stiffer substrates, stronger forces are generated due to the high resistance of the substrate; therefore, weaker focal adhesions are established, lower actin stress fibers are developed, and lower cell proliferation is observed. On PCL membranes, with a tensile modulus (E) of 344.4 ± 48.9 MPa (wet), high cell proliferation and fibronectin deposition were observed, in contrast to CHT membranes with a tensile modulus (E) of 24.4 ± 3.5 MPa (wet), where low cell proliferation and no fibronectin deposition were found. PCL membrane exhibits a tensile modulus more similar to those of in vivo dermal components and, in particular, by collagen fibers, the main structural constituent of the dermis. PCL membranes, owing to their tensile strength and elastic properties, their lower hydrophilic character, and the presence of microstructures as suitable cues on their surface, offered a proper microenvironment that fostered the highest proliferation activity not only for SSCs, but also for human keratinocytes in epidermal and dermal–epidermal systems. Indeed, in the dermal–epidermal membrane systems, the cell–cell contacts and the intercellular communications stimulated the cell proliferation and differentiation as well as the migration of keratinocytes in the dermal layer. Structures that are more complex were formed on the CHT-PCL and PCL membranes. In the dermal–epidermal membrane systems, the fibronectin deposition by SSCs was enhanced and speeded up compared to the SSCs homotypic cultures, more likely to the mutual cell–cell interactions. The fibronectin deposition is stimulated by the presence of keratinocytes, which are, in turn, triggered by SSCs to secrete growth factors, which are known to be involved in the regulation of fibronectin expression (Mimura et al., 2004). SSCs in the dermal membrane systems maintained their multipotency within the time, as highlighted by the expression of specific mesenchymal markers, such as vimentin, which is a specific cytoskeleton protein of MSCs, and of CD90 and CD271, which are MSC-related surface antigens. In the dermal–epidermal membrane systems, the expression of these mesenchymal markers was markedly reduced after 14 days. Simultaneously, in the dermal CHT-PCL and PCL membrane systems, SSCs acquired with time the expression of markers of epidermal cell lineage (e.g., cytokeratins 1, 18, and 19), as proof of their ability to transdifferentiate into keratinocytes. Membranes provided specific environmental and mechanical cues, able to favor the high proliferative and self-renewal abilities of SSCs, and to induce epidermal and dermal transdifferentiation. In this direction, proper mechanical properties of PCL membrane (close to the in vivo dermis), and the presence of topographical nano- and microstructures on both PCL and CHT-PCL membranes, offered a favorable milieu to SSCs for their skin cell lineage differentiation. In the dermal systems, on CHT-PCL and PCL membranes, SSCs after 14 days of culture were simultaneously able to express specific epidermal markers and to deposit fibronectin, as a result of dermal differentiation. In the dermal–epidermal systems, a more suitable microenvironment to promote the SSCs transdifferentiation is established owing to the interactions with keratinocytes, so that, after 2 weeks of co-culture, SSCs reduced the expression of mesenchymal markers (Salerno et al., 2017b).

Abbreviations

CHT	chitosan
DIV	day in vitro
ECM	extracellular matrix
GAP 43	growth-associated protein
HA	hydroxyapatite
hADSCs	human adipose-derived stem cells
HF	hollow fiber
hMCSs	human mesenchymal stem cells
nHA	hydroxyapatite nanoparticles
PCL	polycaprolactone
PEEK-WC-PU	modified polyetheretherketone and polyurethane
PLA	polylactic acid
PLLA	poly-L-lactic acid
PLLA-MTA	poly-L-lactic acid- microtube array
PU	polyurethane
Ra	roughness
RMS	root mean square

References

Ahmed, H. M. M., Salerno, S., Morelli, S., Giorno, L., and De Bartolo, L. (2017). 3D liver membrane system by co-culturing human hepatocytes, sinusoidal endothelial and stellate cells. *Biofabrication*, **9**, 025022. DOI: 10.1088/1758-5090/aa70c7.

Ai, J., Kiasat-Dolatabadi, A., Ebrahimi-Barough, S., Ai, A., Lotfibakhshaiesh, N., Norouzi-Javidan, A., Saberi, H., Arjmand, B., and Aghayan, H. R. (2014). Polymeric scaffolds in neural tissue engineering: A review. *Arch. Neurosci.*, **1**, pp. 15–20.

Bose, S., Roy, M., and Bandyopadhyay, A. (2012). Recent advances in bone tissue engineering scaffolds. *Trends Biotechnol.*, **30**, pp. 546–554.

Cao, G. and Wang, Y. (2011). *Nanostructures and Nanomaterials: Synthesis, Properties, and Applications,* 2nd Ed. (World Scientific Publishing Co Pte Ltd, Singapore).

Chen, N., Tian, L., He, L., and Ramakrishna, S. (2016). Nanobiomaterials for neural regeneration. *Neural Regen. Res.*, **11**, pp. 1372–1374.

Chiono, V. and Tonda-Turo, C. (2015). Trends in the design of nerve guidance channels in peripheral nerve tissue engineering. *Prog. Neurobiol.*, **131**, pp. 87–104.

Ciardelli, G. and Chiono, V. (2006). Materials for peripheral nerve regeneration. *Macromol. Biosci.*, **6**, pp. 13–26.

Curcio, E., Piscioneri, A., Salerno, S., Tasselli, F., Morelli, S., Drioli, E., and De Bartolo, L. (2012). Human lymphocytes cultured in 3-D bioreactors: Influence of configuration on metabolite transport and reactions. *Biomaterials*, **33**, pp. 8296–8303.

Das, S., Carnicer-Lombarte, A., Fawcett, J. W., and Bora, U. (2016). Bio-inspired nano tools for neuroscience. *Prog. Neurobiol.*, **142**, pp. 1–22.

De Bartolo, L., Morelli, S., Gallo, M. C., Campana, C., Statti, G., Rende, M., Salerno, S., and Drioli, E. (2005). Effect of isoliquiritigenin on viability and differentiated functions of human hepatocytes maintained on PEEK-WC-polyurethane membranes. *Biomaterials*, **26**, pp. 6625–6634.

De Bartolo, L., Morelli, S., Rende, M., Campana, C., Salerno, S., Quintiero, N., and Drioli, E. (2007b). Human hepatocyte morphology and functions in a multibore fiber bioreactor. *Macromol. Biosci.*, **7**, pp. 671–680.

De Bartolo, L., Rende, M., Giusi, G., Morelli, S., Piscioneri, A., Canonaco, M., and Drioli, E. (2007a). *Evolutionary Molecular Strategies and Plasticity: Research Signpost*, eds. Canonaco, M. and Facciolo, R. M., Chapter 18 "Membrane bio-hybrid systems: A valuable tool for the study of neuronal activities" (Research Signpost, Kerala, India), pp. 379–396.

De Bartolo, L., Rende, M., Morelli, S., Giusi, G., Salerno, S., Piscioneri, A., Gordano, A., Di Vito, A., Canonaco, M., and Drioli, E. (2008). Influence of membrane surface properties on the growth of neuronal cells isolated form hippocampus. *J. Membr. Sci.*, **325**, pp. 139–149.

De Bartolo, L., Salerno, S., Curcio, E., Piscioneri, A., Rende, M., Morelli, S., Tasselli, F., Bader, A., and Drioli, E. (2009). Human hepatocyte functions in a crossed hollow fiber membrane bioreactor. *Biomaterials*, **30**, pp. 2531–2543.

De Bartolo, L., Salerno, S., Morelli, S., Giorno, L., Rende, M., Memoli, B., Procino, A., Andreucci, V. E., Bader, A., and Drioli, E. (2006). Long-term maintenance of human hepatocytes in oxygen-permeable membrane bioreactor. *Biomaterials*, **27**, pp. 4794–4803.

Di Loreto, S., Sebastiani, P., Benedetti, E., Zimmitti, V., Caracciolo, V., Amicarelli, F., Cimini, A., and Adorno, D. (2006). Transient maintenance in bioreactor improves health of neuronal cells. *In Vitro Cell. Dev. Biol. Anim.*, **42**, pp. 134–142.

Di Vito, A., Mele, M., Piscioneri, A., Morelli, S., De Bartolo, L., Barni, T., Facciolo, R. M., and Canonaco, M. (2014). Overstimulation of glutamate signals leads to hippocampal transcriptional plasticity in hamsters. *Cell. Mol. Neurobiol.*, **34**, pp. 501–509.

Drioli, E. and De Bartolo, L. (2006). Membrane bioreactor for cell tissues and organoids. *Artif. Organs*, **30**, pp. 793–802.

Emerich, D. and Thanos, C. (2003). Nanotechnology and medicine. *Expert Opin. Biol. Ther.*, **3**, pp. 655–663.

Evans, P. J., Mackinnon, S. E., Levi, A. D., Wade, J. A., Hunter, D. A., Nakao, Y., and Midha, R. (1998). Cold preserved nerve allografts: Changes in basement membrane, viability, immunogenicity, and regeneration. *Muscle Nerve*, **21**, pp. 1507–1522.

Fabbro, A., Bosi, S., Ballerini, L., and Prato, M. (2012). Carbon nanotubes: Artificial nanomaterials to engineer single neurons and neuronal networks. *ACS Chem. Neurosci.*, **3**, pp. 611–618.

Garbayo, E., Estella-Hermoso de Mendoza, A., and Blanco-Prietom, M. J. (2014). Diagnostic and therapeutic uses of nanomaterials in the brain. *Curr. Med. Chem.*, **21**, pp. 4100–4131.

Giusi, G., Facciolo, R. M., Rende, M., Alo, R., Di Vito, A., Salerno, S., Morelli, S., De Bartolo, L., Drioli, E., and Canonaco, M. (2009). Distinct α subunits of the GABAA receptor are responsible for early hippocampal silent neuron-related activities. *Hippocampus*, **19**, pp. 1103–1114.

Gu, Z., Aimetti, A. A., Wang, Q., Dang, T. T., Zhang, Y., Veiseh, O., Cheng, H., Langer, R. S., and Anderson, D. G. (2013). Injectable nano-network for glucose-mediated insulin delivery. *ACS Nano*, **7**, pp. 4194–4201.

Jin, S.-C., Kim, S.-G., Oh, J.-S., Lee, S.-Y., Jang, E.-S., Piao, Z.-G., Lim, S.-C., Jeong, M., Kim, J.-S., and You, J.-S. (2013). A comparative study of bone formation following grafting with different ratios of particle dentin and tricalcium phosphate combinations. *J. Biomed. Nanotechnol.*, **9**, pp. 475–478.

Jones, G. L., Motta, A., Marshall, M. J., El Haj, A. J., and Cartmell, S. H. (2009). Osteoblast: Osteoclast co-cultures on silk fibroin, chitosan and PLLA films. *Biomaterials*, **30**, pp. 5376–5384.

Karp, J. M., Ferreira, L. S., Khademhosseini, A., Kwon, A. H., Yeh, J., and Langer, R. S. (2006). Cultivation of human embryonic stem cells without the embryoid body step enhances osteogenesis in vitro. *Stem Cells*, **24**, pp. 835–843.

Khan, F., Tanaka, M., and Ahmad, S. R. (2015). Fabrication of polymeric biomaterials: A strategy for tissue engineering and medical devices. *J. Mat. Chem. B*, **3**, pp. 8224–8249.

Krishna, L., Dhamodaran, K., Jayadev, C., Chatterjee, K., Shetty, R., Khora, S. S., and Das, D. (2016). Nanostructured scaffold as a determinant of stem cell fate. *Stem Cell Res. Ther.*, **7**, 188, DOI: 10.1186/s13287-016-0440-y.

Kung, F. C., Lin, C. C., and Lai, W. F. (2014). Osteogenesis of human adipose-derived stem cells on hydroxyapatite-mineralized poly(lactic acid) nanofiber sheets. *Mat. Sci. Eng. C Mater. Biol. Appl.*, **45**, pp. 578–588.

Kutikov, A. B. and Song, J. (2013). An amphiphilic degradable polymer/hydroxyapatite composite with enhanced handling characteristics promotes osteogenic gene expression in bone marrow stromal cells. *Acta Biomater.*, **9**, pp. 8354–8364.

Lao, L., Wang, Y., Zhu, Y., Zhang, Y., and Gao, C. (2011). Poly(lactide-coglycolide)/hydroxyapatite nanofibrous scaffolds fabricated by electrospinning for bone tissue engineering. *J. Mater. Sci. Mater. Med.*, **22**, pp. 1873–1884.

Lee, J. B., Park, H. N., Ko, W.-K., Bae, M. S., Heo, D. N., Yang, D. H., and Kwon, I. K. (2013). Poly(L-lactic acid)/hydroxyapatite nanocylinders as nanofibrous structure for bone tissue engineering scaffolds. *J. Biomed. Nanotechnol.*, **9**, pp. 424–429.

Li, X. M., Wang, L., Fan, Y. B., Feng, Q. L., Cui, F. Z., and Watari, F. (2013). Nanostructured scaffolds for bone tissue engineering. *J. Biomed. Mater. Res. Part A*, **101**, pp. 2424–2435.

Lin, H. J., O'Shaughnessy, T. J., Kelly, J., and Ma, W. (2004). Neural stem cell differentiation in a cell-collagen-bioreactor culture system. *Dev. Brain Res.*, **153**, pp. 163–173.

McMahon, R. E., Wang, L., Skoracki, R., and Mathur, A. B. (2013). Development of nanomaterials for bone repair and regeneration. *J. Biomed. Mater. Res. Part B*, **101**, pp. 387–397.

Meek, M. F., Varejae, A. S., and Geuna, S. (2004). Use of skeletal muscle tissue in peripheral nerve repair: Review of the literature. *Tissue Eng.*, **10**, pp. 1027–1036.

Mele, M., Morelli, S., Fazzari, G., Avolio, E., Alò, R., Piscioneri, A., De Bartolo, L., Facciolo, R. M., and Canonaco, M. (2016). Application of the co-culture membrane system pointed to a protective role of catestatin on hippocampal plus hypothalamic neurons exposed to oxygen and glucose deprivation. *Mol. Neurobiol.*, **54**, pp. 7369–7381. DOI 10.1007/s12035-016-0240-5.

Messina, A., Morelli, S., Forgacs, G., Barbieri, G., Drioli, E., and De Bartolo, L. (2015). Self-assembly of tissue spheroids on polymeric membranes. *J. Tissue Eng. Regen. Med.*, **11**, pp. 2090–2103. DOI:10.1002/term.2105.

Michalopoulos, G. K. and De Frances, M. C. (1997). Liver regeneration. *Science*, **276**, pp. 60–66.

Midwood, K. S., Williams, L. V., and Schwarzbauer, J. E. (2004). Tissue repair and the dynamics of the extracellular matrix. *Int. J. Biochem. Cell Biol.*, **36**, pp. 1031–1037.

Mimura, Y., Ihn, H., Jinnin, M., Asano, Y., Yamane, K., and Tamaki, K. (2004). Epidermal growth factor induces fibronectin expression in human dermal fibroblasts via protein kinase C delta signaling pathway. *J. Invest. Dermatol.*, **122**, pp. 1390–1398.

Misch, C. E., Qu, Z., and Bidez, M. W. (1999). Mechanical properties of trabecular bone in the human mandible: Implications for dental implant treatment planning and surgical placement. *J. Oral Maxillofac. Surg.*, **57**, pp. 700–706.

Morabito, C., Steimberg, N., Mazzoleni, G., Guarnieri, S., Fanò-Illic, G., and Mariggiò, M. A. (2015). RCCS bioreactor-based modelled microgravity induces significant changes on in vitro 3D neuroglial cell cultures. *BioMed. Res. Int.*, **2015**, 754283. DOI: 10.1155/2015/754283.

Morelli, S. and Piscioneri, A. (2013). *Stem cell therapy: State of the Art and Vision for the Future*, eds. De Bartolo, L and Bader, A. Chapter 6 "Cell-material interactions" (CRC Press, Boca Raton, FL), pp. 153–180.

Morelli, S., Facciolo, D., Messina, A., Piscioneri, A., Salerno, S., Drioli, E., and De Bartolo, L. (2013). Polycaprolactone-hydroxyapatite composite membrane scaffolds for bone tissue engineering. *Mater. Res. Soc. Symp. Proc.*, **1502**. DOI: 10.1557/opl.2013.567.

Morelli, S., Piscioneri, A., Messina, A., Salerno, S., Al-Fageeh, M. B., Drioli, E., and De Bartolo, L. (2015a). Neuronal growth and differentiation on biodegradable membranes. *J. Tissue Eng. Regen. Med.*, **9**, pp. 106–117.

Morelli, S., Piscioneri, A., Salerno, S., Al-Fageeh, M. B., Drioli, E., and De Bartolo, L. (2014). Neuroprotective effect of didymin on H_2O_2-induced injury in neuronal membrane system. *Cells Tissues Organs*, **199**, pp. 184–200.

Morelli, S., Piscioneri, A., Salerno, S., Chen, C. C., Chew, C. H., Giorno, L., Drioli, E., and De Bartolo, L. (2017a). Microtube array membrane bioreactor promotes neuronal differentiation and orientation. *Biofabrication*, **9**, 025018. DOI: 10.1088/1758-5090/aa6f6f.

Morelli, S., Piscioneri, A., Salerno, S., Rende, M., Campana, C., Tasselli, F., Di Vito, A., Giusi, G., Canonaco, M., Drioli, E., and De Bartolo, L. (2012b).

Flat and tubular membrane systems for the reconstruction of hippocampal neuronal network. *J. Tissue Eng. Regen. Med.*, **6**, pp. 299–313.

Morelli, S., Piscioneri, A., Salerno, S., Tasselli, F., Di Vito, A., Giusi, G., Canonaco, M., Drioli, E., and De Bartolo, L. (2012a). PAN hollow fiber membranes elicit functional hippocampal neuronal network. *J. Mater. Sci. Mater. Med.*, **23**, pp. 149–156.

Morelli, S., Salerno, S., Ahmed, H. M. M., Piscioneri, A., and De Bartolo, L. (2016a). Recent strategies combining biomaterials and stem cells for bone, liver and skin regeneration. *Curr. Stem Cell Res. Ther.*, **11**, pp. 676–691.

Morelli, S., Salerno, S., Holopainen, J., Ritala, M., and De Bartolo, L. (2015b). Osteogenic and osteoclastogenic differentiation of co-cultured cells in polylactic acid-nanohydroxyapatite fiber scaffolds. *J. Biotechnol.*, **204**, pp. 53–62.

Morelli, S., Salerno, S., Piscioneri, A., Campana, C., Drioli, E., and De Bartolo, L. (2010). Membrane bioreactors for regenerative medicine: An example of the bioartificial liver. *Asia-Pac. J. Chem. Eng.*, **5**, pp. 146–159.

Morelli, S., Salerno, S., Piscioneri, A., and De Bartolo, L. (2017b). *Comprehensive Membrane Science and Engineering*, 2nd Ed., eds. Drioli, E. and Giorno, L. "Membrane approaches for liver and neuronal tissue engineering" (Academic Press, Oxford) DOI: 10.1016/B978-0-12-409547-2-12260-7.

Morelli, S., Salerno, S., Piscioneri, A., Rende, M., Campana, C., Drioli, E., and De Bartolo, L. (2009). *Membrane Operations: Innovative Separations and Transformations*, eds. Drioli, E. and Giorno, L. Chapter 19 "Membranes in regenerative medicine and tissue engineering" (Wiley-VCH Verlag GmbH & Co. KGoA, Weinheim), pp. 433–446.

Morelli, S., Salerno, S., Piscioneri, A., Tasselli, F., Drioli, E., and De Bartolo, L. (2016b). Neuronal membrane bioreactor as a tool for testing crocin neuroprotective effect in Alzheimer's disease. *Chem. Eng. J.*, **305**, pp. 69–78.

Morelli, S., Salerno, S., Rende, M., Lopez, L. C., Favia, P., Procino, A., Memoli, B., Andreucci, V. E., d'Agostino, R., Drioli, E., and De Bartolo, L. (2007). Human hepatocyte functions in a galactosylated membrane bioreactor. *J. Membr. Sci.*, **302**, pp. 27–35.

Netzlaff, F., Lehr, C. M., Wertz, P. W., and Schaefer, U. F. (2005). The human epidermis models EpiSkin, SkinEthic and EpiDerm: An evaluation of morphology and their suitability for testing phototoxicity, irritancy, corrosivity, and substance transport. *Eur. J. Pharm. Biopharm.*, **60**, pp. 167–178.

Olson, M. J., Mancini, M. A., Venkatachalam, M. A, and Roy, A. K. (1990). *Cell Intercommunication,* ed. De Mello, W. C. "Hepatocyte cytodifferentiation and cell-to-cell communication" (CRC Press, Boca Raton, Florida), pp. 71–92.

Piscioneri, A., Campana, C., Salerno, S., Morelli, S., Bader, A., Giordano, F., Drioli, E., and De Bartolo, L. (2011). Biodegradable and synthetic membranes for the expansion and functional differentiation of rat embryonic liver cells. *Acta Biomater.*, **7**, pp. 171–179.

Piscioneri, A., Morelli, S., Mele, M., Canonaco, M., Bilotta, E., Pantano, P., Drioli, E., and De Bartolo, L. (2015). Neuroprotective effect of human mesenchymal stem cells in a compartmentalized neuronal membrane system. *Acta Biomater.*, **24**, pp. 297–308.

Pucci, D., Crispini, A., Sanz Mendiguchía, B., Pirillo, S., Ghedini, M., Morelli, S., and De Bartolo, L. (2013). Improving the bioactivity of Zn(II)-curcumin based complexes. *Dalton Trans.*, **42**, pp. 9679–9687.

Remya, K., Joseph, J., Mani, S., John, A., Varma, H., and Ramesh, P. J. (2013). Nanohydroxyapatite incorporated electrospun polycaprolactone/polycaprolactone-polyethyleneglycol-polycaprolactone blend scaffold for bone tissue engineering applications. *Biomed. Nanotechnol.*, **9**, pp. 1483–1494.

Salerno, S. and Bartolo, L. (2017). Biohybrid membrane systems and bioreactors as tools for in vitro drug testing. *Curr. Pharm. Des.*, **23**, pp. 319–327.

Salerno, S., Campana, C., Morelli, S., Drioli, E., and De Bartolo, L. (2011). Human hepatocytes and endothelial cells in organotypic membrane systems. *Biomaterials*, **32**, pp. 8848–8859.

Salerno, S., De Bartolo, L., and Drioli, E. (2013a). *Biomaterials for Stem Cell Therapy: State of Art and Vision for the Future,* eds. De Bartolo, L. and Bader, A. Chapter 2 "Membrane systems in liver regenerative medicine" (CRC Press, Boca Raton, FL), pp. 37–64.

Salerno, S., Messina, A., Giordano, F., Bader, A., Drioli, E., and De Bartolo, L. (2017b). Dermal-epidermal membrane systems by using human keratinocytes and mesenchymal stem cells isolated from dermis. *Mater. Sci. Eng. C Mater. Biol. Appl.*, **71**, pp. 943–953.

Salerno, S., Morelli, S., and De Bartolo, L. (2017a). Advanced membrane systems for tissue engineering. *Curr. Org. Chem.*, **21**. DOI: 10.2174/1385272820666160617092944.

Salerno, S., Morelli, S., Drioli, E., Bader, A., and De Bartolo, L. (2013c). Biofabrication of layered membrane systems by using human hepatocytes and endothelial cells: a comparative study. *Curr. Tissue Eng.*, **2**, pp. 109–118.

Salerno, S., Morelli, S., Giordano, F., Gordano, A., and De Bartolo, L. (2016). Polymeric membranes modulate human keratinocyte differentiation in specific epidermal layers. *Colloid. Surf. B*, **146**, pp. 352–362.

Salerno, S., Piscioneri, A., Laera, S., Morelli, S., Favia, P., Bader, A., Drioli, E., and De Bartolo, L. (2009). Improved functions of human hepatocytes on NH_3 plasma-grafted PEEK-WC-PU membrane. *Biomaterials*, **30**, pp. 4348–4356.

Salerno, S., Piscioneri, A., Morelli, S., Al-Fageeh, M. B., Drioli, E., and De Bartolo, L. (2013b). Membrane bioreactor for expansion and differentiation of embryonic liver cells. *Ind. Eng. Chem. Res.*, **52**, pp. 10387–10395.

Sato, M. and Webster, T. (2004). Nanobiotechnology: Implications for the future of nanotechnology in orthopedic applications. *Expert Rev. Med. Devices*, **1**, pp. 105–114.

Shalumon, K., Sowmya, S., Sathish, D., Chennazhi, K., Nair, S. V., and Jayakumar, R. (2013). Effect of incorporation of nanoscale bioactive glass and hydroxyapatite in PCL/chitosan nanofibers for bone and periodontal tissue engineering. *J. Biomed. Nanotechnol.*, **9**, pp. 430–440.

Smith, L. and Ma, P. (2004). Nano-fibrous scaffolds for tissue engineering. *Colloid. Surf. B*, **39**, pp. 125–131.

Sun, F., Kang, H. G., Ryu, S.-C., Kim, J. E., Park, E. Y., Hwang, D. Y., and Lee, J. (2013). Guided bone regeneration using a flexible hydroxyapatite patch. *J. Biomed. Nanotechnol.*, **9**, pp. 1914–1920.

Thomas, R. J., Bhandari, R., Barrett, D. A., Bennett, A. J., Fry, J. R., Powe, D., Thomson, B. J., and Shakesheff, K. M. (2005). The effect of three-dimensional co-culture of hepatocytes and hepatic stellate cells on key hepatocyte functions in vitro. *Cells Tissues Organs*, **181**, pp. 67–79.

Tortelli, F., Pujic, N., Liu, Y., Laroche, N., Vico, L., and Cancedda, R. (2009). Osteoblast and osteoclast differentiation in an in vitro three-dimensional model of bone. *Tissue Eng. Part A*, **15**, pp. 2373–2383.

Tos, P., Battiston, B., Nicolino, S., Raimondo, S., Fornaro, M., Lee, J. M., Chirila, L., Geuna, S., and Perroteau, I. (2007). Comparison of fresh and predegenerated muscle–vein combined guides for the repair of rat median nerve. *Microsurgery*, **27**, pp. 48–55.

Tuzlakoglu, K., Bolgen, N., Salgado, A., Gomes, M., Piskin, E., and Reis, R. (2005). Nano- and micro-fiber combined scaffolds: A new architecture for bone tissue engineering. *J. Mater. Sci. Mater. Med.*, **16**, pp. 1099–1104.

Valmikinathan, C. M., Hoffman, J., and Yu, X. (2011). Impact of scaffold micro- and macro-architecture on Schwann cell proliferation under dynamic conditions in a rotating wall vessel. *Mater. Sci. Eng. C*, **31**, pp. 22–29.

Wang, Z. L. (2002). *Handbook of Nanophase and Nanostructured Materials-Characterization* (Kluwer Academic Plenum Publishers).

Wang-Bennett, L. T. and Coker, N. J. (1990). Analysis of axonal regeneration through the silicone regeneration chamber: A retrograde tracing study in the rabbit facial nerve. *Exp. Neurol.*, **107**, pp. 222–229.

Woo, K., Chen, V., and Ma, P. (2003). Nano-fibrous scaffolding architecture selectively enhances protein adsorption contributing to cell attachment. *J. Biomed. Mater. Res. Part A*, **67**, pp. 531–537.

Zhang, N., Yan, H., and Wen, X. (2005). Tissue-engineering approaches for axonal guidance. *Brain Res. Rev.*, **49**, pp. 48–64.

Index

For Product Safety Concerns and Information please contact our EU representative GPSR@taylorandfrancis.com
Taylor & Francis Verlag GmbH, Kaufingerstraße 24, 80331 München, Germany

www.ingramcontent.com/pod-product-compliance
Ingram Content Group UK Ltd.
Pitfield, Milton Keynes, MK11 3LW, UK
UKHW022200190926
475702UK00003B/126
* 9 7 8 9 8 1 4 7 7 4 7 9 6 *